Space Geodesy and Geodynamics

Space Geodesy and Geodynamics

Edited by
Allen Joel Anderson

Department of Geodesy
Institute of Geophysics
The University of Uppsala
Uppsala, Sweden

Anny Cazenave

Groupe de Recherches de Géodésie Spatiale
Centre National d'Etudes Spatiale
Toulouse, France

1986

ACADEMIC PRESS
Harcourt Brace Jovanovich, Publishers
London Orlando San Diego New York Austin
Boston Sydney Tokyo Toronto

ACADEMIC PRESS INC. (LONDON) LTD
24/28 Oval Road, London NW1 7DX

United States Edition Published by
ACADEMIC PRESS INC.
Orlando, Florida 32887

British Library Cataloguing in Publication Data
Space geodesy and geodynamics.
1. Geodesy
I. Anderson, Allen Joel II. Cazenave, Anny
526'.1 QB 281

ISBN 0-12-057840-9

Printed in Great Britain by
J. W. Arrowsmith Ltd., Bristol

Contributors

R. J. ANDERLE, General Electric Company, Valley Forge Space Center, Box 8555-43A30, Philadelphia, Pennsylvania 19101 USA

ALLEN JOEL ANDERSON, Institute of Geophysics, Department of Geodesy, The University of Uppsala, Hällby, S-755 90 Uppsala, Sweden

G. BALMINO, Bureau Gravimetrique International, 18 Avenue Edouard-Belin, F-31055 Toulouse, France

W. E. CARTER, National Geodetic Survey, Charting and Geodetic Services, National Ocean Service, NOAA, Rockville, Maryland 20852, USA

ANNY CAZENAVE, Groupe de Recherches de Géodésie Spatiale, Centre National d'Etudes Spatiale, 18 Avenue Edouard-Belin, F-31055, Toulouse, France

R. E. CHENEY, National Geodetic Survey, Charting and Geodetic Services, National Ocean Service, NOAA, Rockville, Maryland 20852, USA

J. O. DICKEY, Jet Propulsion Laboratory, California Institute of Technology, 4800 Oak Grove Drive, Pasadena, California 91109, USA

B. C. DOUGLAS, National Geodetic Survey, Charting and Geodetic Services, National Ocean Service, NOAA, Rockville, Maryland 20852, USA

T. M. EUBANKS, Jet Propulsion Laboratory, California Institute of Technology, 4800 Oak Grove Drive, Pasadena, California 91109, USA

J. L. FANSELOW, Jet Propulsion Laboratory, California Institute of Technology, 4800 Oak Grove Drive, Pasadena, California 91109, USA

T. HERRING, Smithsonian Astrophysical Observatory, 50 Garden Street, Cambridge, Massachusetts 02138, USA

EMILE A. OKAL, Department of Geological Sciences, Northwestern University, Evanston, Illinois 60201, USA

W. R. PELTIER, Department of Physics, University of Toronto, 60 St. George Street, Toronto, Ontario M5S 1A7, Canada

D. C. McADOO, National Geodetic Survey, Charting and Geodetic Services, National Ocean Service, NOAA, Rockville, Maryland 20852, USA

RICHARD H. RAPP, Department of Geodetic Science, Ohio State University, 1958 Neil Avenue, Colombus, Ohio 43210, USA

N. A. RENZETTI, Jet Propulsion Laboratory, California Institute of Technology, 4800 Oak Grove Drive, Pasadena, California 91109, USA

D. S. ROBERTSON, National Geodetic Survey, Charting and Geodetic Services, National Ocean Service, NOAA, Rockville, Maryland 20852, USA

D. T. SANDWELL, National Geodetic Survey, Charting and Geodetic Services, National Ocean Service, NOAA, Rockville, Maryland 20852, USA

SETH STEIN, Department of Geological Sciences, Northwestern University, Evanston, Ilinois 60201, USA

C. L. THORNTON, Jet Propulsion Laboratory, California Institute of Technology, 4800 Oak Grove Drive, Pasadena, California 91109, USA

D. L. TURCOTTE, Department of Geological Sciences, Cornell University, 210 Kimball Hall, Ithaca, New York 14853, USA

JOHN M. WAHR, Department of Physics and Cooperative Institute for Research in Environmental Sciences, Campus Box 449, University of Colorado, Boulder, Colorado 80309, USA

J. ZSCHAU, Institut für Geophysik, Christian Albrechts Universität, Olshausenstrasse 40–60, D-2300 Kiel, Federal Republic of Germany

Preface

Space techniques provide a powerful means of studying the earth. This obvious fact has been realized only recently. Many of the most well known of the space techniques, such as remote sensing, only allow measurements of surface phenomena. Space methods do exist, however, which provide unique ways to study the dynamics and deep interiors of the planets. These include the measurement of many diverse components, such as the mapping of the gravity and magnetic fields, surface deformations, changes in the global orientation, ocean circulation, atmosphere/ocean/solid planet interaction, and auxiliary data on the dynamic interior structure of the planet.

An understanding of these physical components and their interrelationship is a prerequisite to understanding the evolution of the earth's surface. Access to the environment of the other planets, coupled with in-depth studies of similar phenomena on earth, is crucial for understanding the delicate balances which are maintaining the dynamic systems on our planet.

Space Geodesy has become one of those unifying subjects for many scientific disciplines. The subject is involved with studies of such diverse phenomena as the radio source structure of objects at the edge of the universe to the surface character and time variability of the Sargasso Sea in the North Atlantic.

That these things are related depends only upon one's point of view. The sixth century Ionian philosopher, Heraclitus of Ephesus, once observed that things are always changing, nothing is constant: "You cannot step twice into the same river, for fresh waters are ever flowing in upon you." Space Geodesy actually tries to measure these quantities on the surface of the earth and planets and to show exactly what it is that they depend upon. In order to do that one needs a cosmic reference system to place small, seemingly imperceptible events, in clear and meaningful perspective.

This book is intended as a review of recent work in Space Geodesy, satellite geophysics and geodynamics which have important bearing on the further understanding of the dynamics of our planet. Some very exciting progress has been achieved in this regard during recent years, and several satellite projects are now being planned that should allow us to probe even deeper and with greater detail into the dynamic properties of the earth.

ALLEN JOEL ANDERSON
Uppsala

ANNY CAZENAVE
Toulouse

Contents

Part 1

Introduction and Summary

1.1 Space Geodesy and Geodynamics: The New Earth Science

Allen Joel Anderson

Department of Geodesy, Institute of Geophysics, The University of Uppsala, Uppsala, Sweden

I. Introduction

In 1975 the late Sir Edward Bullard wrote that the most significant work which led to the discovery of plate tectonics was a study of the geology beneath the oceans—something that had been left out from classical geology earlier and that so limited the understanding of Earth processes as to miss the most important fact about Earth dynamics. This historical account of the facts is in little dispute today.

With plate tectonics now firmly established in the minds of Earth scientists, one often gets the impression that all there is remaining to be done is to put together the details and explain certain remaining disagreements between theory and observation. Actually little is known about the exact form of convection in the mantle, the very mechanism believed to be responsible for plate tectonics. Even less is known about the processes behind the periodic orogeny of the continents and their large-scale uplift and subsidence. Many details concerning the relationship between core–

SPACE GEODESY AND GEODYNAMICS
ISBN 0-12-057840-9

mantle interactions, Earth rotation and polar motion variations and problems concerning the dynamic structure of the lithosphere are left unexplained. Even less is known about the history of plate tectonics which may not have been active during the full time scale of the 4.5 billion year history of the Earth. Indeed, to be honest, we still have but a sketchy picture of what planetary dynamics is really all about.

Without being too brash, it might be suggested that one of the next steps in improving our knowledge of planetary dynamics might come from studies using satellite and space techniques; satellites which observe the dynamic properties, potential fields and changing properties of the Earth and planets. Measurements from space have now become sufficiently precise to allow more accurate and detailed representations in many geophysical parameters than terrestrial methods provide.

Space based methods often provide the systematic coverage and homogeneity to allow for the necessary unification of data to properly interpret relevant quantities related to planetary dynamics. These include the study of potential fields, topography, changes in the global reference frame, surface motions and a large set of parameters which can be derived from satellite based observation of the Earth and planets.

Many geophysicists hope to be able for the first time to actually measure and model the detailed motions predicted by plate tectonic theory. They expect in the near future to model Earth dynamics much as the weather is modelled today. That is within the realm of space geodesy and geodynamics.

II. The Potential Fields

The satellite derived gravity fields have expanded our understanding of the dynamics of the Earth dramatically. In the short wavelengths, they have revealed properties of the crust and lithosphere, structural effects such as the oceanic-continental boundaries, regional properties of the asthenosphere, and they have allowed us to verify the time-dependent thermal age relationship of the oceanic lithosphere. In the long wavelengths, they have pointed the way to understanding the behaviour of the Earth's mantle undergoing three-dimensional convection, and have allowed the study of the forces and properties of the mantle in subduction zones and the ocean ridges.

Large-scale studies of the topographic-gravity admittance functions have produced important results, not only for our planet, but for the moon and terrestrial planets as well. It is possible to probe the thickness of the elastic lithosphere and its response with time to surface loading. Using known or plausible response functions it is possible to determine the formation ages and ancient properties of planetary lithospheres (Lambeck and Phillips, 1980). Although the use of planetary gravity fields to study the dynamic properties of the Earth seems well established today, it is important to point out the rocky road of uncertainty which followed its development (Lambeck, 1983).

The subject began during the early 1960s when the first models of the global gravity field were produced. In less than 20 years, by the late 1970s, satellite global gravity modelling had grown to a fine art, some might say a black art. At each stage in the development, new interpretations about the meaning of the global gravity anomalies were made. In the beginning many people, following Heiskanen and others, felt that the Earth must be in hydrostatic equilibrium below a certain depth. Even though the crust must support some stress differences, mass anomalies could not exceed certain limits. The early satellite results produced global gravity field models which clearly exceeded these limits and something was wrong.

In 1963, W. H. Kaula, using a global gravity model to degree 4 showed that if the mantle was in a static state, then several hundred bars of static stress would be needed in order to produce the observed gravity anomalies. Keith Runcorn, in 1964, argued that the anomalies were caused by mantle convection and did not reflect static stress. The first edition of W. H. Kaula's book "Satellite Geodesy" appeared in 1966. Here he described in detail his derivations of satellite orbit theory, parameter estimation and satellite orbit inversion methods to produce gravity field parameters. The same year the first reliable global gravity field model was published by the Smithsonian Astrophysical Observatory which produced solutions for harmonics through degree and order 8 (Smithsonian Astrophysical Observatory, 1966). This was a remarkable achievement for the time. Some systematic structure was seen, which however did not agree with the then prevailing explanation.

By 1969 Kaula now argued that the gravity field reflected global surface tectonics. He pointed out that plate subduction zones were related to gravity highs; ocean ridges and continental collisions also produced highs. The plate interiors were largely negative. Until recently it has been difficult to fit these observations together with acceptable models of mantle convection.

For the global gravity models themselves, there is reason to be happy. Dramatic improvements were made during the 1970s, and with the advent of satellite sea surface altimetry a whole new science developed. Paradoxically the gravity field over the oceans has now become far better determined than over many continental regions of the Earth. For that reason and others, new satellite projects have been conceived, such as GRAVSAT, GRM and satellite gravity gradiometer missions, GRADIO, to map the detailed gravity field of the Earth independently of whether it be over the oceans or continents.

Let us look at one example of a low flying satellite–satellite mission, NASA's GRM, to see what type of results may be achievable compared with present-day techniques. In measuring the geopotential, there are several advantages to be gained by flying two satellites in a low 160-km polar orbit having a separation of a few hundred kilometers between them (Colombo, 1984). If we assume that a microwave Doppler tracking system, which is used for measuring relative acceleration between the two spacecraft, will have a precision of at least 1.0 μm/s, then we can make several comparisons between the data provided by this mission and other types of geophysical measurements currently available.

The measurement sensitivity (the measurement signal to noise) of the two-satellite GRM system compared with the signal produced by an ocean trench anomaly would be resolved to better than 1 part in 6000 or greater than a 75 dB signal resolution for this geophysical gravity anomaly (see Fig. 1). The resolution of this gravity anomaly is much greater than with

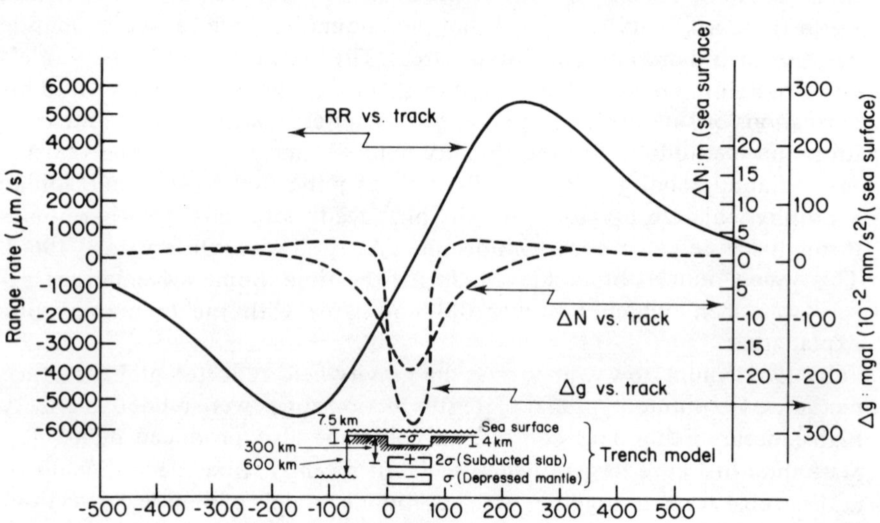

Fig. 1. The modelled effects on GRAVSAT of an oceanic trench geophysical gravity anomaly. The full line indicates the signal as measured by GRAVSAT, the satellite–satellite low orbit configuration (scale on left axis). The dashed lines indicate the signal obtained with sea gravimetry and satellite altimetry (as marked with scales on right axis). The geophysical model used to generate the gravity anomaly is shown at the bottom of the figure. Adapted from Wagner and Sandwell (1984).

other geophysical measurements made here to date. For example, in surface-based ocean gravity measurements, the amplitude of this feature is typically 200 mgal with the ocean gravimeter measurement errors ranging above 2 mgal. This is a useful signal resolution no better than 40 dB, some 100 times less than with the satellite–satellite technique.

If we compare with another more recent space technique, that of satellite sea surface altimetry, we find a similar advantage. Here the ocean trench anomaly produces a signal of some 20 m in the geoid at the ocean surface. Present altimeter errors for these wavelengths are approximately 10 cm. This implies a usable sensitivity for the altimeter measurement of 48 dB, providing 50 times less resolution than with the satellite–satellite concept.

Furthermore by combining satellite-to-satellite tracking data at 160 km above the Earth's surface with sea surface altimetry data, one should be able to fix accurately the depth of features of the gravity anomaly. This data, further combined with bathymetry, topography and some knowledge of surface geology and structure, should give a detailed picture of the structure and dynamic state of the lithosphere, asthenosphere and mantle of the Earth.

III. The Current Situation

As our understanding of the phenomenon increases, there seems always to be ways around the limitations imposed by classical methodologies. In a review of radial orbital variations due to errors in the gravitational model, Wagner (1985) shows that the use of crossover information in satellite altimetry imposes conditions which allow for a method to observe the gravity field variations more accurately. He shows that the linear perturbations can be used to give a substantial improvement in the low-degree geopotential from the indirect use of SEASAT altimetry at track crossovers. Furthermore there are clear indications that a separation of "permanent" sea topography and geoid may be possible from the direct use of the heights along the track.

An example of the present detail in the resolution of the oceanic geoid is illustrated in Fig. 2. Here the global geopotential of spherical harmonic degree and order less than 10 has been removed from the data so as to enhance the short-wavelength variation which have bathymetric, lithospheric, tectonic and asthenospheric origin. One of the large-scale features evident in this type of three-dimensional representation is the large downward slope in the geoid over the Baltic Sea. This has been attributed to the effects of a thickened continental lithosphere, as well as the possibility of revealing effects of the downgoing current of a large-scale convective cell in the upper mantle (Anderson, 1984). Fennoscandia also shows an

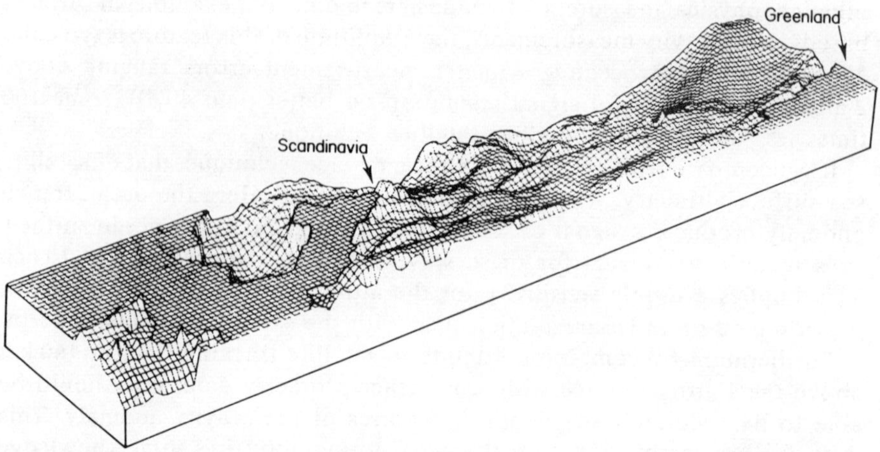

Fig. 2. Three-dimensional representation of the geopotential over the North Atlantic and Arctic Ocean made by using data obtained from SEASAT. The data have been reduced by subtracting out the low-order terms in the geopotential less than or equal to degree 10 to enhance the shorter wavelengths in the projection. The view is from the North Pole looking south over the North Atlantic. The Scandinavian peninsula is on the left and Greenland on the right (flat areas). The area shown is from latitude 55–70° N, longitude 40° W–40° E.

anomalous low topographic height in comparison with other continental crust as shown in Fig. 3. This may imply the existence of a colder, downgoing portion of the mantle residing underneath the old Fennoscandian craton.

One of the problems which have intrigued earth scientists for years has been the question about the structure of density anomalies in the mantle which give rise to long-wavelength features in the geopotential. Clearly these long-wavelength features are related in a rather direct way to the dynamics in the mantle.

Hager and Richards have produced results which show the geoid signatures which can be expected from mantle convection models with differentiating vertical viscosity structure (Hager, 1983, 1984; Richards and Hager, 1984). Even when taking into account possible horizontal variation in viscosity structure, the predicted signatures from these models do not change by more than 50%. In some cases effects are amplified. This gives some hope of actually being able to deduce the viscosity structure within the actively convecting region of the mantle by carefully studying the geoid anomaly after removal of the various lithospheric effects. This area of research should clearly be important to further our understanding of the convective properties of the mantle. Unfortunately the restrictions placed on numerical models of mantle convection in a large computer have often

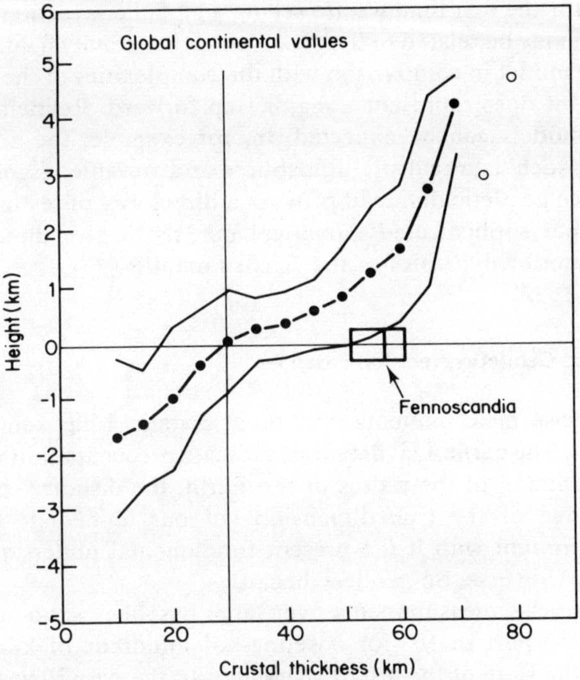

Fig. 3. Continental height as a function of crustal thickness. The class interval adopted for grouping the thickness data is 5 km. The upper and lower lines connect the $+1\sigma$ and -1σ points for each class, respectively; the two thickest classes contain only one estimate each. The relationship for the Fennoscandian lithosphere is shown in the marked box. Here the topography is more than 1 km lower than regions with similar lithospheric thickness on the earth. Adapted from Cogley (1984).

been too severe to allow for a valid generalization to be made about the true nature of convection in the Earth's mantle.

With the increasing access to supercomputers, it has now become feasible to produce three-dimensional mantle convection models which have less departure from realistic Earth mantle convection than was possible earlier. For example, Baumgardner (1985) has produced a series of three-dimensional global Earth convecting models which begin to have the appearance of actual observable global heat flow patterns.

Furthermore these models indicate the tendency for conditions of the present Earth structure to clump mantle convection into large plume-like features, focusing high heat flow patterns into splotches along ridge-like features. The colder sinking portions of mantle material tend to pile up in large arc-like features. These features are reminiscent of the observed global gravity lows in continental or near continental subduction zones of the Earth.

Possibly for the first time we are seeing a global convection model of the Earth which may be related to the observed global gravity field. Even though still a crude model in comparison with the complexities of the actual Earth, nevertheless it does represent a major step forward. Refinements to these numerical models can be expected, in, for example, the adding of new parameters, such as a realistic lithosphere and variable viscosity.

Here space geodetic data will provide a direct way of testing the validity of these super sophisticated numerical models to give direct knowledge about the interior dynamics of the Earth's mantle.

IV. The Space Geodetic Precision Baselines

Making precise measurements over great distances has somehow always been with us. The earliest civilizations indicate preoccupation with this task. The determination of the radius of the Earth, the distance to the sun and the knowledge of the true dimensions of our immediate surroundings eventually brought with it the present fundamental understanding of our place in the Universe. So geodesy began.

Making precise measurements over large baselines is no easy task. Precisions of one part in 10^5 for baselines of hundreds of kilometers were considered the state of the art 20 years ago. In the past 20 years precisions have increased by four orders, from parts in 10^5 to parts in 10^9. A whole new subject has thereby been created—space geodesy and geodynamics. This four order of magnitude improvement in measurement has been made possible through the use of space techniques. The measurement of our world has suddenly become 10,000 times more accurate, and new phenomena, such as global geodynamic phenomena, are measurable for

Table I. Earth dynamics[a]

Phenomenon	Velocity (m/s)
Changes in the geoid, the equipotential surface of the Earth	$\sim 10^{-11}$
Plate tectonic movements, vertical crustal movements	$\sim 10^{-9}$
Movement of the physical pole in response to latest deglaciation event	$\sim 10^{-8}$
Polar motion	$\sim 10^{-6}$
Earth tidal deformations	$\sim 10^{-5}$
Movement of the geomagnetic pole, dipole field motion	$\sim 10^{-3}$

[a] A short list of various Earth phenomena together with their current estimated velocities. Note that 8 orders of magnitude separate the rate of change of the geoid from motions in the geomagnetic dipole field.

the first time. Not only had geodesy changed quantitatively, it had changed qualitatively as well. It now became necessary to deal with the physical methods of observations, and to even explain the observations themselves in terms of physical processes and causes. Thus understanding of geodynamic processes became an important prerequisite.

One type of application, the measurement of variations in interplanetary baseline lengths to parts in 10^{15} over periods of 10,000 s has even brought forth the possibility of detecting gravitational radiation which may emanate from the "big bang" itself (Anderson, 1971, 1977, 1982). Similar Doppler tracking techniques applied to the "Viking" landers on Mars have allowed for precise planetary orbit ephemerides data to be obtained. These data have been able to set upper limits to the amount of possible change in the gravitational coefficient of attraction "G" to less than two parts in 10^{12} per year (Hellings *et al.*, 1984). These are truly remarkable feats which demand exceptional work on the part of literally thousands of committed scientists now involved in these space technology efforts worldwide.

For the Earth, space geodetic baselines now stretch across its face much like intercontinental aircraft routes. A recent publication by Smith *et al.* (1984) looks much like an airline timetable, showing laser ranging baselines obtained during the first four years of the NASA crustal dynamics project (see Fig. 4). It is an impressive list. The majority of the standard deviations of the measured baselines are within 5 cm, some do exceed 30 cm however. The baseline ties are listed in chronological order by year. Unfortunately many have not yet been repeated regularly enough in order to obtain good information about global crustal deformation and dynamics. There is some hope that shorter baseline measurements can soon be made available to complement these by using a global positioning satellite (GPS) interferometric based receiver technique. The already obtainable precisions (Bock *et al.*, 1984; MacDoran *et al.*, 1985) indicate these systems have the capability to measure long-wavelength crustal deformation. Thornton *et al.*, see Chapter 3.3, discuss the GPS system and some of its planned uses for the determination of centimeter quality regional baselines.

V. Earth Orientation

There are also available measurements of Earth rotation and polar motion, the physical motion of the pole. This work is described by Dickey and Eubank in Chapter 4.1. Recent changes have been made in the methods used from those in the past to determine Earth orientation. The purely optical astrolab and PZ-tube measurements began to be replaced some 15 years ago by using solutions obtained through the measurement of satellite

YEAR	FROM		LAT	LON	HGT	SIG(H)	TO		LAT	LON	HGT	SIG(H)	BASELINE	SIG(B)	CHORD	SIG(C)	YEAR
1979	GRASSE	7835	43	6	1315.45	0.06	YARRAGDE	97090	-29	115	234.34	0.04	11157716.11	0.08	11156358.61	0.06	1979
1980	GRASSE	7835	43	6	1315.76	0.02	YARRAGDE	97090	-29	115	234.23	0.01	11157716.02	0.03	11156358.35	0.02	1980
1981	GRASSE	7835	43	6	1316.15	0.09	YARRAGDE	97090	-29	115	234.28	0.01	11157716.79	0.12	11156358.72	0.09	1981
1982	GRASSE	7835	43	6	1316.22	0.02	YARRAGDE	97090	-29	115	234.29	0.01	11157716.51	0.03	11156358.37	0.03	1982
													0.21	0.02	-0.02	0.02	SLOPE
1979	GRASSE	7835	43	6	1315.45	0.06	HAYSTACK	97091	42	288	84.93	0.03	5889525.40	0.10	5888879.85	0.09	1979
1980	GRASSE	7835	43	6	1315.76	0.02	HAYSTACK	97091	42	288	84.91	0.01	5889526.18	0.05	5888880.50	0.05	1980
1981																	1981
1982																	1982
													0.78	0.11	0.65	0.10	SLOPE
1979																	1979
1980	GRASSE	7835	43	6	1315.76	0.02	AM SOMA	97092	9	167	25.82	0.02	11266751.66	0.05	11265568.01	0.04	1980
1981																	1981
1982																	1982
1979	GRASSE	7835	43	6	1315.45	0.06	KWAJELIN	97096	-14	189	42.09	0.04	12329278.55	0.08	12327966.57	0.05	1979
1980	GRASSE	7835	43	6	1315.76	0.02	KWAJELIN	97096	-14	189	41.93	0.02	12329278.72	0.04	12327966.60	0.02	1980
1981																	1981
1982																	1982
													0.17	0.09	0.02	0.06	SLOPE
1979	GRASSE	7835	43	6	1315.45	0.06	GORF 101	97101	39	283	1.81	0.06	6409775.71	0.12	6409114.83	0.12	1979
1980																	1980
1981																	1981
1982	GRASSE	7835	43	6	1316.22	0.02	GORF 101	97101	39	283	1.37	0.93	6409776.12	2.82	6409115.07	2.78	1982
													0.14	0.94	0.08	0.93	SLOPE
1979	GRASSE	7835	43	6	1315.45	0.06	GORF 102	97102	39	283	10.90	0.04	6410221.56	0.10	6409556.07	0.09	1979
1980	GRASSE	7835	43	6	1315.76	0.02	GORF 102	97102	39	283	10.91	0.02	6410222.12	0.05	6409556.47	0.05	1980
1981	GRASSE	7835	43	6	1316.15	0.09	GORF 102	97102	39	283	10.89	0.03	6410221.74	0.18	6409555.90	0.17	1981
1982	GRASSE	7835	43	6	1316.22	0.02	GORF 102	97102	39	283	10.90	0.02	6410222.31	0.06	6409556.43	0.06	1982
													0.17	0.03	0.05	0.03	SLOPE
1979	GRASSE	7835	43	6	1315.45	0.06	GORF 103	97103	39	283	10.98	0.06	6410218.15	0.12	6409552.62	0.11	1979
1980																	1980
1981																	1981
1982	GRASSE	7835	43	6	1316.22	0.02	GORF 103	97103	39	283	10.87	0.03	6410218.86	0.07	6409553.00	0.07	1982
													0.24	0.05	0.13	0.04	SLOPE
1979	GRASSE	7835	43	6	1315.45	0.06	GORF 104	97104	39	283	2.92	0.05	6409866.20	0.11	6409204.74	0.10	1979
1980																	1980
1981																	1981
1982																	1982
1979																	1979
1980																	1980
1981	GRASSE	7835	43	6	1316.15	0.09	GORF 105	97105	39	283	12.10	0.02	6410202.01	0.18	6409535.57	0.17	1981
1982	GRASSE	7835	43	6	1316.22	0.02	GORF 105	97105	39	283	12.12	0.01	6410202.65	0.06	6409536.17	0.06	1982
													0.64	0.19	0.60	0.18	SLOPE

Fig. 4. A document page showing global satellite laser baseline data from Smith *et al.* (1984). The information given (from left to right) is: year baseline was measured; name and position of first site; name and position of second site; baseline length (to nearest cm); standard deviation of measurement; and where four measurements are available a rate of change (slope) estimate is given.

Doppler determined positions using the U.S. navy navigation system (TRANSIT). This is described by Anderle in Chapter 3.1. More recently a dedicated set of large very long baseline radio-interferometric (VLBI) dishes have been used which produce 2–5 day solutions having formal errors of less than 2 cm in pole positions and 0.15 ms in Earth rotation. These methods even allow values of these parameters to be determined each hour on the POLARIS and IRIS network. This system and results are described by Carter and Robertson in Chapter 4.2.

One of the key issues has been the determination of the source of the observable so called irregular "decade fluctuations" in the variations of the length of day (LOD). The linear increase in the length of day, as determined by ancient eclipse records, is estimated to be between 1–2 ms per century. The LOD also has irregular decade fluctuations of about 4–5 ms over time scales of from 20–30 years (Lambeck, 1980). These are believed to be caused by transfer of angular momentum between the liquid core and mantle of the Earth. As such they are a unique set of data regarding the dynamics of the liquid core of the Earth.

Variations in the LOD with periods less than 5 years or so are primarily driven by the exchange of angular momentum within the atmosphere and oceans. Wahr describes, in Chapter 4.3, the implications of Earth orientation observations in modelling the internal constituents of the Earth.

There is also a new lunar laser ranging network which began its operation in 1985. Dickey *et al.* (1985) gives a review of the accomplishments over the past 15 years. An interesting and increasingly important data set is available from the deep space tracking network (DSN) which measures globally using three sites, each on separate continents (Australia, Europe and North America). By measuring Doppler shifts of signals from interplanetary spacecraft, these data provide some of the most accurate information on the dynamics of the planetary system, the dynamics of the Earth and several of the planets as well.

VI. Large-scale Dynamics of the Lithosphere

One of the areas which has been richly provided for with space geodetic data has been in studies of the behavior of the lithosphere under surface loading and its response to the forces of plate tectonics. Here space geodesists have worked closely with seismologists and tectonophysicists.

The story of our subject has now come full circle, reflecting back to the early days of studies of the geology of the sea floor while at the same time producing data which are providing new understanding on the dynamics and structural complexities of the oceanic lithosphere.

One of these areas has been the study of lithospheric flexure. Much of this work has been carried out within the framework of the equation

$$D = EH^3/[12(1 - v^2)]$$

where D is the flexural rigidity, H the elastic thickness, E Young's modulus, and v Poisson's ratio. By comparing observed data (gravity, geoid height, vertical movement, etc.) with theoretical estimates one can derive plausible values for D and H. Recent results on this work are described by Cazenave and Okal in Chapter 5.1. They also review data related to oceanic fracture zones and subduction zones.

Since much of the ocean geoid data is provided from satellite altimetry, it is necessary to distinguish between the influence of variations in the geopotential and those having to do with oceanographic features, currents and time-dependent structures of the ocean surface. Indeed it was the oceanographic applications which in themselves were one of the primary reasons why satellite altimeters where constructed in the first place. Some of these oceanographic applications of space geodetic data are reviewed in Cheney *et al.*; see Chapter 5.2.

The use of space geodetic data can now be incorporated with seismological studies of lithospheric deformation. In Chapter 5.3 Stein and Okal review briefly the seismological methods used for studying the large-scale deformations of the plates. They discuss plate boundary processes, plate-wide seismicity and localized intraplate processes by taking up such areas as the seismicity at passive margins, volcanism and the seismicity at evolving plate boundaries.

The subject would not be complete without a discussion of the various lithospheric compensation mechanisms and how they may be related to the geoid. Turcotte (Chapter 5.4) presents a review of much of the work he has done during the past decade. He summarizes the basic equations relating the geoid and geopotential to dipole structures relevant to plausible geological structure and behavior. He discusses a full set of compensation mechanisms and includes derivations of the equations for modelling the parameters. These include Airy, Pratt, thermal and lithospheric thinning mechanisms, with a discussion of some relevant error sources and some limitations on their use.

VII. Conclusions

During the past two decades space geodesy and geodynamics has emerged as one of the new Earth sciences. Together with seismology and tectonophysics it forms a basis for the study of global tectonics. The subject encompasses areas of physics, celestial mechanics and fundamental research in the areas of gravitation and space time.

By providing large amounts of data based on a "cosmic reference system" it allows a proper description of global geodynamic phenomenon which takes place in an inertial reference frame. One of the subject's great achievements has been to provide a planet-wide gravity field with sufficient detail to show the dynamic effects of convection in the Earth's mantle and the effects of plate tectonics. Precise baselines are now being measured that show the large-scale deformation between the plates.

The use of detailed ocean gravity field data has provided additional information about the structure and tectonics of the oceanic lithosphere. Global gravity field data over the continents should allow similar refinement in our knowledge of the dynamics of continental lithospheres.

Acknowledgments

Much of the work on this book was carried out while the author was at the University of Uppsala in Sweden. He thanks the Swedish Natural Science Research Council for support during this period, and for secretarial help provided by Ingegerd Ohlsson and Susanna Boholm.

References

Anderson, A. J. (1971). *Nature* **229**, 547–548.

Anderson, A. J. (1977). In *Experimental Gravitation* (B. Bertotti, ed.), pp. 235–246. Academy of Sciences, Rome.

Anderson, A. J. (1982). Precise spacecraft tracking using VLBI digital tone extraction for the purposes of gravitational wave detection. *NOAA Tech. Rep. NOS 95/NGS 25*, 285–296.

Anderson, A. J. (1984). *Mar. Geophys. Res.* **7**, 191–203.

Baumgardner, J. R. (1985) "Three-D numerical models of mantle convection in the earth and moon". In *Proc. NATO Adv. Study Inst. Phys. Planet.* Newcastle-upon-Tyne (preprint).

Bock, Y., Abbot, R. I., Counselman, C. C., Gourevitch, S. A., King, R. W. and Paradis, A. R. (1984). Geodetic accuracy of the macrometer model V-1000. In *Proc. Int. Assoc. Geodesy (IAG), Hamburg*, **2**, 223. Ohio State Univ., Columbus.

Bullard, E. (1975). *Ann. Rev. Earth Planet. Sci.* **3**, 1–30.

Cogley, J. G. (1984). *Rev. Geophys. Space Phys.* **22**, 101–122.

Colombo, O. L. (1984). The global mapping of gravity with two satellites. Netherlands Geodetic Commission, Publ. on Geodesy, Vol. 7, No. 3.

Dickey, J. O., Williams, J. G. and Newhall, X. X. (1985). "Fifteen Years of Lunar Laser Ranging: Accomplishments and Future Challenges". JPL Geodesy and Geophysics preprint, No. 121. Also in: *Proc. 5th Int. Workshop on Laser Ranging Instrumentation.* Royal Greenwich Observatory.

Grafarend, E. and Rapp, R. H. (eds) (1984). *Advances in Geodesy.* Selected papers from *Rev. Geophys. Space Phys.*, AGU, Washington D.C.

Hager, B. H. (1983). *Earth Planet. Sci. Lett.* **63**, 97–109.

Hager, B. H. (1984). *J. Geophys. Res.* **89,** (B7), 6003.

Hellings, R. W., Adams, P. J., Anderson, J. D., Keesey, M. S., Lau, E. L., Standish, E. M., Canuto, V. M. and Goldman, I. (1984). "An experimental test of the variability of G using Viking lander ranging data". Jet Propulsion Laboratory (preprint).

Kaula, W. M. (1963). *J. Geophys. Res.* **68**, 4967.

Kaula, W. M. (1966). *Satellite Geodesy.* Ginn (Blaisdell), Boston, Massachusetts.

Kaula, W. M. (1969). In *The Nature of the Solid Earth* (E. Robertson, ed.), McGraw-Hill, New York.

Lambeck, K. (1980). *The Earth's Variable Rotation.* Cambridge Univ. Press, Cambridge.

Lambeck, K. (1983). *Terra Cognita,* **3**, 51–59.

Lambeck, K. and Phillips, R. J. (1980). *Rev. Geophys. Space Phys.* **18**, 27–76.

MacDoran, P. F., Miller, R. B., Buennagel, L. A. and Whitcomb, J. H. (1985). Codeless systems for positioning with NAVSTAR-GPS. In *Proc. GPS Symp.* NGS, Rockville (preprint).

Mörner, N.-A. (ed.) (1980). *Earth Rheology, Isostasy and Eustasy.* Wiley, New York.

Richards, M. A. and Hager, B. H. (1984). *EOS, Trans. Am. Geophys. Union,* **65**, 1092 (abstract).

Runcorn, S. K. (1964). *J. Geophys. Res.* **69**, 4389.

Smith, D. E., Christodoulidis, D. C., Kolenkiewicz, R., Klosko, S. M., Torrence, M. H. and Dunn, P. J. (1984). GSFC Annual Station Positions from SL5. Goddard Space Flight Center Publ., Greenbelt, Maryland.

Smithsonian Astrophysical Observatory (1966). Special Report 200, Cambridge, Massachusetts.

Turcotte. D. L. and Schubert, G. (1982). *Geodynamics*; *Applications of Continuum Physics to Geological Problems.* Wiley, New York.

Wagner, C. A. (1985). *J. Geophys. Res.* **90**, 3027–3036.

Wagner, C. A. and Sandwell, D. T. (1984). *J. Geophys. Res.* **89**, 4419–4426.

Part 2

The Global Gravity Field

2.1 Present Status and Future Improvements in Measuring the Gravity Field of the Earth and Planets

G. Balmino

Bureau Gravimétrique International, Toulouse, France

I. Introduction

There has been a breakthrough in geopotential modeling in recent years thanks to the improvement of satellite laser tracking systems and the availability of the high altitude LAGEOS satellite and to the advent of satellite altimetry. The former has yielded a better knowledge of long-wavelength variations and the latter has brought much higher resolution in the geoid over oceanic parts. There remains however a great deal that needs to be done to improve the global accuracy and resolution of Earth gravity field models for geophysics, oceanography and geodynamics. Ground techniques and conventional satellite methods cannot be improved much further, and

SPACE GEODESY AND GEODYNAMICS
ISBN 0-12-057840-9

new ways of exploring gravity variations in the resolution range of 50–200 km, especially over land, which were proposed more than a decade ago are now close to realization. These methods are satellite-to-satellite tracking (referred to as SST) and satellite gradiometry (SGRAD). From the various concepts and projects which have been proposed, we shall concentrate on the SLALOM and GRM projects in SST and the GRADIO project in SGRAD, actually putting more emphasis on the latter for which the state of development looks very promising.

The same techniques may be applied to other terrestrial bodies, such as the Moon, Mars and Venus about which, although big progress has been made in the last decade, planetologists require higher precision and/or higher resolution, at least over some parts of their surface.

II. The Status of Geopotential Knowledge

We present in Table I the requirements in resolution and accuracy for various studied phenomena of the three major disciplines which need Earth gravity field models: geophysics (which use gravity anomalies), oceanography (which deals with geoid height variations) and geodesy-geodynamics.

Some of these requirements are very stringent in view of the presently achieved capabilities of geopotential modeling, especially over land as far as geophysics and geodesy are concerned. To appreciate this, let us briefly recall the classical methods used up to now to obtain such models, as shown on Fig. 1. They are (i) gravimetry, astrogeodetic measurements (deviation of vertical), and sometimes torsion balance observations (local) on the Earth surface; (ii) perturbation analysis of satellite orbits, which is one of the lengthiest, most tedious and indirect methods; and (iii) satellite altimetry which yields knowledge of an averaged sea surface (assumed to be close enough to the geoid for geophysical investigations). New techniques are also illustrated on this figure and are mentioned in the following. It has to be noted that the potential function decreases with the altitude h with $(R/R+h)^{k+1}$, where k is the wave number ($k \approx 20{,}000/\rho$; ρ is the resolution in kilometers), which explains why classical satellite methods (with h in the 500–5000-km range) only give the low-frequency part of the field, whereas ground-level data allows some high resolution. In between, new space techniques which use one or several free flyers at the lowest possible altitudes (160–220 km for envisaged projects) and which require drag compensation systems or frequent orbit corrections, could bring medium-wavelength information (in the 50–200-km range, typically) all over our planet.

TABLE I. Magnitude of various quantities related to Earth gravity perturbations at different resolutions[a]

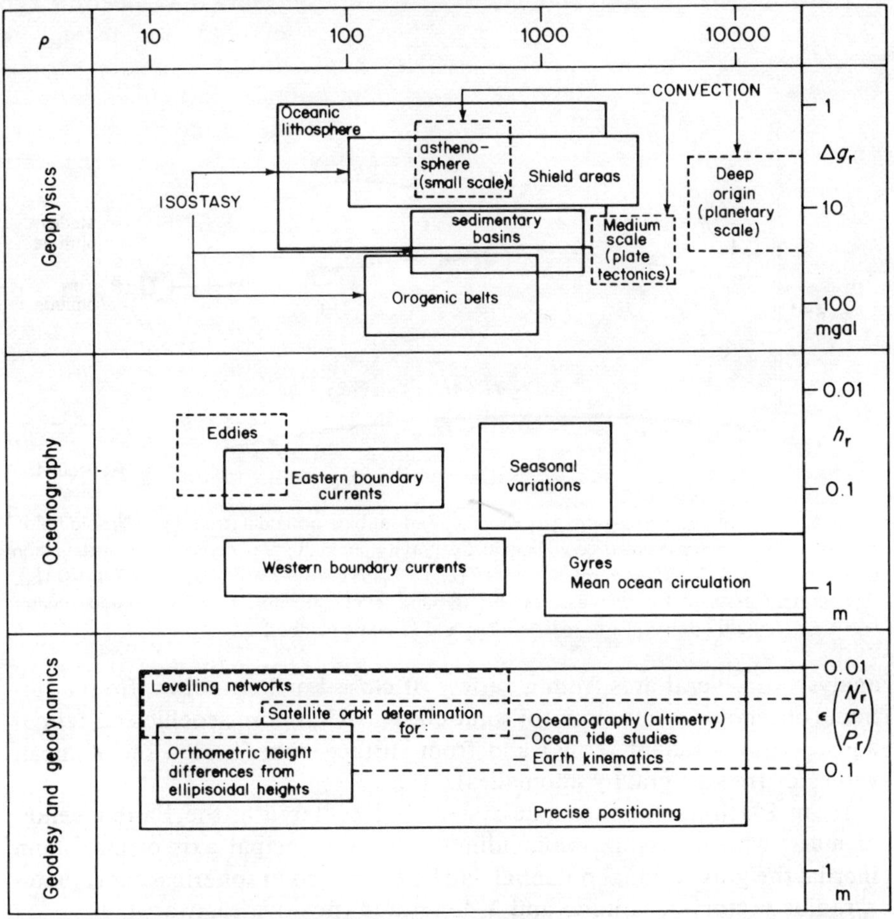

[a] ρ, resolution, or half wavelength of phenomenon (km); Δg_r, relative variation of gravity anomaly (mgal); h_r, relative variation of sea surface height (m); $\varepsilon(N_r)$, accuracy of geoid height relative variation (m) where $N_r \simeq 3.25 \times 10^{-4} \rho \Delta g_r$; $\varepsilon(R)$, accuracy of satellite orbit radial component (m); and $\varepsilon(P_r)$, accuracy of relative geodetic positioning (m).

A. Past methods

1. Basic Aspects

The philosophy behind the method is to combine in the least squares sense, different normal equation systems derived separately from the perturbation

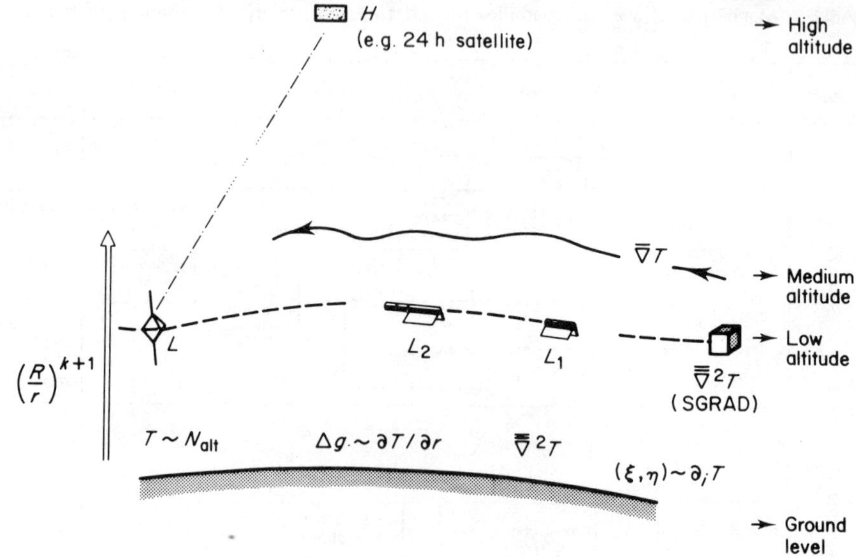

Fig. 1. Ways of determining the geopotential. T, disturbing potential (true field-ellipsoid field); N_{alt}, satellite altimetry derived geoid height; Δg, gravity anomaly; ξ, η, components of deviation of vertical; H–L, SST, high–low mode, yields $T(L)$; L_1–L_2, SST, low–low mode, yields $T(L_2)-T(L_1)$; $\bar{\bar{\nabla}}^2 T$, gravity gradient tensor measurements (ground level, satellite: SGRAD); r, radius vector ($r \approx R+h$, $h=$ altitude); R, Earth mean radius; and k, wave number.

analysis of several arcs from a variety of close Earth satellites, from additional observation equations for some classes of harmonic coefficients (zonal and tesseral resonant terms) and from surface data (here a set of mean values of free-air gravity anomalies).

In an Earth-fixed coordinate system $\{x_i\}$ centered at the Earth's center of mass, with the x_3 axis coinciding with the principal axis of maximum inertia, the gravitational potential will be expressed in spherical coordinates r, radius vector; φ, latitude and λ, longitude (positive eastward), as:

$$U(r, \varphi, \lambda) = \frac{GM}{r}\left[1 + \sum_{l=2}^{\infty} \sum_{m=0}^{l} \left[\frac{R}{r}\right]^{l} (\bar{C}_{lm} \cos m\lambda + \bar{S}_{lm} \sin m\lambda)\bar{P}_{lm}(\sin \varphi)\right] \tag{1}$$

where GM is the product of the gravitational constant G by the Earth mass M, whose adopted value is generally 398,600.5 km^3 s^{-2}; R is the reference length value, which is usually equal to the semi-major axis of a reference ellipsoid (6378.137 km); and $\bar{P}_{lm}(x)$ the fully normalized Legendre functions (and polynomials when $m=0$) such that

$$\frac{1}{4\pi} \iint_{\text{unit sphere}} \bar{P}^2_{lm}(\sin \varphi) \begin{bmatrix} \cos^2 m\lambda \\ \sin^2 m\lambda \end{bmatrix} d\sigma = 1$$

where $\bar{C}_{lo}$ are the zonal harmonic coefficients and $\bar{C}_{lm}\bar{S}_{lm}$, $m>0$ are the tesseral harmonic coefficients. In the adopted reference frame, one has $\bar{C}_{21}=\bar{S}_{21}=0$. If the x_3 axis was different, or if we assume that the main inertia axis moves as it must in the case for a real elastic Earth, these values should be made time dependent by means of a model (Reigber, 1981).

For the Earth, $\bar{C}_{20}\approx -5\times 10^{-4}$, while the higher-degree coefficients appear to decrease in magnitude according to $10^{-5}/l^2$ (Kaula's rule).

The coefficients $\bar{C}_{lm}$, $\bar{S}_{lm}$ are then determined:

a From the analysis of satellite orbit perturbations

In an inertial reference frame $\{X_i\}$, the acceleration of a spacecraft subject to the potential (1) and to other (known) accelerations **F**, is determined by:

$$\mathbf{X}=\nabla\, U(\bar{C}_{lm}, \bar{S}_{lm})+\mathbf{F} \tag{2}$$

(with given initial conditions).

From analytical satellite motion theory (e.g. Kaula, 1966a) it is easy to show that a given mean orbit is mostly sensitive to some classes of harmonics and that, given a certain tracking accuracy, only a finite number of linear combinations of coefficients can be determined. Furthermore it is known that many different orbits of varied inclinations are necessary to then separate these coefficients. To meet the accuracy of actual tracking systems, numerical integration must be used to solve systems like Eq. (2), although the matter remains a classical inverse celestial mechanics problem in which force parameters have to be computed from the knowledge of the orbit (starting from initial values $\bar{C}_{lm}$, $\bar{S}_{lm}$), by linearizing the solution of Eq. (2).

Two aspects of the gravity estimation problem are basic to this approach; both pertain to the arc length of data which can be practically reduced to determine an initial-state vector. First, in the case of our planet, we are definitely limited by the effect of the atmosphere of which the perturbations cannot yet be accurately modeled (Barlier, 1981), and the empirical factor which must be adjusted for each arc to account for all model uncertainties is valid for some days only in some cases. Second, even when long arcs of continuous data are available (which was not the case in the first 12 years or so of the space age), extensive numerical integration and other small but poorly modeled accelerations introduce errors which can lead to erroneous results. Thus it is preferable to divide a long arc into a number of smaller batches in order to enhance accuracy. Therefore the method used here is based upon filtering multiple batches of data for several satellites.

Besides the direct inversion of Eq. (2) through differential correction techniques by using observations in a two-stage process, one can make use of other characteristics of the response of an orbit to some coefficients. As a matter of fact, the dominant orbit perturbations are the secular changes in the angular elements due to $\bar{C}_{20}$ and the other even-degree zonal terms, with superimposed long-period perturbations with frequencies $\dot{\omega}$, $2\dot{\omega}, 3\dot{\omega} \ldots$ (ω being the argument of perigee), that are governed by the odd and even zonal harmonics. Therefore, special methods have been developed (e.g. Kozai, 1966) which use a multi-stage filtering technique yielding expansions of residuals in the orbital elements of a satellite in terms of the zonal coefficients; geopotential model specialists then make use of these final equations as derived by various authors. The same approach has been adopted for tesseral harmonic coefficients corresponding to cases when a given satellite orbit is in shallow resonance with the Earth rotation. Physically it occurs when successive ground paths of the satellite are separated by an interval equal to exactly the wavelength of the geopotential harmonic. After a certain number of revolutions of the satellite, the ground track sequence repeats itself (after the Earth has rotated k times) and the satellite's motion is perturbed in an identical manner, enhancing the earlier perturbations. Cases encountered most frequently correspond to $k=1$ (sometimes $k=2$), and involve harmonic coefficients of order $m=11$, 12, 13, 14 and 15 (sometimes $m=26$, 27, 28, 29, 30, 31).

b From the analysis of surface data

Low-order coefficients can be reliably estimated since they result in perturbations with periods that exceed a few hours and which can be well sampled from tracking stations of a global network, but the higher-degree coefficients become increasingly difficult to estimate because of the $(R/r)^l$ attenuating factor and the fact that $r-R \geqslant 600$ km (at least for all processed orbits due to the atmosphere). To illustrate this situation Fig. 2 indicates the approximate limits of the sensitivity of the satellite to the potential for different accuracies: 10 m (typical for optical observations), 1 m (old laser and classical Doppler systems), 0.20 m (actual laser stations); the shaded regions between the lines indicate those coefficients that cannot be estimated; the coefficients below A_x can be determined from short period perturbation analyses while those above A'_x follow from shallow resonance studies (for a tracking accuracy equal to x m).

Therefore, it is necessary to complement the satellite orbit analysis informations by other data in order to obtain a global model complete up to a certain degree and order $l_{max} = m_{max} = L$. Presently such data consist in:

(i) Surface free-air gravity anomalies, Δg, which are related to U through the well-known equation (Heiskanen and Moritz, 1967), $\Delta g = -\partial T/\partial r - 2T/r$, in spherical approximation, where T, the disturbing potential, equals $U - U^*$, with U^* the gravitational potential of a selected

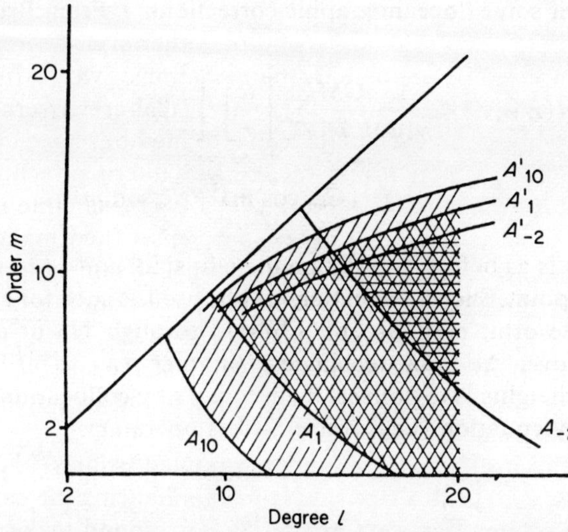

Fig. 2. Schematic representation of the non-zonal coefficients that cannot be determined from satellite orbit perturbation analyses. The coefficients of degree and order (l, m) lying in the shaded area between the lines A_x and A'_x cannot be determined if the tracking accuracy is lower than x meters (from Lambeck, 1979).

reference ellipsoid (expressed as a spherical harmonic expansion involving only $\bar{C}^*_{2n,0}$ terms). This equation yields:

$$\Delta g(r, \varphi, \lambda) = \frac{GM}{r^2} \sum_{l=2}^{\infty} (l-1) \left[\frac{R}{r}\right]^l \left[(\bar{C}_{l,0} - \bar{C}^*_{l,0}) \bar{P}_{l,0}(\sin\varphi) + \sum_{m=1}^{l} (\bar{C}_{lm} \cos m\lambda + \bar{S}_{lm} \sin m\lambda) \bar{P}_{lm}(\sin\varphi) \right] \tag{3}$$

where the point (r, φ, λ) lies on the reference ellipsoid, that is with $r \approx R(1 - f \sin^2\varphi)$, f being the flattening of this ellipsoid and R the mean semi-major axis.

Equation (3) is used for residual mean values of gravity (with respect to the initial field of coefficients $\bar{C}^\circ_{lm}$, $\bar{S}^\circ_{lm}$) over spherical rectangles $\Delta\varphi \times \Delta\lambda$ (here with $\Delta\varphi = \Delta\lambda = 1°$) and the set of these yields normal equations for corrections to the harmonics.

(ii) Geoid heights N with respect to a reference ellipsoid (usually the same as above) measured by satellite altimetry, that is derived as the difference between the satellite altitude H above the ellipsoid (known with some error from a necessary first-stage orbit adjustment), and the observed distance h between the satellite and the sea surface (assumed to be the geoid surface, practically after some "oceanographic corrections"). From Brun's formula, one has:

$$H-h=N(r,\varphi,\lambda)=\frac{1}{\gamma(\varphi)}\frac{GM}{r}\sum_{l=2}^{\infty}\left[\frac{R}{r}\right]^{l}\left[(\bar{C}_{l,0}-\bar{C}^{*}_{l,0})\bar{P}_{l,0}(\sin\varphi)\right.$$
$$\left.+\sum_{m=1}^{l}(\bar{C}_{lm}\cos m\lambda+\bar{S}_{lm}\sin m\lambda)\bar{P}_{lm}(\sin\varphi)\right] \quad (4)$$

where (r, φ, λ) is as before, a point on the ellipsoid, and $\gamma(\varphi)$ is the normal gravity at that point. Such an equation may be used as such for each altimetry data point (the orbit may also be adjusted through H) or also used for mean values over $\Delta\varphi\times\Delta\lambda$ rectangles. Another way to utilize altimetry derived geoid heights is to convert them to Δg's by collocation or by means of Molodensky equation (i.e. inverse Stokes operator).

2. Data Handling

a Satellite data

(*i*) *Processing.* Since the estimation process is basically linear, it is necessary to compute by some means reference orbits for the involved satellites which model all our *a priori* information on the gravitational and non-gravitational forces. This improved orbit determination is based on a preliminary knowledge of the initial-state vector and the observational data which are available from the observing stations.

The observations may take the form of angular data, range or range-rate data, with varying accuracy and reliability. The differential correction process then works as follows:

Let $\bar{u}=(u_1, u_2, \ldots, u_n)$ denote any set of independent parameters, which in the present case will be:

(i) arc-dependent parameters (inner unknowns): $\bar{P}_0$ (set of components of the initial-state vector), k_D and k_P (multiplying factors for the drag and solar pressure, respectively), $\bar{f}_0$ (set of values of frequency offsets per pass, in the case of Doppler observations).

(ii) arc-independent parameters (outer unknowns): $\bar{C}_{lm}$, $\bar{S}_{lm}$ (spherical harmonics of the Earth's gravity field), $\bar{S}$ (set of coordinates of the observing stations).

Let $q=q\ (t,\bar{u})$ denote any type of observation from any station; then if the time t and an approximate set $\bar{u}_0$ is given, a reference orbit can be obtained by integrating the equations of motion, Eq. (2) and a theoretical observation $q_c\ (t,\bar{u}_0)$ can be computed from the orbit ephemeris, approximate station coordinates and from analytical expressions for this type of observation (Balmino, 1974a).

On the other hand, we have observed, with a certain error, a quantity q, denoted q_0. Writing that the true vector $\bar{u}$ is $\bar{u}_0+\Delta\,\bar{u}$ where $\bar{u}_0$ is the initial-(known) vector and $\Delta\bar{u}$ the unknown, we perform a separation between the inner and outer parameters and expand linearly the difference q_0-q_c as:

$$q_0(t,\bar{u})=q_c(t,\bar{u}_0)+\sum_{\text{inner}}\left[\frac{\partial q}{\partial\bar{u}_i}\right]_{\bar{u}=\bar{u}_0}\Delta u_i^{\text{in}}+\sum_{\text{outer}}\left[\frac{\partial q}{\partial\bar{u}_i}\right]\Delta u_i^{\text{out}} \tag{5}$$

or

$$\Delta\bar{q}^{(K)}=\bar{a}_{\text{I}}^{(K)}\,\Delta\bar{u}_{\text{I}}+\bar{a}_{\text{O}}^{(K)}\Delta\bar{u}_{\text{O}} \tag{6}$$

where K stands for the Kth arc of satellite data which are all processed together, $\bar{u}_{\text{I}}$ and $\bar{u}_{\text{O}}$ are the vectors of inner and outer unknowns and $\Delta\bar{q}$ is the residual vector (q_0-q_c).

With all the variables given in Eq. (5), it is, independent of the number of satellite arcs and observations, practically impossible to solve for all unknown parameters at the same time, and the arc-dependent parameters $(\Delta\bar{u}_{\text{I}})$ have to be eliminated before solving for the arc-independent unknowns. To secure, on the other hand, the linear approximation of Eq. (5), the determination of improved satellite orbits is necessary beforehand. Consequently the differential correction procedure is split into two steps:

(i) The differential orbit improvement process, in which the right-hand side member of Eq. (5) is reduced to its first term (for $\Delta\bar{u}_{\text{I}}$). A set of new values for the inner parameters is then obtained after the process of expanding the residuals Δq has been iterated (until convergence), and serves as initial values for the next step.

(ii) The geodetic parameter correction process, where the residual vector is completely re-expanded, as in Eq. (5), yielding a complete observation equation system Eq. (6).

It remains to be said that the satellite motion equations are either solved analytically (in the case of the Smithsonian Astrophysical Observatory (SAO) Standard Earth models), or integrated numerically by means of Cowell-type predictor corrector methods, sometimes regularized by a Sundmann transformation (Moynot, 1971) with additional circular terms (Balmino, 1974b), the stepsize being kept constant and equal to 30–120 s in most cases. The partial derivatives needed in Eq. (5), and which involve the satellite position vector $\mathbf{r}\,(t)$, are computed as:

$$\left[\frac{\partial q(t,\bar{u})}{\partial u_i}\right]_{\bar{u}=\bar{u}_0}=\sum_{j=1}^{3}\frac{\partial q(t,\bar{u})}{\partial r_j(t)}\frac{\partial r_j(t)}{\partial\bar{u}_0} \tag{7}$$

where r_j are the components of **r** and the partials $\partial r_j(t)/\partial \bar{u}_0$ are computed either analytically (in cases where the solution to Eq. (2) is also analytical) or numerically by direct simultaneous integration of the variational equations for the inner parameters or by a Peano–Baker method (Balmino, 1974b) for the $\bar{C}_{lm}$ and $\bar{S}_{lm}$ coefficients.

Forces **F** in Eq. (2) include: gravitational forces due to the Sun and Moon (direct and tidal forces which include oceanic components), non-gravitational forces, namely atmospheric drag (Jacchia 1965 model and Barlier correcting coefficients, Berger-Jaeck, 1976) and solar radiation pressure (direct effect and Earth's albedo); apparent forces due to the choice of the reference system (Balmino, 1974c) and additional terms coming from the pole position motion which results in corrections to $\bar{C}_{21}$, $\bar{S}_{21}$ and $\bar{C}_{20}$ (Reigber, 1981).

Each observation equation being properly weighted, the resulting normal equation system for arc K can be written as:

$$\begin{bmatrix} N_{\mathrm{II}} & N_{\mathrm{IO}}^{\dagger} \\ N_{\mathrm{IO}} & N_{\mathrm{OO}} \end{bmatrix}^{(K)} \begin{bmatrix} \Delta u_{\mathrm{I}}^{(K)} \\ \Delta u_{\mathrm{O}} \end{bmatrix} = \begin{bmatrix} b_{\mathrm{I}} \\ b_{\mathrm{O}} \end{bmatrix}^{(K)} \tag{8}$$

where the normal sub-matrices are derived from the $\bar{a}^{(K)}$'s matrices of Eq. (5) and from the weight matrix of the observations. As we said, in order to reduce the number of unknowns, the $\Delta u_{\mathrm{I}}^{(K)}$ are eliminated, and then

$$N^{*(K)}\,\Delta u_{\mathrm{O}} = \mathrm{b}^{*(K)} \tag{9}$$

with

$$N^{*(K)} = (N_{\mathrm{OO}} - N_{\mathrm{IO}} N_{\mathrm{II}}^{-1} N_{\mathrm{IO}}^{\dagger})^{(K)}$$
$$b^{*(K)} = (b_{\mathrm{O}} - N_{\mathrm{IO}} N_{\mathrm{II}}^{-1} b_{\mathrm{I}})^{(K)}$$

The combination of the normal equations is then simply performed in adding the $N^{*(K)}$ matrices and the $b^{*(K)}$ vectors, respectively, so that:

$$\left(\sum_K N^{*(K)}\right)\Delta u_{\mathrm{O}} = \sum_K b^{*(K)} \tag{10}$$

(*ii*) *The data sets.* Directional (optical, interferometric), range (radar, then laser), range rate or radial velocity (Doppler) measurements collected since the beginning of the space age have been used in different ways, independently or in a combined way, by the various teams engaged in geopotential modeling in the world. As an example, Fig. 3 shows a diagram of all satellite orbits included in the European model GRIM-3B (Reigber *et al.*, 1984), which spans quite a big area around the Earth.

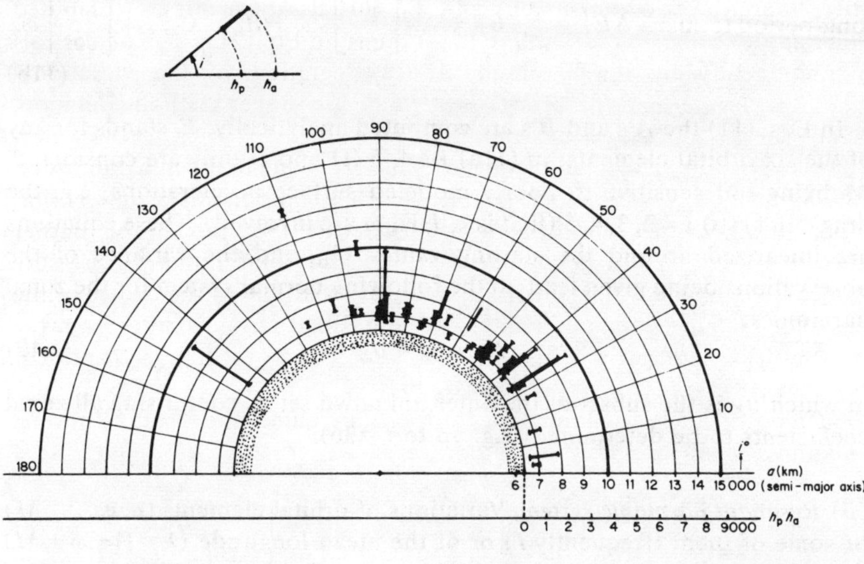

Fig. 3. Diagram of distribution in space of satellite orbits used in GRIM-3B. *i*, inclination; h_p, altitude of perigee; h_a, altitude of apogee.

b Further treatment of special perturbations

In addition to the system of combined reduced normal equations (Eq. 9), additional normal equation subsystems coming from specific studies on zonal and resonant harmonics are combined with the normal system (Eq. 10).

(*i*) *Zonal harmonics systems.* Observation equations for the even and odd zonal harmonics of the geopotential can be derived from secular motions of the angular variables ω and Ω and from amplitudes of long-periodic terms with ω as argument for the orbital elements (e, I, Ω, ω)—eccentricity, inclination, longitude of node and argument of perigee, respectively. Such equation sets may be found in previous works by Kozai (1969), Cazenave *et al.* (1972) and Gaposchkin (1973), with appropriate linear corrections to be made whenever one changes the initial harmonic values.

From the analysis of mean elements of different satellites at various inclinations, one gets observation equation systems of the form:

$$\text{secular terms:}\quad \dot{E}_i^S = \sum_{n=1}^{p} A_{E_i}^n \bar{C}_{2n,0} + A_i^0 \bar{C}_{2,0}^2 \tag{11a}$$

long-period terms: $$\Delta E_j^L = \sum_{n=1}^{q} B_{E_j}^n \bar{C}_{2n,0} \begin{bmatrix} \sin \\ \cos \end{bmatrix} 2\omega + B_{E_j}^{2n+1} \bar{C}_{2n+1,0} \begin{bmatrix} \sin \\ \cos \end{bmatrix} \omega \tag{11b}$$

In Eqs. (11) the A's and B's are computed analytically, E_i stands for any of the six orbital elements; in (11a) $i = 4, 5$ (Ω and ω only are considered, M being too sensitive to poorly modeled surface accelerations, e.g. the drag); in (11b) $i = 2, 3, 4, 5$ (that is e, I, Ω, ω are involved). These equations are linearized around the starting values $\bar{C}_{l,0}^0$ and the variance of the observations being given lead to the following normal system for the zonal harmonics:

$$N_Z \, \Delta u_Z = b_Z \tag{12}$$

in which u_Z is the subset of the outer unknown set u_0 comprising all zonal coefficients to be determined (e.g. up to $l = 36$).

(*ii*) *Resonant harmonic system.* Variations of orbital elements ($a, e, \ldots, M$) or some of them (frequently I) or of the mean longitude ($\lambda = \Omega + \omega + M$) of well-observed resonant satellite orbits, have also been analysed to derive independent sets of observation equations for even and odd degree resonant harmonics of the order $m = 11$, 12, 13, 14, 15. Several methods are used: one of them (Reigber, 1973a) starts from a decomposition of any orbital element ($a, \ldots, M$; or I; or λ) in the form:

$$\Delta E_i^v = \sum_{l=m}^{N} (\gamma_{i,lm}^v \bar{C}_{lm} + \sigma_{i,lm}^v \bar{S}_{lm}) \tag{13}$$

where ΔE_i^v denotes the amplitude of the observed resonance effect (of basic period T), v the frequency number, the coefficients γ and σ being given by:

$$(\gamma, \sigma)_{i,lm}^v = \frac{2T}{v\pi} \int_0^1 \cos v\pi t \sum_{p,q} (f_\gamma, f_\sigma)_{lmpq}^v \tag{14}$$

and computed by Gauss quadrature formulae. The p,q summation and the f terms represent the right-hand sides of the Lagrange equations according to Kaula's expansion of the disturbing potential (Kaula, 1966a).

This technique starts from the computation of a reference orbit as in II.a.(i), where inner parameters have been determined; then by ignoring the resonant terms under consideration, a satellite ephemeris is generated by numerical integration (taking account of all other forces), and boundary values of the corrections in $E_1, \ldots, E_6$ and the amplitudes ΔE_i^v are computed by least squares. All the details can be found in Reigber (1973b).

Various sets of equations may be found in Reigber and Rummel (1979) for $m = 12$, in Balmino and Reigber (1975) for $m = 13$ and Reigber and Balmino (1976) for $m = 14$. The semi numerical (or semi analytical) approach is based on the integration of averaged Lagrange–Gauss planetary equations, so as to filter out the shortest periodic variations in the orbital elements (Borderies, 1976). Here observables are the osculating elements which are also filtered in a first-stage process and then used in a step of differential correction improvement of a mean orbit and of some gravity harmonics. This yields observation equations similar to Eq. (13), which can be found in Klosko and Wagner (1975) for $m = 12$, Dunn (1978) and King-Hele *et al.* (1979) for $m = 14$ or Klokocnick and Kostelecky (1980). A particular approach which uses decaying satellite inclination variation analysis has led several authors to other types of equations. Details can be found in King-Hele *et al.* (1974, 1975a, b) who used the method for $m = 15$, or in Wagner (1974) for $m = 11$ or Klokocnick and Kostelecky (1978).

All these equations, being properly rewritten for the *ad hoc* starting values and weighted according to a variance derived from the observation accuracy and adjustment are used to generate additional sets of normal equations, that is:

$$N_{R,m}\, \Delta u_{R,m} = b_{R,m} \tag{15}$$

where the subscripts R and m stand for resonance and a given order m, respectively.

c Further treatment of station positions

Observation equations for coordinate corrections of particular stations (unknown denoted Δu_S) may come from:

(i) independent works of geometrical geodesy yielding

$$N_S^{(gg)}\, \Delta u_S = b^{(gg)} \tag{16}$$

(ii) semi-dynamical methods in which for a given data arc (H), we process the following equations (analogous to Eq. 5)

$$[A, B]\, [\Delta \tilde{u}_{\mathrm{I}}^{(H)}\, \Delta u_S]^T = \Delta q \tag{17}$$

where $\Delta \tilde{u}_{\mathrm{I}}^{(H)}$ is the set of inner parameters computed for this arc, and where corrections to the force model coefficients are not included. Eliminating the inner unknowns and adding all the arcs pertinent to the considered stations lead to:

$$\left[\sum_H N_S^{*(H)}\right] \Delta u_S = \sum_H b_S^{*(H)} \tag{18}$$

(iii) survey informations which usually have a high accuracy for short station-to-station vectors (up to a few hundred kilometers). The resulting equations may be treated as constraints (in the absolute sense), in which case elimination or Lagrange multiplier techniques can be used, or as additional observation equations with a large weight (an approach which we favor, by experience), yielding other normal systems:

$$N_S^{\text{survey}}\,\Delta u_S = b^{\text{survey}} \tag{19}$$

d Surface data

(*i*) *Gravity anomalies.* Mean values (over rectangles $\Delta\varphi^\circ \times \Delta\lambda^\circ$) of free-air gravity anomalies, Δg, are usually obtained:

(i) by collocation (Wiener–Hopf filtering) by using all data contained in each rectangle and its neighborhood.
(ii) by prediction, using distant data and/or topo-isostatic models (correlations with elevation).
(iii) by inversion of satellite altimetry derived geoid heights (as said above).

Global sets of $1^\circ \times 1^\circ$ data have accordingly been computed, but they do not cover the whole Earth and are unhomogeneous (Rapp 1979, 1983).

(*ii*) *Geoid heights.* So far, no one has yet exploited the point data in their original form for geopotential determination. The tendency is either to use the equivalent inverted gravity anomalies or mean values of the geoid height that are computed by collocation.

(*iii*) *Using the mean values.* Due to the resolution of all geopotential models in the 1970s (about 500 km corresponding to a maximum degree and order equal to 36), many authors chose to take account of surface data in the form of $5^\circ \times 5^\circ$ means, the advantages of which are a smaller computer time and an easier weighting, especially for areas where data were unsatisfactory; the disadvantages consisting of a further step of computation and smoothing. Others, such as the European model team members, are making use of the original mean values. In all cases an additional normal system is generated:

$$N_{\text{surface}}\,\Delta u_0' = b_{\text{surface}} \tag{20}$$

with $\Delta u_0'$ the subset of included harmonics, usually not the whole set, that is those of a minimum degree $1_{min} > 2$ to avoid artifacts due to uneven coverage.

3. The Models

The first harmonic coefficients determined at the beginning of the space age were the zonal terms (e.g. Kozai, 1959, 1964) from optical observations, and the first global models up to degree and order 6 were determined by Guier and Newton (1965) and Anderle (1966), who already used numerical integration and Doppler measurements, and then Izsak (1966) by analytical methods and with optical data only. The first one who combined satellite orbit perturbation analysis and gravity measurements was Kaula (1966b). Numerous models were then computed, improved from time to time, principally by four groups (the first three are in the United States). Each one uses its own methodology, algorithms and software, but still shares a lot of common data.

a The Naval Surface Weapons Center

Previously the Naval Weapons Laboratory, the Naval Surface Weapons Center (NSWC), under the leadership of Anderle, has computed models that are characterized by the following:

(i) numerical integration of the equations of motion,
(ii) utilization of only Doppler data, with an excellent coverage (presently about 40 stations—the DMA network),
(iii) strong weighting of the TRANSIT polar satellites,
(iv) short arcs (2–5 days), and
(v) simultaneous determination of tesseral and zonal harmonics, the latter being obtained from the short-periodic perturbations.

These models (such as NSWC-9 Z.2) are unfortunately classified; only the oldest ones are published. They seem to be of excellent quality, as this can be concluded from published scientific results based on these models, such as the polar motion monitoring.

b The Smithsonian Astrophysical Observatory

The Smithsonian Astrophysical Observatory, under the leadership of Gaposchkin, was the real pioneer in this discipline. Several very good models, named Standard Earth, were published and are characterized by the following:

(i) analytical theories are used to describe the satellite perturbations complemented by some empirical terms (for instance, surface forces),
(ii) optical and laser observations,
(iii) long arcs (15–30 days),
(iv) used in combination with surface gravity data from the SE2 model, and
(v) separate determination of zonal harmonics from secular and long-period orbital perturbations; unfortunately, the SAO stopped these activities in 1980.

c The Goddard Space Flight Center

The Goddard Space Flight Center, under the impulse of many people (Marsh, Wagner, Klosko, Lerch, among others), have computed tens of models and have published 12 (GEM-1 to GEM-10C; see Lerch *et al.*, 1981), plus many solutions tailored to given satellites (SEASAT, STARLETTE, LAGEOS). Their main features are:

(i) numerical integration,
(ii) sophisticated models of surface forces,
(iii) all types of data are included and surface observations are combined in the even-number GEM solutions,
(iv) separate determination of the station positions, using short arcs (semi-dynamical method),
(v) simultaneous computation of the zonal coefficients, and
(vi) addition of *a priori* statistical information to help separating the unknowns.

d The Groupe de Recherches de Géodésie Spatiale (France) and the Deutsches Geodätisches ForschungsInstitut (FRG)

These two European groups have already published four global models (GRIM- 1, 2, 3, 3B) and one tailored solution for LAGEOS (GRIM 3B-L1; Reigber *et al.*, 1984). The characteristics of these works (which started much later than the other groups) come from the experience of the others and have tried to combine the advantages of all previous methods:

(i) numerical integration (stabilized and regularized),
(ii) rigorous treatment of the reference system,

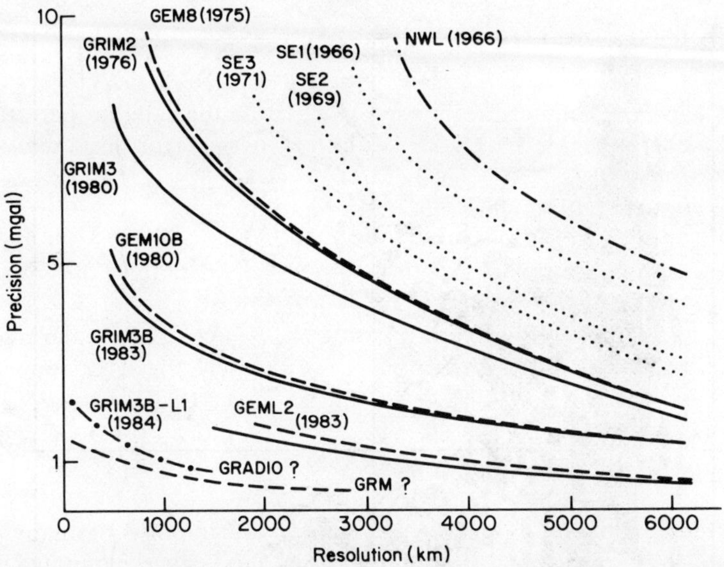

Fig. 4. Progress in geopotential modeling from 1966 to 1984 for the following projects; – – –, GEM (Goddard Earth Models); ——, GRIM (GRGS and SFB78/DGFI models—European); · · · ·, SAO (Smithsonian Astrophysical Observatory Models); – · – · –, NWL (Naval Weapon Laboratory models).

(iii) improvements in the algorithms for the surface forces, the derivation of $1° \times 1°$ ground gravity data equations, and the computation of the solution, etc.,

(iv) all satellite data types are considered,

(v) combination of external normal equation systems for station coordinates, zonal harmonics and resonant terms, and

(vi) simultaneous determination of all parameters, without *a priori* constraints or collocation-type equations.

To briefly summarize the evolution of geopotential modeling efforts and results (truncated spherical harmonic series up to a maximum degree and order $l(\max) = m(\max) = k$, varying from 6 in the earliest models, to 36 in the most recent solutions), we show in Fig. 4 their quality, roughly displayed in terms of the (estimated) precision of the derived free-air gravity anomalies versus the resolution ρ. It may seem overly pessimistic to the nonspecialist who must remember that such conditions (especially in the lowest ρ's) are mostly due to the existence of major gaps in the ground data coverage for land areas, as demonstrated by Fig. 5, where the location of unknown (therefore predicted) anomalies is shown. Obviously, local potential (or

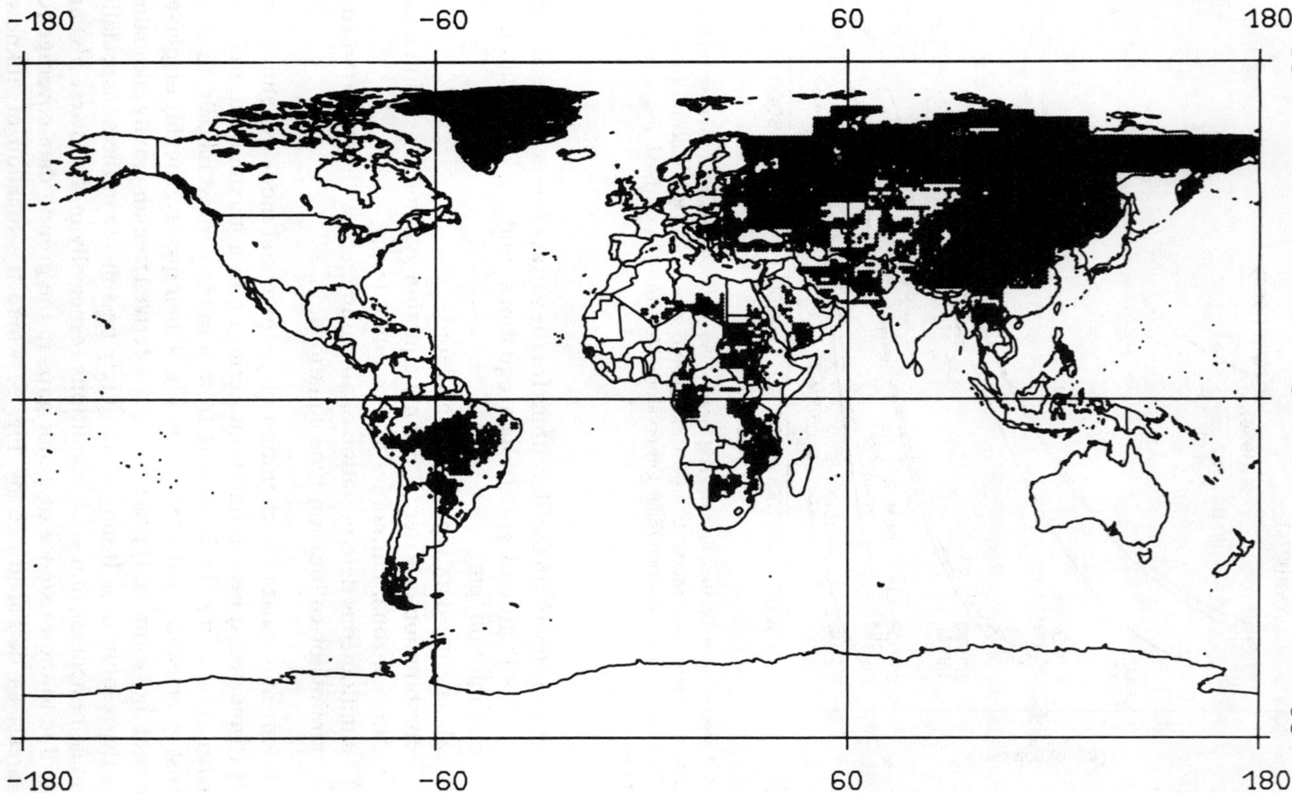

Fig. 5. Location of the 6361 anomalies estimated by geophysical techniques (from Rapp, 1983).

derived gravity) variations are much better known (in the geophysical sense i.e. in equivalencing mean sea surface and geoid) over oceanic areas thanks to the GEOS-3 and SEASAT satellite altimetry experiments, and what geophysicists and geodesists are looking for is a similar knowledge of the field everywhere over land; oceanographers need an independent "true" geoid (reference surface) with respect to which they could represent the sea surface variations at the decimenter level, or better.

B. Continuation of Past Methods

1. With Classical Data

The three major groups mentioned above, still engaged in geopotential modeling activities, plus a group at the University of Texas, will continue to acquire satellite data of high precision (mostly range measurements by third-generation laser tracking stations) to further improve their models. As an example, the determination of a GRIM-4 model has started at GRGS and DGFI and should be completed sometimes in 1986. In parallel, a few additional surface gravity data sets will be slowly obtained and should help fill a (small) part of the gap.

2. With a New Generation of Satellite Data

It is well recognized that the crucial problems of geopotential modeling from space lie in the satellite altitude and the coverage of data. As said before, the Defence Mapping Agency Doppler tracking network is probably the best so far with respect to the latter although its accuracy is still limited. That is why new systems are under study such as the DORIS tracking system, an uplink one-way two-frequency range rate measuring technique of accuracy of about 0.3 mm/s, to be put on board the French satellite SPOT-2 (CNES, 1984) as a test experiment with 50 ground stations in 1987 and to be used further as the major positioning system of geodynamics satellites such as TOPEX–POSEIDON (Born *et al.*, 1984). Such a system will not only ideally serve our community, as far as gravity field studies at long and medium wavelengths are concerned, but will also allow the determination of tectonic plate motions over a decade or so.

In addition, progress will also come from a refinement and better coverage of the satellite altimetry mapping technique thanks to the advent of new satellites equipped with such systems: GEOSAT launched in Spring 1985, ERS-1, a European oceanographic satellite (European Space Agency 1984) and TOPEX–POSEIDON. Higher-resolution maps of the oceanic "geoid" will therefore be available to the geophysicists but will not be fully adequate to oceanographers.

III. Future Projects for the Earth

The requirements and actual limitations of present models have been recognized on many occasions (e.g. National Research Council, 1979) and new methods have been subsequently proposed; SST and SGRAD techniques are now the most popular, although some advanced studies using very low-altitude satellites tethered to the space shuttle are also being made, but will not be technically detailed here.

A. Possible SST and SGRAD Missions—First Generation

1. SST

High–low SST (which theoretically provides the disturbing potential at the lowest point) was performed in an experimental way, though successfully, between the 24-hr satellite ATS-6 and the GEOS-3 altimetric satellite, but yield only some $5° \times 5°$ mean values of surface gravity anomalies due to the medium altitude (≈ 800 km) of the lowest orbiter. The technique could obviously be used again but would need a much lower altitude satellite and an increase in the electronic measurement accuracy made between the two spacecraft, by several orders of magnitude. Groups engaged in SST methods presently favor the low-low mode, in which (at least) two orbiters flying very close circular trajectories perform a radial velocity measurement at the 1 μm/s accuracy level, at an altitude of about 200 km or less, measuring directly the difference of the disturbing potential values at the two satellite points.

Such a concept was foreseen early as the next "big" way to improve geopotential models and the United States GRAVSAT project originated from it (see for instance, Pisacane *et al.*, 1981). Technical difficulties and budget constraints have postponed the real start of the project and even reoriented it somewhat. The present concept, named GRM (Smith *et al.*, 1982), is to map the Earth's gravity and magnetic fields using two 2000-kg drag-free spacecraft orbiting in polar orbit at a mean altitude of 160 km (Fig. 6) and performing: (a) between them double one-way Doppler observations at two frequencies (91 GHz, 42 GHz), resulting in radial velocity change measurements to a value of 10^{-6} m/s; (b) on one spacecraft scalar

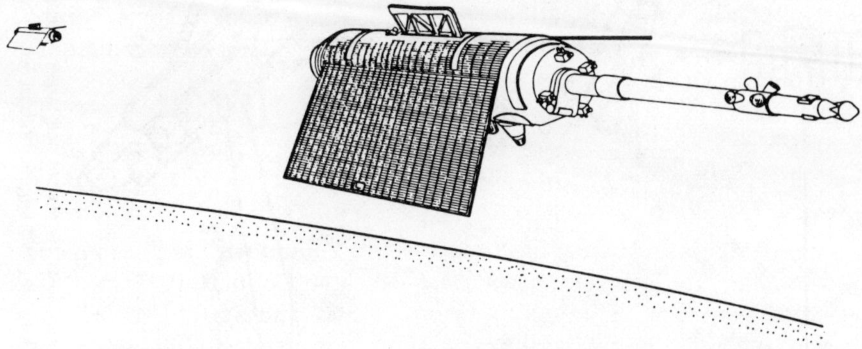

Fig. 6. The GRIM spacecrafts. The one with the boom carries the magnetometers.

and vector measurements of the magnetic field. The drag-free system uses DISCOS (and propulsion subsystems) and the orientation of the satellite is provided by star cameras and sun sensors. Here, the advantages of the low-low configuration are the minimization of the ionispheric effects on the Doppler data and the superior accuracy obtainable from the ground tracking (DMA network).

Another approach has been studied by the Europeans (European Space Agency, 1978) and consists of using the shuttle either as one of the low spacecraft or as an intermediate bus carrying a retrievable platform where the main instruments would be, the second satellite being a fairly simple free-flyer; the measurement of the radial velocity between the shuttle (or the platform) and the free-flyer would be made by laser interferometry (a CO_2 laser would be used) therefore avoiding propagation effects. Laboratory experiments and studies have shown that an accuracy of 1 μm/s can be achieved. Besides the base line configurations (Fig. 7a, b), it is also feasible to use two targets in various modes (i.e. with one or two laser telescopes on the shuttle) in order to get rid of shuttle attitude disturbances. The major drawback in using the shuttle is that a polar-type orbit cannot be flown, but full coverage of the Earth between latitudes 50° N and 50° S can be obtained in several shuttle missions, making the experiment highly valuable and complementary to GRM.

2. SGRAD

A spaceborne gradiometer (SGRAD) in principle measures some components of the difference of the gravity vectors $\bar{g}(A)$, $\bar{g}(B)$ at two points

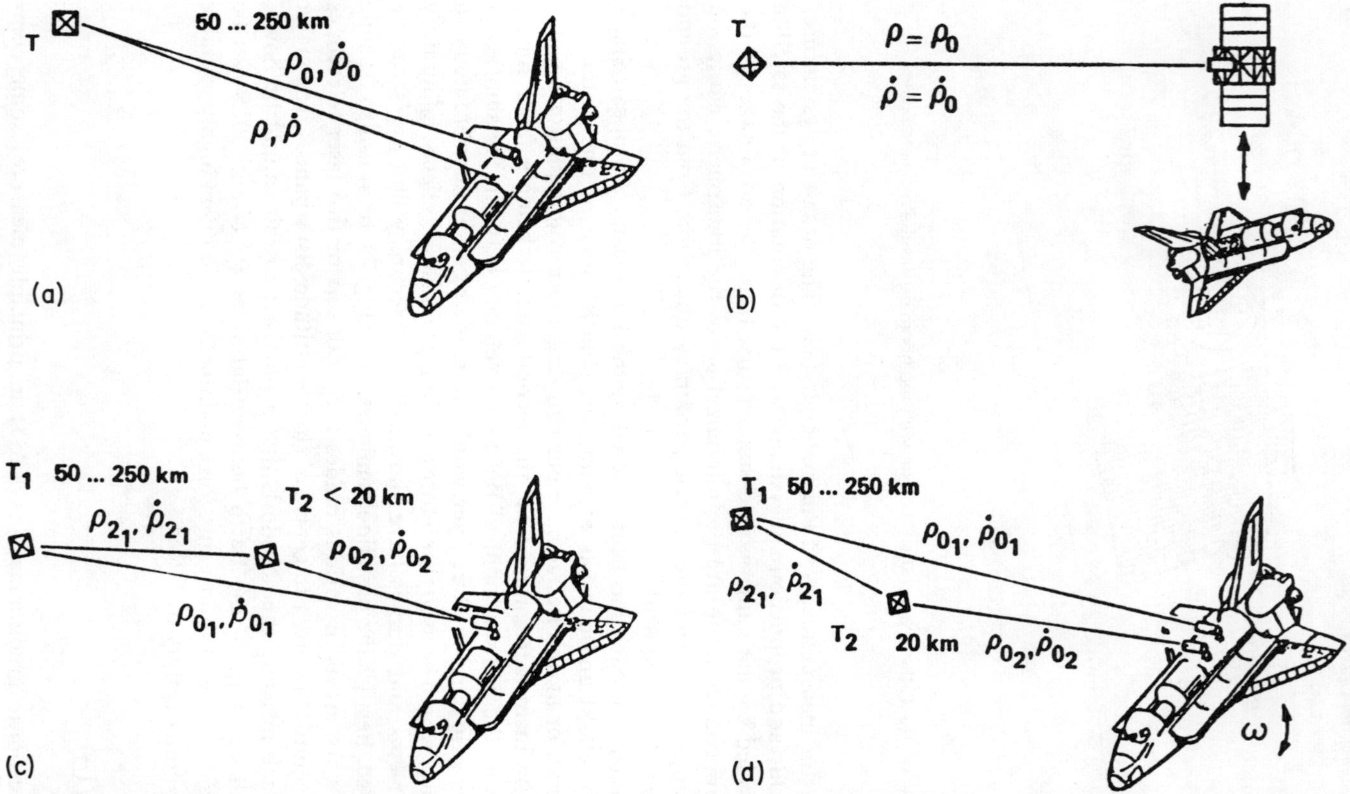

Fig. 7. The SLALOM-SST concept. The possible mission configurations are as follows: (a) one shuttle-based laser telescope, one target equipped with retroreflectors; (b) laser on detached, retrievable platform, one target; (c) one shuttle-based laser telescope, two targets, both of which must remain within the fov of the telescope; (d) two shuttle-based laser telescopes, two targets.

A, *B* fixed in a satellite, yielding components of the gravity gradient tensor $\nabla^2 T$ since $\bar{g}(B) - \bar{g}(A) = [\nabla^2\{T + U(\text{ref})\}] \cdot \overline{AB} + o(\overline{AB}^2)$; here T is the disturbing potential and U (ref) the reference ellipsoid potential. Many types of instruments have been built from this principle, either for inertial navigation or as laboratory test apparatus which were expected to be flown on board Earth orbiting satellites (Forward, 1973) or even lunar orbiters (Wells *et al.* 1981). Components are measured in eötvös ($1\ \text{E} = 0.1\ \mu\text{gal/m}$), the largest component at the Earth's surface being approximately 3000 E. To give an idea of the type of signal we may expect at low satellite altitude, let us consider a surface, square-shaped gravity anomaly of 1-mgal magnitude, of $2u$-km width, repeating itself with period Δ, infinite in the orthogonal direction (Fig. 8). The gravity anomaly function can be expanded as a one-dimension Fourier series from which the disturbing potential at altitude z is obtained by multiplying each spectral component of wave number N by $\exp[-Kz]/K$ with $K = 2\pi N/\Delta$; then $\partial^2 T/\partial z^2$ can be derived in close form. Figure 9 shows the obtained signal at 200-km altitude for various resolutions (u) when $\Delta \to \infty$ (isolated anomaly); amplitudes of several 10^{-3} E for 1 mgal anomaly are experienced in the desired resolution range. This is confirmed by computing (by derivating the Stokes–Pizetti formula) the amplitudes of some components of $\nabla^2 T$ over some terrestrial areas, as shown on Fig. 10: signals of up to 0.5 E are seen for 100-mgal anomalies spread over 100–200km. The deciphering of all spectral components require the inversion and downward continuation of the measurements, a problem which has been given great attention by theoretical geodesists for many years and which is common to both SST and SGRAD techniques (see e.g. Rummel, 1979). Its solution requires: (1) to extrapolate the behavior of the gravity field harmonics at high frequency; (2) to invert an improperly posed

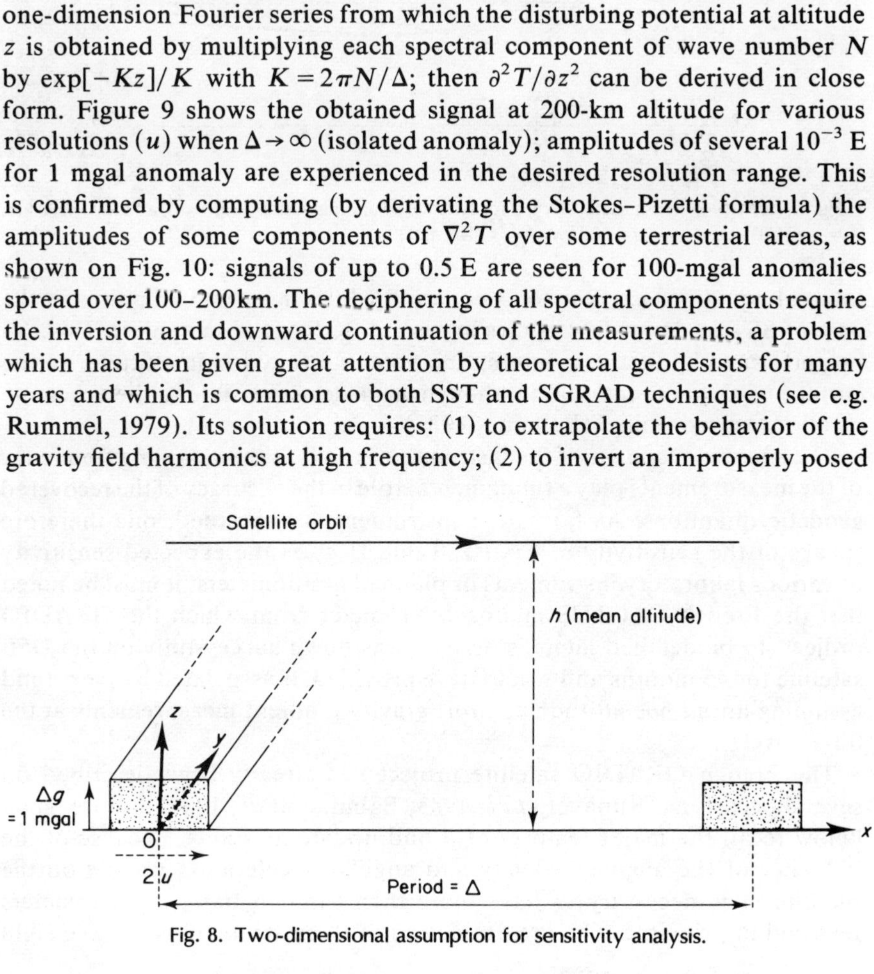

Fig. 8. Two-dimensional assumption for sensitivity analysis.

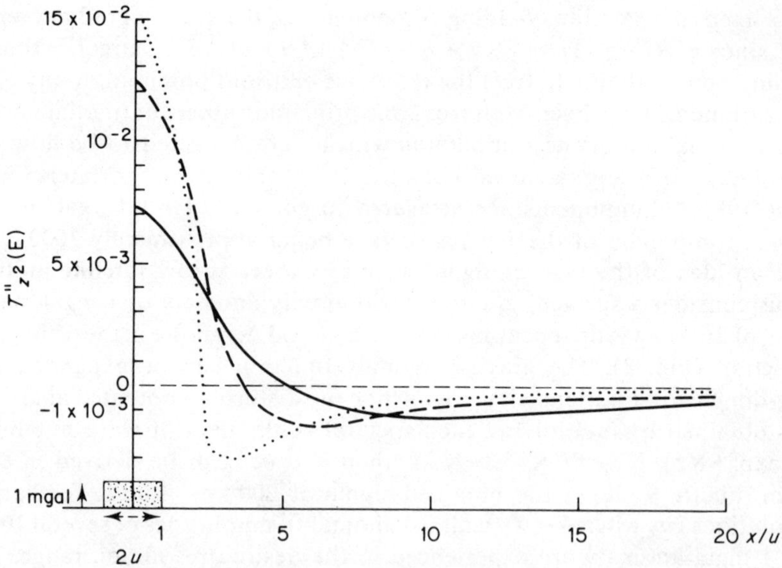

Fig. 9. Gravity gradient signal along z for an isolated anomaly (period $\Delta \to \infty$), altitude = 200 km. ——, $u = 50$ km; — —, $u = 100$ km; · · · ·, $u = 150$ km.

integral equation, e.g. Poisson's equation (differenced at two points in the case of SST or derived two times in case of SGRAD observables). This can be performed by first determining the potential function in altitude, on the surface of a sphere internal to, though as close as possible to the ensemble of all measured points (Balmino, 1982).

The spectral response of the instrument as well as the sampling frequency of the measurements play a fundamental role in the accuracy of the recovered geodetic quantities. As far as the instrument is concerned, one therefore speaks of the sensitivity in $E/\sqrt{Hz}$. Table II gives the expected sensitivity of various laboratory instruments or planned gradiometers; it must be noted that the French CACTUS microaccelerometer from which the GRADIO project (to be detailed later) is derived was flown successfully on the D5B satellite for 45 months and would have provided, if associated by pairs (and assuming an *ad hoc* attitude control) gravity gradient measurements at the 0.1-E level.

The French GRADIO satellite project has already been described on several occasions (Runavot *et al.*, 1983; Balmino *et al.*, 1984) and we shall below recall the major features of it and update its status. Because of the influence of the angular velocity and angular acceleration vectors on the satellite, it is necessary to have more than the two basic accelerometers required in principle. The present design of the instrument is to have eight

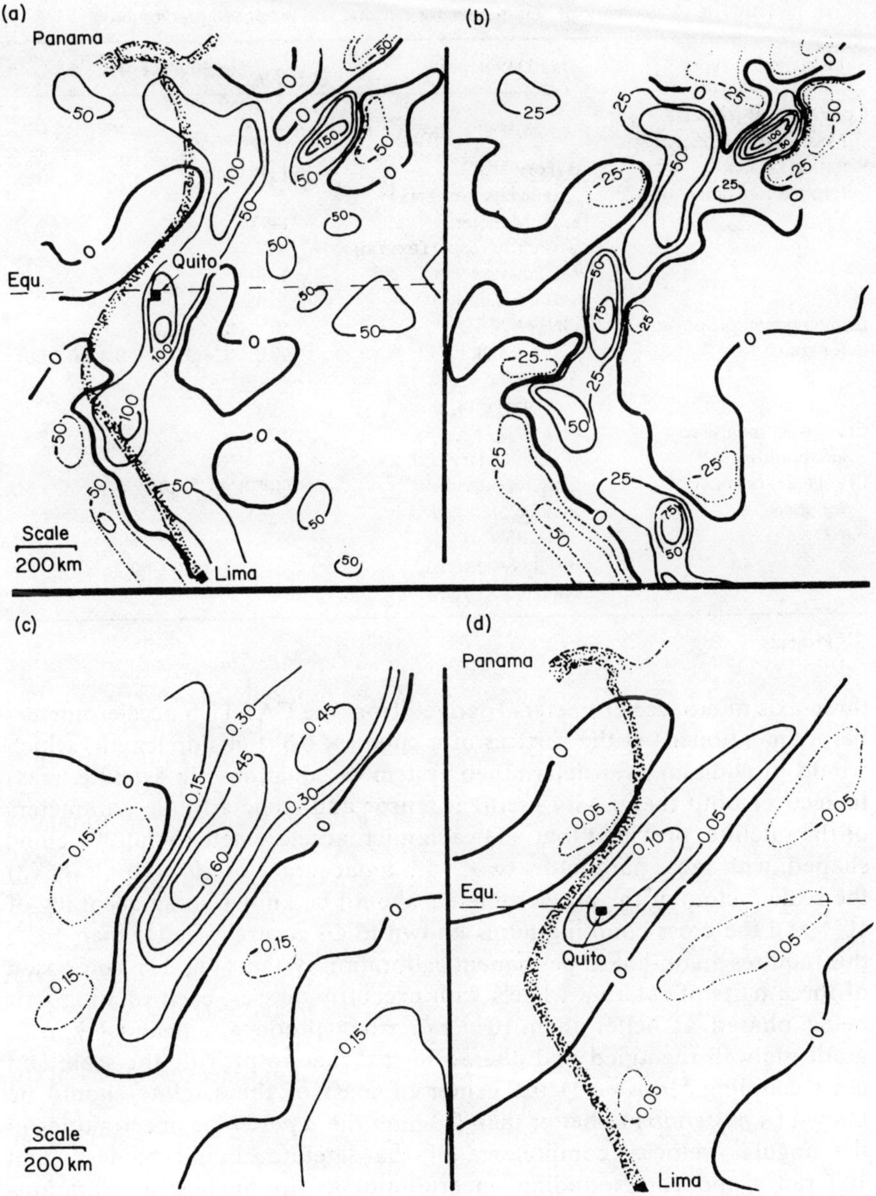

Fig. 10. Example of gravity gradients and anomalies (Ecuador, South America). (a) Gravity anomalies (mgal), altitude = 0 km. (b) Gradiometric tensor (E), component $T''_{X^2} - T''_{Y^2}$, altitude = 0 km. (c) and (d) Gravity gradients (E), altitude = 250 km. (c) Component $T''_{X^2} - T''_{Y^2}$, (d) component T''_{XZ}.

TABLE II. A summary of sensitivities for existing and proposed gradiometers

Instrumental type	Developers	Sensitivity $E\sqrt{Hz}^{-1}$
Conventional (*rt.*) Measured in lab.		
Rotating G.G. (ground instruments)	B. Forward (Hughes Res. Lab.)	1–10 (a)
	E. H. Metzger (Bell Aerosp./Textron)	1–5 (b)
	M. Trageser (C. S. Draper)	1–5 (c)
Conventional, proposed for space	ONERA/CNES ("GRADIO"[a])	10^{-3}–10^{-2} (N.B. "CACTUS": 0.1 on D5B)
	SAO/PSN[a] +above (a, b, c)[a]	2×10^{-2}–1
Cryogenic, measured in laboratory	H. J. Paik (Univ. Maryland)[a]	0.7
Cryogenic, proposed for space	Univ. of Maryland[a]	(10^{-5}) 10^{-4}–10^{-3}
	Bendix/Stanford Univ.[a]	
	SAO/PSN[a]	
	Strathclyde Univ.[a]	
	Sperry[a]	

[a] Projects.

three-axis microaccelerometers (derived from the CACTUS accelerometers before mentioned) at the vertices of a cube (of 0.5–1 m side length) which would provide an over-determined system of equations (in satellite axes) to recover both the gravity gradient tensor and some attitude parameters of the satellite, provided that: (1) each microaccelerometer would be cubic shaped, with faces parallel by twos with an accuracy of 10^{-4}–10^{-5} rad; (2) the scale factors of the accelerometers should be known to an accuracy of 10^{-2} and the cross-coupling terms known to an accuracy better than 10^{-3}; this requires an on-board permanent calibration system (Fig. 11) composed of three pairs of rotating wheels with excentric masses, each rotating pair being phased at better than 10^{-2} rad (they produce sinusoidal known gradients well identified and filtered in real time to provide the scale and cross-coupling terms); (3) the center of mass of the satellite should be known (*a posteriori*) to better than 0.3 mm; the *a posteriori* uncertainties in the angular velocity components of the satellite should be less than 10^{-6} rad/s and corresponding uncertainties in the angular accelerations smaller than 10^{-8} rad/s^2. Provided these specifications are met, which presently seems feasible, one would obtain gravity gradients with an accuracy of a few 10^{-3} $E/\sqrt{Hz}$. The planned satellite, weighing about 1000 kg, could fly in the early 1990s, at a low altitude (200–220 km) orbit, heliosynchronous

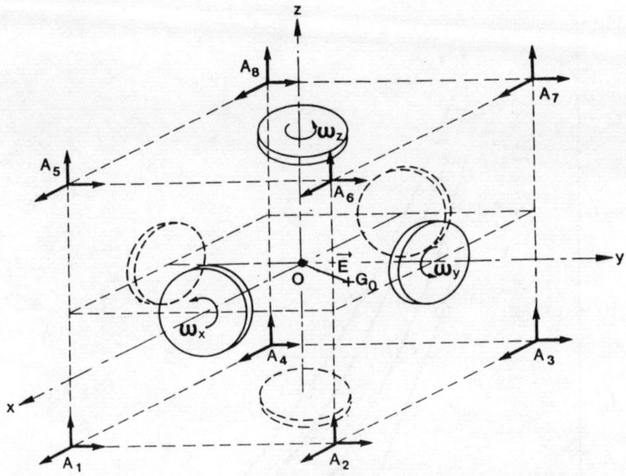

Fig. 11. Basic configuration of the GRADIO project accelerometers with three pairs of calibration device wheels.

($I \approx 98°$), for 3–6 months. A lower altitude, such as 160 km as is foreseen for GRM, would bring its performance to a better level than the competing project, as shown by Fig. 12, the 200-km altitude would make it very much comparable.

B. Second-generation Gradiometer Missions

The next decade will see the start of several new space methods to map globally the gravitational potential of the Earth with an accuracy and resolution needed by all the Earth physics scientists. The most promising techniques are satellite-to-satellite tracking and satellite gradiometry. The GRM on the one hand and GRADIO on the other appear as very complementary planned experiments and should yield comparable results for the geoid heights and gravity anomaly variations, e.g. 7 cm, or equivalently 2 mgals at 100-km half wavelength. This may be sufficient for many presently foreseen investigations in geophysics, but would not yet satisfy the oceanographic community as far as actual studies are concerned. To reach an accuracy of, say, 2 cm at 50-km resolution (required to study the eddies as shown in Table I), an equivalent accuracy of 1 mgal is necessary in the recovery of the gravity anomalies and, from Fig. 12, it is obvious that gradiometers with sensitivity (and stability) of 10^{-4} E at least, if not 10^{-5} E are needed.

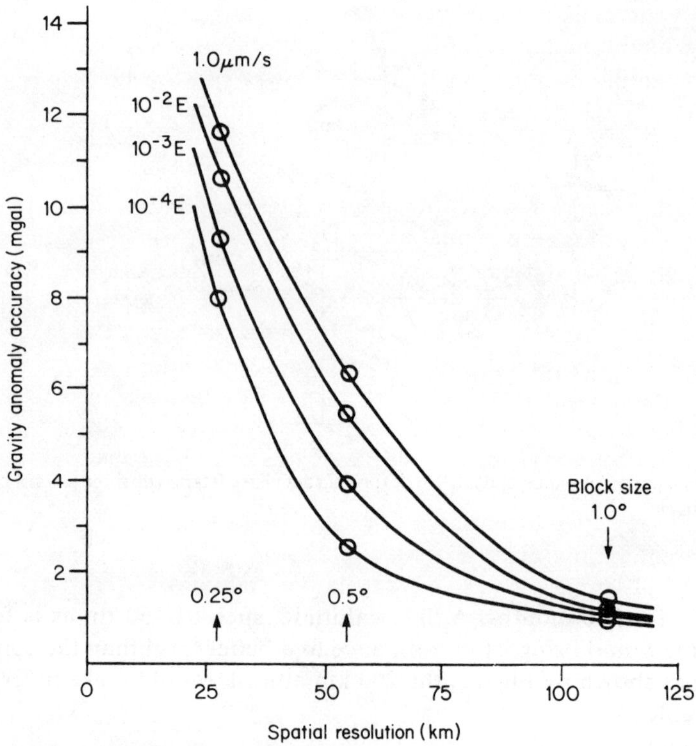

Fig. 12. Expected gravity anomaly accuracy recovery with SST (GRM base line) and SGRAD missions (6-month mission at 160 km altitude).

The fact that we stress this type of technique comes from the possibility of actually building such instruments as it was shown from a recent workshop on the subject (Wells, 1983); in the following list the various envisaged gradiometers with their expected (ultimate) characteristics, are given.

1. University of Maryland Instrument (Paik, 1981): Superconducting Gravity Gradiometer with SQUID readout (Superconducting Quantum Interference Device)

Its components are superconducting accelerometers. Each basic accelerometer is composed of a weakly suspended superconducting proof mass, a superconducting magnetic transducer and a low noise SQUID; the magnetic field produced by the transducer coils is modulated by the motion of the proof mass and detected by the SQUID magnetometer.

A general analysis shows that the minimum detectable gradient amplitude is determined by two fundamental noise sources: the Brownian motion and the amplifier noise, such that the power spectral density is given by:

$$\Gamma_N^2(f) \cong \frac{8\omega_0}{Ml^2}\left[\frac{k_B T}{Q} + k_B T_N \frac{\omega_0}{\omega_S}\right] \tag{21}$$

where M, l, ω_0, Q, T are the mass, baseline, (angular) resonance frequency, quality factor and the temperature of the proof masses, and ω_S, T_N are the (angular) signal frequency and amplifier noise temperature. Here $\omega_0 > \omega_S$ is assumed. For a device with $M = 10$ kg, $\omega_0/2\pi = 1$ Hz, $l = 50$ cm, a power spectral density of less than 10^{-4} E Hz$^{-1/2}$ is expected. Of course, the common mode accelerations must be balanced to a sufficient degree in order to reach this instrument noise limit. It appears that these conditions could be satisfied in a carefully designed orbiting satellite.

Operation at a cryogenic temperature renders additional important advantages. The mechanical stability of materials at low temperature and the stability of supercurrents suggest that an instrument could be built with an extremely low drift, an essential characteristic for a prolonged orbital mission. The excellent low-frequency response of a SQUID permits a compact assembly of a tensor gravity gradiometer around a common center.

A reduced-scale prototype gradiometer could be flown for testing in the shuttle as early as 1987.

2. Smithsonian Astrophysical Observatory/Piano Spaziale Nationale Efforts

Studies are underway to design the best possible tensorial gradiometer, based on capacitive proof masses as a first stage (a project close to GRADIO)—with sensitivities of 10^{-2}–10^{-3} E Hz$^{1/2}$, then on cryogenic microwave cavities in a second stage, with expected sensitivities in the range 10^{-4}–10^{-5} E Hz$^{1/2}$. It is envisaged to fly the first instrument (if not also the second one later) on board a tethered satellite system linked to the shuttle.

3. University of Strathclyde (U.K.) Project

This is a superconducting gravity gradiometer, initially intended for shipboard inertial navigation systems. Its design is based largely on the early work of Paik. Basically, the gradiometer consists of two accelerometers whose outputs are amplified differentially by a SQUID. The accelerometer consists of a Niobium diaphragm whose motion is sensed by the alteration of the inductances of a pair of pancake coils. Since the coils are part of a resistanceless circuit, the changing inductance alters the distribution of currents; this alteration is proportional to the force on the diaphragm and can be amplified by the SQUID.

4. The Bendix concept (Reinhardt et al., *1982) and the Stanford University Superconducting Cavity Oscillator*

Each accelerometer of the proposed gradiometer would consist of a mass spring system, the displacement sensor being a superconducting cavity oscillator (SCO) formed by the system itself. The great advantage of using SCO frequency directly as the observable is that the demonstrated fractional frequency stability of SCOs is better than 3×10^{-15} for periods of 1-1000 s; this implies that an SCO displacement sensor would have a 3×10^{-15} cm resolution and, when used with an SCO reference oscillator, would have inherent limitations in dynamic range due to measurements of the observable. Sources of noise have been investigated and show that a gradiometer with a mass sensor of a few kilograms and a baseline of 50 cm, can have a resolution of 10^{-4} E for a 1 s sampling time; for longer sampling times the resolution may improve.

5. Sperry Defense Systems Design

A levitated balance arm is chosen for the active element of the sensor; if the arm is balanced to one microgram in a total mass of 100 grams, the common mode acceleration signal is reduced by 8 orders of magnitude. The angular position of the balance arm relative to its cylindrical housing is monitored by sensing the position of one of several rigid Niobium disks at the ends of the arm; when the disk moves, current flows in pancake coils placed at the sides of the disk; the current flows through the input coil of a d.c. SQUID, thereby converting the balance arm position to a voltage. A tensorial gradiometer has been designed from this sensor (Hastings, 1983). The planned instrument for operations on an aircraft or other Earth-based moving vehicles, would have a 0.01 E with a 1 s integration time. Minor modifications could provide 10^{-4} E sensitivity for operation in the more benign noise environment of a satellite.

As a conclusion we must note that all planned second-generation gradiometers are open loop instruments, based on measuring the displacement of a mass in a sensor, and operating at cryogenic temperature. It is our feeling that, although the concepts are all well understood and even though some laboratory instruments have been built, it may well take more than a decade from now to perform the full experiment which all scientists are awaiting, but the present design is such that the 10^{-4} E level will then certainly be reached allowing the study of all phenomena listed in Table I.

IV. The Case of the Planets and Solar System Bodies

The identified goals for solar system exploration are so broad, e.g. determination of the origin, evolution and present state of the solar system, that the gravity field and many other observations are logically included within the scope of any mission. The strategy for studying a solar system body follows a logical sequence of steps. The first step, reconnaissance, is usually accomplished by a flyby spacecraft, and planetologists place a high priority on completing the reconnaissance missions to yet unvisited bodies, particularly asteroids and comets. Orbiters, atmospheric probes and surface landers are employed during the exploration phase. Where there is special interest, exploratory missions are being planned to bodies that have already been visited. Here, we concentrate on the exploratory orbiter missions, because global gravity mapping is a natural objective of these missions, and because gradiometers are candidate orbiter instruments. Eventually there may also be an opportunity to use a gravity gradiometer as part of an intensive study mission, e.g. sample return.

A. Present Status

For the Moon, Mars and Venus, the following orbital missions have provided initial gravity models:

Moon:	Lunar Orbiter, Apollo command modules, Apollo subsatellites
Mars:	Mariner 9, Viking
Venus:	Pioneer Venus

The methods are in principle the same as those used in the Earth case (excluding surface data). The models have been determined by analysis of the radio tracking signal due to a change in the radial velocity of the spacecraft relative to the receiving station. Direct analysis of the tracking data also yielded line-of-sight acceleration maps at the orbital altitude caused by gravity anomalies after all other known effects have been accounted for. A review on planetary gravity fields can be found in Balmino (1981).

The accuracy, resolution, and coverage of these maps and models is determined by tracking system performance and the orbit parameters. Independently from the coverage, the maximum degree of the spherical harmonic coefficients to be determined from Doppler tracking data of accuracy ε_D, can be derived from the expectations of the largest Doppler shifts, or equivalently of the largest velocity disturbances $\{\langle \delta V^2 \rangle_l\}^{1/2}$, due to a gravity variation of wave-number (degree) l. Around periapsis, one has:

$$\varepsilon_D = \{\langle \delta V^2 \rangle_l\}^{1/2} = K\sqrt{GM/a}\,[(1+e)(1-e)]^{-1/2}\,[R/(R+h)]^l (2l+1)^{1/2}/l^2 \tag{22}$$

where GM is the product of the gravitational constant by the mass of the planet, R the radius, h the spacecraft altitude at periapsis, (a, e), the semi-major axis and eccentricity of the orbit, K a scale factor derived from the spectral behavior of the gravity harmonics, as compared with the Earth case. For most planetary missions, the radial velocity was measured with an accuracy of about 1 mm/s. For Mars and Venus, the orbits have been elliptical so that altitudes up to 2000 km have been needed to get good coverage. This places global spatial resolution at the 1000-km level, and anomaly accuracy at about 50 mgal. For the Moon, no data are available for the far side; however, good data (20-km resolution and 5-mgal accuracy) were obtained by the Apollo subsatellites for the nearside equatorial region.

Significant gravity anomalies are present. The Moon has "mascons" that are associated with the circular mare; Mars has gravity anomalies over its large volcanoes, and Venus has a few mountains and plains that rise well above an otherwise flat surface. The existing data have been used to estimate the thickness of the mare basalts, and the stresses and thickness of the crusts that support anomalous features.

B. Projects

The future missions which are presently foreseen in the next 10 years are the following:

(i) Venus radar mapper (VRM): decided
(ii) Mars geoscience/climatology orbiter, which (for our interest here) could be coupled with European Kepler emission; SST could be performed between the two spacecraft, the former flying a 350-km altitude circular polar orbit and the second one an elliptical higher one (in most parts).
(iii) Comet rendezvous/asteroid flyby
(iv) Titan probe/radar mapper
(v) Lunar geoscience orbiter
(vi) Saturn orbiter

The list is not complete and the reader should keep in mind that priorities always fluctuate with an unpredictable spectrum (if there is a spectrum at all). Nevertheless the situation seems to be the following.

The first two missions, which are planned to be launched in 1988 and 1990, will orbit bodies for which gravity data are available. More tracking data will be obtained with a probable accuracy of 0.1 mm/s (X band); however, a significant improvement in knowledge of the gravity field is not expected. They may, however, aid interpretation by providing a good topographic map of Venus (1-km surface resolution or better) and a determination of the global surface composition of Mars. The low-cost philosophy being applied to these missions, as well as the elliptical orbit planned in VRM, appears to preclude the use of gravity gradiometers on these missions in an attempt to obtain more meaningful data. The comet rendezvous and Titan probe missions in the core program do not even plan to orbit these very interesting bodies.

Of the other missions, the Lunar geoscience orbiter has the best justification for a gravity gradiometer, namely the acquisition of far-side gravity data without a requirement for a separate communications relay satellite (as it was planned in earlier projects such as LPO or POLO). At 50–100 km altitude, a gradiometer with 1-E accuracy can meet the stated objective of measuring anomalies with 5-mgal precision along 1000-km paths; it may be able to provide 1-mgal resolution over smooth areas, and extend the 5-mgal accuracy to global coverage.

All orbiter (or rendezvous) missions could benefit from use of gradiometry. Complete coverage can be obtained more quickly because gradiometers are not affected by the occultation and ground-station scheduling problems that are encountered with radio tracking. Missions producing smaller amounts of data could reduce communications time significantly. In addition, a gradiometer would permit onboard automation of orbit computation and control; this would be helpful for outer planet orbiters where round trip communication times are long, and for reducing time devoted to communications with any orbiter.

V. Conclusions

Classical methods for determining the geopotential—as well as the field of other planetary objects, have been discussed in some details since they are still going to be in use for some years until quite different, easier techniques (in principle) take place and allow a finer modeling of the functionals (harmonic coefficients, geoid heights, gravity anomalies) needed by scientists in solid Earth and ocean physics. These new approaches consist in mapping directly the variations of the gravity field by performing in orbit (*in situ*) measurements—as any usual sensor does in general. The observed quantities may be readily converted into potential function variations in the case of SST or require some integration procedure in SGRAD experiments where second derivatives of the field are measured. Two generations of possible Earth missions have been considered (i) GRM, a United States low-low SST experiment and GRADIO, an SGRAD mission presently studied in Europe; (ii) an advanced satellite gradiometry mission which could use any of the five presently identified candidate superconducting gradiometers, flying a very low orbit (150 km or less), and from which geoid heights and gravity anomalies could be recovered with an accuracy of 2 cm and 1 mgal, respectively, at 50-km resolution at least. Finally, the satellite gradiometry technique seems to be also the best approach to map the global gravity field of terrestrial bodies and could especially well be applied to the case of the Moon when mankind finds again some interest in his natural satellite.

References

Anderle, R. J. (1966). Geodetic parameters set NWL-SE-6 based on Doppler satellite observations. *Proc. 2nd Int. Symp. Geod. Use Satellites*, Athens (G. Veis, ed.).

Balmino, G. (1974a). Determination of the Earth's gravitational potential by space methods. CNES Summer School: Geodesy and Geodynamics, Lannion.

Balmino, G. (1974b). Numerical methods of orbital dynamics. *COSPAR/IAU-IUIAM Symp., Satellite Dynamics*, pp. 50–97. Springer-Verlag, Berlin and New York.

Balmino, G. (1974c). Coriolis forces in numerical integration of satellite orbits and absolute reference systems. *Cel. Mech.* **10**, 4, 423–436.

Balmino, G. (1981). Gravity field and rotation of planets. A review from the point of view of planetary geodesy. *Ann. Geophys.* **37**, 161–172.

Balmino, G. (1982). Détermination Précise du Géoide. CNES Ecole d'Eté: l'Océanologie Spatiale, Grasse.

Balmino, G. and Reigber, Ch. (1975). 13th-order harmonics in the geopotential from an analysis of four resonant satellites. *Space Res.* **XV**, 53–57.

Balmino, G., Letoquart, D., Barlier, F., Ducasse, M., Bernard, A., Sacleux, B., Bouzat, C., Runavot, J. J., Le Pichon, X. and Souriau, M. (1984). Le projet GRADIO et la détermination à haute résolution due géopotentiel. *Bull. Geod.* **58**, 151–179.

Barlier, F. (1981). Revue sur les résultats obtenus avec l'accéléromètre CACTUS. *Ann. Geophys.* **37**, 223–234.

Berger-Jaeck, C. (1976). Analyses et modélisations statistiques de différents paramètres de l'Atmosphère Neutre. Assymétrie sphérique, Univ. P. & M. Curie, Thèse Doct. d'Etat.

Borderies, N. (1976). La méthode de centrage en géodésie spatiale. Thèse 3ème cycle, U.P.S., Toulouse.

Born, G. H., Wunsch, C. and Yamarone, C. A. (1984). Topex observing the ocean from space. *EOS*, **65**, 433–434.

Cazenave, A., Forestier, F., Nouel, F. and Pieplu, J. L. (1972). Improvement of zonal harmonics by the use of observations of low inclinations satellites: DIAL, SAS and PEOLE. A.G.U. Monograph, no. 15.

CNES (1984). "Determination d'orbite et radiopositionnement intégrés par satellite (DORIS)". Internal Report, Toulouse.

Dunn, P. (1978). 14th order resonant harmonics equations derived from the observation of LANDSAT 1. *Proc. Int. Symp.*, Use Artificial Satellites *Geod. Geodyn. 3rd Lagonissi*, (G. Veis, ed.).

European Space Agency (1978). SLALOM Mission/System Definition Study. *Final Rep.* **MBB-GRGS-SFB78,** ESA Contr. no. 3483/78/F/DK(SC).

European Space Agency (1984). ERS 1 radar altimeter data products. *Proc. ESA Workshop, Frascati, Italy.* ESA-SP. 221.

Forward, R. L. (1973). Review of artificial satellite gravity gradiometer techniques for geodesy. *Int. Symp. Use Artificial Satellites Geod. Geodyn.* Athens.

Gaposchkin, E. M. (1973). Smithsonian standard Earth III. *SAO, Sp, Rep.* no. 353, 262–276.

Guier, W. U. and Newton, R. R. (1965). The Earth's gravity field deduced from the Doppler tracking of five satellites. *J. Geophys. Res.* 4613–4626.

Hastings, R. (1983). Design considerations for a mobile, high sensitivity, cryogenic gravity gradiometer. Sperry Defense Systems, *Rep. no.* **PX14015**.

Heiskanen, W. and Moritz, H. (1967). "Physical Geodesy". Freeman, San Francisco, California.

Izsak, I. G. (1966). A new determination of non-zonal harmonics by satellites. (J. Kovalevsky ed.), Trajectories of Artificial Celestial Bodies as Determined from Observations; Springer-Verlag, Berlin.

Kaula, W. (1966a). "Theory of Satellite Geodesy". Ginn (Blaisdell), Boston, Massachusetts.

Kaula, W. (1966b). Test and combinations of satellite determination of the gravity field with gravimetry. *J. Geophys. Res.* **71**, 5303–5314.

King-Hele, D. G., Walker, D. M. C. and Gooding, R. H. (1974). Geopotential harmonics of order 15 and even degree, from changes in orbital eccentricity at resonance. *RAE Tech. Rep.* **74**, 120.

King-Hele, D. G., Walker, D. M. C. and Gooding, R. H. (1975a). Geopotential harmonics of order 15 and odd degree, from analysis of resonant orbits. *RAE Tech. Rep.* **75006**.

King-Hele, D. G., Walker, D. M. C. and Gooding, R. H. (1975b) *Planet Space Sci.* **23**, 1239.

King-Hele, D. G., Walker, D. M. C. and Gooding, R. H. (1979). Evaluation of 14th order harmonics in the geopotential. *Planet. Space Sci.* **27**, 1–18.

Klokocnik, J. and Kostelecky, J. (1978). *Bull. Astron. Inst.*, Czech. **29**, 10.

Klokocnik, J. and Kostelecky, J. (1980). Possible contribution to the Earth model GRIM 3. *Tech. Rep.* (private communication).

Klosko, S. M. and Wagner, C. A. (1975). Equations for 12th order tesseral harmonics. (Private communication.)

Kozai, Y. (1959). The Earth's gravitational potential derived from the motion of satellite 1958β2. *SAO Sp. Rep.* no. 22.

Kozai, Y. (1964). New determination of zonal harmonic coefficients in the Earth's gravitational potential. *Publ. Astron. Soc.* Jpn. **16**.

Kozai, Y. (1969). Revised zonal harmonics in the geopotential. *SAO Sp. Rep.* **295**.

Lambeck, K. (1979). Methods and geophysical applications of satellite geodesy. *Rep. Prog. Phys.* **42**, 547–628.

Lerch, F. J., Putney, B. H., Wagner, C. A. and Klosko, S. M. (1981). Goddard Earth models for oceanographic applications (GEM 10B and 10C). *Mar. Geod.* **5**, 145–187.

Moynot, B. (1971). Caractéristiques d'un programme de correction différentielle en géodésie dynamique. *Astron. Res.* 27–35.

National Research Council (1979). "Applications of a dedicated gravitational satellite mission". Workshop on a gravitational satellite Mission. Nat. Academy of Science, Washington D.C.

Paik, H. J. (1981). Superconducting tensor gravity gradiometer for satellite geodesy and inertial navigation. *J. Astron. Sci.* **29**, 1–18.

Pisacane, V. L., Ray, J. C., MacArthur, J. L. and Bergeson-Willis, S. E. (1981). Description of the dedicated gravitational satellite mission (GRAVSAT). *Int. Geosci. Remote Sensing Symp.* Washington, D.C.

Rapp, R. H. (1979). Global anomaly and undulation recovery using Geos-3 and terrestrial data. Dept. Geodetic Sci., *Rep.* no. 285, 54.

Rapp, R. H. (1983). The development of the January 1983 $1° \times 1°$ mean free-air anomaly data tape. Dept. of Geodetic Science and Surveying, *Intern. Rep.* Ohio State Univ., Columbus, Ohio.

Reigber, C. (1973a). Fourier analysis of resonant orbits. *Space Res.* **XIII**, 3–12.

Reigber, C. (1973b). Bestimmungsgleichungen für Resonanzparameter der Ordnung 13 ans der Analyse von Bahnen der Satelliten Geos B, BEC und DID. Doctorate Dissertation Tech. Univ. Munich.

Reigber, C. (1981). Representation of orbital elements variations and force function with respect to various reference systems. *Bull. Geod.* **55**, no. 2, 111–131.

Reigber, C. and Balmino, G. (1976). 14th order Harmonics from analysis of mean longitude variations of resonant satellites. *Space Res.* **XIX**.

Reigber, C. and Rummel, R. (1979). 12th order harmonic from resonant perturbations of satellites using non-homogeneous weight functions. *Int. Symp. Use Artificial Satellites Geod. Geodyn. 3rd Lagonissi*, (G. Veis, ed.) pp. 217–240.

Reigber, C., Muller, H., Bosch, W., Balmino, G. and Moynot, B. (1984). GRIM gravity model improvement using LAGEOS. *J. Geophys. Res.* (Submitted.)

Reinhardt, V. S., *et al.* (1982). A supersensitive accelerometer for spacecraft gradiometry. *Proc. IEEE Position, Location Navigation Symp.*

Rummel, R. (1979). Determination of short wavelength components of the gravity field from satellite-to-satellite tracking or satellite gradiometry: an attempt to an identification of problem areas. *Manuscr. Geod.* **4**.

Runavot, J. J., Bouzat, C., Bernard, A. and Sacleux, B. (1983). Global measurement of the Earth gravity field using a gradiometric satellite. *Acta Astronaut.* **10**, no. 9, 599–607.

Smith, D., Langel, R. and Keating T. (1982). Geopotential research mission. *NASA-GSFC Rep.* (Aug.).

Wagner, C. A. (1974). 11th order geopotential resonance on the orbit of Vanguard 3. *J. Geophys. Res.* **79**, no. 23, 3335–3341.

Wells, W. C. (1981). Gravity gradiometers and the POLO mission. Summary of a Workshop at NASA.

Wells, W. C. (1983). Spaceborne gravity gradiometers. *NASA Workshop Proc.* NASA Conf. Pub., no. 2305.

2.2 The Gravity Field of the Earth and Prospects for Improvement of its Determination

Richard H. Rapp

Department of Geodetic Science and Surveying, Ohio State University, Columbus, Ohio, USA

I. Introduction

The need to know the gravity field of the Earth was originally driven by the need to reduce geodetic measurements from the surface of the Earth to the reference ellipsoid on which computations were to be made and geodetic coordinates to be given. Therefore, techniques were developed to determine geoid undulations (the separation between the geoid and the ellipsoid) and deflections of the vertical (the direction differences of the normal to the ellipsoid and the normal to the geoid) from gravity data. These techniques required global gravity data which limited the prospect of the actual application of the famous equations of Stokes and Vening-Meinez.

In the 1930s instruments were developed so that relative gravity measurements could be made by using devices that were easily transportable. Such devices became the mainstay for geophysical prospecting using gravity methods. Although such instruments were primarily developed for use in land areas, they were also used in the measurement of gravity in the ocean areas. However, the progress in this area was very slow due to many technical problems involving ship accelerations and precise navigation.

In the 1960s it became clear that the Earth's gravity field could be described through the analysis of motion of artificial satellites. Through this analysis, potential coefficient parameters in a truncated spherical harmonic series that describe the Earth's gravitational potential could be estimated. The use of this series gave the community an alternate view of the gravity field of the Earth. This view changed from the point concept represented by individual gravity measurements to one where the gravity field was viewed as a continuous function.

The initial determination of the gravity field from satellite data used optical measurements of the satellites against the background of the stars. These measurements enabled the determination of the coefficients in a spherical harmonic series up to degree and order 12 (i.e. about 170 terms). As new data became available, higher-degree solutions have become possible. Today the most complete solutions are those that combine satellite and terrestrial gravity data.

Today not all data is acquired by the conventional techniques of the past. For close to Earth (e.g. 1000 m) measurements, helicopters have become a tool for the fast acquisition of accurate data in areas of limited geographic extent (say on the order of 50 km × 50 km). Inertial equipment can now be used as a precise interpolator for the estimation of the deflection of the vertical and gravity values. New satellite techniques such as satellite-to-satellite tracking have enabled large-scale anomalies to be estimated in areas of the world where no gravity information was previously available. Of particular importance here is the direct sensing of the ocean surface through satellite altimetry. Such information has supplied us with a better picture of the gravity field in ocean areas than was ever available for most land areas.

Today the application of gravity field data ranges from the classical geodetic needs to providing short- and long-wavelength information for geophysical and oceanographic studies. New measurement techniques (Taylor *et al.*, 1983; Jekeli, 1983) have been developed that have the potential for acquiring a substantially improved gravity field in the next decade. But before we look at the improvements to be expected, we will first look at the current status of our knowledge of the gravity field.

II. Surface Gravity Data

Gravity data is collected on land using gravimeters (for gravity differences) in conjunction with known reference-base stations for which values of gravity are usually given in the International Gravity Standardization Net 1971 (IGSN71). Absolute values of gravity can also be determined by using instruments that are becoming more accurate and more portable. However, the number of absolute values is very small (~30) when compared to the very large number of gravity values determined through relative measurements. Gravity data at sea is collected by organizations in the United States such as Woods Hole Oceanographic Institution, Lamont-Doherty Geological Observatory, University of Hawaii, Scripps Institution of Oceanography and the Department of Defense. These data are usually acquired as a continuous profile, so for most applications the profile is sampled at a certain interval to select specific point values.

The gravity values acquired from land or ship measurements are used to compute various types of gravity anomalies such as free-air or Bouguer. Such computations require a precise elevation for the station and a geographic position known to an accuracy sufficient for normal (or theoretical) gravity computations. The data are then usually collected in a center for retrieval purposes.

In the United States the prime center for the worldwide acquisition of gravity data is the Defense Mapping Agency Aerospace Center (DMAAC) in St. Louis. As of 1 November 1983 there were 11,583,246 gravity stations in their automated data system. Not all of these values are releasable for general use as some of the data have been acquired for use only by the Department of Defense.

The National Geophysical Data Center (NGDC) in Boulder serves as a general distribution center for land gravity data in the United States and worldwide oceanic data that have been submitted to it. The land data set contains approximately 605,000 values while there are 5,500,000 values in the ocean areas (Wilcox and Scheibe, 1983). Details on the gravity data held at NGDC are given by Hittleman *et al.* (1982).

The National Geodetic Survey (NGS) also maintains a data base of gravity data in North America that includes NGS data as well as data from DMAAC and other sources.

The Bureau Gravimetrique International (BGI) is an organization in Toulouse, France that operates through the International Association of Geodesy. The Bureau, under the direction of Georges Balmino, has, as its main task, the collection and distribution of gravity data on an international basis. As of June 1982, the BGI has 2,680,000 measurements from 1501 sources. A detailed description of the operations of the BGI (from submitting data to requesting data) is given by Balmino (1982). The BGI regularly publishes a Bulletin D'Information that describes the activities and acquisitions of the Bureau.

Counts of point anomalies that are available are not completely meaningful. For some purposes it is of more importance to know the distribution of the data. In Fig. 1 we show the distribution of point gravity data that was used in the development of the Society of Exploration Geophysicists (Society of Exploration Geophysicists, 1982) Bouguer gravity anomaly map of the United States. The point gravity densities represented in this figure are typical of those available in parts of Canada, Europe and Australia. Coverage is substantially poorer in South America, Africa and in the polar areas. In the Soviet Union and China no recent gravity information is available although such data exists for China (Gu, 1983). This problem typifies the situation where some countries will not release their gravity information for public use. Recent resolutions at scientific meetings strongly encourage countries to release as many gravity data as possible for scientific use, either as point values or as mean values in areas such as $1°\times1°$ equiangular blocks.

For a number of geodetic and geophysical purposes it is necessary to aggregate the point anomaly values into mean values over various size blocks. Although 10° and 5° have been used in the past the most conventional size for global analysis is a $1°\times1°$ equiangular block. The distribution of such data in a compilation (Rapp, 1983a) is shown in Fig. 2, where anomaly estimates based only on direct observations are shown. The accuracy of the data is quite variable. In much of North America, Europe and most of Australia, the accuracy may be at the 2–5-mgal* level. In ocean areas the accuracy may be on the order ± 10 mgal, but it can be considerably poorer in areas of steep gradients. In some areas (both land and ocean) there can be considerable (200 mgal) disagreement between anomaly estimates from various sources (Balmino, 1983).

For some geodetic applications, mean anomalies in smaller size blocks have been estimated. Torge *et al.* (1983) describe the estimation of 103,997 free-air anomalies in $6'\times10'$ blocks for much of Europe including surrounding marine areas. Ganeko (1982) describes a large set of $10'\times10'$ mean free-air anomalies covering much of Japan and surrounding seas. Lachapelle (1978) has developed a $5'\times5'$ data set for North America.

It is clear that in the past few years there has been a substantial increase in the number of gravity measurements that have been made. However, the geographic distribution of the measurements remains quite skewed. It appears that there will be still a number of areas of the world in which

* 1 mgal $= 10^{-5}$ m s^{-2}

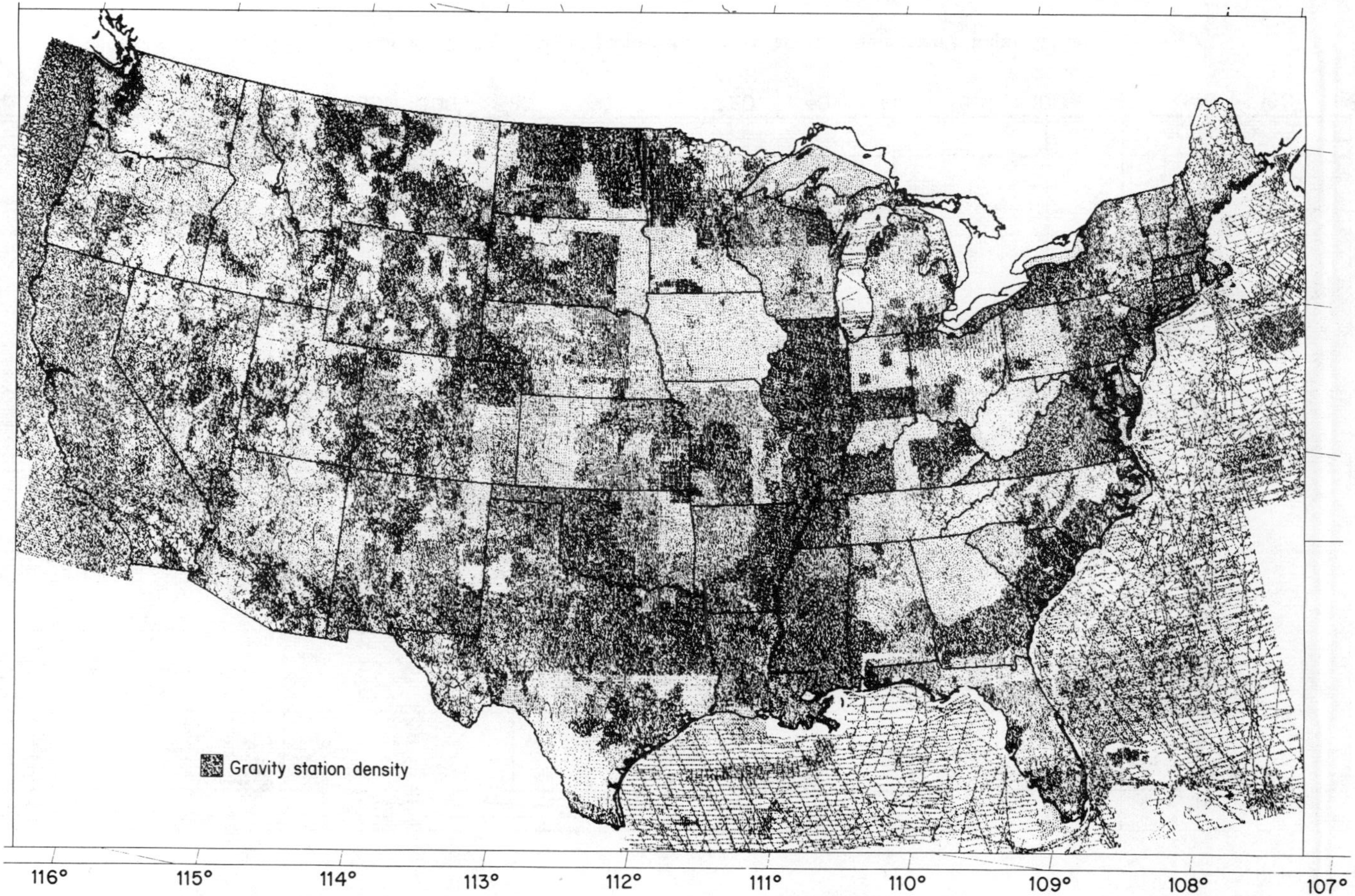

Fig. 1. Gravity data in the United States used in the preparation of the SEG Bouguer anomaly map (1982).

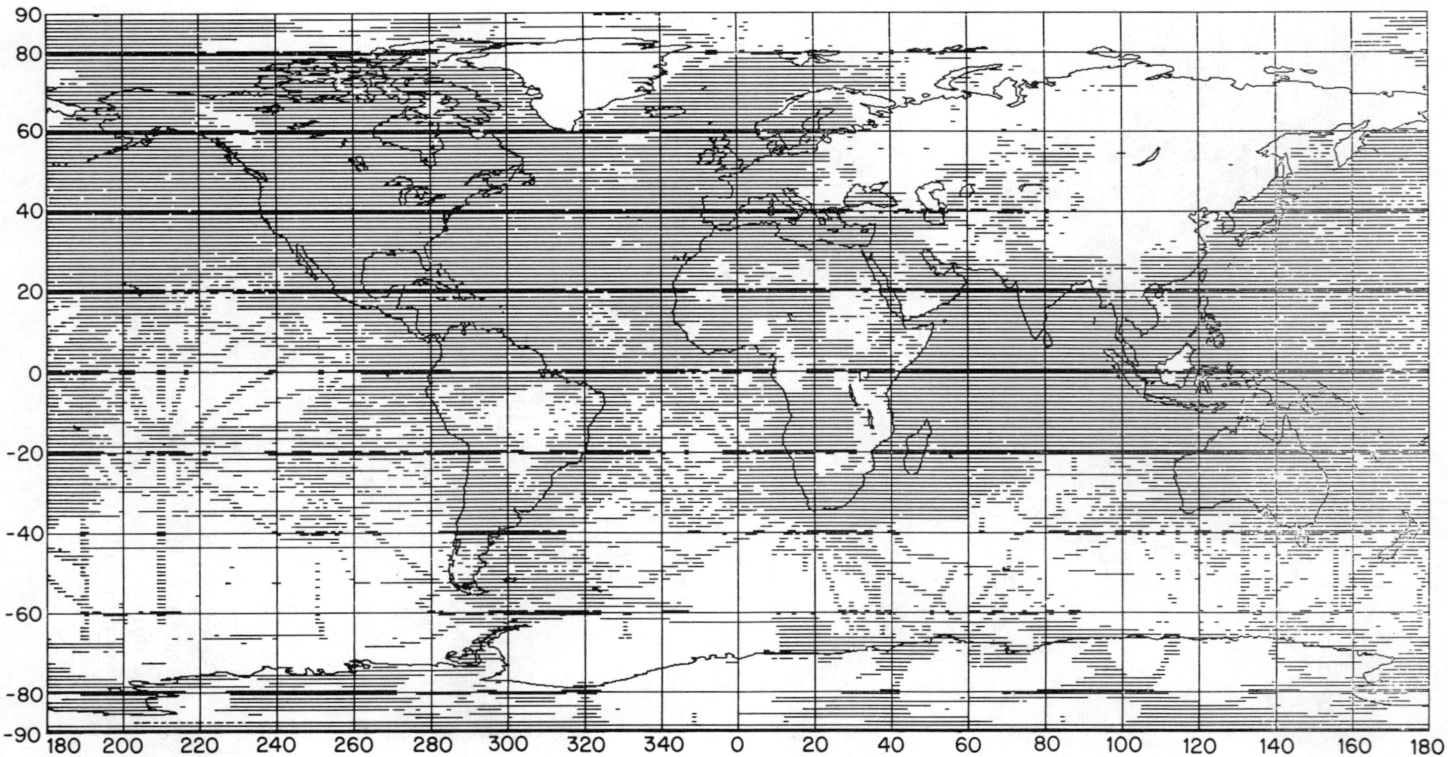

Fig. 2. Location of 38/52 1° × 1° gravity anomalies based on actual measurements (Rapp, 1983a).

gravity data will not be readily available due to political or geographical inaccessability.

III. Gravity Field Information from Satellites

A. Potential Coefficient Models

Since 1958 our knowledge of the gravity field of the Earth has improved through the analysis of the orbits of satellites to deduce spherical harmonic potential coefficients. A review of the progress made in this area since 1958 has been given by Lambeck and Coleman (1983). A summary of activities in the United States has been made by Lerch (1983).

Potential coefficients ($\bar{C}_l$, $\bar{S}_{lm}$) are used to define the geopotential V at a point whose geocentric coordinates are r, $\bar{\phi}$, λ, in the following standard form (Lerch *et al.*, 1979):

$$V(r, \bar{\phi}, \lambda) = \frac{kM}{r}\left[1 + \sum_{l=2}^{\infty}\left(\frac{a}{r}\right)^{l} \sum_{m=0}^{l} (\bar{C}_{lm} \cos m\lambda + \bar{S}_{lm} \sin m\lambda)\bar{P}_{lm}(\sin \bar{\phi})\right] \quad (1)$$

where kM is the geocentric gravitational constant, a a scaling parameter nominally the equatorial radius, $\bar{P}_{lm}$ are the fully normalized associated Legendre functions, $\bar{\phi}$ is the geocentric latitude and λ is the longitude.

Many sets of potential coefficients have been derived in the past years. These coefficient sets vary depending on the maximum degree of the expansion and the types of data used in their determination. Some models use just satellite data by considering orbital variations while other solutions combine the previous data with satellite altimeter data and/or terrestrial gravity data. Table 1 summarizes the status of some current geopotential models.

The GEM solutions are produced at NASA's Goddard Space Flight Center. The GRIM solutions are a cooperative effort of several organizations in Europe (Reigber *et al.*, 1983). Although the maximum degree for some may be 20, the solutions will also contain additional isolated coefficients determined through the effect of resonance perturbations.

Satellite-derived gravity models are of two general types. The first type is a general gravity model which incorporates data from as many different satellites as possible to derive a most representative gravity field. The second type is a tuned gravity field model that is designed to give the best orbit for a specific satellite. The GEML2 model (Lerch *et al.*, 1982a) is in some sense a tuned model for the LAGEOS satellite since substantial amounts of laser tracking data to LAGEOS were used in its determination. However, it is also a general model to the extent that the GEM9 normal equations,

TABLE I. "Current" geopotential models

Model	Date	NMAX	Data used[a]
GEM9	1977	20	S
Rapp	1978	180	S+G(A)+G(T)
SAO	1980	30	S+A+G(T)
GEM10B	1981	36	S+A+G(T)
GEM10C	1981	180	S+A+G(T)
Rapp	1981	180	S+G(A)+G(T)
GEML2	1982	20	S
GRIM3B	1983	36	S+G(A)+G(T)
GRIM3-L1	1984	36	S+G(A)+G(T)

[a] S is satellite data excluding altimetry; A altimeter data; G(A) anomalies derived from altimeter data; and G(T) terrestrial gravity anomalies.

involving data from many satellites, were used in GEML2 development. Tuned models have been derived for other satellites including SEASAT and STARLETTE.

Even though the high-degree expansions to degree 180 contain 32,761 potential coefficients, their use for certain computations has been made quite efficient for such quantities as geoid undulations, gravity anomalies and deflections of the vertical (Tscherning *et al.*, 1983). Efficient techniques to use these high-degree expansions for orbit generation are not yet available.

Potential coefficients can be compared in terms of anomalies, deflections of the vertical and undulations, by value, and by spectrum (or degree l). For example, Fig. 3 shows the undulation difference between the GEML2 coefficient set and several other sets. The GEML2 and the GRIM3-L1 coefficient sets should be the most accurate sets at low degrees since they incorporate a large amount of LAGEOS tracking data in their solution. Also shown in Fig. 3 is the estimated accuracy (1 s.d.), by degree, of the GEML2 solution. The commission error in the computation of geoid undulations, from the GEML2 solution, by considering all the coefficients to degree 20 is ± 1.6 m. At the individual degrees (or wavelengths), the undulation is much more accurately known. For example, at degree 2, the accuracy is ± 1 cm; at degree 4, it is ± 5 cm.

When the comparisons are made at lower degrees some of the differences are still substantial. For example, when comparisons are made up to degree 6 between GEML2 and GRIM3-L1, the RMS undulation difference is ± 40 cm and the RMS anomaly difference is 0.36 mgal. The maximum undulation difference is 1.1 m and the maximum anomaly difference is 0.9 mgal.

One of the advantages of using spherical harmonic expansions to represent the gravity field is that we can obtain information on the spectrum

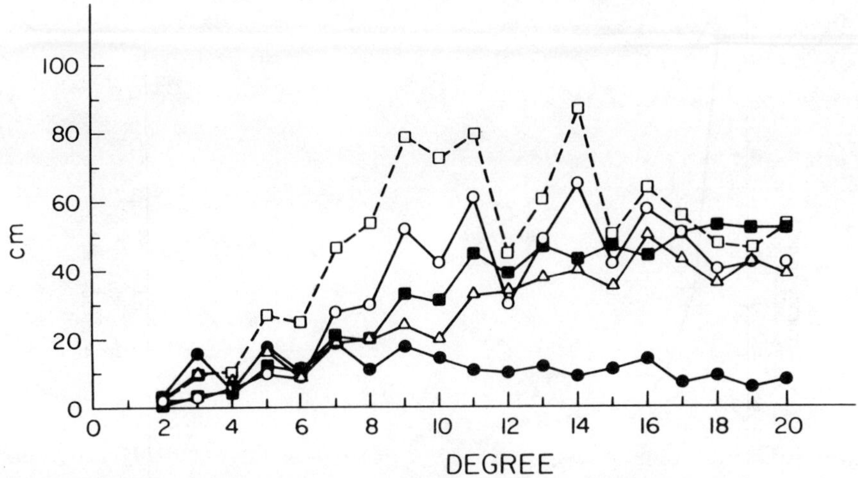

Fig. 3. Geoid undulation differences between the GEML2 model and other selected models. ○ represents OSU (December 1981); ●, GEM9; △, GEM10B; □, GRIM3-L1; and ■, GEML2 accuracy.

of the gravity field. We have already seen in Fig. 3 undulation differences and accuracies by spherical harmonic degree. By knowing the accuracy of the coefficients that have been estimated and a statistical estimate of the behavior of the high-degree components of the field, it is possible to compute the accuracy of mean quantities as a function of the area (or resolution) in which the quantity of interest has been computed. By using the Rapp (1981) 180 solution [assuming an average anomaly accuracy (1 s.d.) of ±10 mgal], the accuracy of a gravity anomaly computed from this field is shown in Fig. 4 as a function of the size (or resolution) of the feature. At the short wavelengths (e.g. 100 km), the global accuracy of today's data is approximately ±10 mgal; at long wavelength (e.g. 2000 km), the accuracy is ±0.6 mgal. These numbers represent global averages and so accuracies in

TABLE II. RMS differences between potential coefficient sets in terms of gravity anomalies and geoid undulations[a]

	GEM9		GEML2		GRIM3-L1	
	Undulation	Anomaly	Undulation	Anomaly	Undulation	Anomaly
OSU 81	±1.8 m	±3.6 mgal[b]	±1.8 m	±3.6 mgal[b]	±1.1 m	±1.9 mgal[b]
GEM9	—	—	±0.5	±0.8	±2.4	±4.6
GEML2	—	—	—	—	±2.4	±4.5

[a] $l_{max} = 20$.
[b] 1 mgal $= 10^{-5}$ m s^{-2}.

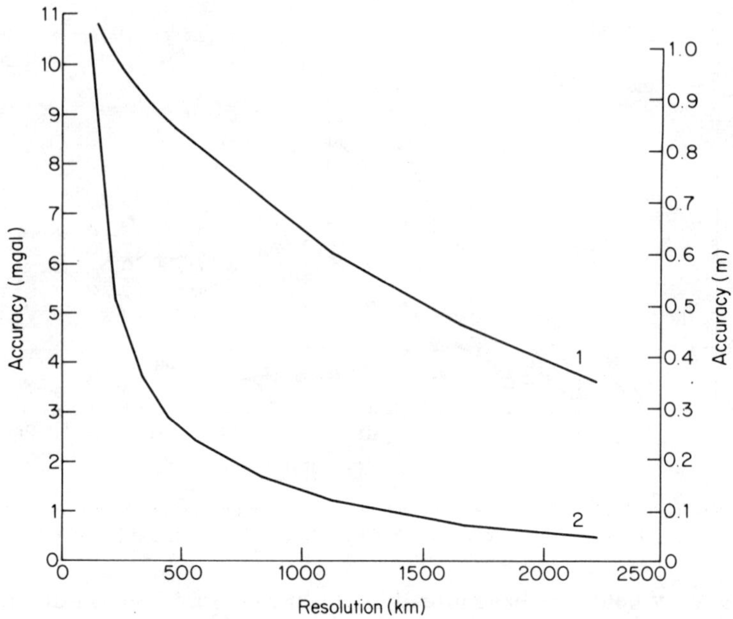

Fig. 4. Accuracy of mean gravity anomalies and geoid undulations by block size or resolution based on 180 fields. Curve (1) undulations and curve (2) anomalies.

specific geographic regions may be considerably worse (or better). Also shown in Fig. 4 are the undulation accuracies. At 100-km resolution the undulation accuracy is ± 1.0 m while at the 2000-km resolution level it is ± 0.38 m.

B. The Estimation of *kM*

The determination of *kM* in Eq. (1) has improved considerably in the past decade primarily through the efforts of laser ranging to the moon and to artificial satellites. Specific results are described by Lerch *et al.* (1982a) and Dickey *et al.* (1983). A current representative value (including the mass of the atmosphere) is $398{,}600.44 \pm 0.01\ \mathrm{km}^3/\mathrm{s}^2$ (Rapp, 1983c). There is some indication that the actual standard deviation is somewhat smaller than quoted.

C. The Role of Satellite Altimetry

Satellite altimeter data obtained from GEOS-3 and SEASAT have played a substantial role in our knowledge of the gravity field of the Earth as well as other quantities of interest such as bathymetry. The altimeter data has been used in at least three different ways for gravity field improvement.

1. Mean Sea Surface Determination

Given the altimeter measurements and orbit adjustment procedures, the elevation of the mean sea surface above a defined reference ellipsoid can be determined. Although classically associated with the geoid, the mean surface is not an equipotential surface because of ocean currents, wind effects, etc. The accuracy of these elevations depends on a number of factors including altimeter measurement type (GEOS-3 or SEASAT), orbit adjustment procedures (orbit mode or crossing arc), tides and so forth. A representative 1-s.d. estimate is ±40 cm. Typical results are described by Marsh and Martin (1982) and by Rapp (1982).

2. Improved Orbit Determinations

The altimeter measurement is incorporated in the actual orbit determination and gravity field improvement process. An example of this is the GEM10B model and a model for SEASAT described by Lerch *et al.* (1982b). The result of the SEASAT analysis was a set of potential coefficients complete to degree and order 36 that enabled the orbit of SEASAT to be computed for a length of 12 days with an rms radial accuracy of about 70 cm.

3. Direct Anomaly Recovery

Gravity anomalies in various size blocks can be recovered directly from the sea surface heights. Since the altimeter data have an almost complete ocean-wide coverage between ±70° latitudes, an ocean-wide recovery of gravity anomalies is possible. Such estimates for 5° and 1°×1° as obtained from SEASAT data are described by Rapp (1983b) Sample estimates for 30′×30′ mean values (i.e. 50 km×50 km) blocks across the Japan trench are described by Rapp (1983d) on the basis of a combined GEOS-3/SEASAT altimeter data set. These latter computations show the ability of the altimeter data to accurately imply anomaly changes of 250 mgal in adjacent 50 km blocks.

The accuracy of mean anomalies derived from altimeter data should be: ±3 mgal for 5° equal area blocks, ±6 mgal for 1°×1° blocks, and ±12 mgal for 30′×30′ blocks. These accuracy estimates vary over the ocean depending on the roughness of the anomaly field and the accuracy of the actual altimetric data considering orbit, tidal and sea surface height corrections. It is also possible to derive point gravity anomalies from the altimeter data but the accuracy will be poorer (± 15 mgal) than obtained from ship measurements. The accuracy spectrum of anomalies derived from the altimeter data is not clear since the error spectrum of such altimeter data has not been defined when the data is processed in a crossing arc adjustment (Wunsch and Zlotnicki, 1984). This implies that such anomalies must be carefully used when long-wavelength implications are being sought.

Altimeter data has also been used for studies indirectly related to gravity field information. Specifically the sea surface signal implied by topographic features has been used to infer such features (Dixon *et al.*, 1983; Dixon and Parke, 1983; White *et al.*, 1983; Haxby *et al.*, 1983; Sandwell, 1984). These studies vary from individual seamount detection to ocean-wide maps of topographic relief implied by the altimeter data.

D. Satellite-to-Satellite Tracking

Gravity field information is also obtained from the analysis of satellite-to-satellite tracking data. Unfortunately the data available is primarily from the GEOS-3/ATS-6 tracking data. The limitation of this data lies in the high altitude (800 km) of GEOS-3 which makes a range rate signal only slightly perturbed by local gravity anomalies.

Despite this, the data has been successfully used by Marsh *et al.* (1981, 1984) to improve the gravity field in the Pacific area and by Kahn *et al.* (1982) for 10° mean anomaly improvement in parts of Africa and the Indian Ocean. Of perhaps the greatest importance of this work is that it demonstrates the value of satellite-to-satellite tracking for gravity field improvement. Such experiments have a great potential for a substantial improvement in the gravity field.

IV. The Gravity Field: Future Developments

The previous discussion was concerned with our current knowledge of the gravity field of the Earth. In the future, substantial improvement is possible in several different areas. We discuss this improvement in two areas; (1) terrestrial-based measurements and (2) satellite-based measurements.

A. Terrestrial-based Measurements

In this section we are primarily concerned with data acquired from sensors that are on or close to the surface of the Earth.

The acquisition of ground-based gravity data is likely to continue to improve our general areal coverage of the world. However, land-based measurements are not expected to be available in many areas because of physical and political inaccessibility problems. This implies that there is no real hope in acquiring terrestrial data in Eastern European countries, China, parts of Asia, South America and Africa.

The acquisition of ship gravity data is likely to continue on cruises of opportunity. This data is very helpful for the high-density, along-track coverage that is provided. Such data can be acquired by almost no other means. However, this data is hampered by its wide-track spacing except in areas where areal coverage is obtained.

As the developments in absolute devices continue, to make them more accurate, more portable and less expensive, we should see these measurements widely used for the measurement of a number of geophysical phenomena. The current accuracy of $\pm 10\ \mu gal$ is sufficient for many studies. If a factor of 10 improvement could be obtained, the interpretation of the time variations in an absolute sense could be quite interesting. For example, general proposals (e.g. Biró, 1983) have been made to form a worldwide net of gravity stations where gravity and height variations are measured.

Another area in which we would expect to see additional development is in the tidal area where new instruments with very low drift give a very accuate long-term record of gravity at a site. Such records can lead to a better understanding of the Earth's interior, ocean tides and perhaps the measurement of polar motion (Richter, 1983).

A major new data source that will become available in the next few years will be that obtained from gradiometers. These measurements of the second derivative of the gravity potential will lead to improved estimates of the high-frequency part of deflections of the vertical and gravity anomalies at the surface of the Earth. Such measurements will be primarily made by aircraft flying at an altitude of (say) 600 m. A description of such a system is given by Jekeli (1983). Using a survey spacing of 5 km, one would expect anomaly accuracies (the high-frequency part) on the order of ± 0.5 mgal and deflections on the order of $\pm 0.06''$.

B. Satellite-based Measurements

In order to improve the knowledge of the gravity field for geodetic, geophysical and oceanographic purposes, satellite-based missions have been discussed for many years (e.g. Committee on Geodesy, 1979).

The most ambitious plan has been developed by NASA for a Geopotential Research Mission (Taylor *et al.*, 1983; Murphy, 1983). If approved as a mission, in the next year or so, GRM could be launched in the 1992/1993 time frame. This mission could be for the improvement of both the gravity field and magnetic field on a global basis. The goals for the gravity field improvement are to obtain gravity anomalies accurate to ±1 mgal and the geoid accurate to ±5 cm with a resolution of 100 km. The main features of the mission would be two spacecraft at an altitude of approximately 160 km operating with a drag compensation system. The spacing between the spacecraft could vary from 100–600 km. The observation to be made would be that of the range rate between the two spacecraft. The accuracy of this measurement is expected to be ±1 μm/s. This data would enable spherical harmonic fields to be determined to very high degree (say 300). The analysis of this data will be a large computational effort. Efficient techniques for processing such data are now being developed.

Figures 5 and 6 show the accuracy to be expected from the mission for anomalies and geoid undulations as a function of resolution or area average. Below 100 km the error rapidly increases although information is present in the measured signal. These results are plotted with current accuracy assessments of the GEML2 field and an 180×180 field from Ohio State University. Since GEML2 is only complete to degree 20, the resolution is primarily limited to about 1000 km.

Another sensor that has great potential for space applications is the gravity gradiometer. The very high gradient accuracies needed for space applications appear to be feasible for the cryogenic gravity gradiometer being developed by Paik (1981). The gradiometer can be designed to measure one or more components of the tensor (Wells, 1984). Jekelli and Rapp (1980), Breakwell (1979), Colombo and Kleusberg (1983) and others have carried out approximate simulation studies to indicate the resolution and accuracy in the determination of the gravity from satellite gradiometer data (primarily the second-order radial component). By using the procedures of Jekeli and Rapp (1980) with a radial component gradiometer having an accuracy of $\pm 10^{-4}$ E (Eötvös), at an altitude of 130 km for a six-month mission, the expected accuracies for anomalies are shown in Fig. 5 and for undulations in Fig. 6. The SST mission improves the current accuracy situation (on a global average) by a factor of about 10. The gradiometer mission is an improvement over the SST mission by about 60% except at the very short wavelengths (i.e. <20 km). The gradiometer mission could obtain anomalies to an accuracy of about ±3 mgal with a resolution of 50 km. This is three times better than that expected from the SST mission.

The missions are primarily global in extent. It is also possible to have a limited mission where only portions of the Earth are covered by a sensor.

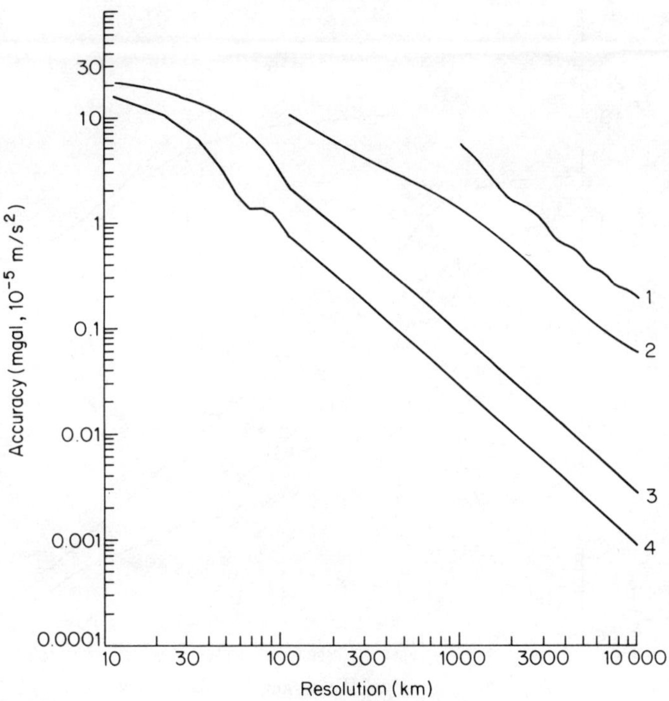

Fig. 5. Gravity anomaly accuracies for current and future gravity field models. Curve (1) GEML2, curve (2) OSU81, curve (3) SST mission (± 1 μm/s, 160 km) and (4) gradiometer mission ($\pm 10^{-4}$ E, 130 km).

An example of such a mission could be a gradiometer in the space shuttle, or a gradiometer (or equivalent) placed in a satellite tethered from the shuttle. Gullahorn *et al.* (1984) describe a system where a gradiometer could be held at an altitude of approximately 130 km, tethered from the shuttle. The launch of such a system could be in the late 1980s.

Another future gradient measurement technique has been described by Balmino *et al.* (1984) as PROJECT GRADIO. In this proposed mission, microaccelerometers would be placed in a satellite to determine the components of the gradient tensor in a satellite-fixed coordinate system. The expected accuracy of the measurement would be on the order of ± 0.01 E. Although this is not as good an accuracy as would be associated with the Paik equipment, it should be sufficient to obtain a substantial improvement in the knowledge of the gravity field. The earliest launch date for this mission is 1990.

Colombo and Kleusberg (1983) have also proposed a mission that would combine an altimeter with a gradiometer that would sense the second radial

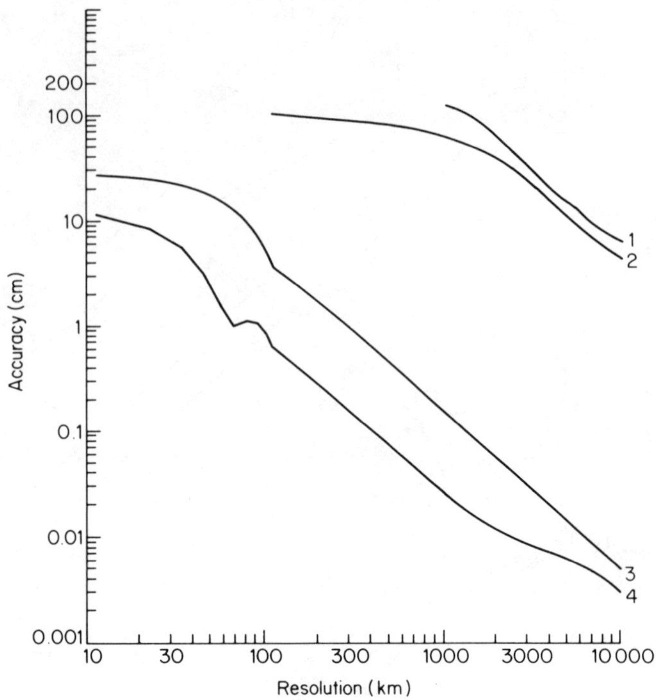

Fig. 6. Geoid undulation accuracies for current and future gravity field models. Curve (1) GEML2, curve (2) OSU81, curve (3) SST mission (±1 μm/s, 160 km) and curve (4) gradiometer mission ($\pm 10^{-4}$ E, 130 km).

derivative of the potential. The basic principle is that the geocentric distance of the satellite could be determined given: (1) the gradient measurement, (2) a fairly good set of low- (<30) degree potential coefficients and (3) an accurate kM value. The accuracy needed for the latter two quantities is well within that expected in the next few years. With this type of information the geocentric distance could be determined to an accuracy approaching 1 cm. If this can be done the need for extensive tracking of the altimeter satellites could be considerably reduced.

The previous discussions have been directed towards improvements in the gravity field that can be considered in the next 10–15 years. However, there are developments in the shorter term that will yield improved gravity field information. The first procedure relates to the development of an interim gravity field model which uses existing data and data that would be acquired through specific tracking campaigns. This interim field model (Murphy, 1983; National Aeronautics and Space Administration, 1982) could provide the basis for the tracking of altimeter satellite TOPEX which may be launched in 1990 (Born *et al.*, 1984). The tracking of TOPEX can also provide additional gravity field information. The accuracy and resolution of the gravity field to be obtained from TOPEX will much depend on the accuracy of the tracking procedures used.

We should also note in the near term developments of the launch of LAGEOS II in 1987. This satellite can be expected to give improved low-degree gravity fields and their time variation. It will not help, because of its high orbit, in the determination of the short-wavelength gravity field.

Future developments for the gravity field determination from space are needed. Since the accuracy to be expected from a single-component gradiometer seems to be only a factor of two better than the SST mission of GRM, more studies need to be made to see the effect of knowing multiple components of the gradient tensor, and to assess other techniques for gravity field improvements from space.

The above discussion has outlined some of the possible future developments related to gravity field improvement. There are more issues involved, however, than just planning new missions. There are theoretical studies needed on the downward continuation problem for an Earth with topography. We need to understand the auxiliary data that could be needed for the reduction and interpretation of the data. For example, improved knowledge of the topography of the Earth should be a future goal. This might be acquired by altimetric missions or space photogrammetry. As our knowledge of the gravity field improves, in terms of long and short wavelengths, temporal variations will play an increasingly important role. For example, polar-cap melting can have effects on sea level as well as on gravity. Will such change be detected by future gravity measurements? At what level of accuracy will we need to define a global gravity field in terms of a specific epoch?

In the near future, dramatic improvements in our knowledge of the gravity field at the long and short wavelengths will be possible. There will be a time at which the need for such determinations on a global basis will stabilize. But there will be continuing interest in the gravity field from the temporal point of view for geodetic, oceanographic and geophysical applications.

Acknowledgments

Much of the paper was originally written for the report "Geodesy: A Look to the Future" which is to be a report of the Committee on Geodesy, Board on Earth Sciences, National Research Council. Some of the material discussed in this report was developed under sponsorship of NASA (grant No. NGR 36-008-161) or the U.S. Air Force (contract F19628-82-K-0022) through The Ohio State University Research Foundation.

References

Balmino, G. (1982). "Bulletin D'Information", No. 50. Bureau Gravimetrique International, Toulouse, France.

Balmino, G. (1983). The surface gravity data available for improvement of the global knowledge of the geopotential. *Proc. Gen. Assem. Int. Assoc. Geod. IUGG MTG, Hamburg.* Dept. of Geodetic Science and Surveying, Ohio State Univ. Columbus.

Balmino, G. *et al.* (1984). Le project gradio et al determination a haute resolution du geopotential. *Bull. Geod.* **58**, 3.

Biro, P. (1983). The role of gravimetry in the geodynamic interpretation of geodetic observations. *Proc. Int. Assoc. Geod. Symp., IUGG Meet., Hamburg*, Dept. of Geodetic Science and Surveying, Ohio State Univ. Columbus.

Breakwell, J. (1979). Satellite determinations of short wavelength gravity variations *J. Astronaut. Sci.* **27**, 4.

Colombo, O. and Kleusberg, A. (1983). Applications of an orbiting gravity gradiometer. *Bull. Geod.* **57**, 83–100.

Committee on Geodesy (1979). "Applications of a Dedicated Gravitational Satellite Mission". National Academy of Sciences, Washington, D.C.

Dickey, J. *et al.* (1983). Geophysical applications of lunar laser ranging. *Proc. IAG/IUGG Symp., Hamburg*, Dept. of Geodetic Science and Surveying, Ohio State Univ. Columbus.

Dixon, T. and Parke, M. (1983). Bathymetry estimates in the southern oceans from SEASAT altimetry. *Nature* **304**, 406–411.

Dixon, T., *et al.* (1983). Bathymetric prediction from SEASAT altimeter data. *J. Geophys. Res.* **88**, C3, 1563–1571.

Engelis, T. and Rapp R. (1983). Global ocean circulation patterns based on SEASAT altimeter data and the GEML2 gravity field. Presented at the IAG/IUGG meeting in Hamburg.

Ganeko, Y. (1982). $10' \times 10'$ detailed gravimetric geoid around Japan. *Proc. Gen. Meet. Int. Assoc. Geod., Tokyo.*

Gu, C.-X. (1983). "National Report on Geodesy". XVIIIth General Assembly of IUGG, Hamburg, China National Committee for the International Union of Geodesy and Geophysics, Beijing, China.

Gullahorn, G., *et al.* (1984). Feasibility of gravity gradient measurements from a tethered subsatellite platform, abstract. *EOS* **65**, 438. 1984 July 10.

Haxby, W. L. *et al.* (1983). Digital images of combined oceanic and continental data sets and their use in tectonic studies. *EOS* **64**, 995–1004.

Hittleman, A. M., Scheibe D. and Goad, C. (1982). "U.S. Land Gravity, Key to Geophysical Records Documentation", No. 18. National Geophysical Data Center, Boulder, Colorado.

Jekeli, C. and Rapp, R. (1980). Accuracy of the determination of mean anomalies and mean geoid undulations from a satellite gravity field mapping mission. Report No. 307. Dept. of Geodetic Science and Surveying. Ohio State Univ. Columbus.

Jekeli, C. (1983). Achievable accuracies for gravity vector estimation from airborne gravity gradiometry. *Proc. Gen. Assem. Int. Assoc. Geod. IUGG Meet., Hamburg.* Dept. of Geodetic Science and Surveying. Ohio State Univ. Columbus.

Lachapelle, G. (1978). Evaluation of $1° \times 1°$ mean free air gravity anomalies in North America, Collected papers. Geodetic Survey of Canada, Surveys and Mapping Branch, 183–213.

Lambeck, K. and Coleman, R. (1983). The Earth's shape and gravity field; A report of progress from 1958 to 1982. *Geophys. J.R. Astron. Soc.* **74**. 25–54.

Lerch, F. (1983). Status of the geopotential. *Rev. Geophys. Space Phys.* **21**, 560–565.

Lerch, F. *et al.* (1979). Gravity model improvement using GEOS-3 (GEM9 and 10). *J. Geophys. Res.* **84**, B8, 3897–3916.

Lerch, F. J., *et al.* (1982*a*). A refined gravity model from LAGEOS (GEM-L2). *Geophys. Res. Lett.* **9**, 1263–1266.

Lerch, F., Marsh, J., Klosko, S. and Williamson, R. (1982b) Gravity model improvement for SEASAT. *J. Geophys. Res.* **87**, C5 3281–3296.

Marsh, J. and Martin, T. (1982). The SEASAT altimeter mean sea surface model. *J. Geophys. Res.* **87**, 3269–3280.

Marsh, J. G., *et al.*, (1981). The gravity field in the central Pacific from satellite to satellite tracking. *J. Geophys. Res.* **86**, 3979–3997.

Marsh, B. D., *et al.* (1984). On gravity from SST, geoid from SEASAT, and plate age and fracture zones in the Pacific. *J. Geophys. Res.* **89**, 6070–6078.

Murphy, J. (1983). Geopotential research program. *Proc. Int. Assoc. Geod. Symp. IUGG Meet. Hamburg, Dept. Geod. Sci. and Surv.* Ohio State Univ. Columbus.

National Aeronautics and Space Administration (1982). Report of the gravity field workshop. Goddard Space Flight Center, NASA TM 84003.

Paik, H. J. (1981). Superconducting tensor gravity gradiometer for satellite geodesy and inertial navigation. *J. Astronaut. Sci.* **29**, 1–16.

Rapp, R. H. (1981). The Earth's gravity field to degree and order 180 using SEASAT altimeter data, terrestrial gravity data, and other data. Report No. 322. Dept. of Geodetic Science and Surveying, Ohio State Univ. Columbus, 1981.

Rapp, R. H. (1982). A Golobal Atlas of the Sea Surface Heights Based on the Adjusted SEASAT Altimeter Data, Report No. 333, Dept. of Geodetic Science and Surveying, Ohio State Univ. Columbus.

Rapp, R. H. (1983a). The development of the January 1983 $1° \times 1°$ mean free-air anomaly data tape. Internal report. Dept. of Geodetic Science and Surveying, Ohio State Univ., 23 pp.

Rapp, R. H. (1983b). The determination of geoid undulations and gravity anomalies from SEASAT altimeter data. *J. Geophys. Res.* **88**, C3, 1552–1562.

Rapp, R. H. (1983c). Report of special study group No. 5.39 of IAG: Fundamental Geodetic Constants. Presented at the XVIII General Assembly of IUGG, Hamburg, August.

Rapp, R. H. (1983d). The role of satellite altimeter data in solving geodetic and oceanograhic problems. In "Geodesy in Transition" (Schwarz and Lachapelle, eds.) University of Calgary, 1983d.

Reigber, C., Muller, H., Rizos, C. Bosch, W., Balmino, G. and Moynot, B. (1983). An improved GRIM3 Earth gravity model (GRIM3B). Presented at the IAG/IUGG General Assembly, Hamburg, August.

Reigber, C., Muller H., Bosch, W., Balmino G. and Moynot, B. (1984). GRIM gravity model improvement using LAGEOS (GRIM3-L1). *J. Geophys. Res.*

Richter, B. (1983). The long periodic tides in the Earth tide spectrum. *Proc. Int. Assoc. Geod. Symp., IUGG Meeting, Hamburg.* Dept. of Geodetic Science and Surveying, Ohio State Univ. Columbus.

Sandwell, D. (1984). A detailed view of the South Pacific geoid from satellite altimetry *J. Geophys. Res.* **89**, B2, 1089–1104.

Society of Exploration Geophysicists, (1982). Gravity anomaly map of the United States, 1;2,500,000 scale. Tulsa, Oklahoma.

Taylor, P., *et al.* (1983). GRM: Observing the terrestrial gravity and magnetic fields in the 1990's. *EOS* **64**, 609–611.

Torge, W., Weber, G. and Wenzel, H.-G. (1983). $6' \times 10'$ free-air gravity anomalies of Europe including marine areas. Presented at the IAG/IUGG meeting, Hamburg, August.

Tscherning, C. C., Rapp, R. and Goad, C. (1983). A comparison of methods for computing gravimetric quantities from high degree spherical harmonic expansions. Manuscr. Geod. **8**, 249–272.

Wells, W. (ed.) (1984). "Spaceborne Gravity Gradiometers." NASA Conference on publication 2305.

White, J. V., *et al.* (1983). Detection of seamount signatures in SEASAT altimeter data using matched filters. *J. Geophys. Res.* **88**, C3, 1541–1551.

Wilcox, L. and Scheibe, D. (1983). Gravimetry. *Rev. Geophys. Space Phys.* **21**, 537–543.

Wunsch, C. and Zlotnicki, V. (1984). The accuracy of altimetric surfaces. *Geophys. J.* **78**, 795–808.

2.3 Slow Changes in the Earth's Shape and Gravitational Field: Constraints on the Glaciation History and Internal Viscoelastic Stratification

W. R. Peltier

Department of Physics, University of Toronto, Toronto, Ontario, Canada, M5S 1A7

I. Introduction

Recent advances in seismic imaging have led to the construction of models of the internal elastic structure of the Earth of ever increasing complexity and, one hopes, realism. In the past several years attention has shifted from an exclusive concentration on the problem of inferring the best "terrestrial monopole" (e.g. Gilbert and Dziewonski, 1975; Dziewonski and Anderson, 1981) to the more difficult issue of attempting to resolve, as a function of depth, the existing lateral heterogeneity of elastic properties (e.g. Dziewonski *et al.*, 1977; Masters *et al.*, 1982; Woodhouse and Dziewonski, 1984; Dziewonski, 1984). The hope in this work is to obtain direct images

SPACE GEODESY AND GEODYNAMICS
ISBN 0-12-057840-9

of the thermally induced circulation in the mantle which is responsible for continental drift and seafloor spreading.

The existence of a thermally induced convective circulation in the mantle is of course possible only if this region of the Earth behaves as a viscous fluid when it is subjected to a shear stress which is of sufficient temporal duration. The vigour of this circulation is determined essentially by mantle viscosity if other parameters in the system, Rayleigh number, are considered to be fixed on the basis of laboratory measurements (e.g. Peltier, 1980, 1985). For a variety of reasons, not the least of which is connected with the very low strain rates involved in the convection process, direct laboratory measurement of the effective viscosity of mantle material presents difficult problems. It is therefore of some interest to enquire as to whether there might be observable phenomena apart from mantle convection itself to which one might appeal to constrain this Earth property directly.

The observations which have most often, and successfully, been employed for this purpose are those connected with the phenomena of glacial isostasy. These phenomena are each in some direct way connected with the response of the planet to the glaciation–deglaciation cycle to which it has been subjected for the last few million years which comprise the Pleistocene geological epoch of Earth history. On the basis of $\delta^{18}O$ data from deep-sea sedimentary cores, it has been well established (Hays *et al.*, 1976; Imbrie *et al.*, 1984) that the dominant cryospheric oscillation has a time scale near 10^5 yr and is characterized by a slow buildup of continental ice mass on a time scale near 9×10^4 yr and a fast "termination" lasting about 10^4 yr. In the last 800,000 years this oscillation has been a durable feature of the climate record and is understood to exist because of the small changes in seasonal solar insolation received by the Earth because of temporal variations in the parameters which govern its orbit around the sun (e.g. eccentricity, obliquity, precession). The physics required to understand the dominance of the 10^5-yr cycle in the record of ice-sheet volume fluctuations appears to be intimately dependent on the same process of glacial isostatic adjustment from which one can infer the internal viscoelastic stratification of the planet (Peltier, 1982; Peltier and Hyde, 1984; Hyde and Peltier, 1985).

The last glacial maximum of the current ice age occurred about 18,000 years ago, and deglaciation was complete by 6000–8000 years ago. Three major continental ice masses were involved in this large-scale melting event. The most important of these was the Laurentide ice sheet which covered all of Canada and parts of the northern United States (when the Cordilleran and Franklin-Innuitian complexes of western Canada and the Canadian Arctic, respectively, are included). The volume of this ice sheet was such that its melting alone was responsible for a global sea-level rise of approximately 60 m. The remaining centres of deglaciation were in Fennoscandia, where an ice complex was centred on the Gulf of Bothnia whose melting produced a sea-level rise of about 20 m, and in Antarctica, where the melting of the west Antarctic ice sheet contributed a further sea-level increase near 20 m. The melting of the ice sheets in these three widely separated geographical regions was essentially contemporaneous, a fact which until recently has been difficult to understand on the basis of the Milankovitch (1941) hypothesis in which all such fluctuations are attributed to orbitally controlled summer seasonal insolation fluctuations. The reason for this difficulty is that the seasonal insolation variations are out of phase in the Southern and Northern Hemispheres so that one should not expect the approximately inphase oscillations which are in fact observed. The currently accepted explanation for this paradox is that the oscillation of the west Antarctic ice sheet is forced by the sea-level variations caused by the accumulation and disintegration of the Northern Hemisphere ice masses. This hypothesis is reasonable given the fact that the west Antarctic ice sheet is marine (e.g. Denton and Hughes, 1980). The Southern Hemisphere "tail" is wagged by the Northern Hemisphere "dog".

Given the magnitude of the continental ice volume fluctuations involved in the ice-age cycle, it should not be too surprising to learn that these fluctuations of surface load have been responsible for inducing large deformations in the shape of the underlying planet which have left easily discernible signatures in the geological record. The most important of these signatures are the histories of relative sea-level change which are recorded in sequences of raised and submerged beach horizons both at locations which were once under the ice sheets and at locations peripheral to them. The time interval since deglaciation began at 18 KBP (kiloyears before present) is such that ^{14}C methods provide a direct means of dating the individual beach horizons in a sequence. Four examples of such relative sea-level data are shown in Fig. 1 for a site which was under the Laurentian ice sheet (the Ottawa Islands), a site which was at the edge of the ice sheet (Boston), a site south of the ice-sheet margin on the United States East Coast (Brigantine), and a site in the far field (Recife, Brazil). Inspection of these data shows that in the region which was once under the ice, the sea-level record is characterized by monotonic emergence, at the edge by nonmonotonic emergence followed by submergence, and further afield by monotonic submergence. In the far field at sites like Recife, the relative sea-level (RSL) record is characterized by the emergence of a raised beach at 6 KBP, the time at which melting effectively ceases.

Over the past decade, beginning with the paper by Peltier (1974), a complete linear viscoelastic field theory has been developed to explain the global pattern of observed changes in relative sea level and to predict other signatures of the isostatic adjustment process which may also be employed

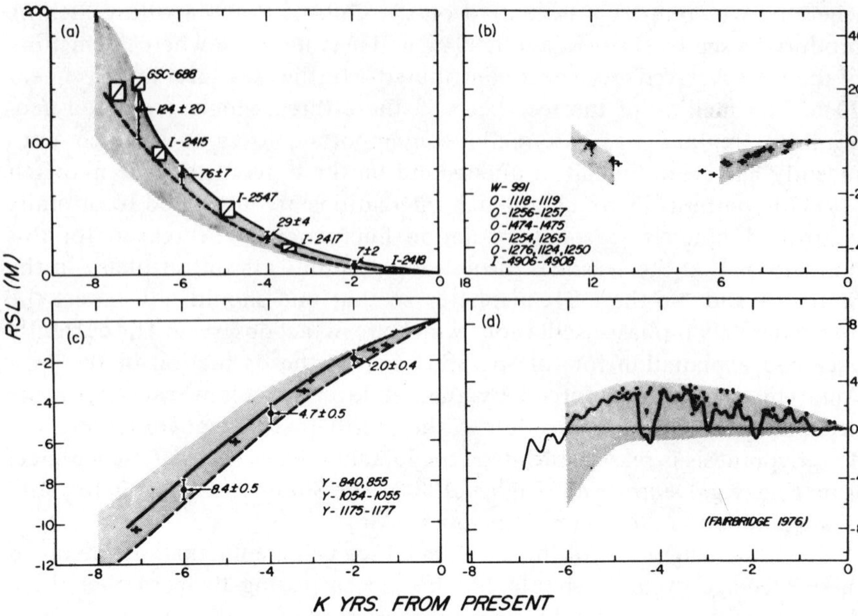

Fig. 1. Four examples of radiocarbon controlled relative sea-level curves from (a) the Ottawa Islands in central Hudson Bay, (b) the vicinity of Boston, Massachusetts which was near the margin of the Laurentide ice sheet, (c) the vicinity of the coastal city of Clinton, Connecticut and (d) the vicinity of Recife, Brazil, South America.

to constrain the internal viscoelastic layering of the planet. Examples of these other signatures of the adjustment process include free-air gravity anomalies over the present day centres of rebound, variations in the Earth's rate of rotation and in the wander of its rotation pole with respect to the surface geography. This theory will be reviewed and extended in the following sections and examples will be shown of its utility as a means of inferring the parameters of Earth's internal viscoelastic stratification. As will also be demonstrated, modern space-based geodetic techniques are having a profound impact on the use to which these observations may be put because they are significantly improving the accuracies to which they are determined.

II. The Theory of Glacial Isostasy

The model which has been developed to describe the Earth's response to surface ice loads is one in which the rheology of the interior is assumed to be linearly viscoelastic. The field equations of the model consist of the equations of momentum conservation and for the gravitational potential which respectively take the form (e.g. Backus, 1967):

$$0 = -\rho_0 \nabla \tilde{\phi}_1 - \tilde{\rho}_1 g_0 \mathbf{e}_r - \nabla(\tilde{\mathbf{u}} \cdot \rho_0 g_0 \mathbf{e}_r) + \nabla \cdot \tilde{\boldsymbol{\tau}} \tag{1a}$$

$$4\pi G \tilde{\rho}_i = \nabla^2 \tilde{\phi}_i \tag{1b}$$

In (1b) the density perturbation $\tilde{\rho}_1$ is determined by the linearized form of the continuity equation as:

$$\tilde{\rho}_1 = -\rho_0 \nabla \cdot \tilde{\mathbf{u}} - \tilde{\mathbf{u}} \cdot (\partial_r \rho_0) \mathbf{e}_r \tag{1c}$$

where the momentum equation (1a) has been obtained by linearizing about a hydrostatic equilibrium configuration (ρ_0, p_0, g_0) which satisfies:

$$\nabla p_0 = -\rho_0 g_0 \mathbf{e}_r \tag{1d}$$

$$\nabla^2 \phi_0 = 4\pi G \rho_0 \tag{1e}$$

In Eqs. (1), ρ_0 and g_0 are density and radial basic state gravitational acceleration, $\tilde{\mathbf{u}}$ is the displacement vector, $\tilde{\phi}$ the perturbation of the gravitational potential, $\tilde{\boldsymbol{\tau}}$ the stress tensor and $\mathbf{e}_r$ a unit vector in the radial direction. The tildas above the perturbation quantities in Eqs. (1) signify that the linear field equations have been transformed from the time domain to the domain of the Laplace transform variable s, and the inertial force has been dropped from the left-hand side of (1a) because of the long time scale of the phenomena which are of interest to us. In general, the perturbation geopotential ϕ_1 (tildas will be dropped in all that follows) is the sum of two parts, ϕ_2 and ϕ_3, which are respectively the potential of the applied gravitational force field (the load) and that due to the internal redistribution of mass effected by the load induced deformation. Inside the Earth it thus follows that $\nabla^2\phi_3 = 4\pi G\rho_1$ whereas $\nabla^2\phi_2 = 0$. Because Eqs. (1a) and (1b) are fully coupled, their solution will include the full effects of self-gravitation. The following subsections will separately describe the most important ingredients in the field theory embodied in these equations.

A. The Rheological Constitutive Relation

The constitutive equation which is employed in the model to connect the elements of the stress tensor τ_{kl} to the elements of the strain tensor e_{kl} is that for a linear Maxwell solid. Since the Maxwell solid appears Hookean elastic in the limit of small time and Newtonian viscous in the limit of long time, it is asymptotically realistic in the sense that it agrees with the requirements of seismology in the former case and with the requirements of mantle convection in the latter. In fact it is not exactly correct in the seismic limit because it cannot correctly predict the observed variations of seismic Q with frequency. This can be rectified, however, by extending the model to what I have previously called the "generalized" Burgers' body (e.g. Peltier, 1982) but this extension is not required for the understanding of glacial isostatic adjustment. The time domain form of the constitutive relation for the Maxwell solid is:

$$\dot{\tau}_{kl}+\mu/\nu(\tau_{kl}-\tfrac{1}{3}\tau_{kk}\delta_{kl})=2\mu\dot{e}_{kl}+\lambda\dot{e}_{kk}\delta_{kl} \tag{2}$$

In (2) λ and μ are the usual elastic Lamé parameters, ν is the Newtonian molecular viscosity, and δ_{kl} is the unit diagonal tensor. In the domain of the Laplace transform variable s, Eq. (2) has the form:

$$\tilde{\tau}_{kl}=\lambda(s)\tilde{e}_{ii}\delta_{kl}+2\mu(s)\tilde{\varepsilon}_{kl} \tag{3}$$

where the viscoelastic compliances $\lambda(s)$ and $\mu(s)$ are:

$$\lambda(s)=\frac{\lambda s+\mu K/\nu}{s+\mu/\nu} \tag{4a}$$

$$\mu(s)=\frac{\mu s}{s+\mu/\nu} \tag{4b}$$

$$K=\lambda+\tfrac{2}{3}\mu=\lambda(s)+\tfrac{2}{3}\mu(s) \tag{4c}$$

Because the bulk modulus K is independent of s, the model has no bulk dissipation. The quantity $\nu/\mu = T_{\mathrm{m}}$ is called the Maxwell time and is clearly the time scale which must be exceeded in order to effect the transition from Hookean elastic to Newtonian viscous behaviour in response to an applied shear stress.

The technique which is employed to obtain time domain solutions of the field equations (1) is called the correspondence principle. This principle (e.g. Biot, 1954) assures us that if we solve the equivalent elastic loading problem by using the constitutive relation (3) for all possible values of the Laplace transform variable s, we will have constructed the Laplace transform of the solution to the viscoelastic problem of interest. The only problem remaining is then to invert the solution so obtained back into the time domain. There are two absolutely crucial ingredients which are required to make this method work in the context of the isostatic adjustment problem. These concern the way in which the Archimedes force [the gravitational restoring force (buoyancy) due to the load-induced surface deflection] is introduced into the problem, and the way in which the Laplace transform domain solutions are inverted into the time domain. These ingredients are discussed below.

B. Prestress Advection and the Archimedes Force

The fundamental physical concept involved in the process of isostatic adjustment is that a finite load placed on the free surface of a viscous fluid will eventually induce a steady finite deformation of this free surface, in which state the weight of the load will be supported by the buoyancy force due to the displacement of material by the deformation. In every analysis of the isostatic adjustment process prior to that in Peltier (1974), this buoyancy force was introduced through the boundary conditions by adding an extra component of the normal stress proportional to the surface displacement (e.g. Haskell, 1935, 1936, 1937). In the viscoelastic theory of isostatic adjustment which employs the correspondence principle, however, this buoyancy force is introduced [following Love's (1927) analysis of the elastic problem] by the addition of an extra body force to the equation of momentum conservation. This extra body force is that described by the third term on the right-hand side of Eq. (1a) which has the form of the gradient of the radially advected prestress $\mathbf{u} \cdot (\partial p_0/\partial r)\mathbf{e}_r$. Integration of Eq. (1a) over a small pill-box centred on an interface across which there exists a discontinuity of ρ_0 shows that the presence of this term introduces a normal stress discontinuity in the amount $\Delta\rho_0 g_0 u_r$ which is just the buoyancy force per unit area due to the radial displacement of the interface by the amount u_r.

Although prestress advection in the elastic problems of conventional geodynamic interest (ocean tidal loading, elastic gravitational free oscillation, etc.) provides at most a small modification to the results which would otherwise be obtained, this body force is *the* crucial ingredient in the description of isostatic adjustment using the correspondence principle. This fact apparently is not generally understood. At least two recent attempts to understand and employ the analysis of Peltier (1974) have erroneously concluded that there were serious flaws in the correspondence principle approach. However, the prestress advection term was not included in either of these authors' viscoelastic momentum conservation equations (e.g. Svensson, 1980; Nakiboglu and Lambeck, 1980), and their difficulties are traceable to this cause.

In order to demonstrate explicitly the importance of prestress advection in the isostatic adjustment problem, we will briefly discuss in this subsection the solution of Boussinesq's problem for a Maxwell viscoelastic half space using the correspondence principle and compare the result to Haskell's

(1934, 1936) solution based upon the Navier–Stokes equations of conventional hydrodynamics. We begin by neglecting self-gravitation and assuming incompressibility. Both (1b) and the first two terms of (1a) can then be neglected and the equations of equilibrium in the $(x-z)$ plane reduce to the following, in which the superscipt "ve" is used to denote the viscoelastic component of the stress tensor:

$$\partial(\tau_{xx}^{\text{ve}}+\rho_0 g w)/\partial x+\partial\tau_{xz}^{\text{ve}}/\partial z=0 \tag{5a}$$

$$\partial(\tau_{zz}^{\text{ve}}+\rho_0 g w)/\partial z+\partial\tau_{xz}^{\text{ve}}/\partial x=0 \tag{5b}$$

The half space has been assumed to have the constant density ρ_0. Now the incompressible equivalent of (3) requires:

$$\tau_{xx}^{\text{ve}}=-p^{\text{ve}}+2\mu\ \partial u/\partial x \tag{6a}$$

$$\tau_{zz}^{\text{ve}}=-p^{\text{ve}}+2\mu\ \partial w/\partial z \tag{6b}$$

$$\tau_{xz}^{\text{ve}}=\mu(\partial u/\partial z+\partial w/\partial x) \tag{6c}$$

and inspection of (5) shows that they may be rewritten by defining a total stress tensor $\tau_{ij}=\tau_{ij}^{\text{ve}}+\rho_0 g w\delta_{ij}$ (see discussion in Wu and Peltier, 1982, p. 438) as

$$\partial\tau_{xx}/\partial x+\partial\tau_{xz}/\partial z=0 \tag{7a}$$

$$\partial\tau_{xz}/\partial x+\partial\tau_{zz}/\partial z=0 \tag{7b}$$

with (6) in the form

$$\tau_{xx}=-p+2\mu\ \partial u/\partial x \tag{8a}$$

$$\tau_{zz}=-p+2\mu\ \partial w/\partial z \tag{8b}$$

$$\tau_{xz}=\mu(\partial u/\partial z+\partial w/\partial x) \tag{8c}$$

where p denotes the total pressure fluctuation (i.e. the sum of the viscoelastic contribution and the advected contribution). A simple solution to (8) may be constructed in terms of displacement potentials (e.g. Malvern, 1969, pp. 552–554). With the boundary conditions in the domain of the horizontal wave number k as

$$\bar{\tau}_{zz}^{\text{ve}}(k,0)=-\bar{\pi}(k) \tag{9a}$$

$$\bar{\tau}_{xz}^{\text{ve}}(k,0)=0 \tag{9b}$$

the solution for the displacements is simply

$$\begin{bmatrix}\bar{u}(k,z)\\ \bar{w}(k,z)\end{bmatrix}=\frac{\bar{\pi}}{2\mu k+\rho_0 g}\begin{bmatrix}-ikz\\ 1+kz\end{bmatrix}\exp(-kz) \tag{10}$$

while the solution for the stresses is

$$\begin{bmatrix} \bar{\tau}_{xx}(k, z) \\ \bar{\tau}_{zz}(k, z) \\ \bar{\tau}_{xz}(k, z) \end{bmatrix} = \frac{2\mu k \bar{\pi}}{2\mu k + \rho_0 g} \begin{pmatrix} 1 - kz \\ 1 + kz \\ -ikz \end{pmatrix} \exp(-kz) \tag{11}$$

The incorporation of prestress advection into the momentum balance equation is responsible for introducing the additive $\rho_0 g$ factors in the denominators of (10) and (11). To transform these solutions to the time domain we simply substitute for μ from (4b) and transform the resulting expressions back into the time domain. Substituting $\mu(s)$ into (10) and (11) gives

$$\begin{bmatrix} \bar{u}(k, z, s) \\ \bar{w}(k, z, s) \end{bmatrix} = \frac{\bar{\pi}}{2\mu k + \rho_0 g} \left(1 + \frac{\mu/\nu - s_1}{s + s_1}\right) \begin{pmatrix} -ikz \\ 1 + kz \end{pmatrix} \exp(-kz) \tag{12}$$

and

$$\begin{bmatrix} \bar{\tau}_{xx}(k, z, s) \\ \bar{\tau}_{zz}(k, z, s) \\ \bar{\tau}_{xz}(k, z, s) \end{bmatrix} = \frac{-2\mu k \bar{\pi}}{2\mu k + \rho_0 g} \left(1 - \frac{s_1}{s + s_1}\right) \begin{bmatrix} 1 - kz \\ 1 + kz \\ -ikz \end{bmatrix} \exp(-kz) \tag{13}$$

where the parameter s_1 is such that

$$1/s_1 = 2\nu k/\rho_0 g + \nu/\mu \tag{14}$$

The time domain forms of (12) and (13) are then simply:

$$\begin{bmatrix} \bar{u}(k, z, t) \\ \bar{w}(k, z, t) \end{bmatrix} = \frac{\bar{\pi}(k)}{2\mu k + \rho_0 g} \left[\delta(t) + \left(\frac{\mu}{\nu} - s_1\right) e^{-s_1 t} \right] \begin{bmatrix} -ikz \\ 1 + kz \end{bmatrix} \exp(-kz) \tag{15}$$

$$\begin{bmatrix} \bar{\tau}_{xx}(k, z, t) \\ \bar{\tau}_{zz}(k, z, t) \\ \bar{\tau}_{xz}(k, z, t) \end{bmatrix} = \frac{-2\mu k \bar{\pi}}{2\mu k + \rho_0 g} \left[\delta(t) - s_1 e^{-s_1 t} \right] \begin{bmatrix} 1 - kz \\ 1 + kz \\ -ikz \end{bmatrix} \exp(-kz) \tag{16}$$

These are interpreted as the solutions for $\delta(t)$ (impulsive) forcing since the applied pressure field $\pi(k)$ in (9a) was assumed independent of the Laplace transform variable s. The response to Heaviside forcing is obtained from (15) and (16) by convolving these impulse response forms with a Heaviside step function. At $z = 0$ (the surface) these solutions are just

$$\begin{bmatrix} \bar{u}^{\mathrm{H}}(k, 0, t) \\ \bar{w}^{\mathrm{H}}(k, 0, t) \end{bmatrix} = \frac{\bar{\pi}(k)}{2\mu k + \rho_0 g} \left[1 - \frac{(\mu/\nu - s_1)}{s_1} (e^{-s_1 t} - 1) \right] \begin{bmatrix} 0 \\ 1 \end{bmatrix} \tag{17}$$

and

$$\begin{bmatrix} \bar{\tau}^{\mathrm{H}}_{xx}(k, 0, t) \\ \bar{\tau}^{\mathrm{H}}_{zz}(k, 0, t) \\ \bar{\tau}^{\mathrm{H}}_{xz}(k, 0, t) \end{bmatrix} = \frac{-2\mu k}{2\mu k + \rho_0 g} \left[1 + (e^{-s_1 t} - 1) \right] \begin{bmatrix} 1 \\ 1 \\ 0 \end{bmatrix} \tag{18}$$

Since the total and viscoelastic components of the normal stress are related as

$$\tau_{zz}^{\mathrm{ve}} = \tau_{zz} - \rho_0 g w \tag{19}$$

substitution from (17) and (18) shows that

$$\bar{\tau}_{zz}^{\mathrm{ve}}(k, 0) \equiv -\bar{\pi}(k) \tag{20}$$

so that the viscoelastic portion of the stress tensor balances the applied load for all $t > 0$. The advected part of the prestress is

$$\rho_0 g \bar{w}(k, 0) = \left[1 - \frac{2\mu k}{2\mu k + \rho_0 g} e^{-s_1 t}\right] \bar{\pi}(k) \tag{21}$$

which in the $t \to \infty$, inviscid limit reduces to

$$\rho_0 g \bar{w}(k, 0) = \bar{\pi}(k) \tag{22}$$

Thus the advected contribution to the total stress increases until it exactly cancels the surface load. The left-hand side of (22) is of course just the Archimedes force (buoyancy force) due to the vertical deflection of the free surface by the amount w. This simple viscoelastic analysis from Wolf (1985a, b) shows that the adjustment process occurs on the time scale given by (14) which differs from the time scale predicted by the Newtonian viscous analysis of Haskell only by the small correction of the viscous relaxation time $2\nu k/\rho_0 g$ by the Maxwell time ν/μ. This slight increase of the relaxation time was also found in the analysis of the uniform viscous sphere by Wu and Peltier (1982, p. 454), thus demonstrating that the influence of geometry in this regard is unimportant. All of the essential ingredients of the isostatic adjustment process are embodied in the simple example discussed here. The complications introduced by sphericity, self-gravitation, compressibility and radial layering which are included in the general formulation of the next subsection greatly increase the mathematical complexity of the problem, and are very important if the model is to be employed to fit actual geophysical observations, but they are not physically fundamental as is true of the constitutive relation discussed previously and the prestress advection term discussed here.

C. Viscous Gravitational Relaxation of a Layered Compressible Sphere

The general solution of (1) for arbitrary (ρ_0, p_0, ϕ_0) is best expressed in terms of the triplet of nondimensional viscoelastic Love numbers (h_n, k_n, l_n)

which are defined as

$$\begin{bmatrix} U_n(r, s) \\ V_n(r, s) \\ \Phi_{3,n}(r, s) \end{bmatrix} = \Phi_{2,n}(r) \begin{bmatrix} h_n(r, s)/g_0 \\ l_n(r, s)/g_0 \\ k_n(r, s) \end{bmatrix} \tag{23}$$

where U_n, V_n, $\Phi_{3,n}$ are the coefficients in the spherical harmonic expansions for the radial and tangential displacements and for the potential perturbation produced by the response of the layered viscoelastic sphere to excitation by a point mass load added to its surface at $t=0$ and instantaneously removed (Peltier, 1974). The boundary conditions subject to which this solution is obtained are (e.g. Peltier, 1982):

$$T_{rn}(a) = -g\Gamma_n \tag{24a}$$

$$T_{\theta n}(a) = 0 \tag{24b}$$

$$Q_n(a) = -4\pi G\Gamma_n \tag{24c}$$

where the functions on the left-hand sides are coefficients in the spherical harmonic expansions of the radial and tangential stress components and of a variable related to the radial derivative of the potential perturbation (e.g. Peltier, 1974) and a is the Earth's radius. Peltier (1974, 1976) has shown that each of the Love numbers h_n, l_n and k_n has a time domain normal mode expansion entirely analogous to (17) for Heaviside forcing in the form:

$$h_n^{\mathrm{H}}(a, t) = h_n^{\mathrm{HE}}(a)\delta(t) + \sum_{j=1}^{M} \frac{r_j}{s_j}(1 - e^{-s_j t}) \tag{25}$$

where M is the number of normal modes of viscous gravitational relaxation required to represent the time domain solution. Very recently, a new and highly accurate method for the computation of the normal mode amplitudes r_j and inverse relaxation times s_j has been developed (Peltier, 1985) to replace the collocation inverse which I have previously employed. Since this is an important new ingredient of the theory, I will sketch its rudiments here.

Expansion of all unknown fields in spherical harmonics reduces (1) to the form

$$d\mathbf{Y}/dr = \mathbf{A}\mathbf{Y} \tag{26}$$

where the solution 6-vector $\mathbf{Y}$ has elements $(U_n, V_n, T_n, \Theta_n, \Phi_n, Q_n)$ and $\mathbf{A}$ is the 6×6 matrix of coupling coefficients given in Peltier (1974) (see Peltier, 1982 for a recent review). Equation (26) is solved using a standard shooting algorithm which generates solutions in the form of a linear combination of three linearly independent solutions as (suppressing the n dependence)

$$\mathbf{Y}(r, s) = C_1\mathbf{T}_1(r, s) + C_2\mathbf{T}_2(r, s) + C_3\mathbf{T}_3(r, s) \tag{27}$$

The coefficients of combination C_i are selected to satisfy (24) as

$$\mathbf{B}(s) = \mathbf{Y}(a, s) = C_1(s)\mathbf{T}_1(a, s) + C_2\mathbf{T}_2(a, s) + C_3\mathbf{T}_3(a, s), \tag{28}$$

or

$$\begin{aligned} \mathbf{B} &= \mathbf{MC} \\ \mathbf{C} &= \mathbf{M}^{-1}\mathbf{B} \end{aligned} \tag{29}$$

where $\mathbf{M}$ is the matrix whose column vectors are the independent solutions $\mathbf{T}_i(a, s)$. Since

$$\mathbf{M}^{-1} = \mathbf{M}^*/\det \mathbf{M} \tag{30}$$

where $\mathbf{M}^*$ is the transpose of the matrix composed of the cofactors M^*_{ij}. Therefore we have

$$C_i = \sum_{j=1}^{3} \frac{M^*_{ij}(s)B_j}{\det \mathbf{M}} \tag{31}$$

and the general solution of (26) satisfying the boundary conditions is

$$\mathbf{Y}(r, s) = \sum_{i,j} \frac{M^*_{ij}B_j}{\det \mathbf{M}} \mathbf{T}_i(r, s) \tag{32}$$

$$= \mathbf{Q}/\det \mathbf{M} \tag{33}$$

where $\mathbf{Q} = \sum_{ij} M^*_{ij}B_j\mathbf{T}_i$. Analysis in Peltier (1985) then shows that

$$\mathbf{Y}(r, t) = \sum_{j=1}^{M} \frac{\mathbf{Q}^v(r, s_j)\, e^{s_j t}}{(d/ds)\,(\det \mathbf{M})|_{s=s_j}} + \mathbf{Y}^E\delta(t), \tag{34}$$

where

$$Q^v(r, s) = [\mathbf{Y}(r, s) - \mathbf{Y}^E(r)] \det \mathbf{M} \tag{35}$$

and

$$\mathbf{Y}^E(r) = \lim_{s\to\infty} \mathbf{Y}(r, s) \tag{36}$$

The normal mode solution (34) shows that the modal amplitudes r_j can be determined from the slope of the secular function det $\mathbf{M}$ as it passes through zero on the negative real axis of the complex s-plane at the location of the normal mode pole with inverse relaxation time s_j. This turns out (Peltier,

1985) to be a much more accurate method of determining the r_j than the collocation inverse previously employed. Figure 2(a) illustrates the relaxation spectrum for a typical viscoelastic Earth model which has its elastic structure fixed to that of model 1066B of Gilbert and Dziewonski (1975), a lithosphere of thickness 120.7 km, and a constant mantle viscosity of 10^{21} Pa s. The inverse relaxation times s_j have been plotted, with time nondimensionalized with a characteristic time of 10^3 yr, as a function of spherical harmonic degree l and inspection of the figure shows that the relaxation diagram possesses five different branches which are labelled M0, M1, M2, L0 and C0 in the figure. The modes along each of these branches are supported by specific characteristics of the radial viscoelastic structure of the planet. Those along the M0 branch are supported by buoyancy across the free outer surface of the model, M1 by buoyancy across the seismic discontinuity at 670-km depth in the Earth, and M2 by buoyancy across the 420-km discontinuity. Those along the C0 branch are supported by the density jump across the core–mantle boundary and those along L0 exist because of the infinite viscosity contrast between the mantle and lithosphere. Figure 2(b) plots r_j/s_j as a function of l and thus the amplitude of the total viscous relaxation [see Eq. (25)] delivered by each mode in the spectrum to the adjustment of spherical harmonic degree l. This demonstrates that M0, M1, 2 and C0 each carry a significant fraction of the total relaxation for

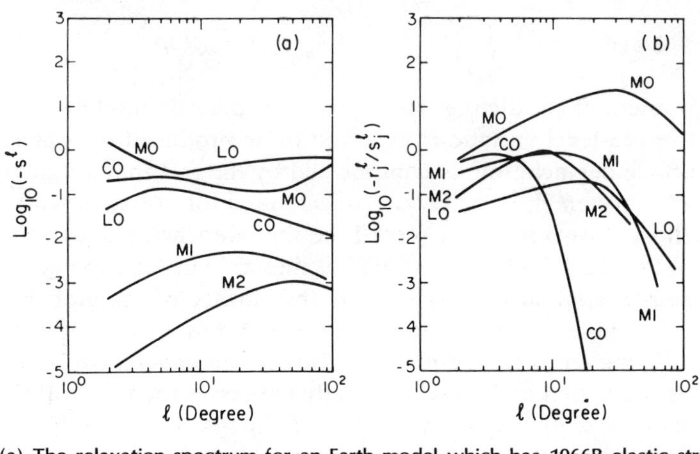

Fig. 2. (a) The relaxation spectrum for an Earth model which has 1066B elastic structure, a constant mantle viscosity of 10^{21} Pa s, and a lithospheric thickness of 120.7 km. The ordinate is inverse modal relaxation time s_j^n, with time nondimensionalized with a characteristic time of 10^3 yr, while the abscissa is spherical harmonic degree n. (b) Modal strength r_j^n as a function of degree n with the individual strengths normalized with the associated modal inverse relaxation time s_j^n. The individual modal branches are named for the features of the Earth model which supports, as discussed in the text.

small l but that for large l (small wavelength) only the fundamental M0 mode is significant. As discussed in Peltier (1980), the effect of the lithosphere on the M0 mode is substantial and in particular is responsible for causing the decrease of relaxation time with increasing l for $l \lesssim 30$ which is observed in Fig. 2a.

D. The Gravitationally Self-consistent Computation of Relative Sea-level Variations

Given the impulse response amplitudes (34) one may synthesize space–time Green's functions for the response of the planet to gravitational interaction with a point-mass load by extending the formulation for the elastic problem of Longman (1962, 1963) and Farrell (1972). Green's functions for several different signatures of the response may be computed in terms of the Love numbers defined in (23). For radial displacement, the free-air gravity anomaly, and the perturbation of the gravitational potential field, these functions (Peltier 1974, 1976) take the respective forms:

$$\Delta R(\theta, t) = \frac{a}{m_e} \sum_{n=0}^{\infty} h_n(t) P_n(\cos\theta) \tag{37a}$$

$$\Delta g(\theta, t) = \frac{g}{m_e} \sum_{n=0}^{\infty} [n + 2h_n(t) - (n+1)k_n(t)] P_n(\cos)\,\theta) \tag{37b}$$

$$\Delta\phi(\theta, t) = \frac{ag}{m_e} \sum_{n=0}^{\infty} [1 - h_n(t) + k_n(t)] P_n(\cos\theta) \tag{37c}$$

The problem of predicting, for a given viscoelastic model of the Earth, the relative sea-level variations expected to be produced by a given model of the surface deglaciation, is complicated by the fact that all gravitational effects in the system must be properly accounted for. Relative sea-level data such as those shown in Fig. 1 record the time-dependent separation of the surface of the solid Earth (on which beaches are cut) and the geoid, which is the equipotential surface defined by the surface of the global ocean. It is, in fact, possible to directly compute this separation but the computation involves solving an integral equation in which the Green's function defined in (37c) is the kernel and which basically expresses the physical constraint that the meltwater produced by ice sheet disintegration must be distributed over the global ocean in such a way that the ocean surface remains an equipotential. The structure and solution of this integral equation have been reviewed and employed in Wu and Peltier (1983). The reader interested in the numerical details of the method by which this equation is solved is referred to this paper and to Peltier *et al.* (1978). The inputs to this theory consist of a model of the radial viscoelastic structure of the Earth which is embodied in the Green's functions (37) and a model of the surface deglaciation history such as the ICE-2 model tabulated in Wu and Peltier (1983). The output from the theory consists of a prediction of the relative variations of sea level everywhere in the global ocean. An example of a solution of the integral equation for relative sea-level rise is shown in Fig. 3 from a calculation based on the ICE-2 melting history (minus Antarctica) and an Earth model with 1066B elastic structure, 10^{21} Pa s constant mantle viscosity, and a lithospheric thickness of 120.7 km. This result is from the analysis reported in Wu and Peltier (1983) and shows the time-dependent rise, in metres, of ocean bathymetry produced by the melting of the Northern Hemisphere ice sheets which began about 18,000 years ago. To obtain predictions from the theory to compare with observations such as those shown on Fig. 1, we simply pierce the stack of such maps output by the solution of the integral equation (which are separated by 10^3-yr intervals) at a location from which we have data and compare the predicted relative sea-level history with the observations.

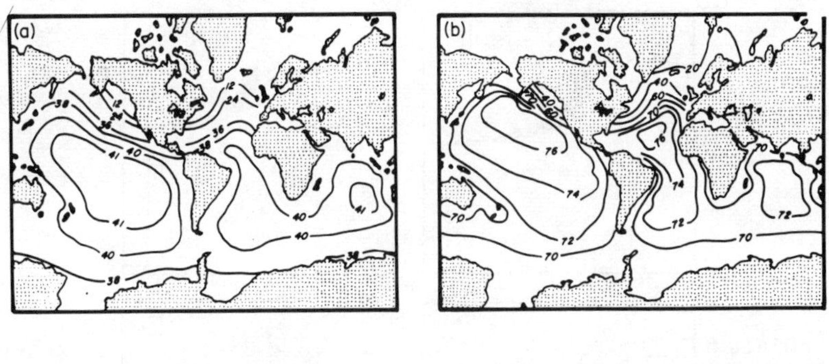

Fig. 3. Four time slices through a representative solution of the sea-level equation; (a) 12,000 B.P., (b) 8000 B.P., (c) 4000 B.P. and (d) present. In the ocean basins the contours of the RSL rise are marked in meters. The calculation which produced this output employed the ICE-2 melting chronology of Wu and Peltier (1983), although Antarctic meltback was neglected, and the same viscoelastic Earth model as that employed in the construction of Fig. 2.

A sample of the results from such comparisons is shown on Fig. 4 for six sites in the high Canadian Arctic and for three different Earth models which differ from one another only in the viscosity of the mantle beneath 670-km depth. Inspection of these comparisons establishes that the model with a uniform mantle viscosity of 10^{21} Pa s (short dashed lines) is the model best able to simultaneously satisfy the observed amplitudes of emergence and present-day uplift rates. Although the model with lower mantle

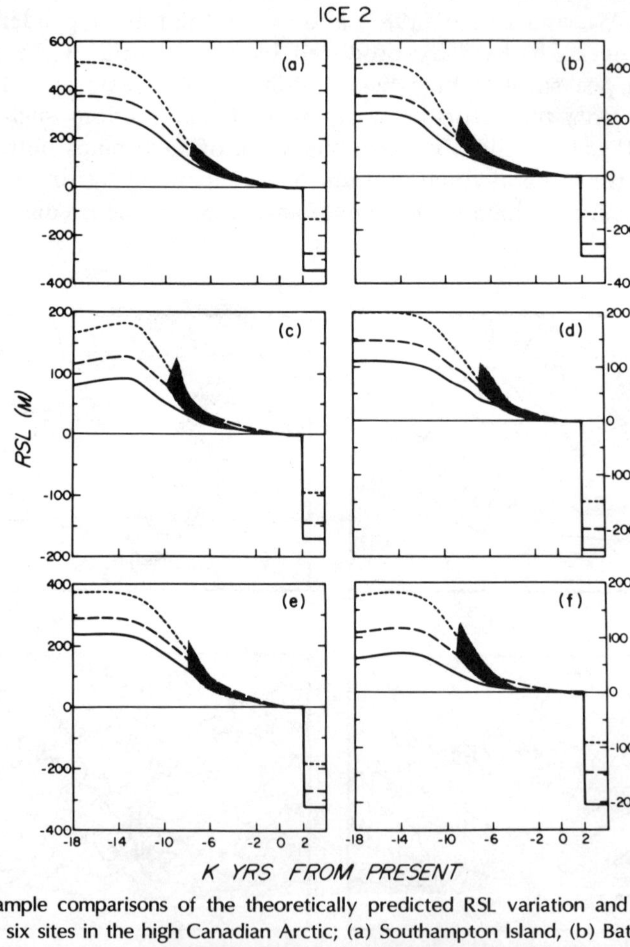

Fig. 4. Sample comparisons of the theoretically predicted RSL variation and the observed variation at six sites in the high Canadian Arctic; (a) Southampton Island, (b) Bathurst Inlet, (c) Sommerset Island, (d) Ipik Bay, (e) Igloolik Island and (f) Cape Tanfield. The melting history is ICE-2 minus Antarctic melting, while the Earth models employed all have 1066B elastic structure, 120.6-km thick lithospheres, and upper mantle viscosities of 10^{21} Pa s. The three curves shown against each plate are for Earth models which have lower mantle viscosity of 10^{21} Pa s (the short dashed curve), 10^{22} Pa s (the long dashed curves), and 5×10^{22} Pa s (the solid curves).

viscosity of 10^{22} Pa s (long dashed lines) fits the observed amplitudes of emergence equally well it generally predicts far too rapid rates of present-day emergence. Finally, the model with lower mantle viscosity of 5×10^{22} Pa s (solid curves) does fit the present-day observed emergence rates but predicts far too little emergence over the past 8000 years of Earth history. In the next section, I will review the extent to which other geodynamic observables may be invoked to further reinforce this apparent requirement of the sea-level record that the viscosity of the Earth's mantle is essentially uniform.

III. Large-scale Signatures of Glacial Isostasy

The additional observables of glacial isostasy which I will discuss in the following subsections consist of (i) the free-air gravity anomalies over present-day centres of post-glacial rebound, (ii) the nontidal component of the acceleration of planetary rotation which has recently been extracted through the analysis of laser-ranging data for the LAGEOS satellite and (iii) the true wander of the rotation pole with respect to the surface geography which is visible in the ILS-IPMS record of polar motion since about 1900 A.D.

A. Free-air Gravity Anomalies

The free-air gravity anomalies associated with the Laurentian and Fennoscandian depressions are shown in Fig. 5. The map for the Hudson Bay region of Canada is based on data of Innes *et al.* (1968) as analysed by

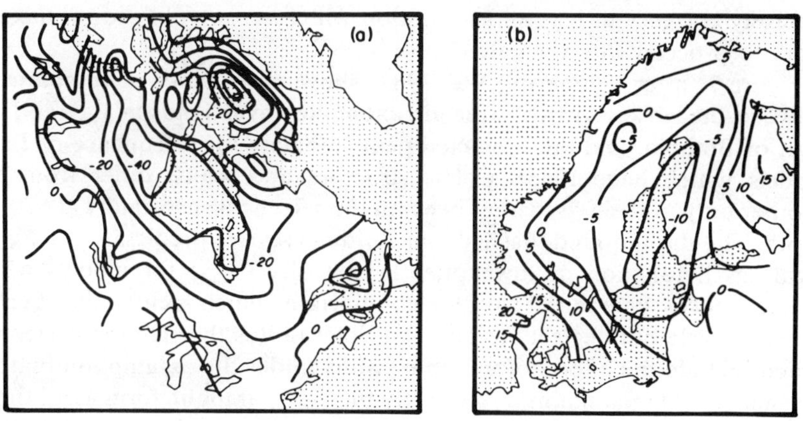

Fig. 5. (a) Free-air gravity anomaly over Hudson Bay with contours in mgals (adapted from Walcott, 1972). (b) Same field over the Gulf of Bothnia with contours in mgals (adapted from Balling, 1980).

Walcott (1980) while that for the Gulf of Bothnia region of northwestern Europe is based on the data of Honkasolo (1963) as analysed by Balling (1980). Although the most recent satellite data agree extremely well with the location and extent of the free-air low over Hudson Bay, the analysis by Rubincam (1984) based on GEM10B data suggests a peak anomaly Δg such that $-50 \leq \Delta g \leq -40$ mgal which is somewhat deeper than that suggested by Walcott's analysis of Inne's data. The same analysis of the GEM10B data also reveals the negative free-air anomaly over the Gulf of Bothnia and with an amplitude equal to that accepted in the present paper (i.e. $-20 < \Delta g < -10$ mgal). To make a theoretical prediction of these observations, we simply convolve the history of surface loading and unloading (including both the ice and water components) with the Green's function for the free-air anomaly given in (37b). As pointed out in Peltier and Wu (1982), the accurate prediction of free-air gravity anomalies is complicated by the fact that this signature of the isostatic adjustment process is extremely sensitive to the history of loading and unloading of the surface for times which extend back into the past for several cycles of the dominant 10^5-year oscillation. Since $\delta\,^{18}O$ data from deep-sea sedimentary cores may be employed to constrain the history of loading, however, one can do calculations which approximately incorporate the memory of the load cycle. Figure 6(a) shows a typical $\delta\,^{18}O$ record along with a simple sawtooth function which may be taken to be a reasonable approximation to it. In order to incorporate the effect of the load history on the free-air gravity anomaly prediction, we are obliged to assume that the average history of the entire cryosphere which is recorded in the sea-core data is an accurate approximation to each of the local histories which comprise it. This is certainly not precisely correct but it is an assumption which we are obliged to make in order to proceed with the calculation.

Figure 7 (after Peltier and Wu, 1982) shows a sequence of comparisons between the observed peak free-air anomalies over Hudson Bay and the Gulf of Bothnia, and the predicted peak anomaly for a sequence of Earth models which have 1066B elastic structure and which differ from one another only in their viscosities beneath the seismic discontinuity at 670-km depth. The dash-dotted, dashed and solid curves are predictions of Δg (in mgal) as a function of lower mantle viscosity (ν_{LM}) for models which respectively assume initial isostatic equilibrium, which include the effect of the load cycle but neglect variations of load radius through the cycle, and which include the effects of variable radius under the assumption that the ice sheet profile maintains a "perfectly plastic" parabolic form at all times. The solid curves are thought to represent the most accurate predictions.

Inspection of this figure shows that the effect of initial isostatic disequilibrium on the predicted free-air gravity anomalies is extremely important

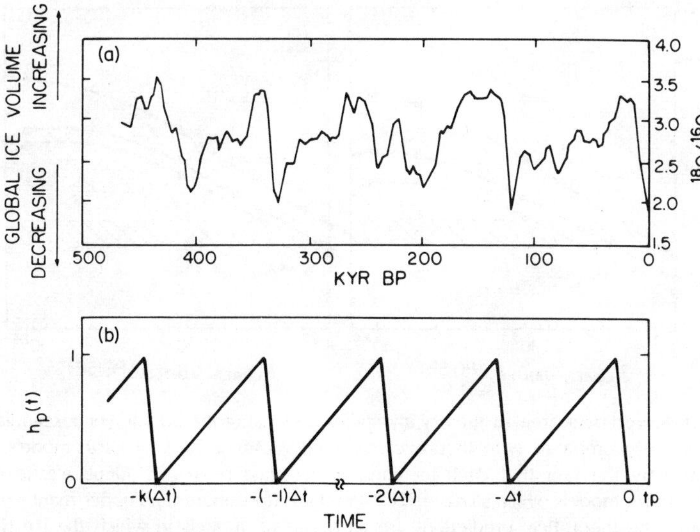

Fig. 6. (a) Typical δ ^{18}O time series of variations of the ratio of concentrations $^{18}O/^{16}O$ deduced from a sedimentary core taken from a deep ocean basin. (b) Simple sawtooth mathematical model of the observed concentration variations which emphasized the 10^5-year periodic pulsation which dominates such records.

and such as to reduce the predicted anomalies from those which are based on the assumption of initial equilibrium. According to these comparisons the viscosity of the lower mantle is bounded below by 10^{21} Pa s and above by about 5×10^{21} Pa s if the upper mantle value is held fixed to the number 10^{21} Pa s preferred by Haskell (1936) on the basis of his analysis of the Fennoscandia data. This is the same range as that required by the relative sea-level data discussed in the last section so that both data sets are reconcilable by the same mantle viscosity profile.

As pointed out in Peltier (1982) and Peltier and Wu (1982), the ability of the model to fit the free-air gravity anomalies with the same viscosity profile demanded by the RSL data completely resolves a long-standing argument in the geophysical literature to the effect that the two types of observation were fundamentally conflicting in their requirements (e.g. Walcott, 1980), with the free-air data seeming to require high lower mantle viscosity and the RSL record requiring the opposite. The reason why the theoretical model discussed in the last section has been able to reconcile these data is because of the new modes of viscous gravitational relaxation which they contain that are supported by the near discontinuities of density and Lamé parameters which occur across the seismic discontinuities at 420- and 670-km depth in the Earth. These discontinuities are most likely

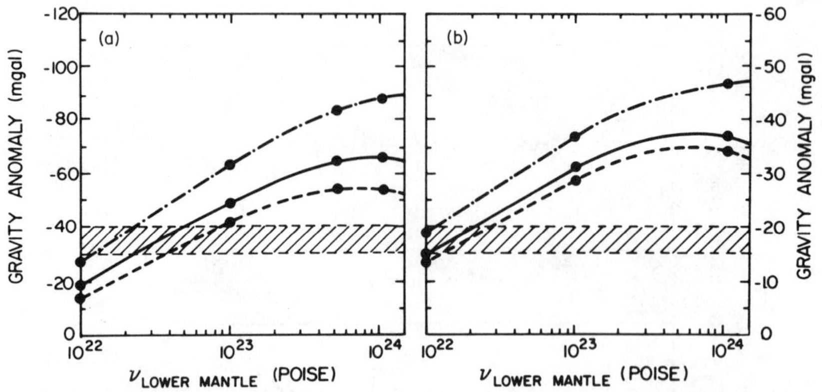

Fig. 7. Observed peak free-air gravity anomalies over Laurentia (a) and Fennoscandia (b) (the hatched regions) compared with theoretical predictions for a class of Earth models which is characterized by the fact that each member of the class possesses 1066B elastic structure. Predictions are for models which all have 120.7-km thick lithospheres and upper-mantle viscosities of 10^{21} Pa s. The theoretical predictions are all based on models in which the ice sheets are approximated by circular disk loads with parabolic cross-sections, and differ only in the assumed time variations of this load. The dash–dotted curves show the results for calculations based upon the assumption that the ice sheets are initially in isostatic equilibrium, while the dashed curves include the effect due to a load cycle of the form illustrated in Fig. 6. In the latter calculations the ice sheet radius is assumed to be fixed while the volume fluctuates. The solid curves should be the most accurate set of calculations because they include the effect of variable load radius. The instantaneous radius during the loading cycle is determined by the assumption that the ice sheet maintains an equilibrium plastic profile while the volume changes.

due to the presence of equilibrium solid–solid phase transitions (e.g. Jeanloz and Thomson, 1983). Because the isostatic adjustment time scale is short compared to the thermal diffusion time scale on which material can actually change phase in response to the changes in ambient pressure associated with surface loading, these density discontinuities behave as if they were non-adiabatic chemical discontinuities. They therefore produce buoyant restoring forces when they are deflected from their equilibrium depths by ice sheet loading, and these buoyancy forces are responsible for supporting the M1 and M2 modes (from the 670-km and 420-km discontinuities, respectively) which are shown on Fig. 2(a). The buoyancy forces are introduced in the theory using the prestress advection formalism illustrated in subsection II.B.

The fact that it *is* this extra source of buoyancy which allows us to reconcile the two kinds of data is shown explicitly by the new results plotted in Fig. 8. We show results from a set of calculations identical to those performed to construct Fig. 7. Here, however, the different curves are for a sequence of models which differs from one another in the parameterization

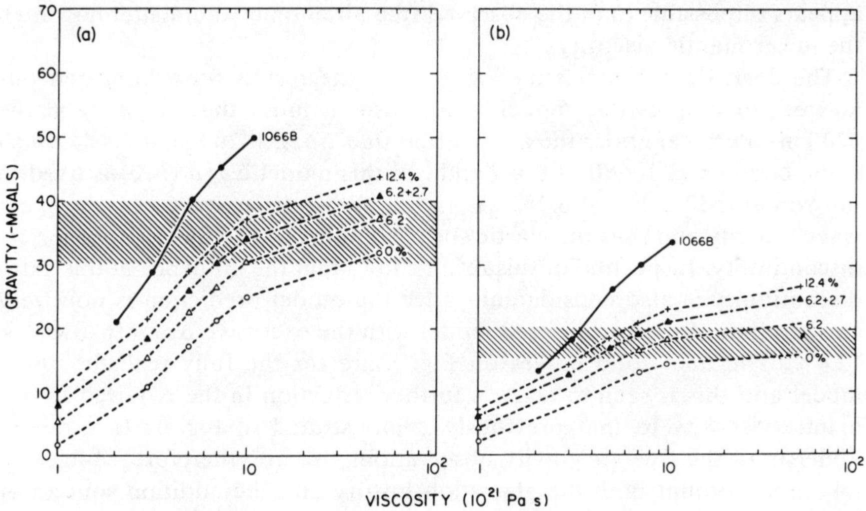

Fig. 8. A sequence of comparisons of observed and predicted peak free-air gravity anomalies for Hudson Bay (Laurentia) (a) and the Gulf of Bothnia (Fennoscandia) (b). The three dashed curves are for simple homogeneous plain-layered approximations to model 1066B with one density discontinuity in the mantle at 670-km depth. All three models have a lithospheric thickness of 120.6 km and an upper-mantle viscosity of 10^{21} Pa s. The density increase in percent is shown adjacent to each curve. The dash–dotted line shows peak free-air gravity anomalies in mgal for a model which includes the density jump at 420-km depth and at 670 km. The percent increases of density across these boundaries are fixed to the values in 1066B. The solid curve for the 1066B model shows that when all of the density variation in this model is considered non-adiabatic the inferred lower-mantle viscosity is further decreased.

employed for the elastic structure of the planetary interior. All calculations employ the constant load radius approximation to the glaciation history of each region (15° and 9.5° for Laurentia and Fennoscandia, respectively) with maximum thickness of the Laurentian and Fennoscandian loads of 3.5 km and 2.5 km, respectively, as for Fig. 7. The three dashed curves, which are labelled with $\Delta\rho/\rho$ (670 km) in per cent, show predicted peak Δg as a function of ν_{LM} for a sequence of models of the homogeneous-incompressible-layered type discussed in Wu and Peltier (1984). For these calculations the mantle is assumed to have only two layers which meet at the 670-km boundary, and results are shown for $\Delta\rho/\rho$ (670 km) = 0.0, in which case both layers have identical properties, for $\Delta\rho/\rho = 6.2\%$ (which is the density contrast in model 1066B at this depth) and for $\Delta\rho/\rho = 12.4\%$. Clearly the incorporation of the buoyant restoring force across this boundary allows one to fit the observation with a lower deep mantle viscosity than would otherwise be required. Indeed, without such additional buoyancy it appears impossible to fit the observed free-air anomaly no matter how high the lower mantle viscosity.

The dash–dotted curves on Fig. 8 show the results for a homogeneous layered, incompressible model which also includes the buoyancy at the 420-km interface, under the assumption that $\Delta\rho/\rho$ (420 km) = 2.7% which is the contrast in 1066B at this depth. In this model $\Delta\rho/\rho$ (670) is fixed to the "observed" value at 6.2%. As in the above discussed cases, the shear wave velocity and thus the elastic shear modulus μ also has an appropriate discontinuity. Inspection of this result shows that the influence of this extra discontinuity is also considerable since the model prediction is now very close to that of the two-layered model with the excessive $\Delta\rho/\rho$ (670 km) = 12.4%. The final solid curves on Fig. 8 are for the fully realistic 1066B model and this is seen to allow a further reduction in the required lower-mantle viscosity to that previously demonstrated in Fig. 7. In order to understand the free-air gravity observations, we are, therefore, obliged to take into account both the glaciation history and the addition sources of buoyancy in the mantle due to the presence of near discontinuities of physical properties due to the phase transitions which bound the transition region both from above and below.

B. The Nontidal Acceleration of Planetary Rotation

A fairly complete theory for the prediction of ice age induced variations of planetary rotation has recently been presented in Peltier (1982) and Wu and Peltier (1984). The first such signal which we will discuss here is the deglaciation-induced acceleration of rotation and thus the deglaciation-forced decrease in the length of day (LOD). Until very recently this residual component of the change in LOD had been inferred only on the basis of the analysis of ancient eclipse observations (e.g. Müller and Stephenson, 1975). Basically what one did was to estimate the present-day lunar-tidal torque based upon the observed rate of recession of the moon from the Earth (using, say, lunar laser ranging). Assuming that this torque has remained constant throughout the past millenia, on the basis of the fact that the configuration of the oceans does not change significantly on this time scale, one can make precise predictions of where and when in the distant past total or near total eclipses of the sun should have occurred. When one does this one finds a systematic error such as to imply the action over the past few thousand years of an acceleration of rotation in the amount $(\dot{\omega}/\Omega)_{NT} = (6.9 \pm 2.6) \times 10^{-11}/\text{yr}$ (e.g. see Lambeck, 1980 for a review). It seems that Dicke (1969) was the first to suggest that this effect might be due to the influence of Pleistocene deglaciation. Both he and subsequently

O'Connell (1971) argued that this observation might be invoked to constrain mantle viscosity, and O'Connell suggested on this basis that the viscosity of the lower mantle was near that of the upper (his argument was wrong, however, as shown below).

The theory developed in Peltier (1982) shows that the predicted acceleration of rotation for an arbitrary glaciation history on a radially layered compressible Earth takes the form:

$$\frac{\dot{\omega}_3}{\Omega}(t) = \frac{-I_{33}^R}{C}\left[D_i\dot{f}(t) + \sum_{j=1}^{M} r_j \frac{d}{dt}(f * e^{-s_j t})\right] \tag{38}$$

where I_{33}^R is the change in the axial component of the inertia tensor of the planet which would be produced at ice age maximum if the earth were rigid and $f(t)$ is the history function which is +1 at glacial maximum and 0 during an interglacial. Assuming the same sawtooth approximation to the load history shown in Fig. 5b, the predictions (38), in which r_j and s_j are the amplitudes and inverse relaxation times for the M modes required to describe the time history of the Love number k_2 in (23), have been calculated using elastic model 1066B with all parameters fixed but the lower-mantle viscosity ν_{LM}. The results of these analyses are shown in Fig. 9 (Peltier, 1983; Wu and Peltier, 1984) for models which include only the Laurentian

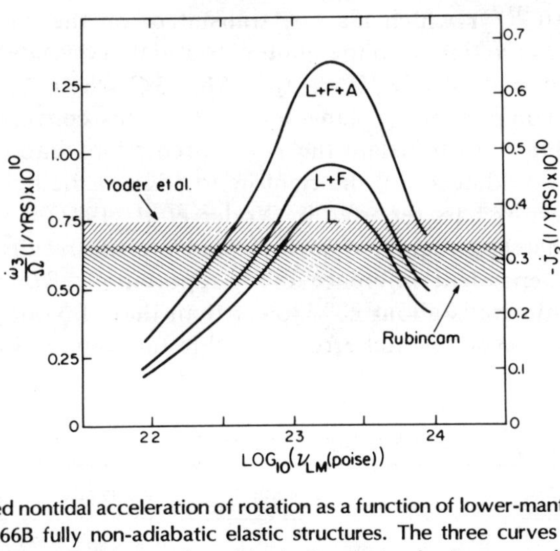

Fig. 9. Predicted nontidal acceleration of rotation as a function of lower-mantle viscosity based on the use of 1066B fully non-adiabatic elastic structures. The three curves are for a loading model including the Laurentian ice sheet (L), the Laurentia plus Fennoscandia (L+F), and the Laurentian, Fennoscandia and the deglaciation of Antarctica (L+F+A). The hatched regions show the observed nontidal acceleration (and $\dot{J}_2$) inferences based upon LAGEOS laser-ranging observations.

ice sheet (the curve marked L), for a model including both the Laurentian and Fennoscandian ice sheets and for a model which also includes the deglaciation of West Antarctica (e.g. Denton and Hughes, 1980). As discussed in detail in Wu and Peltier (1984), in these calculations the ice sheets have been approximated by circular disk loads, and the oceans represented by the ocean function (e.g. Lambeck, 1980). With these approximations I_{33}^R is given by

$$I_{33}^R = \sum_{i=1}^{N} M_i a^2 \left[\frac{2}{15} \frac{a_{20}}{a_{00}} - \tfrac{1}{3} \cos \alpha_i (1 - \cos \alpha_i) P_2^0(\cos \theta_i) \right] \tag{39}$$

where a_{20} and a_{00} are coefficients in the spherical harmonic expansion of the ocean function, M_i are the individual ice sheet masses, α_i their radii and θ_i the latitudinal locations of their centroids. For Laurentia, Fennoscandia and Antarctica, the latter parameters are listed in Table I.

Results for these predictions of present day $\dot{\omega}_3/\Omega$ are shown on Fig. 9, where they are compared with an observed value of $(7.1 \pm 0.6) \times 10^{-11}$/yr. This value differs slightly from that deduced from the eclipse data and is the number recently obtained by Yoder *et al.* (1983) through the analysis of laser-ranging data from the LAGEOS satellite which reveal an acceleration in the node of its orbit which requires a secular variation of the J_2 component of the Earth's gravitational potential field in the amount $\dot{J}_2 = (-3.5 \pm 0.3) \times 10^{-11}$/yr which may be translated, on the basis of angular momentum conservation, into the quoted nontidal acceleration of rotation. The conversion is simply $\dot{\omega}_3/\Omega = -\dot{J}_2 \cdot 2\, Ma^2/3C$ where C, as in (38), is just the axial component of planetary inertia. This conversion has been employed in Fig. 9 to translate the $\dot{\omega}_3/\Omega$ predictions into $\dot{J}_2$ predictions which are shown adjacent on the right-hand edge of the figure. The value of $\dot{J}_2$ with its errors bars reported by Yoder *et al.* (1983) is shown on this figure as the hatched region. Also indicated are the results of the more recent and independent inference of $\dot{J}_2$ reported by Rubincam (1984). Rubincam's estimate is about 20% lower than that of Yoder *et al.* (1983) and his reported experimental error is double that reported by the former

TABLE I. Ice sheet parameters

	Laurentia	Fennoscandia	Antarctica
Mass, M_i (kg)	2.0×10^{19}	0.56×10^{19}	0.7×10^{19}
Radius, α_i (deg.)	15.0	9.5	20.0
Colatitude ϕ_i (deg.)	30.0	25.5	180.0
E. Longitude θ_i (deg.)	270.0	25.0	—

authors. The reasons for the difference in these independent analyses is not known.

Inspection of the comparisons on Fig. 9, for the loading history which contains all the ice sheets, shows that it is apparently possible to fit the observed $\dot{J}_2$ with either one or the other of two widely separated values of the lower-mantle viscosity when the upper-mantle value is fixed to the value of 10^{21} Pa s required by the sea-level data. Therefore the $\dot{J}_2$ datum is inherently ambiguous in regards to the ratio of the viscosity of the lower mantle to that of the upper mantle. As discussed in Peltier (1983), however, sea-level and free-air gravity data may be invoked to completely reject the higher of the two allowed lower-mantle viscosities. The LAGEOS data (as interpreted by Yoder *et al.*, 1983) therefore require a lower-mantle viscosity ν_{LM} such that 2.7×10^{21} Pa s $\leqslant \nu_{\text{LM}} \leqslant 4.4 \times 10^{21}$ Pa s, and thus reinforce the view that the Earth's mantle has no large-scale contrasts in its radial variation of viscosity. This is of crucial importance in understanding the thermal convective circulation which is responsible for plate creation and destruction.

The nontidal acceleration of rotation turns out to be a datum which is only moderately sensitive to the internal density and elastic stratifications, unlike the free-air gravity data discussed in the last subsection. It is also only moderately sensitive to the thickness of the lithosphere. Explicit sensitivity analyses are discussed in Peltier (1985). This strong sensitivity to the viscosity stratification and relatively weak sensitivity to the elastic component of the structure is a characteristic not shared by the true polar wander observation discussed in the following subsection.

C. Deglaciation-induced Wander of the Pole of Rotation

The solution for the true polar wander forced by the glacial cycle is much more complicated than for the acceleration of rotation described previously. If we approximate the history of glaciation and deglaciation in the same way as before, using circular disk loads to represent the ice sheets and the ocean function to describe the ocean basins, then the forced polar wander is given by the following expression (Peltier, 1982; Wu and Peltier, 1984) for the instantaneous polar wander speed:

$$\frac{\dot{\omega}_j}{\Omega} = \frac{\Omega}{A\sigma_0} I_{j3}^R \left[D_i \dot{f}(t) + D_2 f(t) + \sum_{i=1}^{M-1} E_i \frac{d}{dt} (f * \exp[-\lambda_i t]) \right] \tag{39}$$

where

$$D_1 = 1 + k_2^{LE} = l_s - \sum_{j=1}^{M} \frac{r_j}{s_j} \qquad \text{where} \qquad l_s = \lim_{s \to 0} [1 + k_2^L(s)] \tag{40a}$$

$$D_2 = -l_s q(0) \Big/ \prod_{i=k}^{M-1} \lambda_i \tag{40b}$$

$$E_i = \left[\frac{l_s q(-\lambda_i)}{\lambda_i} + \sum_{j=1}^{M} \frac{r_j R_j(-\lambda_i)}{s_j} \right] \Big/ \prod_{k \neq i}^{M-1} (\lambda_k - \lambda_i) \tag{40c}$$

In this solution the $M-1$ inverse relaxation times λ_i are the roots of the degree $M-1$ polynomial

$$Q_{M-1}(s) = \sum_{j=1}^{M} g_j \left[\prod_{i \neq j} (s + s_i) \right] = \prod_{i=1}^{M-1} (s + \lambda_i) \tag{41}$$

while the polynomials $q(s)$ and $R_j(s)$ are

$$q(s) = s \prod_{i=1}^{M-1} (s + \lambda_i) - \prod_{i=1}^{M} (s + s_i) \tag{42a}$$

$$R_j(s) = \prod_{i=1}^{M-1} (s + \lambda_i) - \prod_{i \neq j} (s + s_i) \tag{42b}$$

The main difference between the form of the solution (40), which describes the polar wander forced by the load cycle $f(t)$ with rigid Earth inertia perturbations at glacial maximum I_{j3}^R, and the solution (39) for the forced variation of the LOD is contained in the form of the history dependent terms which involve convolution integrals of the form $f * \exp[-\lambda_i t]$. A comparison of these two equations shows that the nontidal acceleration depends upon the same M inverse relaxation times s_i as govern the relaxation of the degree-two component of the surface deformation. On the other hand, the solution (40) for the polar motion depends on the $M-1$ relaxation times λ_i which are determined as roots of the degree $M-1$ polynomial Q_{M-1} defined in (41). In (41) the numbers g_j are defined by the expression

$$g_j = \frac{t_j / s_j}{\sum_{i=1}^{M} (t_i / s_i)} \tag{43}$$

where the t_j are the amplitudes for impulsive forcing in the spectral representation of the *tidal* Love number k_2^{T} as

$$k_2^{\mathrm{T}}(t) = k_2^{\mathrm{TE}} \, \delta(t) + \sum_{i=1}^{M} t_j \exp[-s_j t] \tag{44}$$

These amplitudes are determined exactly as were those for the equivalent *surface load* Love number except that the appropriate boundary condition at the surface of the Earth is one of vanishing normal stress rather than the finite normal stress which obtains in the surface load case. The reason why the tidal Love number (44) enters the solution for the forced polar wander is that this element of the rotational response depends not only upon the component of the surface deformation which is directly forced by the surface ice and water load, but also upon the deformation due to the changing rotation itself (e.g. see Lambeck, 1980). It is the latter component of the deformation which requires the tidal Love number to describe it. In fact these two components of the polar wander "excitation function" compete with one another in that during free relaxation the rotational deformation amplifies the moment of inertia with respect to the original rotation axis, whereas isostatic adjustment of the load diminishes it. As demonstrated in Peltier (1982), an Earth model which supports only one mode of exponential relaxation (as is the case when the model is a homogeneous one) has a polar wander response which is identically zero whenever the surface is not actively glaciating or deglaciating. This physical fact appears in the solution (40) as a reduction by one of the number of modes which govern the history dependent term (the third term in square brackets) since this term is the one which prevents the polar wander from decaying to zero immediately, the surface load ceases to vary (i.e. when $f=\dot{f}=0$). This is an extremely important point in attempting to understand the polar wander data shown on Fig. 10.

These data show the variation of the position of the Earth's pole of rotation relative to the surface geography, resolved into components x and y along the axes shown on the inset polar projection. These data cover the time range 1900–1975 A.D. and are predominantly those collected by the International Latitude Service (ILS) and discussed by Vincente and Yumi (1969, 1970) and augmented by the data of the later International Polar Motion Service (IPMS). The data have recently been discussed by Dickman (1977). The dominant quasi-periodic variation which these data reveal is the 7-year beat due to the superposition of the 12-month annual and 14-month Chandler wobbles. The maximum amplitude of these combined oscillations is near 0.3 arc s or roughly 10 m in terms of distance along the Earth's surface. Our interest in these data is by virtue of the fact that the dominant oscillatory response is superimposed on a slow secular drift of the pole in the direction shown by the arrow on the inset polar projection. The analysis in Dickman (1977) shows that the speed of polar wander implied by the secular drift is at a rate near $0.95° \pm 0.15°/10^6$ yr. Although previous analyses have attempted to ascribe this drift to concurrent melting of the Greenland and Antarctic ice sheets (Munk and Revelle, 1952) or to

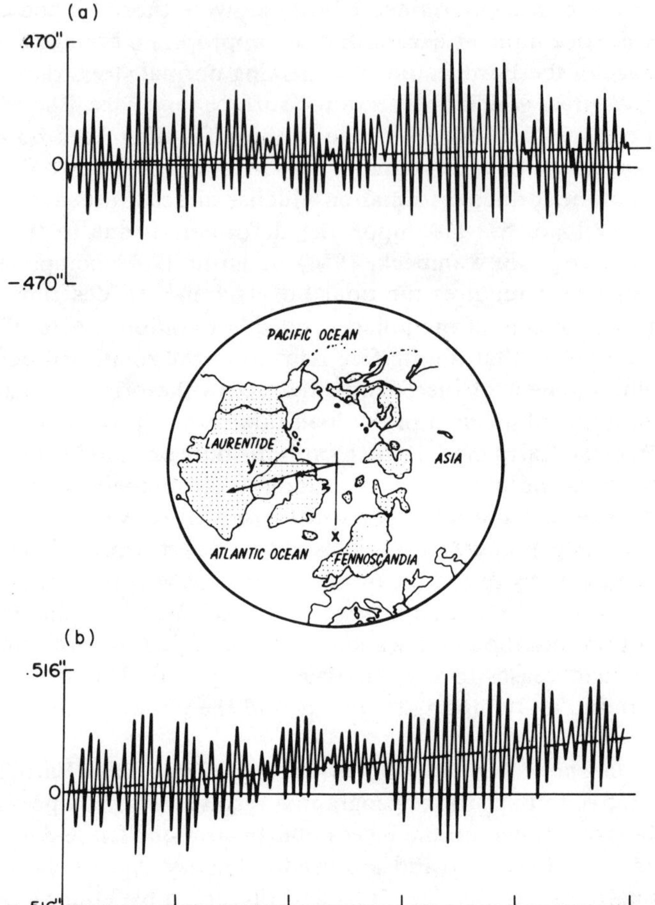

Fig. 10. The ILS–IPMS record of the motion of the rotation pole since 1900 A.D. The x and y co-ordinate directions (a) and (b), respectively, are those shown on the inset polar projection.

continental drift (Dickman, 1977), recent results based on application of the solution (40) have demonstrated that the observed polar wander is fully explicable as a memory of a glacial cycle which had been continuous until the end of the last deglaciation event at about 6000 years BP and after which it has apparently ceased.

This result is an important one since the argument of Munk and Revelle (1952) concerning the necessity of having some presently occurring wastage of the major continental ice masses in order to explain the polar wander observation has recently been resurrected in the climatology literature (e.g. Gornitz *et al.*, 1982; Barnett, 1983). Analyses of a global array of tide-gauge data seem to reveal the necessity of invoking some current increase in water *volume* but the data have not been properly corrected to remove the influence of remaining glacial isostatic disequilibrium as I suggest should be done using the theory for RSL variations discussed previously in this chapter. It is also unclear whether the effect of thermal expansion of the oceans might not be able to completely account for the observed relative sea-level rise, thus obviating the necessity of invoking any current mass loss from the Earth's major (or minor) ice masses. Meier (1984) has nevertheless shown that the small glaciers of the world could, of themselves, account for the apparently required increase of water mass when the effect of thermal expansion is estimated using the best presently available data (Gornitz *et al.*, 1982).

The rate of RSL rise extracted from the tide-gauge records from stations along the East Coast of the United States, and preliminary estimates of the rates which should be observed due to glacial isostatic disequilibrium based on a new set of calculations with the gravitationally self-consistent sea-level model, are listed in Table II. The RSL data from this region have been

TABLE II. Theoretically predicted and observed trends in the relative sea-level record at sites along the East Coast of the continental United States

Location	Theory[0] (mm/a)	Observed Trend[1]	Obs. Trend[2] (1940–1980)	Deviation (2–0)
Woods Hole, Ma	0.72	2.7	2.3	1.6
Newport, RI	0.66	2.6	2.0	1.3
Providence, RI	0.50	1.8	1.8	1.3
New London, CT	0.60	2.2	2.2	1.6
Willets Pt., NY	0.65	2.2	1.6	0.95
Sandy Hook, NJ	0.75	4.2	4.0	3.2
Atlantic City, NJ	1.1	4.0	3.9	2.8
Philadelphia, PA	0.81	2.6	2.3	1.5
Lewes, DE	1.1	3.0	2.0	0.90
Baltimore, MD	0.90	3.2	2.5	1.6
Annapolis, MD	1.00	3.7	3.0	2.0
Washington D.C.	1.00	3.0	2.8	1.6
Charleston, SC	1.10	3.4	2.4	1.3
Fort Pulanski, GA	1.0	2.7	2.5	1.5
Mayport Beach, FL	0.84	2.3	1.5	1.5
Miami Beach, FL	0.44	2.3	1.9	1.5
Key West, FL	0.33	2.2	1.6	1.3
Pensacola, FL	0.76	2.4	1.6	0.8

Average deviation Δ from East Coast sites = 1.5 mm/a.

discussed by Peltier (1984) in connection with an assessment of their sensitivity to variations of lithospheric thickness. Also shown are the residuals resulting from the correction of the tide-gauge data with the predictions of the RSL model. Along this "passive margin" the average residual is near 1.5 mm/yr which is close to the global rate of sea-level rise inferred in the analysis of Gornitz *et al.* (1982) and assumed in the calculations of Meier (1984). This rate should be compared to that which obtained during the last glacial–interglacial transition in which a 100 m eustatic water rise occurred over a time scale of about 10^4 yr to effect an average rate of rise of 10 mm/yr. This is less than one order of magnitude greater than that apparently required by the tide-gauge data to be presently occurring if *no* allowance is made for thermal expansion of the oceans. This would be expected to produce a large influence on the Earth's rotation rate if the source of meltwater were at either of the poles, or a large influence on polar wander if the source were in lower latitudes as Meier (1984) has proposed. In order to appreciate the possibilities, we will proceed to discuss the relation between the rate of true polar wander revealed by the ILS data in Fig. 10 and that which would be expected, even if no current ice-sheet disintegration were occurring, due solely to the residual effects (memory) of an ice age cycle which is assumed to have ended 6000 years ago.

Any true polar wander that should be occurring at present due to this cause is entirely a consequence of the history term in (40) which itself exists only because of the internal viscoelastic stratification of the planet. Figure 11 shows the results of a sequence of calculations of the rate of polar wander which should be occurring at present if the only forcing active is the residual influence of a glacial cycle which ended 6000 years ago and which consisted of the same sequence of 10^5-yr periodic sawtooth oscillations previously employed in analysis of the free-air gravity signals over present-day centres of postglacial rebound. Figure 11 shows the region of $(\nu_{LM}, \bar{L})$ parameter space in which one may fit the observed polar wander speed within the observational limits of $0.95 \pm 0.15°/10^6$ yr. Here ν_{LM} is the lower-mantle viscosity while the upper-mantle value is fixed to the value of 10^{21} Pa s and $\bar{L}$ is the average thickness of the planetary lithosphere. Since $\bar{L}$ is about 120 km and ν_{LM} is near 2×10^{21} Pa s, the predicted polar wander speed, based on the assumption that it is a residual effect of a previously active but now inactive glaciation cycle, is equal to the observed speed. Figures 11(b) and 11(c) show the same tradeoff, but for two different choices of the time since the last deglaciation event corresponding to the present epoch. The best data available at present suggest that Fig. 11(b) provides the most accurate estimate of the allowed region in $(\nu_{LM}, \bar{L})$ parameter space.

The results shown in Fig. 11 establish quite clearly that the observed speed of present-day polar wander is precisely what one would expect if it

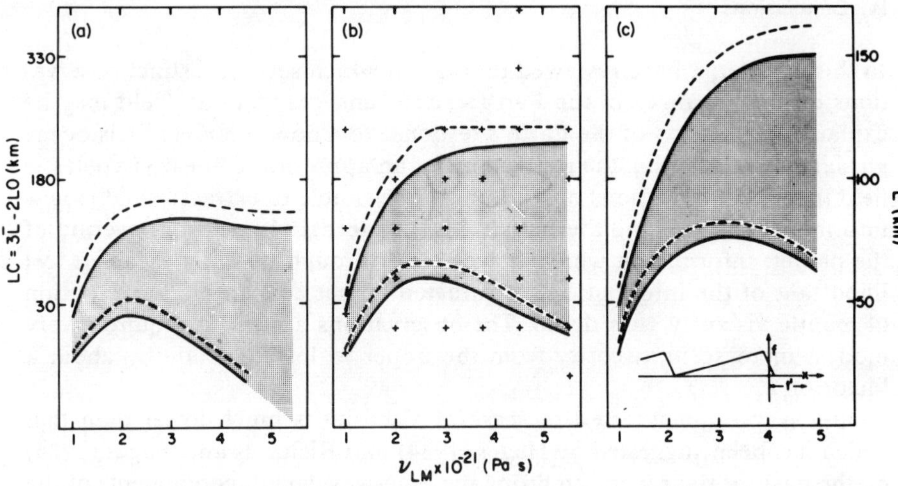

Fig. 11. Trade-off curves in ($\bar{L}$, ν_{LM}) parameter space for models which fit the observed polar wander speed of 0.95°/10^6 yr within the standard error of 0.05°/10^6 yr. The boundaries for load models for 5×10^5 yr and 7×10^5 yr glaciation cycles are represented respectively by (– – –) and (——). The three time midpoints of the last glaciation cycle from different frames are for three different choices for the present; (a) 10 KBP, (b) 11 KBP and (c) 12 KBP. The preferred time based on δ ^{18}O data is near 11 kyr.

were *entirely* due to the influence of a glacial cycle which ended about 6000 years ago. As demonstrated in Wu and Peltier (1984), the observed direction of polar wander also fits beautifully the prediction of a model based on this assumption. Clearly then, no substantial *off-polar* glacial sources of meltwater could be active in the system at present otherwise the present-day polar wander would have to be different than it is observed to be. However, any *near-polar* source of meltwater would cause an appreciable change of the rate of Earth rotation, but this too has also been shown to be completely explicable as a residual effect of a now complete glacial cycle. This would appear to rule out an explanation of the apparent tide-gauge requirement for an increase in the mass of water in the oceans in terms of either an Antarctic or a Greenland source, which is in accord with the *a priori* glaciological evidence (e.g. Meier, 1984) to the effect that both these ice sheets are presently near equilibrium in terms of their mass balances. It will require a new sequence of rotation calculations by using the theoretical model discussed here to determine whether the "many-small glacier" and, thus, distributed-melt model suggested by Meier (1984) may be allowed by the rotational constraints. Such calculations will be reported in the near future.

IV. Conclusions

In this chapter, I have reviewed the way in which several distinct observations of slow changes in the Earth's shape and gravitational field may be explained as aspects of the Earth's response to a now complete Pleistocene glacial cycle. By fitting the predictions of an appropriate linear viscoelastic field theory to the various observations, one is able to extract valuable new information concerning the nature of the internal viscoelastic layering of the planet; information which is accessible through no other means. Most important of the information so extracted is that concerning the variation of mantle viscosity with depth. The observations appear to require a very modest increase in viscosity from the upper to lower mantle by about a factor of 2.

This very slight inferred increase of viscosity is much lower than that which has been suggested by Hager (1984) and Richards and Hager (1984) on the basis of their work on fitting the long-wavelength components of the global field of geoid heights based on the internal lateral heterogeneity of density extracted from the new seismic results on whole Earth tomography (Dziewonski, 1984; Woodhouse and Dziewonski, 1984). This analysis is in fact based on a simple Newtonian viscous implementation of the isostatic adjustment theory of Peltier (1974). The only difference is in the fact that the loads responsible for inducing the gravitational signal are not surface loads but are rather internal to the body of the mantle itself. Their calculations appear to require an increase in viscosity from the upper to lower mantle by more than a factor of 10, which is clearly incompatible with all of the data discussed here. There does exist a rather serious potential flaw in their model, however, which is probably responsible for the difference in the inferred viscosity. This flaw is connected with the fact that they employ a model in which only radial variations of viscosity are allowed, whereas there must exist intense lateral variations of viscosity in the mantle which are perfectly correlated with the lateral variations of density which are assumed to force the gravitational signal observed. This is because the density variations are ultimately related to the lateral variations of temperature which are in turn responsible for driving the mantle convective circulation, and because fluid viscosity, being the consequence of a thermally activated process, depends exponentially on the temperature. This correlation between lateral variations of the load responsible for producing the gravitational signature and the lateral variations of the physical property inferred by fitting a radially stratified model to the observations, strongly undermines the validity of inferred viscosity structure. Their analysis of the long-wavelength constituents of the gravitational field should therefore be repeated by using a model which explicitly incorporates the lateral heterogeneity of viscosity in order to test the above conjecture.

Our analyses of the rotational response of the planet to the Pleistocene glacial cycle have demonstrated that both the present rate and direction of true polar wander and the nontidal component of the acceleration of planetary rotation are simply explicable in terms of the glacial hypothesis. More accurate observations of these slow changes using satellite laser ranging and long base line interferometric measurements will inevitably make them observations which are of even greater use geodynamically. Laser ranging to the LAGEOS satellite has already delivered a very useful new measurement of the nontidal acceleration. It could well be that these rotational observations will eventually provide us with the best way (albeit a rather indirect one!) of detecting the early effect of the amelioration of climate which is expected to occur due to the increase of the atmospheric load of carbon dioxide. This will doubtless lead to an increased rate of mass loss from the planet's glaciers and ice sheets, which may already have been detected in tide-gauge data, and these will cause variations of planetary inertia and thus polar wander and variations of LOD. The signals forced by the last deglaciation event constitute the noise in which this new evidence of climate variation will be embedded.

References

Balling, N. (1980). The land uplift in Fennoscandia, gravity field anomalies and isostasy. In *Earth Rheology, Isostasy, and Eustasy* (N.-A. Mörner, ed.), pp. 297–321. Wiley, New York.

Barnett, T. P. (1983). Recent changes in sea level and their possible causes. *Clim. Change* **5**, 15–38.

Biot, M. A. (1954). Theory of stress–strain relations in anisotropic viscoelasticity and relaxation phenomena. *J. Appl. Phys.* **25** (11), 1385–1393.

Dicke, R. H. (1969). Average acceleration of the Earth's rotation and the viscosity of the deep mantle. *J. Geophys. Res.* **74**, 5895–5902.

Dickman, S. R. (1977). Secular trend of the Earth's rotation pole: Consideration of motion of the latitude observatories. *Geophys. J. Roy. Astron. Soc.* **57**, 41–50.

Denton, G. and Hughes, T. H., eds. (1980). *The Last Great Ice Sheets*, pp. 437–467. Wiley, New York.

Dziewonski, A. M. (1984). Mapping the lower mantle: determination of lateral heterogeneity in P velocity up to degree and order 6. *J. Geophys. Res.* **89**, 5929–5952.

Dziewonski, A. M. and Anderson, D. L. (1981). Preliminary reference earth model. *Phys. Earth Planet. Inter.* **25**, 297–356.

Dziewonski, A. M. Hager, B. H. and O'Connell, R. J. (1977). Large scale heterogeneities in the lower mantle. *J. Geophys. Res.* **82**, 239–245.

Farrell, W. E. (1972). Deformation of the Earth by surface loads. *Rev. Geophys. Space Phys.* **10**, 761–797.

Gilbert, F. and Dziewonski, A. M. (1975). An application of normal mode theory to the retrieval of structural parameters and source mechanisms from seismic spectra. *Philos. Trans. R. Soc. London, A.* **278**, 187–269.

Gornitz, V., Lebedeff, S. and Hansen, J. (1982). Global sea level trend in the past century. *Science* **215**, 1611–1614.

Hager, B. H. (1984). Subducted slabs and the geoid: Constraints on mantle rheology and flow. *J. Geophys. Res.* **89**, 6003–6015.

Haskell, N. A. (1935). The motion of a viscous fluid under a surface load. *Physics* **6**, 265–269.

Haskell, N. A. (1936). The motion of a viscous fluid under a surface load. *Physics* **1**, 56–61.

Haskell, N. A. (1937). The viscosity of the asthenosphere. *Am. J. Sci.* **33**, 22–28.

Hays, J. D., Imbrie, J. and Shackleton, N. J. (1976). Variations in the Earth's orbit: Pacemaker of the ice ages. *Science* **194**, 1121–1132.

Honkasolo, T. (1963). On the use of gravity measurements for investigation of the land upheaval in Fennoscandia. *Fennia* **89**, 21–23.

Hyde, W. T. and Peltier, W. R. (1985). Sensitivity experiments with a model of the ice age cycle: the response to harmonic forcing. *J. Atmos. Sci.*, **42**, 2170–2188.

Imbrie, J., Hays, J. D., Martisen, D. G., McIntyre, A., Mix, A. C., Morley, J. J., Pisias, N. G., Prell, W. L. and Shackleton, N. J. (1984). The orbital theory of Pleistocene climate: support from a revised chronology of the marine $\delta\ ^{18}O$ record. In *Milankovitch and Climate* (A. Berger *et al.*, eds.) pp. 269–306. Reidel Publ. Boston, Massachusetts.

Jeanloz, R. and Thompson, A. B. (1983). Phase transitions and mantle discontinuities. *Rev. Geophys. Space Phys.* **21**, 51–74.

Lambeck, K. (1980). *The Earth's Variable Rotation: Geophysical Causes and Consequences.* Cambridge Univ. Press, London and New York.

Longman, I. M. (1962). A Green function for determining the deformation of the Earth under surface mass loads: Theory. *J. Geophys. Res.* **67**, 845.

Longman, I. M. (1963). A Green's function for determining the deformation of the Earth under surface mass loads: Computations and numerical results. *J. Geophys. Res.* **68**, 485–496.

Love, A. E. H. (1927). *A Treatise on the Mathematical Theory of Elasticity*, 4th ed. Dover, New York.

Malvern, L. E. (1969). *Introduction to the Mechanics of a Continuous Medium.* Prentice-Hall, Englewood Cliffs, New Jersey.

Masters, G., Jordan, T. H., Silver, P. G. and Gilbert, F. (1982). Aspherical Earth structure from fundamental spheroidal mode data. *Nature* **298**, 609–613.

Meier, Mark F. (1984). Contribution of small glaciers to global sea level. *Science* (in press).

Milankovitch, M. (1941). *Canon of Insolation and the Ice Age Problem* (translated from the German) U.S. Department of Commerce, Israel Program for Scientific Translations (distr. by Davey, Hartford, Connecticut, 1969).

Müller, P. M. and Stephenson, F. R. (1975). The acceleration of the Earth and moon from early observations. In *Growth Rhythms and History of the Earth's Rotation* (G. D. Rosenburg and S. K. Runcorn, eds.) pp. 459–534. Wiley, New York.

Munk, W. H. and Revelle, R. (1952). On the geophysical interpretation of irregularities in the rotation of the Earth. *Mon. Not. R. Astron. Soc., Geophys. Supp.* **6**, 331–347.

Nakiboglu, S. M. and Lambeck, K. (1980). Deglaciation effects upon the rotation of the Earth. *Geophys. J.R. Astron. Soc.* **62**, 49–58.

O'Connell, R. J. (1971). Pleistocene glaciation and the viscosity of the lower mantle. *Geophys. J.R. Astron. Soc.* **23**, 299–327.

Peltier, W. R. (1974). The impulse response of a Maxwell Earth. *Rev. Geophys. Space Phys.* **12**, 649–669.

Peltier, W. R. (1976). Glacial isostatic adjustment II. The inverse problem. *Geophys. J.R. Astron. Soc.* **46**, 669–706.

Peltier, W. R. (1980). Ice sheets, oceans, and the Earth's shape. In *Earth Rheology, Isostasy and Eustasy* (N.-A. Mörner, ed.), pp. 45–63. Wiley, New York.

Peltier, W. R. (1982). Dynamics of the ice age Earth. *Adv. Geophys.* **24**, 1–146.

Peltier, W. R. (1983). Constraint on deep mantle viscosity from LAGEOS acceleration data. *Nature* **304**, 434–436.

Peltier, W. R. (1984). The thickness of the continental lithosphere. *J. Geophys. Res.* (in press).

Peltier, W. R. (1985). The LAGEOS constraint on deep mantle viscosity: results from a new normal mode method for the inversion of viscoelastic relaxation spectra. *J. Geophys. Res.* (in press).

Peltier, W. R., Farrell, W. E. and Clark, J. A. (1978). Glacial isostasy and relative sea level: a global finite element model. *Tectonophysics* **50**, 81–110.

Peltier, W. R. and Hyde, W. (1984). A model of the ice age cycle. In *Milankovitch and Climate* (A. Berger *et al.*, eds.) Vol. 2, pp. 565–580. Reidel Publ. Boston, Massachusetts.

Peltier, W. R. and Wu, P. (1982). Mantle phase transitions and the free-air gravity anomalies over Fennoscandia and Laurentia. *Geophys. Res. Lett.* **9**(7), 731–734.

Richards, M. H. and Hager, B. H. (1984). Geoid anomalies in a dynamic Earth. *J. Geophys. Res.* **89**, 6016.

Rubincam, D. P. (1984). Postglacial rebound observed by LAGEOS and the effective viscosity of the lower mantle. *J. Geophys. Res.* **89**, 1077–1087.

Svensen, S. L. (1980). A model for glacial uplift. In *Earth Rheology Isostasy and Eustasy* (N.-A. Mörner, ed.), pp. 65–72. Wiley, New York.

Vincente, R. O. and Yumi, S. (1969). Co-ordinates of the pole (1899–1968), returned to the conventional international origin. *Publ. Int. Latit. Obs. Mizusawa* **1**, 41–50.

Vincente, R. O. and Yumi, S. (1970). Revised values (1941–1961) of the co-ordinates of the pole referred to the CIO. *Publ. Int. Latit. Obs. Mizusawa* **7**, 109–112.

Walcott, R. I. (1980). Rheological models and observational data of glacio-isostatic rebound. In *Earth Rheology, Isostasy, and Eustasy* (N.-A. Mörner, ed.), pp. 3–10. Wiley, New York.

Wolf, D. (1985a). "The Dynamics of the Continental Lithosphere". Unpublished Ph.D. thesis, Department of Physics, University of Toronto.

Wolf, D. (1985b). On Boussinesq's problem for Maxwell continua subject to an external gravity field. *Geophys. J.R. Astron. Soc.* (in press).

Woodhouse, J. H. and Dziewonski, A. M. (1984). Mapping the upper mantle: three dimensional modeling of Earth structure by inversion of seismic waveforms. *J. Geophys. Res.* **89**, 5953–5986.

Wu, P. and Peltier, W. R. (1982). Viscous gravitational relaxation. *Geophys. J. Roy. Astron. Soc.* **70**, 435–486.

Wu, P. and Peltier, W. R. (1983). Glacial isostatic adjustment and the free-air gravity anomaly as a constraint on deep mantle viscosity. *Geophys. J. R. Astron. Soc.* **74**, 377–450.

Wu, P. and Peltier, W. R. (1984). Pleistocene deglaciation and the Earth's rotation; a new analysis. *Geophys. J. R. Astron. Soc.* **76**, 753–792.

Yoder, C. F., Williams, J. G., Dickey, J. O., Schultz, B. E., Eanes, R. J. and Tapley, B. D. (1983). *Nature* **303**, 757–762.

Part 3

Space Geodetic Baselines

3.1 Doppler Satellite Measurements and Their Interpretation

R. J. Anderle

General Electric Company, Valley Forge Space Technology Center, King of Prussia, Pennsylvania, USA

I. Introduction

A. History of NAVSAT

NAVSAT is the acronym for the U.S. Navy Navigation Satellite System developed by the Applied Physics Laboratory of Johns Hopkins University (APL/JHU) to allow navigators aboard ships to determine their position to 100 m accuracy about once an hour. The system became operational in 1963 and is used by tens of thousands of receivers throughout the world. In order to meet the accuracy specification, it was necessary to improve the accuracy of geodetic parameters of the Earth from the levels existing at the start of the development program in 1958. Since the Navy department decided that this effort would be a difficult and critical task, it assigned the Naval Surface Weapons Center (NSWC) (earlier called, in turn, the Naval Proving Ground and the Naval Weapons Laboratory) the responsibility of refining the Earth's geodetic parameters independently of the APL/JHU effort. The two laboratories pursued the research in parallel until 1965 when, the navigation accuracies having been achieved, the NSWC was given the new mission of exploiting the NAVSAT program to meet the mapping, charting and geodesy requirements of the Navy department initially, and then of the Defense Mapping Agency.

Prior to the start of the space program, the Earth's gravitational field was assumed to correspond to that of an ellipsoid, and distances between sites on different continents and between sites and the center of mass of the earth were uncertain to 100–200 m. The first geodetic result of the NAVSAT program was the confirmation of the existence of the third-order zonal gravitational harmonic for the Earth (Cohen and Anderle, 1960) followed by the refinement of other zonal harmonics and the first determination of tesseral harmonics from NAVSAT data (Newton, 1963). Routine geodetic applications of the NAVSAT system started in 1963 with the determination of the positions of LORAN-C navigation beacons on isolated islands in the Pacific Ocean. These computations were performed with a gravitational field complete only through fourth degree and order. Although 10-m accuracies had been achieved for distances between stations up to a few hundred kilometers, errors in the gravitational field resulted in uncertainties in the position of isolated sites of 50–100 m at this time. These were still useful results since astronomical positions used up to this time for isolated sites were uncertain to 300 m. The evaluation of "resonant gravitational coefficients", those corresponding to near-commensurability between orbit period and Earth's rotational period (Anderle, 1965), improved the accuracy of the determination of position to 15–25 m. Including these gravitational harmonics (of 13th and 14th order) in solutions otherwise complete to 7th degree (Anderle, 1967) or 8th degree (Guier and Newton, 1965) allowed the NAVSAT system to eventually reach operational status and caused the Defense Mapping Agency to adopt the NAVSAT system as its principal space technique for geodetic applications in 1965. The NAVSAT results made a significant contribution to the Department of Defense World Geodetic System 1966 (DOD WGS-66) and were a major constituent of DOD WGS-72 which has been widely used internationally (Anderle, 1974; Seppelin, 1974). Repeatability in the determination of the positions of isolated sites to 1–2 m is achieved using orbits computed with a gravitational field complete through 19th degree and order supplemented by higher-degree zonal and resonant coefficients. Another significant error source was eliminated just before the adoption of this geodetic system when NAVSAT data began to be used routinely to determine polar motion, the direction of the Earth's spin axis with respect to its crust (Anderle and Beuglass, 1970). Relatively minor changes were made in NAVSAT computations in the period 1972–1984.

B. Contributions of Doppler Data to Geodesy and Geodynamics

The NAVSAT system is based on the measurement of the Doppler effect on frequencies transmitted by the satellites as described in the next section. The measurement system has also been used with other satellite systems, such as satellites equipped with radar altimeters for use in determining the mean ocean surface, which is an approximation of the geoid, and with the NAVSTAR global positioning system (GPS). However, the altimetric results are a major topic discussed elsewhere in this book, and techniques other than Doppler will be the principal methods of obtaining geodetic and geophysical data from the GPS (Anderle, 1981). Apart from the determination of the orbits of altimetric satellites, Doppler satellites have been used extensively to strengthen and extend terrestrial networks and connect them to an Earth-centered coordinate system, to provide polar motion data and to determine the Earth's gravity field for use in the determination of satellite orbits and, to some extent, for geophysical interpretation. Marginally useful results have been obtained on Earth's rotation and on plate tectonics. While it is theoretically possible to determine the direction of the Earth's inertia axis from analysis of Doppler observations, the accuracy of the results is too low to be useful. The use of the new DOD WGS-84 and other improvements in the Doppler system discussed later will permit higher levels of accuracies to be achieved than those obtained to date. However, it is expected that the GPS system will rapidly replace NAVSAT for use in precise determination of the relative positions of sites and perhaps in other applications.

II. System Description

A. Navy Navigation Satellites

The Navy navigation satellites are in near circular polar orbits at altitudes of about 1000 km. There are typically four satellites operating at any time that transmit continuous wave signals at frequencies 80 parts per million below 150 and 400 MHz (14,998,800 Hz and 399,968,000 Hz). These frequencies are synthesized from the same oscillator so that the signals received on the ground may be combined to eliminate first-order ionospheric effects which are inversely proportional to frequency. The signals are modulated to provide time signals used to synchronize the clocks at the various ground stations. Two classes of satellites, OSCAR and NOVA, are in operational use. Both types are gravity gradient stabilized and can be observed with the same ground receivers. The OSCAR-type satellites have been in use since the initial operation of the system, several performing with better than 99% reliability for over 20 years. The NOVA is a new design incorporating improved oscillators, providing higher signal strength, and equipped with a DISCOS (Disturbance Compensation System) which senses and corrects for nonconservative forces (drag and solar radiation) along the direction of motion of the satellite. Both types of satellites are equipped with memories into which ground stations periodically (about daily) inject predicted satellite positions; the satellites then broadcast their current positions as modulations on the signals so that navigators can determine their position from the satellite positions and the Doppler effect on the broadcast signal.

B. Navy Navigation Ground Receivers

Ground receivers can be considered to be in three classes according to the functions they perform. All receivers use a local oscillator to synthesize coherent frequencies at 150 and 400 MHz which are mixed with the incoming signals. The receivers then, depending on their design, either measure the time required to count the number of beat cycles between the received signals and the ground signals, or measure the number of beat cycles over a preset interval. The OPNET receivers are an integral part of the Navy Navigation System; measurements from the four OPNET sites in Maine, Minnesota, California and Hawaii are used to compute the orbit of each satellite, and to make predictions of its future position which are injected into its memory. These receivers normally use high-gain antennas and make frequency measurements at 4-s intervals, in which each measurement is the time required to count a given number of beat cycles between the received signal and a ground signal. The number of cycles to be counted is set so that the time for the count lasts a little less than 1 s. TRANET receivers are deployed at sites around the world (Fig. 1) either by the U.S. Defense Mapping Agency or by other nations cooperating with the Defense Mapping Agency for the purpose of determining more precise satellite positions for use in geodetic surveys and various scientific studies. Most of these receivers have been at the sites shown in the figure for over 20 years; the international observatories have been operating for periods of time varying from 1–10 yr. Prior to 1971, TRANET receivers made measurements in the same way that the OPNET receivers do but with omni-directional antennas. The receivers were then modified to count Doppler signals in a non-destruct mode. That is, CCID (Continuous Count Integrated Doppler) measurements are made

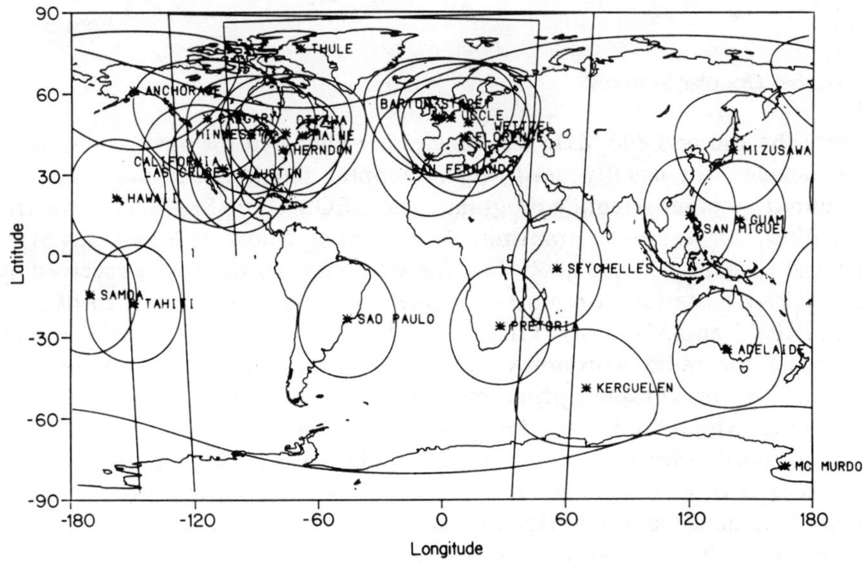

Fig. 1. Ten degree visibility circles at Doppler sites (four near vertical lines show ground track for two revolutions of polar orbiting satellite). Height, 1000 km.

by allowing counters to continue the Doppler cycle count when the time to count the given number of Doppler cycles is recorded. The number of cycles to be counted is typically set so that measurements are recorded at approximately 20-s intervals. The original equipment, TRANET-1, was replaced by TRANET-2 equipment in 1983 to utilize modern electronic components and further automate the stations. The equipment is functionally the same, except that two satellites may be observed continuously with the new equipment. A number of types of portable equipment are available from various manufacturers and used by most nations around the world. A dozen or so of these receivers are periodically deployed by the Defense Mapping Agency to supplement the TRANET station array when more dense observations are required to obtain higher-precision orbits. For example, high-precision orbits were required for the GEOS-C and SEASAT satellites which carried radar altimeters to determine the mean ocean surface as an approximation of the geoid and variations in the surface due to oceanographic effects. Tens of thousands of these receivers are used for ship navigation and many hundreds are used for geodetic surveys. The receivers generally record the beat cycles measured over given intervals, varying from a few seconds to 2 minutes depending on the receiver, in a non-destruct manner. All receivers except the OPNET receivers normally use omni-directional antennas, giving lower signal strength but reduced size over the antennas employed at the OPNET sites.

C. Other Doppler Systems

Since the 150 and 400 MHz frequencies are allocated for navigation, other frequencies are usually used when Doppler satellite measurements are required for other scientific programs. The GEOS-C and SEASAT altimetric satellites, for example, transmitted at 50 ppm below 162 and 324 MHz (161,991,900 and 323,983,800 Hz). These frequencies cannot be received by the OPNET stations or most portable receivers, but were recorded by TRANET 1 and Magnavox GEOCEIVER receivers. Since 162 MHz is now reserved for radio astronomy, other frequencies had to be chosen for nonnavigation Doppler programs. Frequencies 80 ppm above 150 and 400 MHz, which do not interfere with navigation receivers were chosen for this purpose. TRANET 2 and MAGNAVOX MX1102 portable receivers can receive these signals and can in fact operate in a dual acquisition mode recording data from a navigation satellite and a geodetic satellite simultaneously. This system was chosen as the primary orbit determination technique for the TOPEX (topographic experiment) satellite planned for launch in 1990 to obtain radar altimetry data for use in studying global ocean circulation. Orbit accuracies better than 10 cm in height are required for this purpose. Doppler systems other than the navigation systems generally do not broadcast ephemeris data or time signals. Therefore portable receivers deployed to obtain data on such systems usually also acquire observations on one or more navigation satellites for use in time calibration, and the data are used in calculating the ephemeris of the navigation satellites.

III. Information Content of Doppler Data

A. Fundamental Measurement

The frequency received at a ground station may be written

$$f_R = f_T - f_T \dot{r}/c + A/f_T \tag{1}$$

where f_T is the transmitted frequency (say 80 ppm below 150 or 400 MHz), c the velocity of light, $\dot{r}$ the rate of change of range from the ground station to the satellite, and A a function of the electron density along the path of the signal from the satellite to the receiver.

The equation is simplified to omit tropospheric refraction, relativistic effects and higher-order effects of ionospheric refraction due to electron density. If the path from the satellite to the ground is assumed to be the same for two frequencies (another first-order assumption), the signals received at two frequencies can be combined to eliminate the ionospheric refraction term involving A. Let us assume this has been done and the results have been scaled back to measurement 80 ppm below 400 MHz which is free of ionospheric effects. If the resulting signal is mixed with a ground signal, $f_{G'}$ at 400 MHz, the beat cycles of the output N_c can be counted over the time interval t_1 to t_2

$$N_c = \int_{t_1}^{t_2} (f_G - f_R)\, dt = (f_G - f_T)(t_2 - t_1) + \frac{f_T}{c} \int_{t_1}^{t_2} \dot{r}\, dt \tag{2}$$

All receivers operate on the above general principal, although various scaling factors are used in mixing the signals and the ionospheric refraction correction may be made at different stages in the processing. The interval t_1 to t_2 can be intermittent intervals during the pass of a satellite over a station such as in the case of OPNET receivers, or the count can extend from the first measurement on the pass t_1 to sequences of times t_i occurring during the pass as in the case of most other receivers. The values N_c are the basic measurements which could be compared with theoretical values for use in navigating stations or computing orbits. More frequently, the equation is rearranged to provide a more generally recognizable observational quantity, range difference, ΔR, over the interval $t_2 - t_1$

$$\Delta R = \int_{t_1}^{t_2} \dot{r}\, dt = \frac{c}{f_T}[N_c - (f_G - f_T)(t_2 - t_1)] \tag{3}$$

It is clear that the range difference ΔR is dependent on both the satellite position and the receiver position, so that it can be used to attempt to determine either of these quantities, or other parameters on which these quantities depend, and upon the transmitted and ground frequencies f_T and f_G which are not known *a priori* to the precision of the measurement. But before discussing the methods actually used, it is useful to consider the information content in a pass of Doppler data in order to understand the data in a more physically meaningful context.

B. Equivalent Information

If we assume the satellite orbit is known, two geometric quantities which can be inferred from measurements of frequencies received during the pass of a polar-orbiting satellite over a station are (a) the range to the satellite and (b) the position of the station and the station parallel to the track of the satellite at the time the satellite is closest to the receiver. Figure 2 shows that distinctive residual patterns are formed if (a) the station position along the track of the satellite was erroneous, which is equivalent to a fixed time error in the observations, or (b) the range to the satellite was wrong, which would alter the rate of change of the frequency. These are two components of the relative position of the station and satellite in a reference frame defined by the range vector to the satellite and the velocity vector of the satellite relative to the station at the time of closest approach to the station. The third component of position is ill-defined since for a nonrotating Earth, rotation of the satellite path at fixed distance from the station would give the same Doppler signal. The third geometric quantity for which useful information is obtained is the velocity of the satellite; an error in velocity would increase or decrease the received frequency at the beginning and end of the satellite pass more strongly than at the center of the pass. The other two components of velocity have relatively small effects on the received satellite frequency, considering typical errors in velocity uncertainty and measurement accuracy. In addition to the geometric information contained in the Doppler measurements made during a satellite pass, the relative

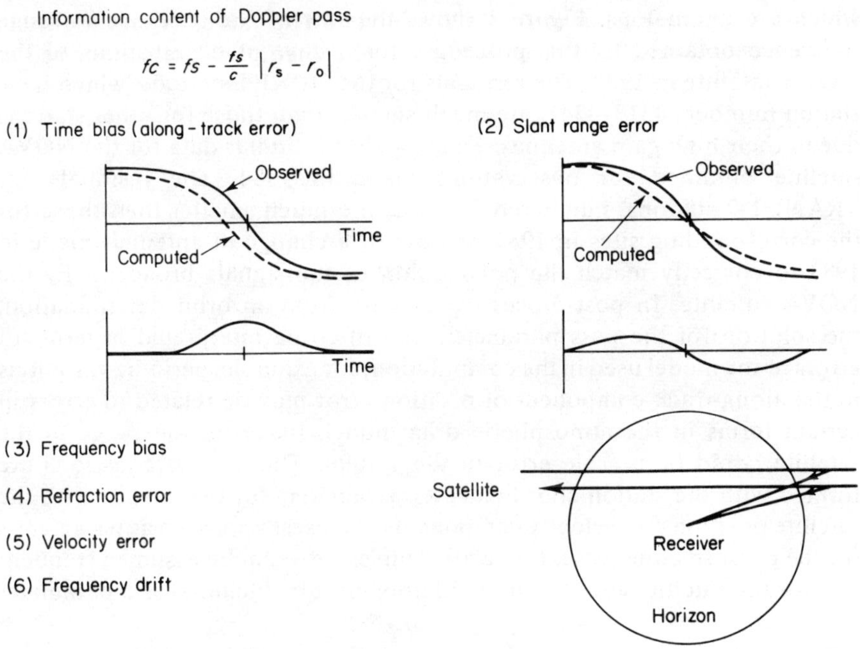

Fig. 2. Information content of Doppler pass.

frequency offset between the transmitted and ground frequencies is well determined because one expects the Doppler effects at the start and end of the pass to be equal and opposite in sign when effects of geometric differences are accounted for. Lastly, a scale factor for the tropospheric refraction corrections can be inferred because the tropospheric effects cause a sharp change in the signal at the beginning and ends of the pass when the satellite is at low elevation angles.

C. Application of Pass Parameter Representation

During pre-processing and post-processing of Doppler data, normal equations for each satellite pass are formed and solved to determine the pass parameters discussed previously. During pre-processing, the process is utilized to produce residuals which are nearly Gaussian which allows strong filtering of individual data points to remove anomalous data. In this sense, the technique provides a high-order filter which is dependent on a small number of physically meaningful parameters. The size of the parameter corrections for various passes are also compared to determine entire passes which are anomalous. Figure 3 shows the rms of the residuals in range difference obtained by this procedure for 8 days of observations of the NOVA satellite in 1982. The residuals for the OPNET stations, which have station numbers 4313–4341, are much smaller than those for other stations due to their high gain antennas. Figure 4 shows similar data for the NOVA satellite obtained for observations made in 1984. The residuals for TRANET-2 stations, numbered 545–562, are much smaller than those for the corresponding sites in 1982 because of a change in antennas made in 1983 to correctly match the polarization of the signals broadcast by the NOVA satellite. In post-processing of data from an orbit determination, the solution for the pass parameters can often be interpreted in terms of errors of the model used in the computation; for example, periodic variations in the along-track component of position error may be related to errors in certain terms in the atmospheric drag model, biases in the range to the satellite could be a scale error in the system. The normal equations are formed with the station coordinates as parameters for the pass or with the satellite position and velocity components at closest approach as parameters for the pass. In either case, the relative offset between the assumed frequencies for the satellite and station oscillators must be included as a parameter

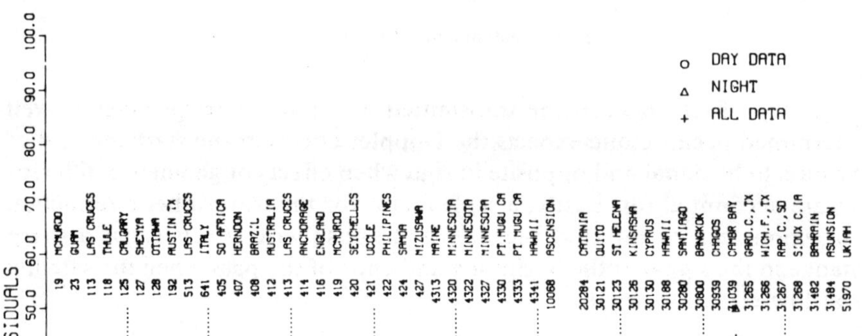

Fig. 3. Random error of Doppler observations of NOVA satellite day 208 1982. O, Day data; Δ, night; +, all data.

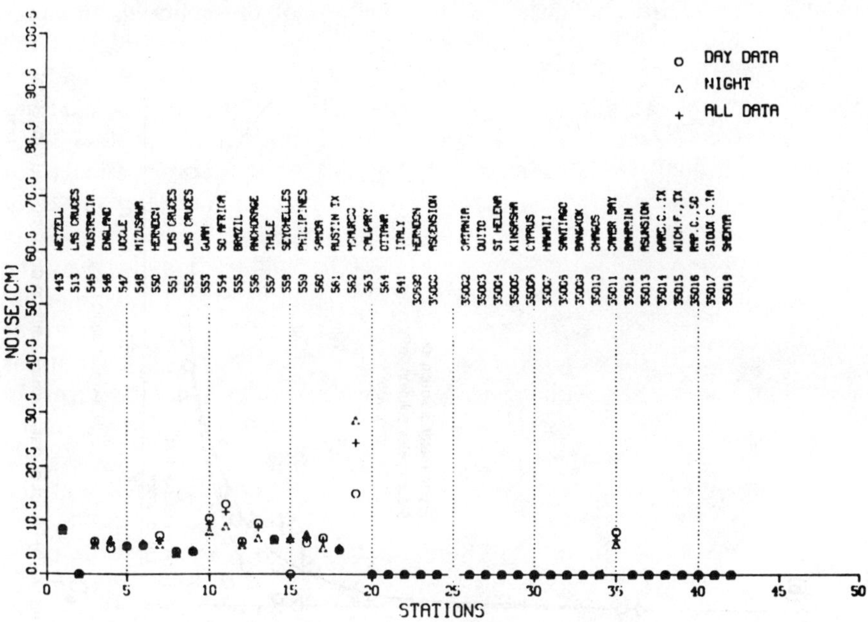

Fig. 4. Random error of Doppler observations of NOVA satellite day 208 1983. ○, Day data; Δ, night data; +, all data.

and a scale factor for tropospheric refraction may be included as a parameter. The normal equations must be constrained to prevent solution for the ill-determined third component of position and to include, at most, only one component of velocity to avoid near-singularity.

D. Eigenvalue Analysis

The effect of uncertainties in the ill-determined component of position on the solution for the two principal components of the relative position of the satellite with respect to the station is shown in Fig. 5(a) (b) as a function of the maximum elevation angle to the satellite during satellite pass. The range error is more sensitive to errors in the ill-determined component than in the along-track component of error, but both are reasonably well determined. Another method of analyzing the separation of the parameters is to compute the eigenvalues and eigenvectors of the normal matrix and resolve the eigenvectors in a reference frame defined by the range vector to the satellite and the velocity vector of the satellite at closest approach. If the normal matrix also contains parameters for the satellite velocity, some correspondence must be established for the relationship between position

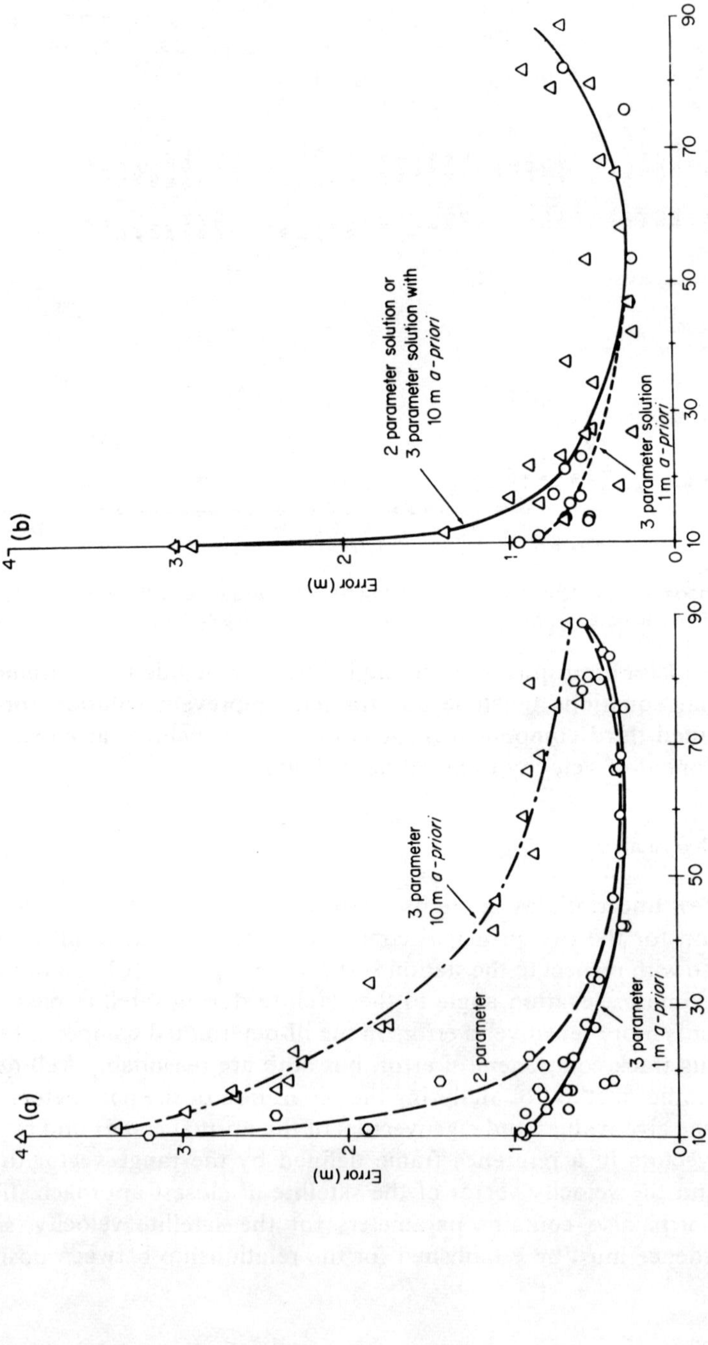

Fig. 5. (a) Effect of parameter selection on computed radial standard error. (b) Effect of parameter selection on computed tangential error.

and velocity errors in order to allow a meaningful interpretation of the results of the analysis. In the example which follows, the normal matrix was scaled on the basis that a meter in position errors is equivalent to a millimeter per-second velocity error. This relationship was established on the basis of a typical pass length of 1000 s and was confirmed for typical errors in satellite orbits by comparison of position and velocity of orbits computed using different parameters. (In the case of positioning a fixed ground station, the effects of the error in the relative velocity of the satellite and station might be small compared to the error in station location.)

TABLE I. Sample eigenvalues and eigenvectors[a]

		Eigenvector components					
		Position			Velocity		
Zenith angle	Eigenvalue	Radial	Tangential	Transverse	Radial	Tangential	Transverse
69°	0.17E07	0.71	0.44	0.08	0.16	−0.51	0.04
	0.11E07	0.38	−0.83	0.04	−0.30	−0.27	−0.07
	0.13E04	0.58	0.00	−0.16	−0.01	0.80	−0.04
	0.43E02	−0.02	0.19	0.03	−0.30	−0.03	−0.93
40°	0.98E07	0.61	0.73	−0.04	0.16	−0.25	−0.03
	0.35E07	0.64	−0.64	−0.06	−0.14	−0.40	0.02
	0.16E05	0.47	−0.07	0.14	−0.04	0.87	0.02
	0.14E04	0.01	−0.19	−0.02	0.73	0.03	−0.66
24°	0.96E07	0.81	−0.45	0.03	−0.09	−0.37	−0.01
	0.36E07	0.40	0.87	0.02	0.18	−0.23	0.01
	0.43E05	0.44	0.03	−0.09	0.04	0.89	0.01
	0.49E03	−0.01	−0.19	0.05	0.88	−0.03	0.43
17°	0.70E07	0.81	0.41	−0.03	0.09	−0.41	−0.01
	0.40E07	0.36	−0.89	−0.01	−0.20	−0.21	0.01
	0.56E05	0.46	−0.01	0.07	−0.05	0.89	0.02
	0.11E04	−0.02	0.21	−0.01	−0.92	−0.05	0.31
50°	0.12E08	0.85	0.13	−0.08	0.04	−0.51	−0.01
	0.60E07	0.12	−0.95	−0.01	−0.27	−0.08	0.05
	0.38E05	0.52	0.00	0.15	0.00	0.84	−0.01
	0.57E03	−0.01	−0.21	0.00	0.59	0.00	−0.78
39°	0.11E08	0.03	0.97	0.00	0.24	−0.01	−0.03
	0.64E07	0.88	−0.03	−0.06	0.00	−0.47	0.00
	0.28E05	0.47	0.01	0.13	−0.03	0.87	0.02
	0.17E04	0.01	−0.21	−0.01	0.74	0.04	−0.64

[a] First four eigenvalues and corresponding eigenvectors for six satellite passes (rows and columns of velocity components were multiplied by 1000 after formal elimination of refraction and frequency parameters).

The effects of the frequency and refraction parameters on the coupling among the geometric parameters are considered by forming the eliminated matrix (Anderle, 1967)

$$B_G^1 = B_G - A * B^{-1} A$$

where B_G is the 6×6 partition of the normal matrix pertaining to geometric parameters, B_B the 2×2 partition of the normal matrix referring to the bias parameters frequency and refraction and A the 2×4 partition of the normal matrix relating the interaction of geometric and bias parameters.

Eigenvectors and eigenvalues for normal matrices formed in this way are shown in Table I for each of the six satellite passes. It can be seen that the two largest eigenvalues, which represent the two strongest parameters, are either along the range vector or along-track direction, or a mixture of both. The third strongest eigenvalue is two or three orders of magnitude lower and lies along the tangential component of velocity. The fourth eigenvalue is one or two orders of magnitude still smaller and the last two eigenvalues are not tabulated. This table provides mathematical justification for declaring that two components of position contain the principal information content in a pass Doppler data and that the next bit of information is contained in the velocity of the satellite.

IV. Computational Methods

A. Ephemeris Computations

Usually both an operational ephemeris and a precise ephemeris are computed for Navy Navigation Satellites. The operational ephemeris is computed by the Naval Astronautics Group at Point Mugu, California on the basis of the observations made at the four OPNET sites. Extrapolations of this ephemeris are injected in the memory of the satellites which then broadcast their positions in real time (the term "broadcast ephemeris" is also used for these satellite positions). The precise ephemeris is computed on a delayed basis by the Defense Mapping Agency Hydrographic Topographic Center using observations at the 20 or so TRANET stations and any additional portable receivers deployed for use in ephemeris computations. The methods of computing the broadcast and precise ephemeris are very similar; differences in the results are primarily due to the fact that the former is a predicted ephemeris and the latter is a post-fit ephemeris. The effects of differences in the number of stations used in the computation and the differences in method of computation are relatively small. Satellite positions are computed by numerically integrating the equations of motion of the satellite arising from the gravitational force of the Earth, sun, and moon, the tidal potential arising from the sun and moon, the force atmospheric drag and solar radiation forces. The initial components of position and velocity for the integration are usually taken from a prediction made for a preceding orbit computation, and are parameters of a linear least squares solution to minimize the differences between observed range differences and those computed for the computed orbit. Because of the variations in atmospheric density due to changes in solar radiation and magnetic flux with time, a scale factor for atmospheric drag is included as a parameter of the solution, usually an independent scale factor for each day. Since the orientation of the Earth's spin axis with respect to the crust varies in a somewhat unpredictable way, the components of pole position are also parameters of the solution. The least squares adjustment is usually made for the observations made during consecutive 48-hr intervals. For each pass of the satellite over each station, the uncertainty in the scale factor for tropospheric refraction and in the relative frequency of the satellite and station are also considered by the formal elimination process described in Section III.D. In the case of the NOVA satellite, the tangential effects of solar radiation force and atmospheric drag are set to zero, but an along-track acceleration is included as a parameter to accommodate a bias in the DISCOS measurement system. Although the system of equations to be solved is highly nonlinear, the initial conditions predicted from the preceding contiguous orbit fit are usually accurate enough so that iteration of the solution is not required.

The formation of the normal equations requires partial derivatives for the computed data with respect to the parameters of the solution. Some of these partial derivatives, such as those with respect to frequency, refraction and pole positions, can be obtained directly by differentiating the range difference equation with respect to the parameters. However, those which depend on dynamic parameters such as initial position and velocity or atmospheric drag require the partial derivative of satellite position at the time of observation with respect to the dynamic parameter. These partial derivatives are obtained by numerically integrating the variational equations, which are the partial derivatives of the satellite acceleration with respect to the dynamic parameters. The equations used are given by Anderle (1974) and a detailed account is given by O'Toole (1976).

B. General Geodetic Solutions

The foregoing orbit computations assume that values are available for the coordinates of most of the stations observing the satellite and that a definition of the Earth's gravity field is available. These data are determined on an infrequent basis by a procedure similar to that used in ephemeris computations. In such a general geodetic solution, variational equations must also be numerically integrated for coefficients of the Legendre polynomial representing the Earth's gravity field, derivatives of data with respect to station coordinates obtained, and variational equations are also carried for additional parameters such as those for the coefficient for the effect of solar radiation force on the satellite and a coefficient for the effect of solid Earth tides on the gravitational potential. In order to separate these terms, data on satellites having different orbital inclinations are used and longer spans of observation are used for each set of orbit constants considered. Other types of data are also included such as surface gravimetry. A discussion of the technique is given by Seppelin (1974) and Anderle (1974).

C. Geodetic Positioning Techniques

Although the principal parameters to be determined in positioning a receiver are the three components of position, the relative frequency offset between satellite and receiver must also be determined for each pass and a tropospheric refraction parameter is usually determined for each pass, where the *a priori* refraction model is assigned a standard error of 10%. In the point positioning technique, the orbit is assumed to be known perfectly and no additional parameters are carried in solving the normal equations for station coordinates. However, uncertainties in the gravity field and atmospheric drag are a significant source of error in such solutions, probably about 1–2 m. The errors probably have correlation distances of 7000 km in the direction of motion of the satellite and 500-km normal to the direction of motion (Anderle and Hoskins, 1977). While these errors cannot be removed from the computed absolute position of a receiver, studies indicate that some improvement is obtained in the repeatability of results if parameters are carried in the normal matrix for each satellite pass to represent the orbit error (Anderle, 1982). Considerable success has also been achieved in reducing the error in the relative positions of sites occupied by different equipment during the same time period by considering orbit errors common to the sites (Kouba, 1983).

V. Error Sources

The principal sources of errors in the application of the Doppler technique to geodesy and geodynamics are the oscillators in the satellite and in the receiver, higher-order ionospheric effects neglected in computing first-order effects from measurements at two frequencies, and orbit error due to uncertainties in the gravity field. The latter two errors can produce systematic differences in results computed at different locations or times, or can be largely eliminated in determining the relative positions of neighboring sites computed from observations made at nearly the same time. The oscillators cause systematic errors during a satellite pass, but the errors will average out after a large number of satellite passes are acquired. The nature of these and other measurement, environmental and orbit errors are discussed in the following sections.

A. Measurement Errors

The effects of random errors of measurement were discussed earlier and it was shown in Fig. 4 that, with sufficient radiated signal strength and properly matched transmitting and receiving antennas, random errors in range difference measurement of 5 cm can be achieved even with omni-directional antennas. These errors were random over Doppler measurement intervals of 20–30 s. For longer periods of time, the stability of the satellite and ground station oscillators can produce significant systematic errors, depending on the quality of the oscillator. Figure 6 shows the measured stability of two quartz oscillators (QZ (SN 7444) and QZ (SN 1222)), of the upper

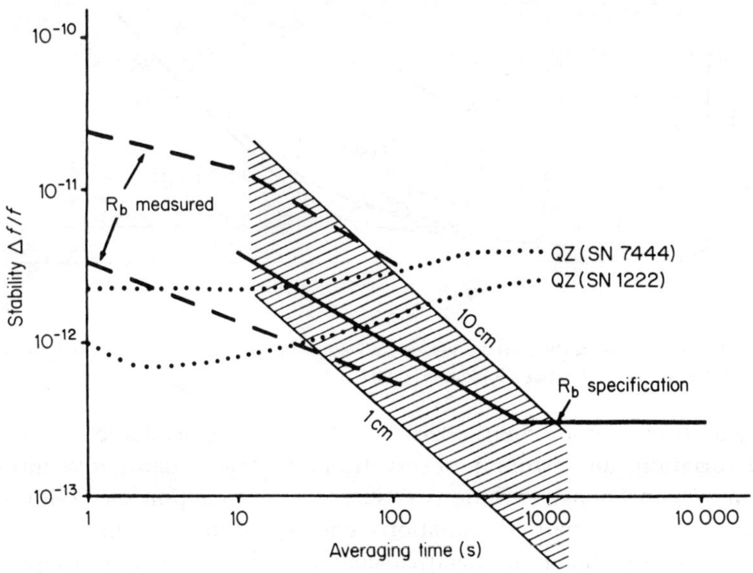

Fig. 6. Oscillator stability (diagonally hatched area indicates desired performance level).

and lower bounds of measurements on a series of rubidium oscillators (Rb measured), and for the specifications of a rubidium oscillator (Rb specification) as a function of measurement interval. The shaded area indicates the range of performance required such that the integrated error due to oscillator stability would lie between 1–10 cm for time spans ranging from measurement intervals of 10 s to the length of a satellite pass of a little over 1000 s. Note that the quartz oscillators generally used with mobile equipment fail to meet the requirement at the longer averaging times and that the poorer performing rubidium oscillators also fail to meet the tolerance. For this reason, it has been general practice to consider the basic observations to be range differences made at the 20–30 s measurement intervals rather than range differences from the start of the pass to each measurement time. For precision applications, such as the TOPEX altimetry satellite program, care will be taken to ensure that oscillator stability is not a significant contributor to the error budget. Figure 7, drawn from data given by Gouldman (1982), shows that range and tangential components of position errors will be below 10 cm provided that the oscillator stability is better than 5 parts in 10^{13} for the duration of the satellite pass (random errors no worse than 5 cm are also assumed).

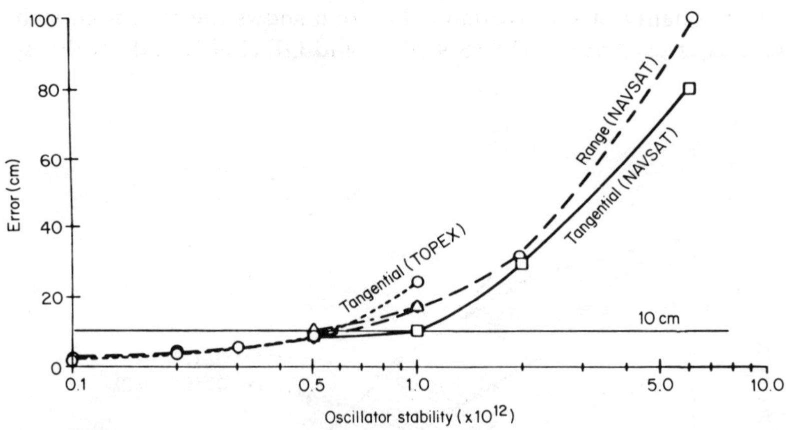

Fig. 7. Single-pass navigation errors versus oscillator stability based on seven-sample Monte Carlo simulations. [After Spence-Gouldman, (1982).]

Two additional equipment errors which must be considered are antenna phase variations and errors in observation time tags. Quantitative information on antenna phase variations is scarce. Manipulation of Taylor series expansions of the pair of equations corresponding to Eq. (2) for two frequencies shows that if the electrical centers of the antennas for frequencies A and B (say $f_A=150$ MHz and $f_B=400$ MHz) are at locations $\bar{x}_A$ and $\bar{x}_B$, then the effective geodetic location of the measurement is at $\bar{x}=\bar{x}_B+(\bar{x}_B-\bar{x}_A)R^2/(1-R^2)$ where $R=f_A/f_B$ (Anderle, 1978). (Note that the effective center is not between the two antennas.) However, the electrical phase centers depend on the nature of the ground plane which could conceivably depend on ground moisture, for example.

Motion of the satellite antenna can be neglected since the satellites are stabilized to about a degree and the antenna is only a meter below the center of gravity of the spacecraft. However, circularly polarized antennas are subject to nearly a wavelength change in phase during the course of a pass which must be corrected for in the data processing algorithm.

The observations are initially time tagged according to a clock operated from the local station oscillator. However, the Navy Navigation Satellites broadcast time signals which are recorded by each station and which are calibrated by the signals recorded at a few stations equipped with high-performance oscillators which have coordinated universal time. Since the satellite is moving about 7 km/s, an uncertainty of 5-μs in time tag would produce a 3.5-cm error in the computed satellite position. A 5-μs accuracy for a 24-hr period implies an oscillator stability of 5 parts in 10^{11}. The satellite oscillators are sufficiently stable to transfer the calibration for at least a 24-hr period, and the time signals can be recovered to the required accuracy if sufficient care is exercised in the processing. Figure 8 shows the satellite oscillator variation inferred from stations equipped with cesium oscillators, and Fig. 9 shows the satellite clock calibrations performed by these stations.

B. Environmental Error Sources

The principal error sources in this category are ionospheric and tropospheric errors. One might consider errors in Earth tides and ocean tidal loading as environmental, but even the main effects are small compared to other error sources. Tropospheric refraction is of no concern if the atmosphere has no horizontal gradients. Figure 10 shows that a refraction-scale parameter is recovered to a precision of 1–5% from the satellite observations if observations are made down to an elevation angle of 5°. Figure 11 shows that the solution for the scale parameter causes little degradation in the precision of measurement of the range component of the position of the satellite if the satellite is seen at elevation angles of 25° or higher. Effects of horizontal gradients in the atmosphere have not been determined, but can be expected to average out in the course of time in most cases. (It would be interesting to simulate weather patterns for coastal stations to determine whether seasonal effects or bias effects are present in such locations.)

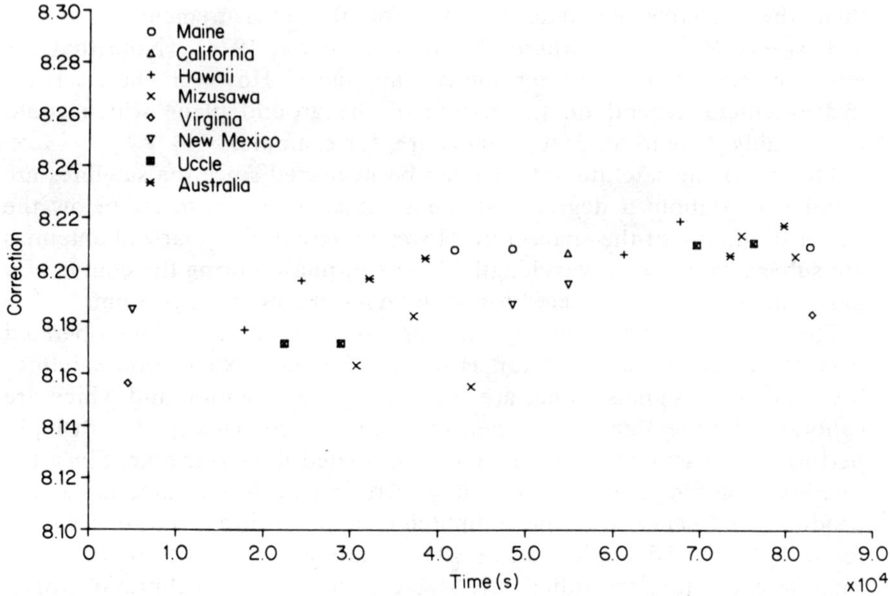

Fig. 8. Frequency corrections in parts per billion observed for NOVA satellite on day 171 1984.

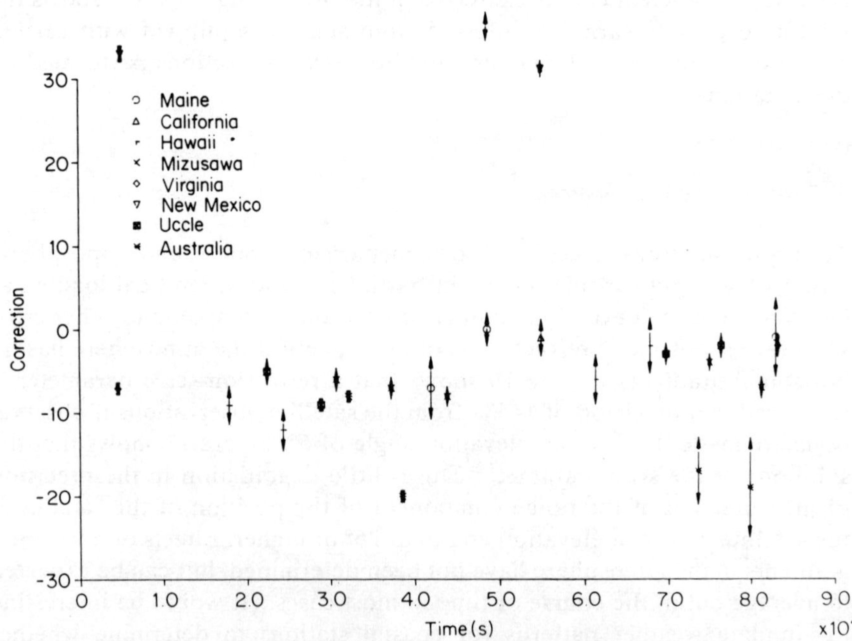

Fig. 9. Time corrections in microseconds observed for NOVA satellite on day 171 1984.

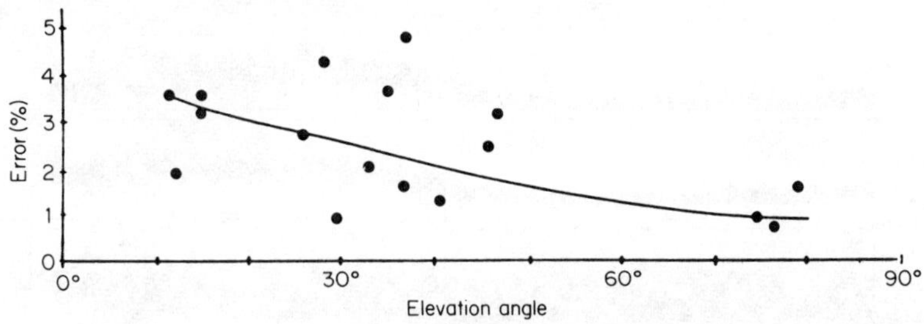

Fig. 10. Standard error of refraction bias versus elevation angle at closest approach (from single-pass navigation normal equations where data were typically observed from 10° elevation at rise to 5° at set).

Residual ionospheric errors which arise after elimination of first-order effects by the use of two frequencies are a significant problem. The ionosphere is excited by solar activity which can be gauged by variations in sun spot number or solar flux measurements. Figure 12, for example, shows that over the course of time (a 13-yr solar cycle primarily), the sun spot number and the solar flux both range from values which are below 100 to values over 200. There are seasonal variations evident in the graph, and the geographical effect will depend on the direction of the sun with respect to the line of nodes of the satellite, which has a nearly annual variation. Simulations of the effect on computed station height due to the neglect of third-order ionospheric errors arising from solar flux levels of 100 and 200 are shown in Fig. 13(a) for various sites. At the lower flux level, the effects are about 10 cm or less, but at the higher level the effects can be several

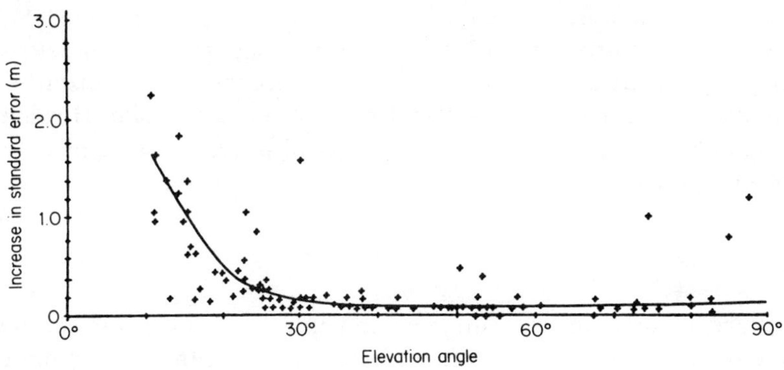

Fig. 11. Increase in standard error in slant range due to use of refraction bias parameters; 10% standard error in refraction and 5% minimum elevation angle for observations.

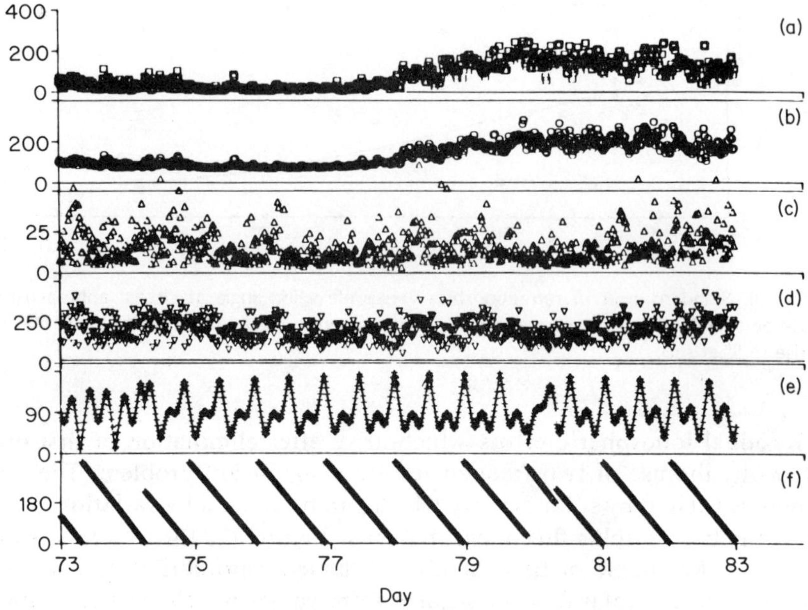

Fig. 12. Solar activity from 1983–1984; (a) sun spot number, (b) 10.7 MHz solar flux, (c) geomagnetic index A_p, (d) geomagnetic index k_p, (e) angle between sun and perigee of satellite used for point positioning and (f) angle between sun and line of nodes of satellite.

meters. The effects are largest in the vicinity of the geomagnetic equator, where the electron density is highest. The effects on the computed latitude and longitude of sites are shown in Fig. 13(b). The magnitude of the effects on the computed horizontal coordinates are similar to those for height, but the direction is fairly random. For polar-orbiting satellites, the effect on longitude probably averages out over a year since ionospheric disturbances are probably equally likely east and west of any given site. However, a north–south gradient in the ionosphere could produce a systematic error in the computed latitude of some sites.

C. Orbit Errors

Although the foregoing measurement and environmental errors could contribute to uncertainties in computed satellite orbits, the large number of observations from the globally distributed stations tend to make these errors negligible when considering the dynamic constraints inherent in the orbit

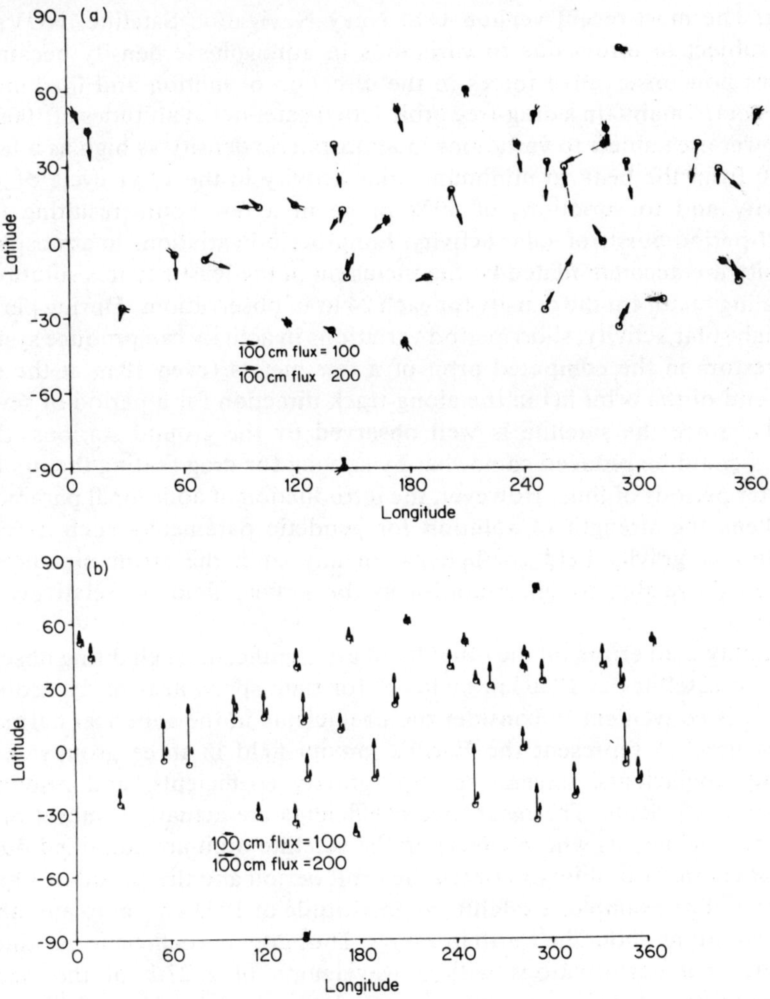

Fig. 13. Simulated effects of third-order ionospheric effect on horizontal coordinates of sites computed from six days of Doppler observations of satellite 77 on days 210–215 at solar flux levels of 100 and 200, (a) horizontal coordinates, and (b) height.

computation. Similarly, errors in tidal-generated potential and in the solar radiation model could produce significant errors in satellite orbits computed for long periods of time, but are largely absorbed in the orbit constants when orbit fits are made to time spans of data of 48-hr or shorter.

For most satellites, variations in atmospheric density cause significant errors in the computed orbits due to unmodeled variations in the force of drag. The most recent version U.S. Navy Navigation Satellite, NOVA, is not subject to errors due to variations in atmospheric density because it senses nonconservative forces in the direction of motion and fires micro-thrusters to maintain a drag-free orbit. Other satellites at altitudes of 1000 km or lower are subject to variations in atmospheric density as high as a factor of 10 from the peak to minimum solar activity in the 13-yr cycle of such activity and to variations of 50% or so in a few hours resulting from short-period bursts of solar activity. Long-period variations in atmospheric density are accommodated by the inclusion in the least-squares solution of a scaling factor for the density for each 24 hr of observations. During periods of high solar activity, short-period variations in activity can produce systematic errors in the computed orbit of a few meters (even 10 m at the start and end of the orbit fit) in the along-track direction for a period of several hours. Since the satellite is well observed by the ground stations, these errors could be reduced somewhat by solving for drag-scaling factors over shorter periods of time. However, the introduction of additional parameters weakens the strength of solution for geodetic parameters such as polar motion or gravity field coefficients. In any case, the errors significantly exceed those due to uncertainties in the gravity field on relatively few occasions.

Gravity field errors on the other hand are significant when fitting observations of satellites at 1000 km or lower for time spans near or exceeding a day. It is convenient to consider the coefficients in the spherical harmonic series used to represent the Earth's gravity field in three groups: zonal gravity coefficients, tesseral/sectoral gravity coefficients, and resonance gravity coefficients. The resonance coefficients are actually a subset of the tesseral coefficients whose effects on the satellite orbit are amplified due to a near commensurability between the orbit period and the period of Earth's rotation. For example, a satellite at an altitude of 1000 km completes about 27 revolutions about the Earth in 2 days. Thus gravity coefficients producing longitudinal perturbations with a wavelength of 1/27th of the Earth's circumference (coefficients of order 27) have effects on the satellite orbit which repeat every 2 days. Such effects would grow in magnitude indefinitely except that the effects disturb the orbit period removing it for exact commensurability. However, exact commensurability is not necessary for a near-resonance condition to produce significant effects on the satellite orbit. For the 1000-km satellite, 13th- and 14th-order coefficients also produce effects on the satellite which are several hundred meters or more in amplitude with a period of 2 or more days. Such long-period effects are easily measured so that precise values can be obtained for the gravity coefficients producing the effects, provided atmospheric drag variations have not produced similar effects. However, the resonance coefficients also produce short-period effects which are modulated at the resonance period. While these effects are smaller, a dense tracking network is required to measure them. Zonal gravity coefficients produce the largest effects on the satellite motion. The largest effects of even-order zonal coefficients are secular motions of the node and perigee of the satellite while the largest effect of the odd-order coefficients is periodic variation in orbital eccentricity with a period corresponding to perigee motion or integer fractions thereof. However, precise values for these coefficients can be obtained by observing the satellite motion over periods of several weeks so that uncertainties in the coefficients will produce small effects on orbits computed for 1 or 2 days. The most troublesome coefficients are the large group of tesseral and sectoral coefficients. These coefficients have periodic effects with periods which are integer fractions of 24 hr or effects at the orbit period but with amplitudes modulated at periods which are integer fractions of 24 hr. Amplitudes of the effects are 2 km or lower. Because of the large number of coefficients which have effects of many different periods, a large station network is also required to separate these effects. It should be noted that all gravity coefficients produce effects at the orbit period and integer fractions of the orbit period. The largest amplitude effect is 10 km produced by oblateness. But the zonal coefficients corresponding to oblateness are very well known and all other coefficients are three orders of magnitude smaller. Therefore the short-period effects would be expected to be 10 m or smaller so that relatively low precision for the values of the coefficients would reduce errors in the short-period effects to negligible levels. The current status of the knowledge of the Earth's gravity field is such that errors in computed orbits of 1000-km satellites arising from this source are about 1 m. The most troubling aspect of the errors is that those arising from sectoral and tesseral coefficients are geostationary, so that the orbit is biased at a level of nearly a meter over regions of the Earth which are 500-km square (Anderle and Hoskins, 1977).

VI. Evaluation of Systematic Errors

A. Introduction

Until the late 1970s, there were few external standards by which to measure the accuracy of the Doppler satellite system. During this time period, lower bounds for the accuracy were evaluated on the basis of residuals of fit to computed orbits, the consistency of predicted orbits with orbits fit to data during the prediction interval, the repeatability of computed station coordinates and pole positions, and simulations on experiments of the effects of variations in the assumed physical model. Now there are a wide variety of external sources wth accuracies comparable to or better than the Doppler system for use in evaluation. These include terrestrial geodimeter measurements of distance between sites and NAVSTAR global positioning system (GPS) interferometric measurements which are better than satellite Doppler measurements for distances up to a few hundred kilometers, very long baseline interferometric (VLBI) measurements of stellar radio sources which provide distances between sites to better accuracy over any distance, satellite laser measurements which give origin and distances between sites with at least comparable accuracy to the Doppler system, and VLBI, lunar laser and satellite laser measurement of pole position and Earth's rotation which have accuracies comparable to or better than the Doppler system. Other special evaluation techniques are the repeatability of radar altimetry measurements to the ocean surface from orbital positions computed from Doppler data at different times and the difference between Doppler positions of mean sea level and mean sea level determined by various expressions for the gravity field or from altimetric data. Nevertheless, the self-consistency tests are also still used extensively either because of certain deficiencies in the external tests or to isolate the sources of errors.

B. Simulated Orbit Errors

Effects of uncertainties in the gravity field on computed orbits are difficult to evaluate due to the scarcity of independent data sources and the large variations in the magnitude of the effects at different orbital conditions. The previous section describes characteristic periods of effects on the orbit which can be observed for gravity coefficients of a given order. However, once these effects are observed, they can be attributed to a linear combination of coefficients of different degrees at that order. Any set of coefficients satisfying that linear combination will significantly reduce the residuals of fit for that orbital condition, but different sets will produce different effects on computed orbits at any other condition, particularly at a different orbital inclination. For this reason, gravity coefficients are usually computed from observations of satellites at different inclinations. But the variety of error sources in the computations makes it difficult to determine whether enough coefficients are being used to represent real gravity effects or whether an excessive number of coefficients are being computed resulting in an over-fit to the observational data and corresponding excessive errors at other orbital conditions or in geographic regions lacking sites which can observe satellites included in the solution.

The effect of neglecting a coefficient of degree n on a computed orbit can be evaluated because it has been observed that the rms of normalized coefficients of degree n is $10^{-5}/n^2$ (Kaula, 1966; Anderle and Smith, 1967). Generating gravity fields which satisfy this decay law, Malyevac (private communication) performed a Monte Carlo simulation to determine the magnitude of the orbit errors to be expected if the gravity coefficients were truncated at various values of n. The results for a 600-nautical-mile polar satellite are shown in Fig. 14. The errors decay slowly as the number of coefficients considered increases, decreasing from 20 m at 12th-degree truncation to 5 m at 19th-degree truncation. However, an aliasing effect occurs during the solution for the coefficients which causes some of the effects of the truncated coefficients to be absorbed in the values of the coefficients which are computed. Anderle *et al.* (1969) carried out a large-scale single-sample Monte Carlo simulation of the effects of gravity field truncation on the accuracy of computed orbits which showed that the 20 m for the polar-orbiting satellite reduced to 5 m as a result of favorable aliasing,

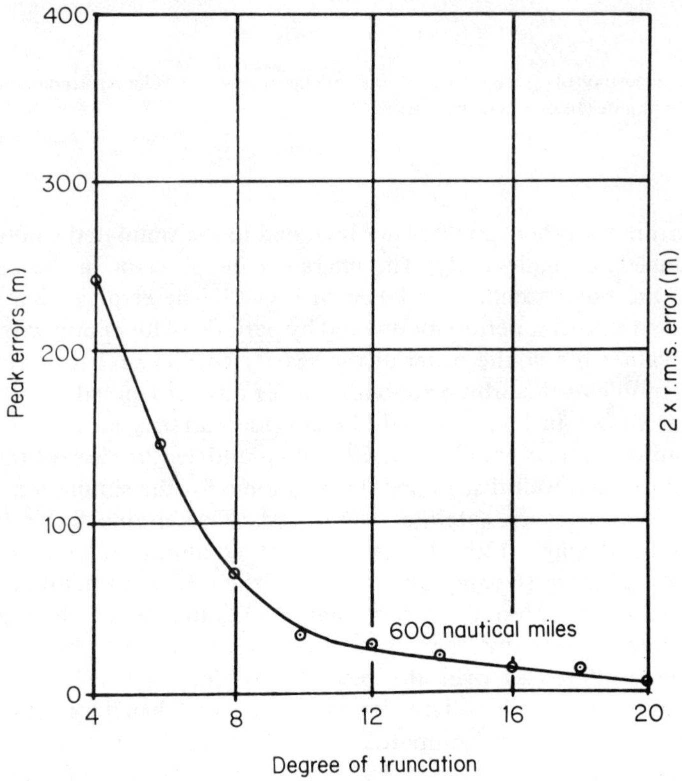

Fig. 14. Effect of truncation of gravity field on computed position of satellite at a height of 600 nautical miles.

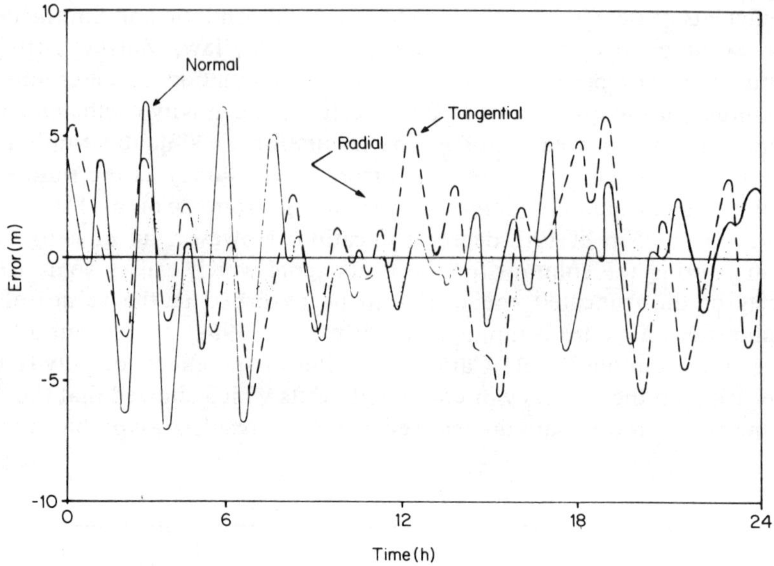

Fig. 15. Simulation of effects of coefficients neglected in geodetic solution on computed position of satellite at a height of 600 nautical miles.

while errors for other satellites not included in the simulated solution were not reduced so significantly. The character of the error in the computed orbit of the polar satellite is shown in Fig. 15. The error is characterized by errors at the orbit period modulated by periods 1440/m min with various integer values for m, the order of the gravity coefficient. The variations in the magnitude of the orbit error with time of day (or equivalent geographic location) shown in Fig. 15 should be considered fairly modest. Such relatively uniform errors are characteristic of an undetermination of the gravity coefficients from well-distributed observations: for the simulation Doppler observations of seven satellites were used to determine fewer than 150 coefficients through 12th degree. Current solutions involve over 1500 coefficients. Figure 16 compares an orbit of the GEOS-3 satellite computed from coefficients determined from alternate fits to a set of observations of the satellite; a large difference occurs in a very localized geographic area and small differences over the rest of the world (Tanenbaum, private communication). Such differences, which occur when large numbers of gravity coefficients are computed from different source data make the characterization of orbit error by a single rms number of questionable value.

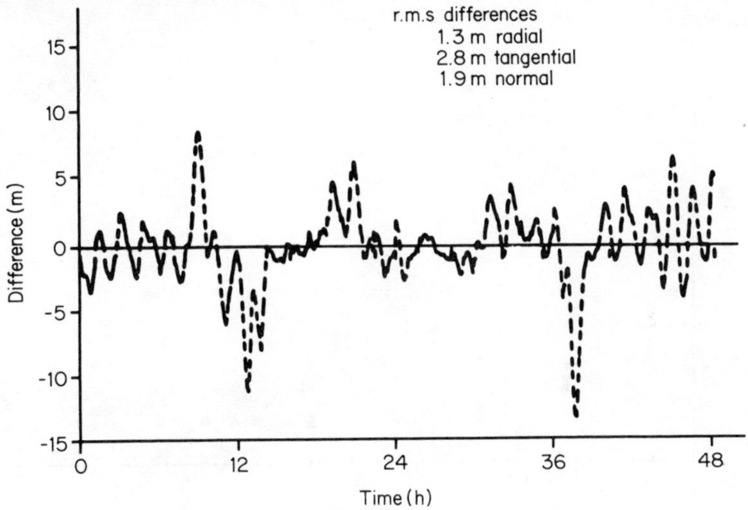

Fig. 16. Tangential orbit differences for geodetic solutions found in GEOS-3 gravity evaluations.

C. Residuals of Fit

As implied by the foregoing, residuals of fit give an optimistic view of orbit accuracy for geographic regions lacking observing sites. Even in regions where the satellite is observed, the orbit error may be partially compensated for by errors in the station coordinates which must be obtained from the observational data. Nevertheless, the residuals provide a lower bound for the orbit error provided that the effects of instrument errors are not a major constituent of the residuals. In order to de-emphasize the effects of instrument error, the residual error can be weighted inversely by the square of the standard error of the observation corresponding to the random error of observation and the root of the weighted squares of the observation error (rws) computed rather than a simple root of mean sums of squares (rms) of observation error. A plot of such rws errors in the range to the Nova satellite is given in Fig. 17 as a function of the maximum elevation angle in the pass. The larger errors at low elevation angle are due to larger effects of instrument errors at lower elevation angles so that the 1-m error at elevation angles above 30° is more representative of the orbit error, subject to the qualifications given earlier. This error level is close to that expected from oscillator instability so that the actual orbit errors could be somewhat smaller. However, results under the best conditions and expected effects of uncertainties in the gravity field make it highly improbable that the accuracy

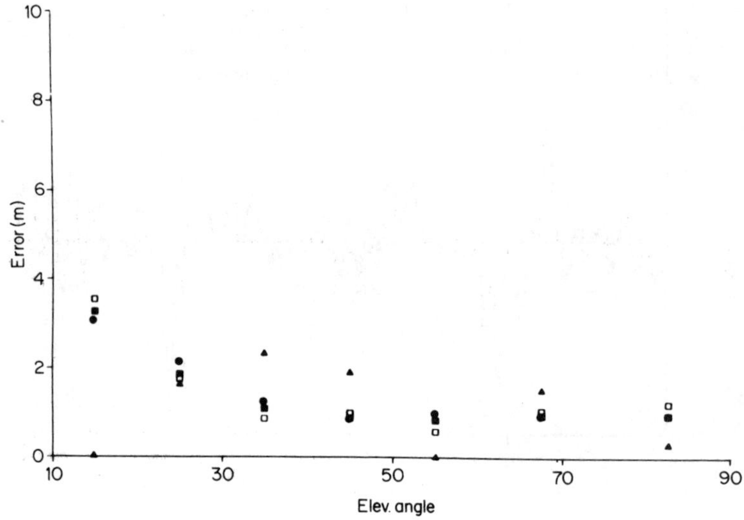

Fig. 17. Root weighted square range navigation errors for NOVA satellite. Receiver comparisons include ■ 378 passes, all stations; □ 122 TRANET-2 passes; ● 208 GEOCEIVER passes and ▲ 33 OPNET passes.

could be significantly better than 1 m, while the potential for errors in unobserved regions and for compensation between effects of errors in station coordinates and effects of errors in the gravity coefficients make it likely that the errors are somewhat greater.

D. Systematic Errors in the Doppler System

In this section, the errors in the coordinate system used to establish geodetic positions will be summarized. Systematic errors in station height of the order of 2 m during periods of high solar activity due to the neglect of higher-order ionospheric refraction effects on signal propagation were described in the section on environmental error sources, and systematic errors in the orbit which are correlated over geographic distances of 500 km were mentioned in the section on orbit errors that arise from uncertainties in the gravity coefficients. The principal ionospheric errors are long period due to the 13-yr period in solar activity and will be discussed further in the section on physical changes in the coordinate system that arises from plate tectonic motions. The correlated orbit error due to gravity effects can be considered a distortion of the coordinate system whose magnitude is not precisely known, but is believed to be about a half meter. The systematic errors in the scale, origin and orientation of the NWL 9 Doppler system will be discussed next. The NWL-9D system is the system in which the precise ephemeris has been computed from 1972 to 1984. The WGS-72 system which was derived from this system and is in extensive use is defined as longitude λ_{WGS}, $\lambda_{9D}+0.26''$; latitude ϕ_{WGS}, $\phi_{9D}+0.26$ and radius r_{WGS}, $r_{9D}-5.27$ m. The WGS-84 system which will replace these systems is related to the NWL-9D system by the biases discussed below.

1. Scale of the Doppler System

When the NWL-9 system was developed, the value used for the central gravitational constant, *GM*, was 398,601 km^3/s^2, the displacement of the satellite antenna from the center of mass of the satellite was neglected, the velocity of light used in the computations was 299,792.5 km/s. Anderle (1980) reported that the effects of the first two assumptions were to yield coordinates which were 1.7 and 0.7 m too high, respectively (with respect to a GM of 398,600.5), and Leroy (1982) subsequently refined these values and computed a correction to a velocity of light of 299,792.465 km/s, obtaining a net correction of 2.97 m rather than Anderle's 2.4 m. The validity of this correction may be evaluated on specific base lines by comparing Doppler derived chord distances with (1) very long base line interferometric derived chord distances, (2) satellite laser derived chord distances (Boucher and Feissel, 1984; Hothem *et al.*, 1982) and (3) terrestrial surveyed distances. Many more sites may be used in the evaluation by comparing equipotential surfaces fit through mean sea level points derived at Doppler sites with a corresponding fit derived for laser sites (Grappo, 1980) although the equipotential surfaces used for such fits lack high-frequency variations which introduces an error in the comparison. A fifth method also involves fitting an equipotential surface to mean sea level derived at Doppler sites and comparing the surface with the geoid defined by satellite altimetry in ocean areas (West, 1982). Doppler-derived heights of mean sea level may also be compared with satellite altimetric derived heights at Doppler sites close enough to the ocean to allow extrapolation of the altimetric geoid (Neil Simmons, Defense Mapping Agency Aerospace Center, private communication). However, there are uncertainties in the magnitude of systematic errors in the calibration sources. Finally, a preliminary solution of satellite and surface gravity normal equations which were included in the DOD WGS-84 solution yielded a best estimate of the scale correction of −1.9 m along with a value of GM which was 398,600.66 km^3/s^2. A likely cause of the discrepancy in GM is the neglected higher-order ionospheric effects which probably also affected the solution for scale. A summary of these corrections to the NWL-9D scale is theoretical, −2.97 m or −0.47 ppm; VLBI, −3.2 m to −4.5 m or −0.5 to −0.7 ppm, Doppler versus laser sites,

−3.5 to 3.8 m or −0.55 to −0.6 ppm and new general geodetic solution, −1.9 m or −0.3 ppm. The neglect of higher-order ionospheric in the theoretical correction can be expected to result in a scale which is too large. Considering the uncertainty in each of these evaluations, the best estimate of the NWL-9D scale correction is considered to be −3.8 m or −0.6 ppm.

2. Origin of the Coordinate System

Many of the data sources discussed above may also be used to evaluate any discrepancy in the origin of the Doppler system from the center of mass of the Earth. The discrepancy in the plane of the Earth's equator (i.e. in the x and y directions) is within the scatter of 0.5–1 m in the results. However, the results generally indicate the Doppler origin lies above the plane of the equator by about 4 m (Hothem *et al.*, 1982; West, 1982; Marsh and Williamson, 1980; Boucher and Feissel, 1984). (In interpreting this Z bias by using altimetric data with Doppler-derived ephemerides, experiments have shown that a 4.5-m Z bias in the Doppler system will result in computed geoid heights which are too low by 2.25 sin ϕm, where ϕ is the latitude of the evaluation point.) Laubscher (1983) has evaluated the effect of the neglect of a relativistic correction in the processing of the Doppler data on the computed Z coordinate. He finds the effect to be periodic with the argument of perigee of the satellite reaching a peak of 5 m for the Navy navigation satellite having maximum eccentricity. Since a variety of satellites and arguments of perigee were used in deriving the Doppler system, it is unlikely that the lack of a relativistic correction is a significant factor in causing the bias. The bias is apparently real and unexplained: it does not exist in general geodetic solutions processed recently and is therefore likely to have been a processing anomaly in the 1972 solution. Doppler data currently yield a correction of 5.7 m. A summary of these comparisons for the Z bias is:

Laser	4.5 m
Gravimetric geoid	2.7–4.8 m
Altimetric geoid	4.1–4.5 m[a]
General geodetic solution	5.7 m

[a] Obtained by scaling the −2.47 m–−2.67 m discrepancy in the geoid by the experimental ratio of 5 m for station coordinate offset to 3 m for corresponding geoid offset.

The best estimate is that 4.5 m should be added to the NWL-9D coordination.

3. Orientation of the Coordinate System

Comparison of rotations about axes in the plane of the Earth's equator show no significant differences between the Doppler system and either the VLBI or the satellite-laser system (Hothem *et al.*, 1982; Boucher and Feissel, 1984). For rotations about the Z axis, only the VLBI and terrestrial gravimetric/astrogeodetic data can be used for evaluation. (The direction is indeterminate in analysis of satellite-laser or satellite-Doppler data.) The VLBI data indicates 0.80 arcsec should be added to the longitudes of Doppler sites to bring them in conformance with the BIH mean meridian, while gravimetric/astrogeodetic data in North America indicates only about 0.5 arcsec should be added by theodolites used in astronomic longitude determinations with a mobile Danjon astrolab similar to the type used extensively in establishing the BIH conventional longitude. The tests revealed a 0.2 arcsec bias between the instruments in a direction which would decrease the NWL-9D discrepancy from the terrestrial longitude. A summary of these results is

VLBI	+0.80 arcsec
NAD	0.50 arcsec
ED	0.50 arcsec
NAD or ED Corrected	0.30 arcsec

The best estimate is taken to be 0.50 arcsec.

4. Summary of NWL-9D Coordinate System Corrections

(i) Decrease scale by 0.6 ppm or station radii by 3.8 m.
(ii) Add 4.5 m to Z coordinates.
(iii) Add 0.5 arcsec to station longitudes.

The corrected system would be uncertain in about 0.1 ppm or 0.6 m in scale, 1 m in origin, 0.3 arcsec in longitude and 0.02 arcsec in pole position.

VII. Geodetic Surveys

The Doppler satellite system has had its greatest impact in geodetic surveys to strengthen, extend and connect terrestrial systems in an Earth-centered astronomically oriented coordinate system. It has been used by most nations on all continents ranging from the most developed and accessible areas to isolated islands, mountainous, jungle, tropical, and sub-zero environments.

Major programs include the multi-national program to develop a New North American Datum and surveys in South America, the Southwest Pacific, India and the People's Republic of China. Since a detailed accounting of these and other programs would become repetitive and statistical, just two examples of programs operating under different conditions will be described. One is the European Doppler Campaign (EDOC-2) which involved many countries already connected in a relatively modern terrestrial network, the European Datum 1950. The other is the African Doppler Survey, ADOS, in which 55 African countries are establishing and connecting terrestrial networks with assistance from a number of European countries, Japan and the United States.

The EDOC-2 campaign involves the 37 Doppler sites in 15 countries shown in Fig. 18. Data from four of these sites are currently used to compute

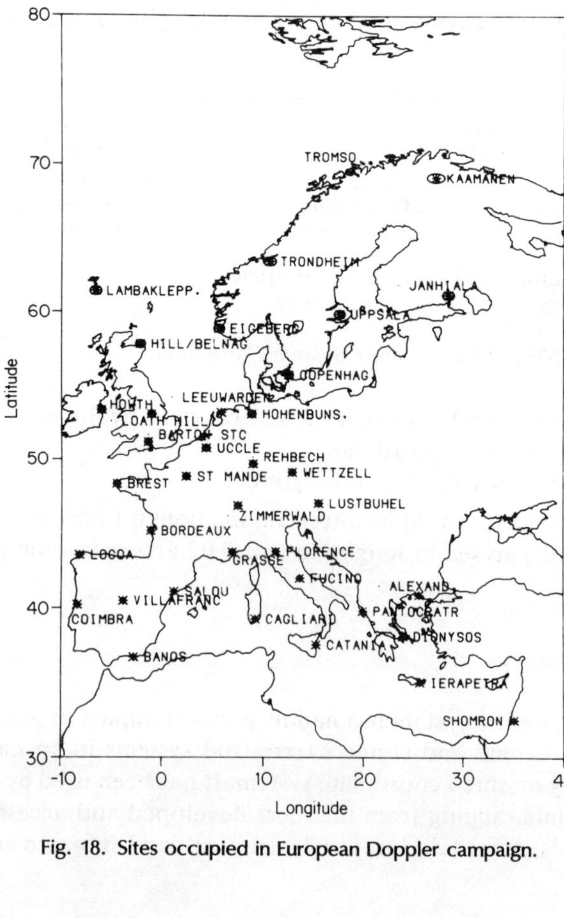

Fig. 18. Sites occupied in European Doppler campaign.

precise ephemerides, although at the time of the campaign the data from Wettzell, FRG were not yet in regular use for this purpose. The other three semi-permanent sites are Barton Stacey in the United Kingdom, Uccle in Belgium and Florence in Italy. Cagliari in Italy observed with similar equipment while the remaining 32 sites observed with portable receivers. The analysis of the results conducted independently by three computing centers is described by Boucher *et al.* (1979). They found distortions in the European Datum of 10 m in the Iberian Peninsula, Scandanavia and elsewhere. The results obtained by different computational methods agreed to 10 cm to 1 m, with the larger discrepancies occurring for sites with the smallest amount or poorest quality data. The accuracy to which Doppler positions can be achieved routinely, at least over relatively small distance (700 km), was reported by Seeger *et al.* (1979) in a report on the German-Austrian Campaign (DODOC). For 15 sites in or near the Federal Republic of Germany, Doppler positions agreed with European Datum 1977 coordinates to 20–40 cm after transformation for differences in the reference systems.

The African Doppler Survey was undertaken under the auspices of the International Association for Geodesy Commission on Geodesy for Africa and of the International Coordination of Space Techniques for Geodesy and Geodynamics (CSTG), a commission of the International Association of Geodesy and a sub-commission of the Committee for Space Research. The purposes of the program (Mueller, 1983) were

(i) to provide zero-order control for future geodetic networks which could support mapping,
(ii) provide control for datum unification and strengthening,
(iii) provide an improved geoid for Africa, and
(iv) provide coastal calibration points for satellite altimetry.

Over 1000 geodetic positions established by the Doppler satellite system will be available to meet these objectives. The number of positions available or to be established in each country is shown in Fig. 19. Depending on the wishes and requirements of each country, equipment, observing teams or computational facilities are made available from a "sister" country having the facility and selected by the host African country. This campaign has been an example of remarkable cooperation among countries in diverse political, economic and technical conditions; its success would not have been possible without the determined efforts of host and sister countries to apply scientific advances to provide a technical base which contributes to the improvement of the economic conditions of the populace, or without the indefatigable efforts of Professor Ivan Mueller in foreseeing and resolving problems during his term as president of CSTG.

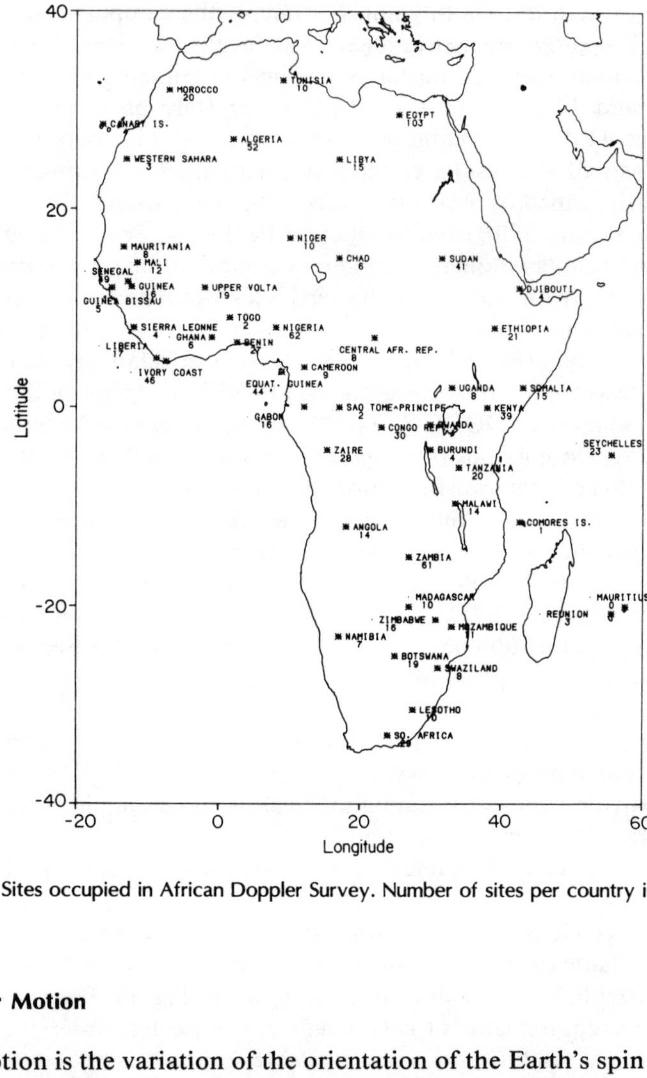

Fig. 19. Sites occupied in African Doppler Survey. Number of sites per country indicated.

VIII. Polar Motion

Polar motion is the variation of the orientation of the Earth's spin axis with respect to the crust. The two primary causes of the motion are the natural Eulerian motion which occurs for a body which is initially set to spin about an axis other than its principal moment of inertia (Vincente, 1970) and the annual motion induced by atmospheric motion and changes in vegetation and snowfall (Jeffreys, 1970). The Eulerian motion, predicted by Chandler (and hence called Chandlerian motion), would be 315 days for a solid earth but is calculated to be 400 days for an elastic earth with liquid core, in reasonable agreement with observations. The Chandlerian and annual motions are each 5–10 m in amplitude at the surface of the Earth, and the amplitudes may add or cancel depending on the phase of the effects. The motions, which have been studied since 1900, are of geophysical interest since:

(i) the damping of the Chandlerian motion is a measure of the elasticity of the Earth,
(ii) excitation of the Chandlerian motion may be caused by earthquakes or other geophysical phenomena,
(iii) drift in the mean pole of rotation may be caused by crustal rebound from glaciation, or
(iv) variations in the linear drift could signify variations in such geophysical phenomena and
(v) disturbance in the polar motion may be associated with earthquakes.

Theoretical studies in these areas have been made by considering various models for the Earth's structure and magnetic or mechanical coupling between the core and mantle of the Earth, and observational data have been compared with the theoretical results. However, the random and systematic errors in the astronomical observations, which are the only ones with a long enough time history to be used for most of these studies, are generally too large to give conclusive information about the validity of the theories. A recent and thorough discussion of the theories and observations is given by Lambeck (1980).

Regular computations of the polar motion from Doppler satellite observations began in 1970 (Anderle and Beuglass, 1970). The initial results were based on residuals of fit to orbits based on a pole predicted from astronomical data. Since August 1971, components of pole position have been derived in a simultaneous fit to observations along with orbit parameters and other parameters of the solution (see Section IV A), and occasional values of pole position have been computed as far back as 1964 in the course of recomputations made for other purposes. Pole positions are normally computed as part of a least squares solution to Doppler observations of a satellite made at 20–40 sites over a period of 2 days. Other parameters of the solution include orbit constants and one or two atmospheric drag-scaling factors and applicable to the 2-day period and a frequency correction and tropospheric refraction scaling parameter for each satellite pass over each station. The pole position results for two or more satellites during 5-day intervals are averaged for comparison with astronomic results.

Over the course of time, improvements have been made to the ground equipment and computational model, and the locations of receiving stations have been changed. Oesterwinter (1979) summarizes the data available from

1964 to 1977 and analyzes results obtained with that data set. While changes in the computational model and in the location of observing sites can introduce systematic changes in pole position results, the effects of such changes have been relatively small since 1972. However, there have been changes which affected the precision of the results (Colquitt *et al.*, 1984). The rise in the 13-yr cycle in solar activity to its peak level in 1981 caused large rapid variations in atmospheric drag which were not accurately modeled in the orbit computations and therefore resulted in larger random errors in the computed pole positions during this time period. On the other hand, results for the NOVA satellite, which senses and compensates for nonconservative forces on the satellite motion, have had significantly smaller random errors. Recent results may also be improved due to improved oscillators in the ground stations, higher signal-to-noise ratios for the NOVA satellites because of new antennas at the ground site, computation of results for more satellites than the number used early in the program, and compensation for biases between results for different satellites; future results may be further improved with the introduction of improved force models.

The most precise measurements of pole position are now being made by very long base line interferometric (VLBI) observations of radio stars. The difference between the x component of position observed by the VLBI technique and the Doppler method is shown in Fig. 20 for the period 1981–1984. The 56-cm bias between the two results is due to arbitrary origins

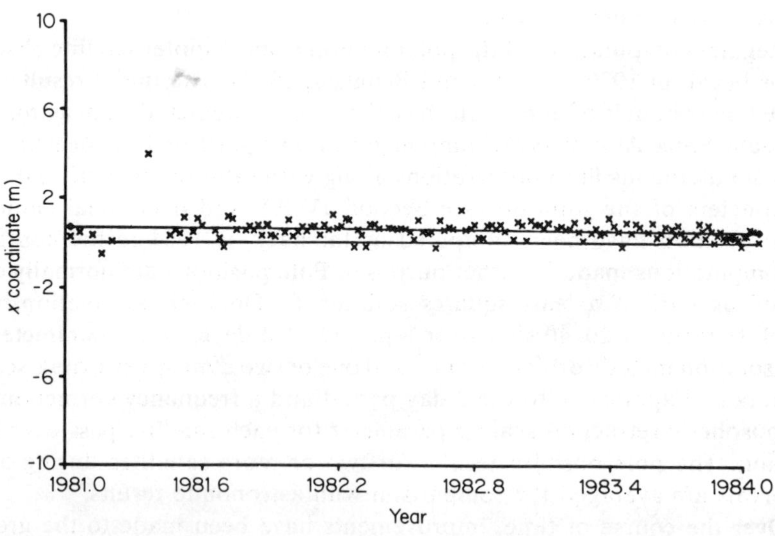

Fig. 20. Difference in computed x component of pole position between Doppler and interferometry data.

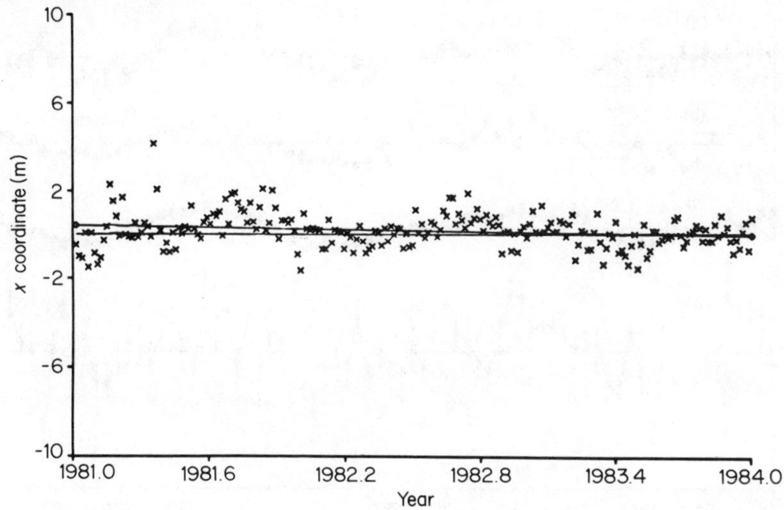

Fig. 21. Difference in computed *x* component of pole position between Doppler and BIH astrometric data.

for each system, and the scatter in the early results is partly due to sparse, less accurate VLBI results and partly due to the larger atmospheric drag effects on the Doppler results. The results cluster very closely at the end of the interval. On the other hand, the astronomic results, which were the only source of pole position data prior to the introduction of the Doppler system in 1970, show large systematic variations and larger random errors in comparison with the Doppler results (Fig. 21). One method of evaluating the random error of observation is the computation of pole-position residuals with respect to a solution for bias, annual and semiannual terms best fitting the data. Such results for the x and y coordinates are given in Figs. 22 and 23, respectively, for four currently available sources of pole position data: astronomic (labelled BIH ASTR), satellite laser (LAGEOS), VLBI, and Doppler. (Dopplermixed refers to the average value of pole positions computed from data for different satellites after adjustment for biases between satellites.) The Doppler results are still competitive in precision with those from more recent techniques, and no disturbances in the results are evident which are common to any two data sources other than long-period variations. Since the Doppler results are the only precise ones which are available for a 12-yr period, which is a useful period for separation of annual and Chandler effects, it would be timely to refine the analysis for the Chandler period (Oesterwinter, 1979) and for secular drift of the pole (Anderle, 1976a).

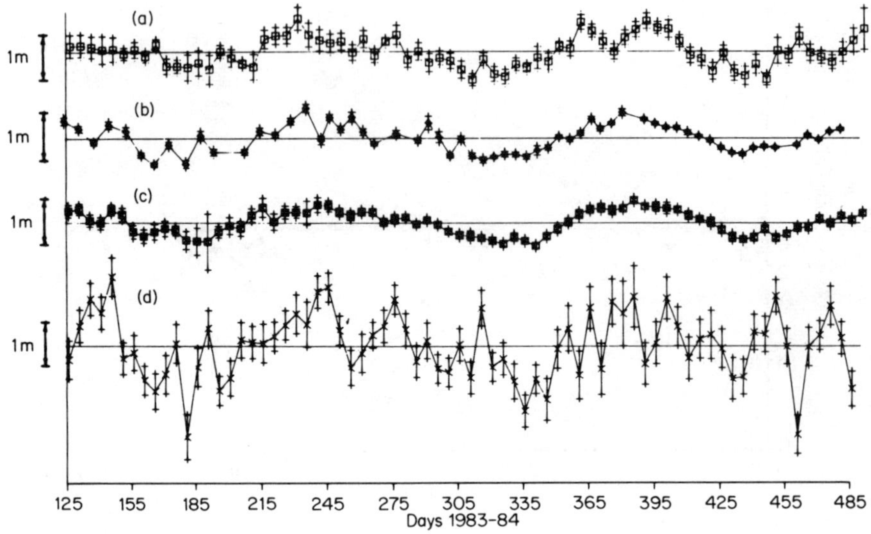

Fig. 22. Residuals of fit of x component of pole position from five parameters fit to the (a) Doppler, rms 28 cm ($0.009''$); (b) VLBI, rms, 26 cm ($0.008''$); (c) LAGEOS, rms 23 cm ($0.007''$); and (d) BIHASTR, rms, 63 cm ($0.020''$).

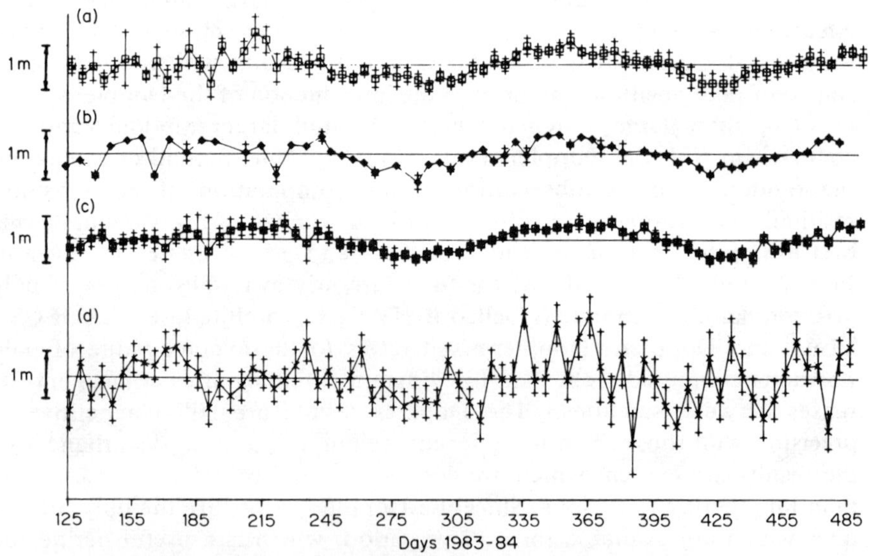

Fig. 23. Residuals of fit of y component of pole position from five parameters fit to (a) Doppler, rms 23 cm ($0.007''$); (b) VLBI, rms 23 cm ($0.007''$); (c) LAGEOS, rms 20 cm ($0.006''$); and (d) BIHASTR, rms 50 cm ($0.016''$).

IX. Earth's Rotation

The rate of Earth's rotation is affected by most of the geophysical effects mentioned in the preceding section and discussed equally fully by Lambeck (1980). The variations are dominated by annual atmospheric variations and long-period changes due to Earth–moon tidal coupling. Although the rate of Earth's rotation could be treated as a parameter of fit to Doppler observations, it has not been the practice to do so because the errors in the values provided by astronomic techniques have been adequate over the 2-day span of data used in the orbit computation. Nevertheless it is possible to compute a correction to the assumed value of Earth's rotation since errors in the value will produce an error in the longitude of the node of the satellite-orbit plane which can be evaluated from the orbit fit to a succeeding span of data. Such errors were computed for the period 1969–1972 (Buonaguro, 1972), but were found to be due to model errors in the computation, particularly the lack of correction for the gravitational effects of ocean tides. The lack of such a correction and in fact even the errors in knowledge of the tidal effects and other force model errors precludes useful determination of long-period changes in Earth's rotation from Doppler observations. However, meaningful results for changes in Earth's rotation for periods of 10–100 days can be made from Doppler data.

It is convenient to accumulate the corrections to Earth's rotation rate since the result is a correction to the value (UT1–UTC) which is directly measured by astronomical techniques. In order to remove long-period systematic errors from the Doppler results, a six-parameter fit is made to the data to remove bias, rate, annual and semiannual variations. Residuals of fit for 1981 are shown in Fig. 24 for two Doppler satellites, for BIH astronomic data, and for LAGEOS laser satellite observations. Variations in (UT1–UTC) of 4 ms for periods of the order of 90 days are apparent in each of the data sources, although any one (most often one of the Doppler curves) shows occasional disagreement. Such variations given for Doppler data in 1975–1979 (Anderle, 1980) and 1980–1981 (Colquitt *et al.*, 1984) have been closely correlated with atmospheric variations in studies of such effects from VLBI and satellite-laser data. The high precision of VLBI data makes the Doppler measurements of UT1 of little significance.

X. Plate Tectonics

One of the challenging objectives of space technology is the measurement of the motions of the Earth tectonic plates for comparison with results of geophysical investigations (Minster and Jordan, 1978). Since the geophysical

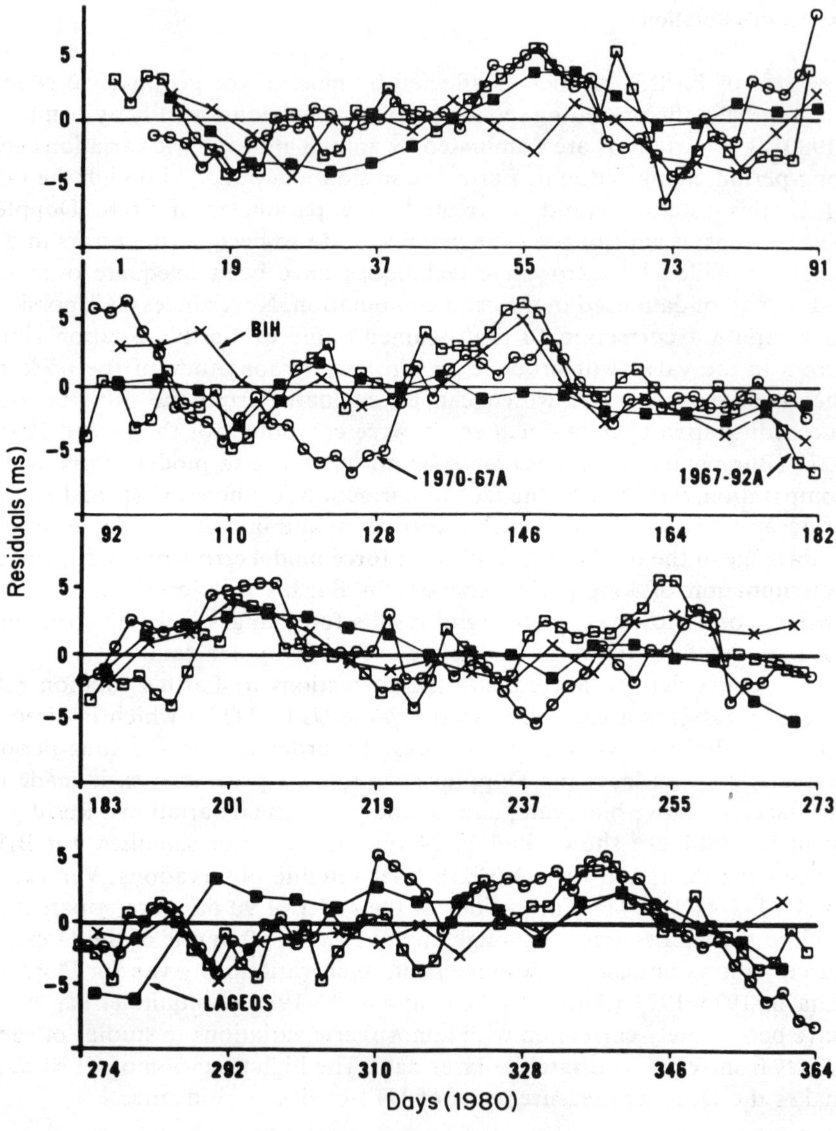

Fig. 24. UTI-UTC residuals from 6 parameters fit to Doppler, BIH astrometric and LAGEOS data.

and satellite measurements pertain to different time periods, and satellite measurements can be made at various locations on the plates, they should provide new insights into the mechanism of the processes involved in the motions. One of the earliest attempts to apply Doppler satellite measurements to the problem was made in 1968 (abstract in Anderle, 1968; text reprinted in Anderle, 1978). The results gave fairly reasonable results for the northward motion of Australia, but were dominated by unreasonably large motions of other sites, including inconsistencies for sites on the same plate. The early results, made from measurements intended to provide 10–20 m accuracy in geodetic positions, were subject to uncertainties in the surveys when antennas were replaced at sites from time to time. Casual survey measurements can be made for a 10-m system, but can produce disastrous errors in measurements requiring 0.1–1-cm/yr precision. Even recent antenna changes can produce results which are suspect since the effective ground plane may be altered changing the effective electrical center for the measurements on the 75-cm wavelength signal. Nevertheless, recent results are tantalizingly close to the accuracy required as will be shown. Although laser and VLBI systems have the potential for significantly higher precision in such measurements, the portability and all-weather capability of the Doppler satellite system allows measurements to be made at a lower cost than may be feasible with the more precise systems.

The source data for the computation of plate motions from Doppler observations are obtained using ephemerides computed at 2-day intervals for use in determining the geodetic positions of mobile equipment, and.the observations made at the base stations which were used to compute the ephemeris. The position of each base station is computed from the same data used in the orbit determination, except that a span of 5 days of data is used. (For computational efficiency, a longer span of 13 days is used when reprocessing older data.) The coordinates obtained in these solutions are plotted versus time since the repeatability of the computed positions are a measure of the stability of the coordinate system in which the positions of neighboring sites at which portable equipment has been deployed.

During the 10-yr period 1973–1983 relatively few changes were made in computational model or station array used in the orbit computation. The changes made in the computation include a revision of the 27-order resonance coefficients and a change in the weighting of the observations to better represent the effect of orbit error on the residuals of observation. The station array used in the ephemeris computation sometimes included portable receivers deployed periodically and antennas were changed at the following sites during the period:

Anchorage	1976
Las Cruces	1976
Ottawa	1976
Thule	1977
Calgary	1978
Maine	1979

Although the surveyed distances between the new and old antenna locations were available, the displacement was treated as an unknown in a least squares fit for station motion because of the possibility that errors might be introduced due to changes in type of antenna, rearrangement of antenna components or changes in the effective ground plane at the new physical locations. Three additional sites (Minnesota, California and Hawaii) provided data from more than one antenna during the period considered, but the bulk of the observations at each site was made using the same antenna. The locations of sites in Japan and England were also changed, but only data from the new sites were used in the computations.

Table II shows the rates of change of the coordinates of 22 sites on 8 tectonic plates based on Doppler observations made during the period 1973–mid 1983. Only the data from one Navy navigation satellite was used in the computations as shown below:

Time period	Satellite
Before 1974.24	1967–92A
1974.24–1979.88	1970–67A
1979.88–1980.30	1967–92A
After 1980.30	1970–67A

(It is not expected that the use of a different satellite would produce different computed relative position changes for the stations.) Since there are a number of inconsistencies in the horizontal rates within the same tectonic plates and unreasonably large rates of change in the heights of most of the stations, an additional parameter was introduced into the least-squares solution for each component of station position and rate of change of position in an attempt to absorb biases caused by the neglect of third-order ionospheric effects on the Doppler observations. The parameter considered in the analysis is the critical frequency, *fc*, which is the minimum frequency which will penetrate the ionosphere at a given time and location. Figure 25 shows contour maps of the critical frequency for a selected time. The values are highest along the geomagnetic equator and at the longitude of the sun. The critical frequency was calculated for each station considered in this report at 5-day intervals. Two values were computed for each day, one when the station longitude was in the plane of the satellite orbit when the satellite was traveling south to north, and one approximately 12 hours earlier or later when the satellite was traveling north to south. The values obtained

TABLE II. Doppler station motions[a]

Site	Rates (cm/yr)			Standard Error (cm/yr)		
	Long.	Lat.	Ht.	Long.	Lat.	Ht.
North American						
Las Cruces	−7.1	3.9	14.5	2.0	1.4	1.9
Austin	−5.1	1.5	1.7	2.4	2.0	2.3
Maine	−12.0	6.5	6.0	3.0	2.1	2.7
Minnesota	−6.5	2.0	−4.0	1.7	1.0	1.4
Ottawa	1.6	−0.4	6.1	3.2	2.2	2.7
Calgary	−18.0	6.5	12.4	6.2	4.8	5.5
Anchorage	0.9	−1.1	5.0	1.8	1.3	2.0
Thule	−0.5	−7.1	−6.0	5.2	5.6	6.5
European						
Uccle	4.4	1.1	9.4	1.3	1.0	1.1
Florence	7.0	0.8	9.6	2.4	2.2	1.8
England	3.5	−1.8	15.3	2.5	1.7	1.9
Mizusawa	4.4	−5.8	8.4	1.9	1.7	2.0
African						
Seychelles	−1.1	−4.7	20.0	1.9	1.6	2.1
Pretoria	−0.5	6.8	8.9	1.1	1.0	1.0
Pacific						
California	−5.9	8.2	1.6	1.2	0.9	1.2
Hawaii	−10.5	0.9	20.3	2.0	1.4	1.7
Samoa	−11.0	2.7	13.9	2.5	1.9	2.0
Philippines						
San Miguel	3.5	0.5	23.2	1.6	1.7	1.8
Guam	−5.6	0.3	21.4	2.1	1.7	2.0
Souch America						
Brazil	−2.0	4.3	8.7	2.5	1.7	1.7
Antarctic						
McMurdo	−2.9	8.8	7.9	4.1	4.0	4.0
Australia						
Australia	−3.8	16.5	5.5	1.1	1.0	1.0

[a] No refraction parameter.

for fc^2 for three of the sites are shown by the symbols (×) and (+) in Fig. 26 while the sums of squares of the values are connected through the symbol Δ. Figure 27 shows the variations in the heights of these stations which appear to correlate quite well with the variations in the critical frequency.

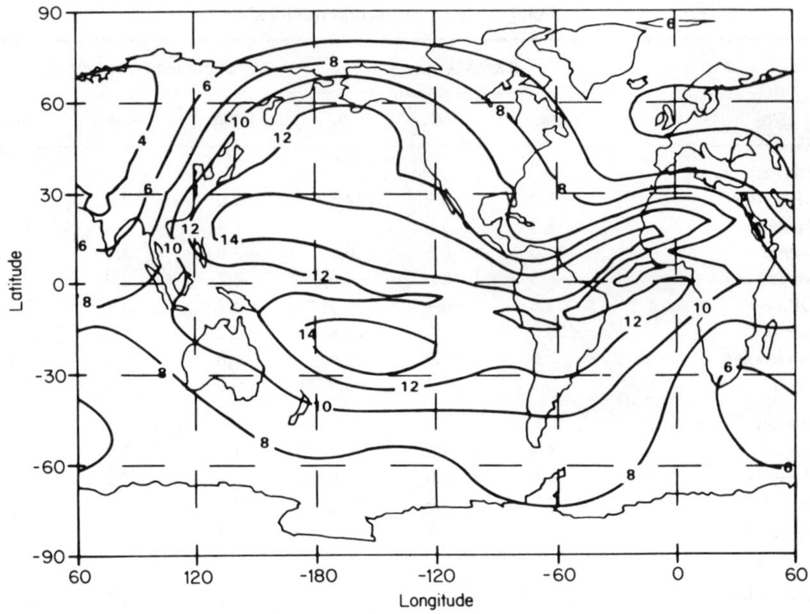

Fig. 25. Critical frequency for sunspot number of 160.

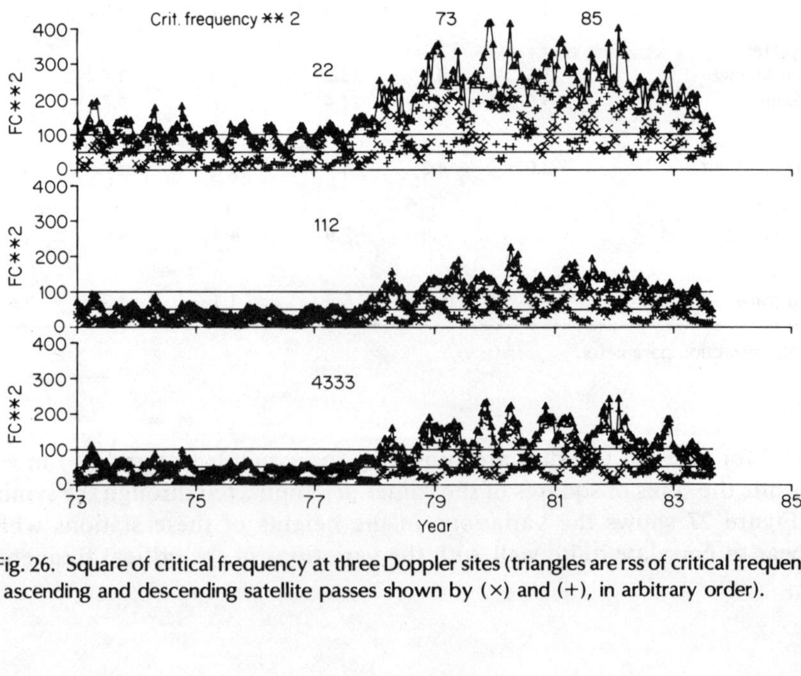

Fig. 26. Square of critical frequency at three Doppler sites (triangles are rss of critical frequencies on ascending and descending satellite passes shown by (×) and (+), in arbitrary order).

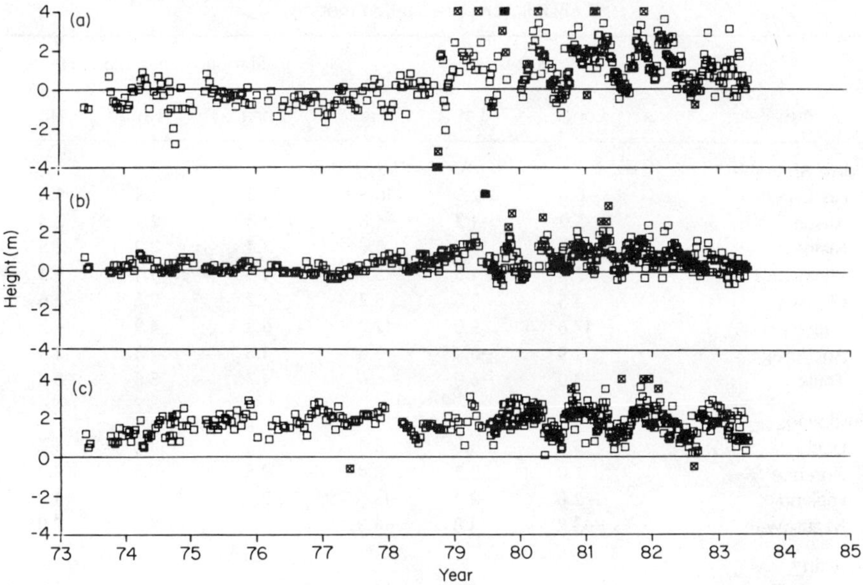

Fig. 27. Computed heights of three Doppler stations; (a) Philippines, (b) Australia and (c) California.

Various functional forms were considered for the critical frequency parameter without a clear preference emerging for any tested. For the results given here, departures above a critical frequency of 5 MHz were squared and added for the two directions of satellite motion and a coefficient of the resulting parameter was included in the least-squares solution for each component of position and rate of change of position. Independent parameters were used for each component since the station height variations would depend primarily on the magnitude of the critical frequency while the latitude and longitude variations would depend primarily on the north–south and east–west gradients of the critical frequency, respectively. There might be some hope that the coefficient for the height variations would be the same for different stations, if the functional form were chosen optimally; however, there would be no reason to expect that the coefficients for horizontal motion would be similar for different sites since differences in horizontal gradients were not considered in the analysis.

The station motions obtained by using the critical frequency parameter are shown in Table III. The station height variations are generally improved, although several still show variations significant, but probably unrealistic, compared to the standard errors of the solution. Only a few horizontal variations showed marked changes: the latitude rate for Mizusawa is in

TABLE III. Doppler station motions[a]

Site	Rates (cm/yr)			Standard Error (cm/yr)		
	Long.	Lat.	Ht.	Long.	Lat.	Ht.
North American						
Las Cruces	−6.6	2.5	10.4	2.0	1.4	1.7
Austin	−5.2	−1.2	−4.8	2.8	2.3	2.5
Maine	−13.0	6.7	8.6	3.1	2.2	2.5
Minnesota	−7.3	1.5	−5.4	1.8	1.1	1.4
Ottawa	1.5	−0.6	5.2	3.2	2.2	2.6
Calgary	−17.8	5.0	12.7	6.2	4.9	4.7
Anchorage	1.9	−0.1	3.6	1.8	1.3	1.9
Thule	−3.0	−6.0	−6.0	4.7	5.6	6.8
European						
Uccle	5.3	1.9	8.1	1.3	1.0	1.2
Florence	7.0	0.2	8.0	2.8	2.5	2.1
England	2.8	−2.1	13.5	2.6	1.8	1.9
Mizusawa	3.8	1.8	−1.9	2.2	1.8	2.0
African						
Seychelles	−1.0	−2.4	12.6	2.4	2.0	2.6
Pretoria	1.2	7.0	6.0	2.4	2.0	2.6
Pacific						
California	−5.5	8.3	−3.6	1.3	0.9	1.1
Hawaii	−12.1	6.8	7.9	2.3	1.5	1.4
Samoa	−2.0	10.2	6.6	3.1	2.3	2.4
Philippines						
San Miguel	0.4	−2.5	10.9	2.1	2.1	2.0
Guam	−6.3	0.2	4.7	2.6	2.0	2.1
South America						
Brazil	−5.7	2.4	3.5	3.2	2.1	2.0
Antarctic						
McMurdo	−5.3	7.1	5.8	4.3	4.3	4.1
Australia						
Australia	−3.3	14.5	−0.3	1.2	1.1	1.0

[a] Solving of coefficient of (F_c-5), F_c=critical frequency.

better agreement with the other Eurasian sites, the Samoan rates are in worse agreement with the other Pacific plate sites, and the Hawaiian rates are neither clearly better or worse. The changes in the rates for the remaining sites, although important geophysically, are not significant compared to the standard errors. The solutions and the standard errors for the parameter

TABLE IV. Coefficient of $(F_c-5)^2$ where F_c=critical frequency in MHz

Site	Coefficient $\times 10^5$			Standard Error$\times 10^5$		
	Long.	Lat.	Ht.	Long.	Lat.	Ht.
North American Datum						
Las Cruces	5	227	14.7	113	79	2.7
Austin	−89	181	15.7	62	37	1.4
Maine	−209	−3	12.2	95	48	1.7
Minnesota	139	60	5.3	113	48	1.9
Ottawa	79	52	9.7	117	57	2.1
Calgary	−157	−506	24.3	189	94	2.8
Anchorage	−575	−235	13.2	123	42	1.9
Thule	19400	−194	−156.0	5970	1650	62.3
European						
Uccle	−170	−87	5.3	82	40	1.4
Florence	5	46	3.6	128	84	2.1
England	156	53	8.5	125	54	1.8
Mizusawa	33	−571	24.6	99	63	2.2
Africa						
Seychelles	−7	−77	7.3	49	40	1.6
Pretoria	−187	−18	6.6	63	55	1.5
Pacific						
California	−43	−5	14.3	62	36	1.3
Hawaii	69	−282	19.6	62	38	1.1
Samoa	−252	−203	7.0	55	39	1.3
Philippines						
San Miguel	1288	12	14.6	51	48	1.5
Guam	41	9	17.8	55	42	1.3
South America						
Brazil	133	62	5.3	68	42	1.2
Antarctic						
McMurdo	−97	26	21.7	94	71	2.0
Australia						
Australia	9880	1540	56.8	5920	1230	37.2

for critical frequency are shown in Table IV. Note that the coefficient for height is significant for most sites; it is not significant for sites at extreme latitudes such as Thule and McMurdo, where the ionospheric levels are always relatively low, but it did not affect the solution of the height rates at these sites. The solution for the coefficient of critical frequency for the horizontal coordinates was rarely statistically significant. Although statistically significant coefficients were obtained for the Mizusawa and Hawaii latitude coefficients, the Pretoria longitude coefficient, and the Samoa coefficients, the standard errors of the coefficient for these sites were of the same magnitude as those for other sites. It is therefore concluded that the treatment of the critical frequency parameter given in this report is not adequate to remove the neglected effects of ionospheric refraction variations on the computed rates of change of horizontal coordinates.

The best estimate of the rates of change of horizontal coordinates are those from Table II, without the critical frequency parameter, while the best height rates are those from Table III. The composite set is given in Table V along with those given by Minster and Jordan (1978) based on geologic records. The geologic rates are absolute as defined by the stations observing the satellite over the 10-yr period. In an attempt to obtain a

TABLE V. Best Doppler estimate of Doppler station rate

	Doppler solution (cm/yr)						Geologic absolute rate	
	Rate			Standard error				
Site	Long.	Lat.	Ht.	Long.	Lat.	Ht.	Long.	Lat.
North American								
Las Cruces	−7.1	3.9	10.4	2.0	1.4	1.7	−2.3	−1.3
Austin	−5.1	1.5	−4.8	2.4	2.0	2.5	−2.4	−1.2
Maine	−12.0	6.5	8.6	3.0	2.1	2.5	−2.6	−0.7
Minnesota	−6.5	2.0	−5.4	1.7	1.0	1.4		
Ottawa	1.6	−0.4	5.2	3.2	2.2	2.6	−2.5	−0.8
Calgary	−18.0	6.5	12.7	6.2	4.8	4.7	−1.8	−1.4
Anchorage	0.9	−1.1	3.6	1.8	1.3	1.9	−0.7	−1.4
Thule	0.5	−7.1	−6.0	5.2	5.6	6.8	−1.8	−0.7
Wtd Mean			1.5					
Wtd Residuals			3.1					
Scaled Std Err			2.5					
European								
Uccle	4.4	1.1	8.1	1.3	1.0	1.2	−0.3	0.2
Florence	7.0	0.8	8.0	2.4	2.2	2.1	−0.3	0.2
England	3.5	−1.8	13.5	2.5	1.7	1.9	−0.3	0.2
Mizusawa	4.4	−5.8	−1.9	1.9	1.7	2.0	−0.3	0.2
Wtd Mean			7.4					
Wtd Residuals			2.8					
Scaled Std Err			2.4					

TABLE V. (*continued*)

Site	Doppler solution (cm/yr) Rate Long.	Lat.	Ht.	Standard error Long.	Lat.	Ht.	Geologic absolute rate Long.	Lat.
African								
Seychelles	−1.1	−4.7	12.6	1.9	1.6	2.6	0.5	1.4
Pretoria	−0.5	6.8	6.0	1.1	1.0	1.1	0.9	1.1
Wtd Mean			7.0					
Wtd Residuals			1.7					
Scaled Std Err			1.7					
Pacific								
California	−5.9	8.2	−3.6	1.2	0.9	1.1	−5.5	3.0
Hawaii	−10.5	0.9	7.9	2.0	1.4	1.4	−8.3	4.9
Samoa	−11.0	2.7	6.6	2.5	1.9	2.4	−9.2	5.1
Wtd Mean			1.5					
Wtd Residuals			3.9					
Scaled Std Err			3.2					
Philippines								
San Miguel	3.5	0.5	10.9	1.6	1.7	2.0	—	—
Guam	−5.6	0.3	4.7	2.1	1.7	2.1	—	—
Wtd Mean			7.8					
Wtd Residuals			1.5					
Scaled Std Err			2.2					
South American								
Brazil	−2.0	4.3	3.5	2.5	1.7	2.0	−3.0	−0.4
Antarctic								
McMurdo	−2.9	8.8	5.8	4.1	4.0	4.1	0.0	0.6
Australia								
Australia	−3.8	16.5	−0.3	1.1	1.0	1.0	1.2	7.3

common reference, the rates for the Minnesota site were subtracted from the Doppler and geologic rates for the other sites and the results were plotted in Figs. 28 and 29, respectively. The mean rates for each of the plates are shown schematically in Fig. 30 where the open arrows are the Doppler rates with their standard errors indicated by the ellipses and the geologic rates, which have negligible standard errors comparatively, are

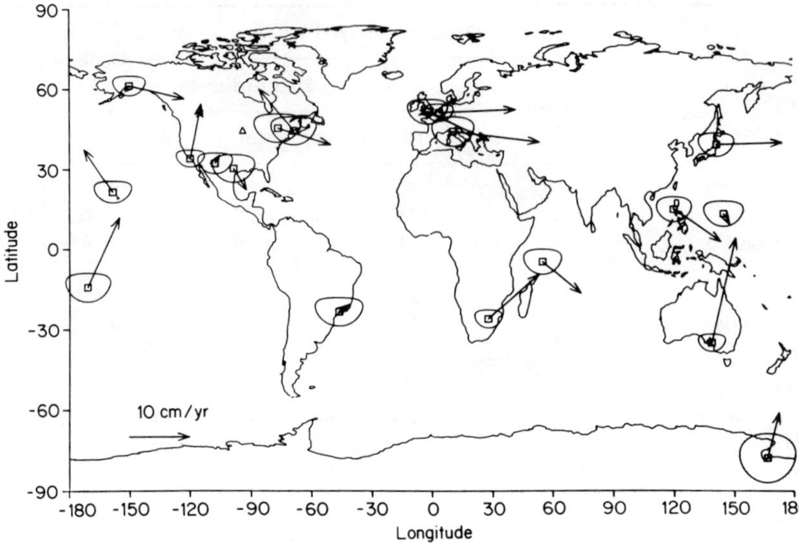

Fig. 28. Station motions relative to Minnesota; Doppler-NAVSAT solution NSWC840912CF.

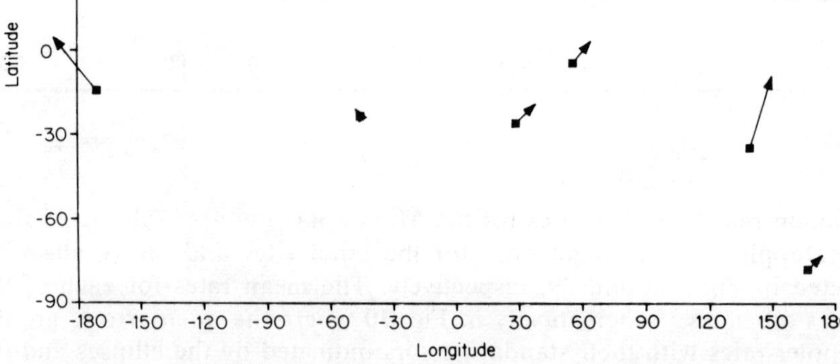

Fig. 29. Station motions relative to Minnesota; geologic motions (Minster and Jordan, 1978.)

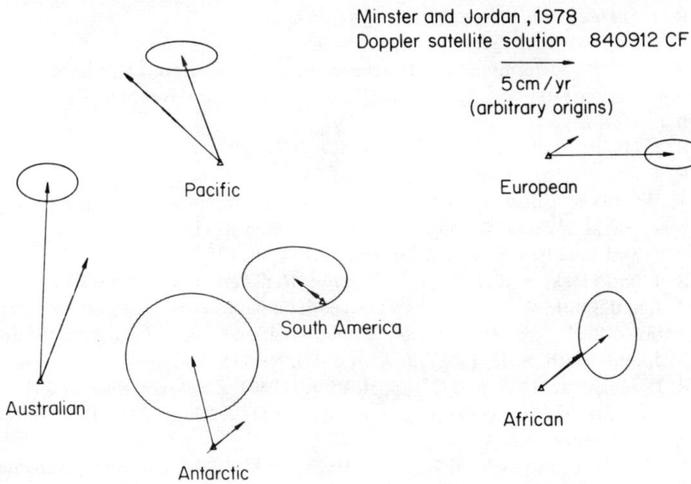

Fig. 30. Plate motions relative to North America (ellipses show standard errors of Doppler rates).

shown by the solid arrowheads. Although averaging latitude and longitude rates on rotating plates is not a geophysically sensible analysis, it is useful for analysis of the consistency of the two solutions since the same number and location of sites is used for both the Doppler and geologic data. The directions of motion which are significant in Fig. 30 are generally consistent and the agreement in magnitude is fair for most of the plates, but the Doppler solution for the separation of the European and North American plates is considerably larger than that given by the geologic data. This is a curious result since there are more Doppler sites on either of these two plates than on any other. No comparisons are given for the Philippine plate since it was not included in the Minster–Jordan solution.

REFERENCES

Anderle, R. J. (1965). *J. Geophys. Res.* **70** (10), 2453–2458.

Anderle, R. J. (1966). In *Trajectories of Artificial Celestial Bodies* (Jean Kovalevsky, ed.), pp. 178–193. Springer-Verlag, Berlin and New York.

Anderle, R. J. (1967). In *The Use of Artificial Satellites for Geodesy* (G. Veis, ed.), pp. 197–220. National Technical University, Athens, Greece.

Anderle, R. J. (1968). *Trans. Am. Geophys. Union* **49** (1).

Anderle, R. J. (1974). *J. Geophys. Res.* **79** (35), 5319–5331.

Anderle, R. J. (1976a). *Bull. Geod.* **50** (1), 43-77.
Anderle, R. J. (1976b). *Bull. Geod.* **50** (4), 377-390.
Anderle, R. J. (1978). Determination of plate tectonic motion from Doppler observations of Navy navigation satellites. *Tech. Rep. 3884.* Naval Surface Weapons Center, Dahlgren, Virginia.
Anderle, R. J. (1980). *Bull. Geod.* **54** (4), 521-527.
Anderle, R. J. (1981). *Ann. Geop.* **37** (1), 37-46.
Anderle, R. J. (1982). Effect of satellite orbit error on computed coordinates of Doppler receivers, Naval Surface Weapons Center. *Tech. Rep.* 82-143.
Anderle, R. J. and Beuglass, L. K. (1970). *Bull. Geod.* **96**, 125-141.
Anderle, R. J. and Hoskins, R. L. (1977). *Geophys. Res. Lett.* **4** (10), 421-423.
Anderle, R. J. and Smith, S. J. (1967). NWL-8 geodetic parameters based on Doppler satellite observations. *Tech. Rep. 2106.* Naval Surface Weapon Center, Dahlgren, Virginia.
Anderle, R. J. and Smith, S. J. (1968). *J. Astron. Sci.* **XV** (1), 1-4.
Anderle, R. J., Malyevac, C. A. and Green, H. L. Jr. (1969). *J. Spacecr. Rockets* **6** (8), 951-954.
Boucher, C. and Feissel, M. (1984). In *Annual Report for 1983*, pp. D79-D83, Bureau International de l'Heure.
Boucher, C., Paquet, P. and Wilson, P. (1979). In *Proc. Third Int. Geod. Symp. Satellite Doppler Positioning, 3rd*, 819-859. Physical Science Laboratory, New Mexico State University.
Buonaguro, A. (1972). Out-of-plane prediction errors in navigation satellite orbits. *Tech. Rep. 2880.* Naval Surface Weapons Center, Dahlgren, Virginia.
Cohen, C. J. and Anderle, R. J. (1960). *Science* **132**, 807-808.
Colquitt, E. S. Anderle, R. J. and Malyevac, C. A. (1980). *J. Astron. Sci.* **XXVIII**(4), 391-403.
Colquitt, E. S., Stein, W. L. and Anderle, R. J. (1984). *Proc. Int. Symp. Space Tech. Geodyn.*
Gouldman, M. W. S. (1982). *Proc. Int. Geod. Symp. Satellite Doppler Positioning, 3rd*, 67-69. Physical Science Laboratory, New Mexico State University.
Grappo, G. A. (1980). *Manuscripta Geodaetica* **5** (3), 201-216.
Guier, W. H. and Newton, R. R. (1965). *J. Geophys. Res.* **70** (18), 4613-4626.
Hothem, L. D., Vincenty, T. and Moose, R. E. (1982). *Proc. Int. Geodetic Symp. Satellite Doppler Positioning, 3rd*, 108-128. Physical Science Laboratory, New Mexico State University.
Jeffreys, Sir H. (1970). *The Earth.* Cambridge Univ. Press, London and New York.
Kaula, W. M. (1966). *Theory of Satellite Geodesy.* Ginn (Blaisdell), Boston, Massachusetts.
Kouba, J. (1983). *Rev. Geophys. Space Phys.* **21** (1), 41-50.
Lambeck, K. (1980). *The Earth's Variable Rotation: Geophysical Causes and Consequences.* Cambridge Univ. Press, London and New York.
Leroy, C. F. (1982). *Proc. Int. Geodetic Symp. Satellite Doppler Positioning, 3rd*, 129-150. Physical Science Laboratory, New Mexico State University.
Malyevac, C. A. and Anderle, R. J. (1985). In *The Use of Artificial Earth Satellites for Geodesy and Geodynamics*, Vol. 3, National Technical University, Athens, Greece.
Minster, J. B. and Jordan, T. H. (1978). *J. Geophys. Res.* **83** (B11), 5331-5354.
Mueller, I. I. (1983). International coordination of space techniques to geodesy and geodynamics. Bulletin No. 6, 4-6. Ohio State Univ. Columbus.
Newton, R. R. (1963). In *The Use of Artificial Satellites for Geodesy* (G. Veis, ed.), pp. 323-328. National Technical University, Athens, Greece.
Oesterwinter, C. (1979). In *Time and Earth's Rotation* (D. D. McCarthy and J. D. Pilkington, eds), pp. 263-278. Reidel Publ., Boston, Massachusetts.
O'Toole, J. W. (1976). Celest computer program for computing satellite orbits. *Tech. Rep. 3565.* Naval Surface Weapons Center, Dahlgren, Virginia.

Seegar, H., Schuluter, W., Herzberger, K., Kramer, F. and Soltau, G. (1979). *Proc. Int. Geodetic Symp. Satellite Doppler Positioning, 2nd,* 889–924. Applied Research Laboratories, University of Texas at Austin.

Seppelin, T. O. (1974). *Can. Surveyor* **28** (5), 496–506.

Vicente, R. (1970). *Ciel Terre* **86** (3).

West, G. B. (1982). *J. Geophys. Res.* **87** (B7), 5538–5540.

3.2 Very Long Baseline Interferometry and its Contributions to Geodynamics

Thomas A. Herring

Harvard College Observatory, Cambridge, Massachusetts, USA

I. Introduction

Very long baseline interferometry (VLBI) has been used since 1969 to measure the positions of radio telescopes and radio sources (Hinteregger *et al.*, 1972). These first VLBI experiments measured the distances between the radio telescopes with precisions of about 1 m. During the past 15 years the improvements in VLBI equipment and data analysis techniques have been impressive (Shapiro *et al.*, 1974, 1976; Ong *et al.*, 1976; Thomas *et al.*, 1976; Cannon *et al.*, 1979; Niell *et al.*, 1979; Robertson *et al.*, 1979; Herring *et al.*, 1981; Robertson and Carter, 1982; Rogers *et al.*, 1983; Sovers *et al.*, 1984). The precision of VLBI measurements using the dual-frequency band Mark III system (Rogers *et al.*, 1983) has now reached the level where geophysically interesting results can be obtained with this system. In this chapter we will analyze the VLBI data obtained by the National Aeronautic and Space Administration (NASA) Geodynamics Program and the National Geodetic Survey's (NGS) project POLARIS between July 1980 and December 1983.

We will examine the temporal behavior of two of the baselines which have been measured in these programs; the tidal parameters of the Earth

SPACE GEODESY AND GEODYNAMICS
ISBN 0-12-057840-9

which can be estimated from these data; and the nutations of the Earth's body axis.

Even after 15 years of development there are still many aspects of VLBI measurements which are not fully understood. Therefore, in this chapter we will not only discuss the geophysical results which can be obtained with VLBI but we will also discuss the influence of systematic errors on the results. We start by outlining the procedures used in analyzing VLBI data.

II. VLBI Data Analysis

Most of the results we will present in this chapter will be derived from the VLBI observations contained in the NASA Geodynamics Data Information System (DIS) and the NGS POLARIS data base. Consequently in this section we will outline the procedures and the mathematical models which have been used in analyzing these data.

The data analyzed consisted of dual-frequency-band interferometric group delays (Shapiro, 1976; Rogers *et al.*, 1983) obtained in observing sessions each of approximately 24-hr duration. The frequency bands used were centered at ~8.3 GHz (X-band) and ~2.3 GHz (S-band). Within each frequency band, data were obtained in four to eight channels, each of 2-MHz bandwidth. The total bandwidth spanned by these channels varied between 300 and 360 MHz at X-band, and 50 and 85 MHz at S-band, thus allowing accurate estimates of the group delays (Whitney *et al.*, 1976).

These observations were analyzed within the framework described in Herring *et al.* (1981). However, the following models were used in place of their counterparts described there: (i) the IAU 1980 nutation series (Wahr, 1981a); (ii) the Greenwich sidereal time formula of Aoki *et al.* (1982) and (iii) the atomic time-coordinate time relationship of Moyer (1981). In addition, we added corrections for the ionospheric propagation delay, based on the dual-frequency-band observations, and for the neutral atmospheric propagation delay, based on surface weather data (Marini, 1974).

We assumed that the observation noise was an independent, zero mean, Gaussian random process. The variance of the noise for each observation was computed as the sum of two variances. The first variance was computed from the signal-to-noise ratio (SNR) of the observation and represents the uncertainty in the observation due predominately to instrumental noise (Whitney *et al.*, 1976). The second variance was calculated from the condition that the χ^2 per degree of freedom of the postfit residuals for each baseline be near unity. This second variance is used to account for the pseudo-noise contribution from the inadequate models used to represent the data.

The observations were analyzed using a sequential, weighted-least-squares technique (see, for examples, Kaula, 1966 and Morrison, 1969), with the weights being taken as the inverses of the variances of the noise contributions to the observations. The following classes of parameters were estimated in the solutions to be discussed in this chapter; (i) coefficients of one to three polynomials in time, each of first or second order, and, in some cases, coefficients of diurnal sine and cosine functions. These polynomial and trigonometric functions were used to represent the relative behavior of the hydrogen maser "clocks" at the different sites; (ii) a daily bias correction for each site to the neutral atmosphere propagation delay calculated from surface weather data; (iii) all crust-fixed site coordinates (assumed time independent), except for those of one site which were used to define the origin of the coordinate system (see Herring *et al.*, 1981); (iv) all radio source positions, save for one parameter—the mean correction to the right ascensions of seven radio sources which served to define the right ascension origin. (The seven radio sources chosen were 0851+202, 0923+392, 1226+023, 1641+399, 2134+004, 2200+420 and 2251+158); (v) daily corrections to the pole position and UT1 given by BIH Circular D and (vi) daily corrections to the nutation angles given by the IAU 1980 nutation series.

In addition to these parameters, we sometimes estimated corrections to the Earth tide parameters (see discussion in Section IV).

The VLBI data analysis involves a number of subjective aspects: (i) the parameterization of the relative behaviors of the hydrogen-maser clocks at the radio telescopes, (ii) the parameterization of the atmospheric delay behavior at each site, (iii) data editing and (iv) data reweighting (see discussion below). The order of the polynomial and the epochs of the boundaries between estimates of the polynomial coefficients for the relative clock behaviors are decided on by the analyst after a visual inspection of the postfit residuals from a solution which utilizes relatively simple parameterizations of the clock behavior. Trends in the residuals which are radio-source independent are usually associated with clock variations. Similarly, trends which are elevation-angle dependent are usually associated with atmospheric variations. In our analysis a single atmospheric parameter was estimated each day for each site and served to remove any constant biases in the surface weather model calibration of the atmospheric delays. Data editing, i.e. the elimination of observations which appear to be defective, is also subjective in nature. (The Mark III correlator has built-in quality checks which now detect most of the defective observations before the analyst examines the observations. These quality checks were still being developed when much of the data we have analyzed were being correlated. Consequently, we felt it was necessary to include additional editing conditions in our analysis.) The editing criterion used deleted from the analysis all observations which had postfit residuals greater than three reweighted standard deviations (see further discussion below). Data reweighting is the final subjective aspect of the data analysis. The scatter of the postfit residuals from least-squares solutions which use Mark III data is nearly always greater than the standard deviations of the observations would indicate. This greater scatter occurs because the observation standard deviations, which are calculated from the signal-to-noise ratio (SNR) of the observations, do not take into account the errors in the mathematical models used to analyze the data, and the random variations of the frequency standards. The contribution of these additional uncertainties can be accounted for phenomenologically by adding a constant variance to the SNR-calculated variances and by using this new variance to calculate the weights used in the weighted-least-squares solution. We call this operation "reweighting". Both the editing and the reweighting criteria we have used depend on postfit residuals, and hence, on the solution used to generate those postfit residuals. To help ensure that the editing and the reweighting operations were not biased by the solution on which they were based, we iterated the previously discussed procedures. Each iteration involved generating a solution using all the data, and then applying the editing and reweighting procedures to the postfit residuals from this solution. Three iterations of this procedure were necessary for the editing and the reweighting to converge.

The final data set consisted of 42,604 group delay observations of which 31,091 were used in the sequential weighted-least-squares solution. The data which were not used either failed the quality checks during correlation (9144 observations), or were rejected by the editing criterion discussed (2383 observations). The large number of observations rejected by the "three-sigma" criterion is probably due to instrumental errors which were not detected during correlation (especially in the early observing sessions before the development of an extensive set of correlator checks), and to large postfit residuals caused by inadequate models.

III. Baseline Length Results

The use of VLBI measurements to study global plate tectonics currently relies on the study of the behaviors of the distances between the radio telescopes participating in each observing session. If a network of VLBI sites were established on a single, stable tectonic plate, thus allowing Earth orientation parameters to be monitored with very high accuracy, then the three-dimensional site coordinates could be used to monitor plate motion. However, such a network of sites does not yet exist, and hence, in this section we will study variations of baseline lengths (which are insensitive to orientation changes).

The first set of baseline length results we will examine is a composite of the single-frequency band Mark I VLBI system and the dual-frequency Mark III system measurements of the ~4000 km baseline between radio telescopes in Westford, Massachusetts and Owens Valley, California. These results are shown in Fig. 1. The Mark I results were extracted from Ryan *et al.*, 1986. The Mark III results were obtained by analyzing each observing session separately with the source positions constrained to the values obtained from the analysis of the complete data set as described in Section II. The weighted-least-squares estimate of the rate of change of this baseline length is -0.4 ± 0.2 cm/yr, where the standard deviation is the statistical estimate of the uncertainty scaled by the square root of the χ^2 per degree of freedom of the individual estimates of the baseline length about the best-fit slope. (We call the square root of the χ^2 per degree of freedom the normalized-root-mean-square (nrms) scatter. If the scatter of the baseline length estimates was in accord with the statistics of the estimates, the nrms scatter would be unity. The nrms scatter being significantly greater than unity is attributed to the effects of systematic errors.) For these results the nrms scatter with 40 degrees of freedom was 1.8. The weighted-root-mean-square (wrms) scatter of the 42 measurements about the slope was 2.1 cm. The estimate of the slope along with its standard deviation would indicate that, over the six years spanned by this data set, there has been no motion between these sites greater than 1 cm/yr (99% confidence interval). The small negative slope of the baseline length estimates is most probably due to the positive bias caused by the effects of the ionospheric delay on the single frequency-band Mark I measurements (Ryan *et al.*, 1986).

When the rate of change of the baseline length is computed using only the Mark III measurements, a rate of 0.2 ± 0.4 cm/yr is obtained. The wrms scatter of the Mark III baseline length measurements about the slope is 1.5 cm. The nrms scatter is 1.6. Again the slope from this subset of data is insignificant. (This latter slope estimate should be more reliable than the estimate from the combined data set because of the possible bias in the Mark I measurements.)

These results indicate that at least part of the North American plate has been stable at the 1-cm/yr level for the last six years. The estimates of the current day rate of expansion of the Basin and Range province (which lies between Haystack and Owens Valley) vary between 0.3 and 2.0 cm/yr (Hamilton and Myers, 1966; Stewart, 1977), with more recent estimates of the order of 0.1 cm/yr (Doser and Smith, 1982). The lower bound of this rate would not yet be detectable by the VLBI measurements presented here. The upper bound seems to be contradicted by the VLBI measurements,

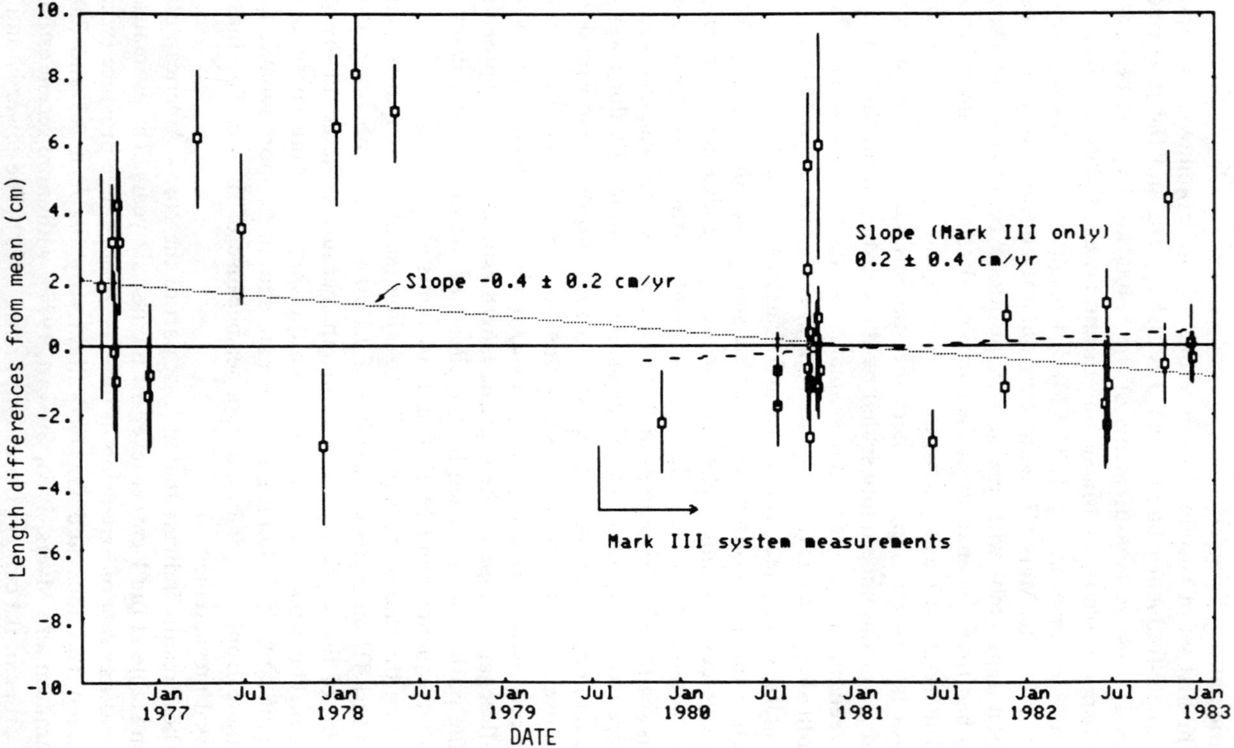

Length differences from mean (cm)
10.
8.
6.
4.
2.
0.
-2.
-4.
-6.
-8.
-10.
Slope -0.4 ± 0.2 cm/yr
Slope (Mark III only)
0.2 ± 0.4 cm/yr
Mark III system measurements
Jan 1977
Jul
Jan 1978
Jul
Jan 1979
Jul
Jan 1980
Jul
Jan 1981
Jul
Jan 1982
Jul
Jan 1983
DATE

unless, as suggested by Allenby (1979), the effects of extensional and compressional regimes in North America tend to cancel. The NASA Geodynamics Program has established mobile VLBI sites in central North America to address the question of the spreading rate in the Basin and Range province. In several years VLBI measurements should be able to answer questions about the internal as well as the overall stability of the North American plate.

The second set of VLBI measurements we will examine are the results obtained for the Haystack and Westford, Massachusetts to Onsala, Sweden baselines. We have used measurements to these two sites in Massachusetts in order to maximize the length of the time history of the baseline length measurements to Onsala. We believe the combination of Haystack–Onsala results ("mapped" to the Westford–Onsala baseline) and the Westford–Onsala results is valid because of the accurately known relative positions of the Haystack and Westford antennas. These antennas are separated by ~1.24 km and their relative positions have been determined with a wrms scatter of the vertical and two horizontal components of 7, 5 and 3 mm, respectively, from 12 independent Mark I VLBI experiments (Rogers *et al.*, 1978; Carter *et al.*, 1980). The VLBI measurements differed from the results of a conventional survey between these antennas by −6, 2 and 4 mm, for the vertical and horizontal components of the baseline after the vertical component was corrected for a known, repeatable effect of deformation of

---

Fig. 1. Length differences from a mean value of 3,928,881.661 m for 42 measurements of the Haystack Observatory to Owens Valley Radio Observatory baseline. The measurements prior to 1979 were made with the single-frequency band Mark I VLBI system. The values plotted were obtained from Ryan *et al.* (1984). The measurements after 1979 were made with the dual-frequency band Mark III VLBI system. The values plotted were obtained from the analysis described in the text. The error bars represent one statistical standard deviation. The weighted-root-mean-square (wrms) scatter for the complete set of measurements is 2.1 cm. For the Mark III measurements the wrms scatter is 1.5 cm. The normalized-root-mean-square (nrms) scatters for these two data sets are 1.8 and 1.6, respectively. The wrms scatter was computed using:

$$\text{wrms} = \left[\left(\sum_{j=1}^{n} \delta b_j^2 \Big/ \sigma_j^2\right) \Big/ \left(\sum_{j=1}^{n} 1 \Big/ \sigma_j^2\right)\right]^{1/2}$$

where δb_j is the baseline length residual from the best-fit sloped line; σ_j is the standard deviation of the baseline length estimate; and n is the number of measurements used to compute the wrms scatter. The nrms scatter was computed using:

$$\text{nrms} = \left[\left(\sum_{j=1}^{n} \delta b_j^2 \Big/ \sigma_j^2\right) \Big/ f\right]^{1/2}$$

where f is the number of degrees of freedom; $f = n - 2$ when a slope and intercept are estimated from the measurements. The uncertainties of the slope estimates are one statistical standard deviation multiplied by the nrms scatter of the baseline length residuals.

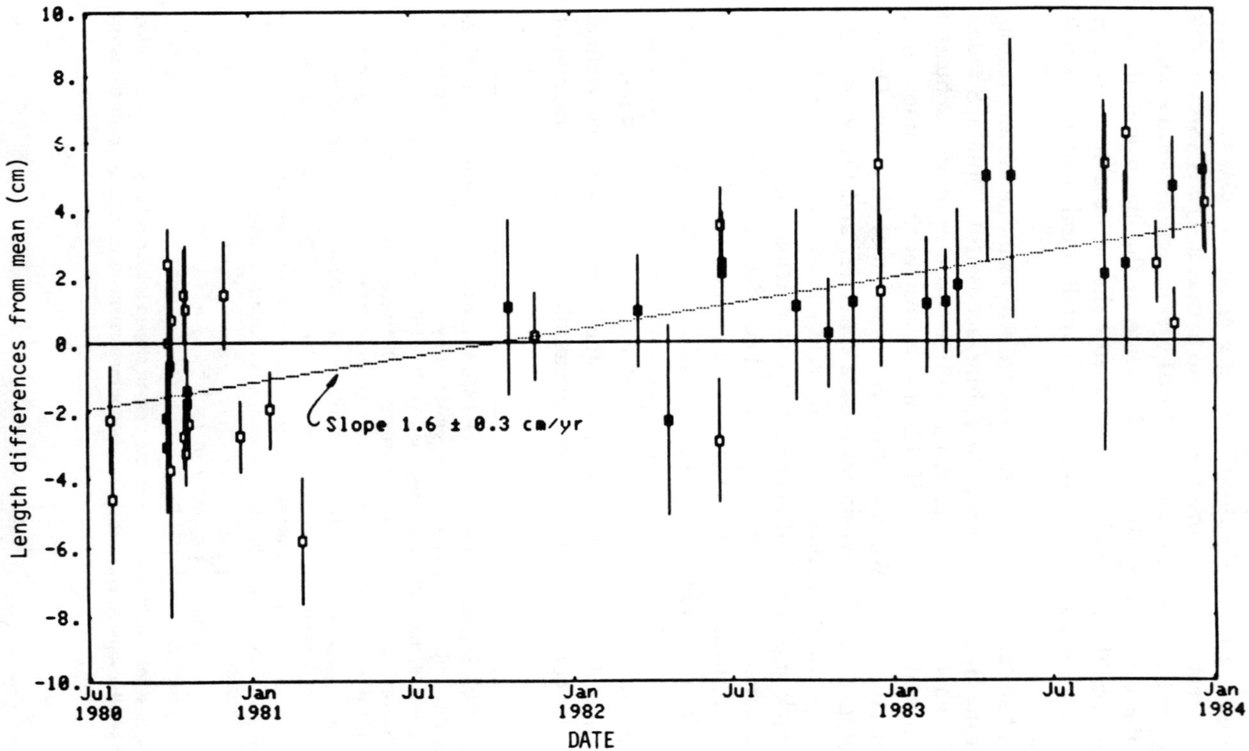

Fig. 2. Length differences from a mean value of 5,599,714.426 m for 47 measurements of the Westford–Onsala baseline length. The data shown with open squares used the Haystack antenna. These results were mapped to the Westford–Onsala baseline using the relative positions of the Haystack and Westford antennas. The solid squares are Westford–Onsala measurements. The wrms and nrms scatters of the baseline length residuals about the best-fit sloped line are 2.0 cm and 1.4, respectively. (See Fig. 1 for definitions.)

the Haystack antenna (see Carter *et al.*, 1980 for details). We expect that the Westford to Onsala baseline lengths inferred from measurements which used the Haystack Observatory should be in error by only a few millimeters due to the uncertainty of the relative positions of the two telescopes. The experiments which used the Haystack Observatory are plotted with open squares in Fig. 2. There does not appear to be any large bias between the Haystack and Westford results.

The sites in Massachusetts are located on the North American plate and the site in Sweden on the Eurasian plate. The expected motion between these sites, projected on to the baseline vector, is 1.7 ± 0.3 cm/yr (Minster and Jordan, 1978). The weighted-least-squares estimate of the slope of the Westford–Onsala baseline length is 1.6 ± 0.3 cm/yr where the standard deviation of the slope is computed using the method discussed for the Haystack–Owens Valley results. The wrms scatter of the individual measurements about the slope is 2.0 cm. The nrms scatter is 1.4 with 45 degrees of freedom. We can not yet claim to have detected the relative motions of these two sites because the systematic errors in the VLBI models could influence the results by amounts greater than the quoted standard deviations. We estimate that these effects could be up to 4 cm for this baseline mainly due to the inadequacies of our current atmospheric model. The 4 cm corresponds to possible site radial errors of 10 cm projected on to the Westford–Onsala baseline vector. If we assume that the 4-cm systematic error contribution varies linearly with time over the 3.5 years spanned by our data, then the realistic uncertainty of the slope estimate is 1 cm/yr. Within this uncertainty our result is consistent with the Minster and Jordan (1978) value. However, it is premature to claim that these results confirm the Minster and Jordan North American–Eurasian spreading rate. Continued measurement of this baseline is required for a reliable estimate of the spreading rate to be obtained.

IV. Earth Tide Parameters

The VLBI group delays are affected by variations of the site positions during an observing session, and hence, are sensitive to the tidal parameters of the Earth. We have estimated the tidal parameters of the Earth, using the data set described in Section II, in three separate analyses. The analyses consisted of (i) estimating the Love numbers h and l, and the tidal lag angle ϕ separately for each observing session; (ii) estimating a single set of these tidal parameters in a weighted-least-squares solution using all of the VLBI observing sessions between July 1980 and June 1983; and (iii) estimating site-dependent values of the radial Love number h with the combined data set. (We only estimated this single Love number in this analysis because of our insensitivity to the Love number l and the lag angle.)

The tidal model we used has been discussed in Herring *et al.* (1983). This model is a "vector" tide model which uses the total tidal potential and its derivatives, rather than the Fourier decomposition of the potential (see, for example, Melchior, 1978), to compute the site displacements due to the solid-earth tides. The advantage of the vector tide model is that the potential of the disturbing bodies (the sun and the moon) are calculated very accurately from planetary ephemeris data. The disadvantage is that any frequency dependence of the Love numbers is suppressed in this model.

The estimates of the Love numbers and tidal lag angle for each observing session are shown in Figs. 3–5. The weighted means and their uncertainties of the tidal parameters are $h = 0.611 \pm 0.003$, $l = 0.071 \pm 0.008$, and $\phi = 0.1° \pm 0.1°$ in which the uncertainties are the statistical estimates of the standard deviations scaled by the nrms scatter of the individual estimates about their means. The nrms scatters were 1.9, 2.8, 2.9, for h, l, ϕ, respectively. These nrms scatters are much greater than we should expect from the random errors in the individual determinations. This additional scatter is probably due to the effects of systematic errors in the models used to analyze the VLBI data. For mid-latitude sites (as used in this study), the contribution of the semi-diurnal and the diurnal tides are about equal (each being ~15 cm). We expect that the diurnal component in the atmospheric delay, which is not modeled correctly by the surface weather model, could disturb the diurnal component of the Earth tide results. Average (spurious) diurnal signatures in the radial site positions of only 1.3 cm could account for the observed scatter of 0.026 in the values of h. (The tidal potential at mid-latitudes (expressed as a displacement) is ~50 cm; an approximate estimate of the error in h due to a radial displacement of 1.3 cm would be ~1.3 cm/50 cm.)

The scatter of the estimates of the Love number l and the tidal lag angle ϕ are 2.8 and 2.9 times the amounts expected from the standard deviations of the individual estimates. These large scatters are probably due to the geometry of the VLBI observations. For the horizontal position (in a local topocentric coordinate system) to be well determined, low elevation angle observations are required and a uniform distribution of observations in azimuth is also desirable. Unfortunately, this latter condition is rarely fulfilled because our western antennas tend to observe in an easterly direction (because of the requirement of mutual visibility of a source from all antennas), and the eastern sites observe in a westerly direction. This nonsymmetric geometry probably makes the horizontal Love number estimates sensitive to inadequacies in atmospheric delay models.

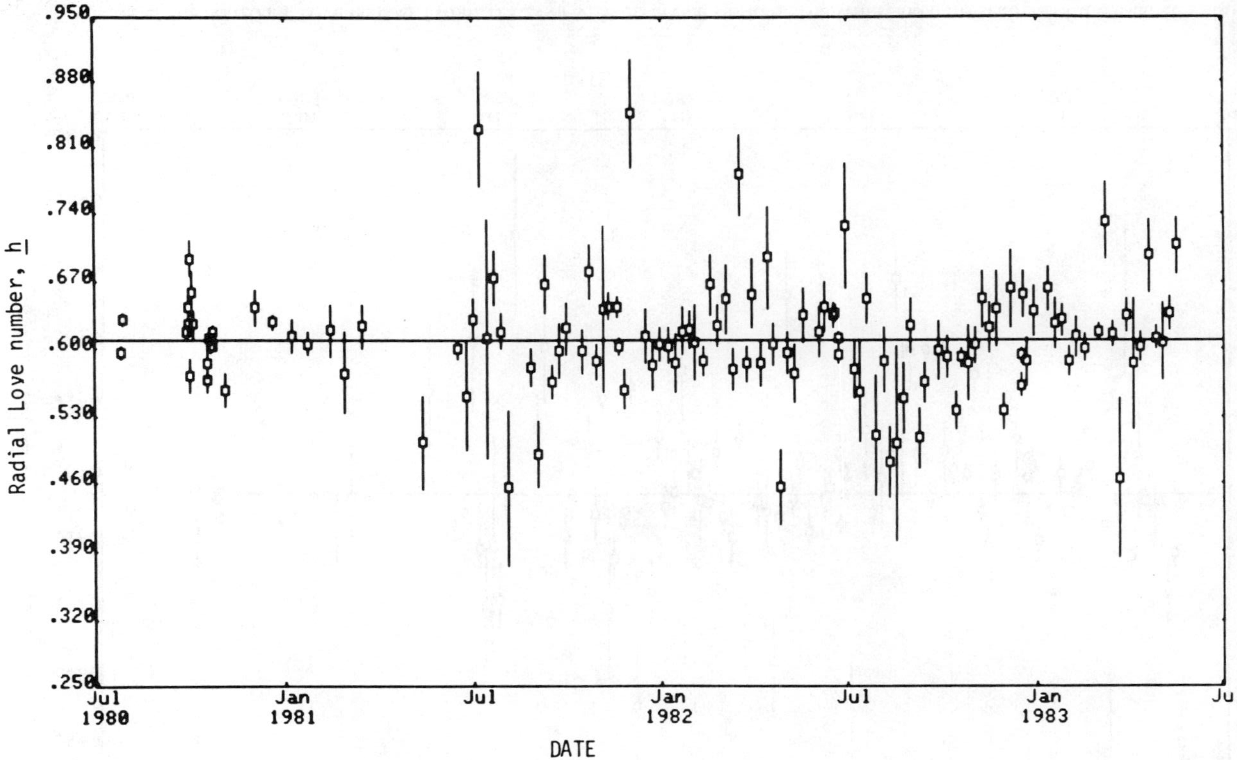

Fig. 3. Estimates of the radial Love number *h* from 123 24-hr VLBI observing sessions. The wrms and nrms scatters of the estimates about the mean are 0.026 and 1.9, respectively (see Fig. 1).

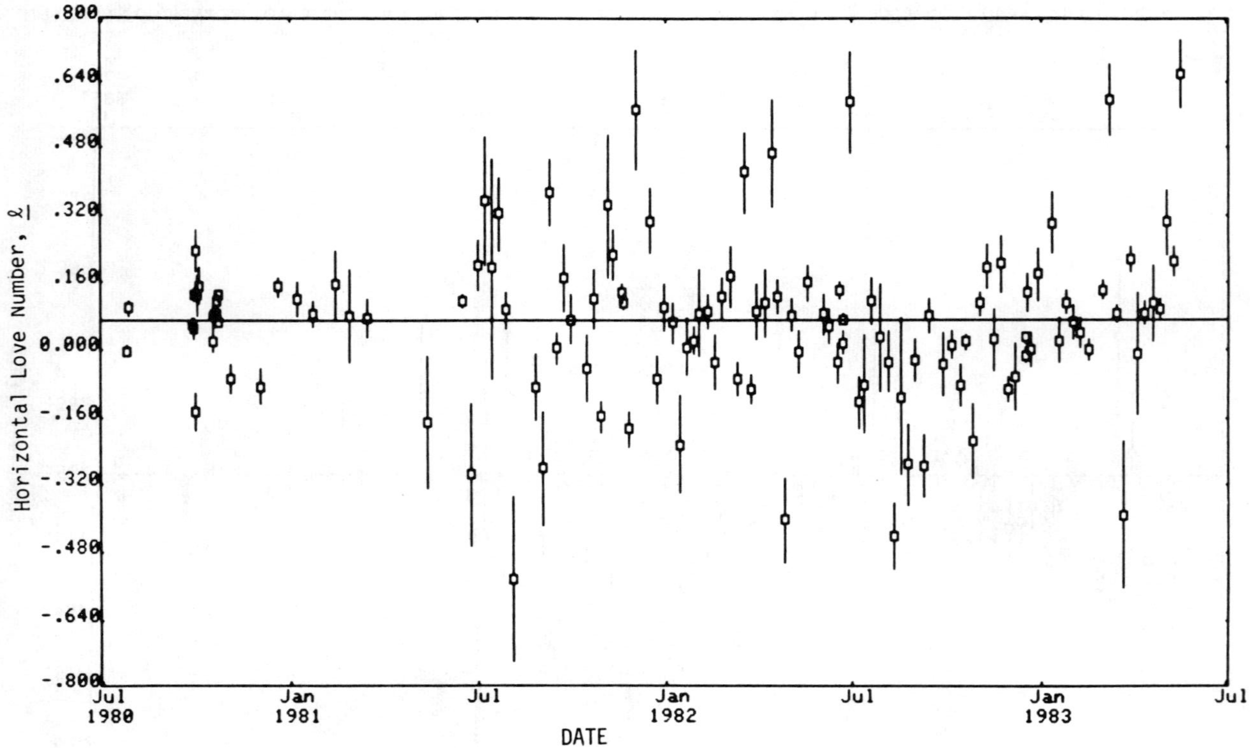

Fig. 4. Estimates of the horizontal Love number *l* from 123 24-hr VLBI observing sessions. The wrms and nrms scatters of the estimates about the mean are 0.081 and 2.7, respectively (see Fig. 1).

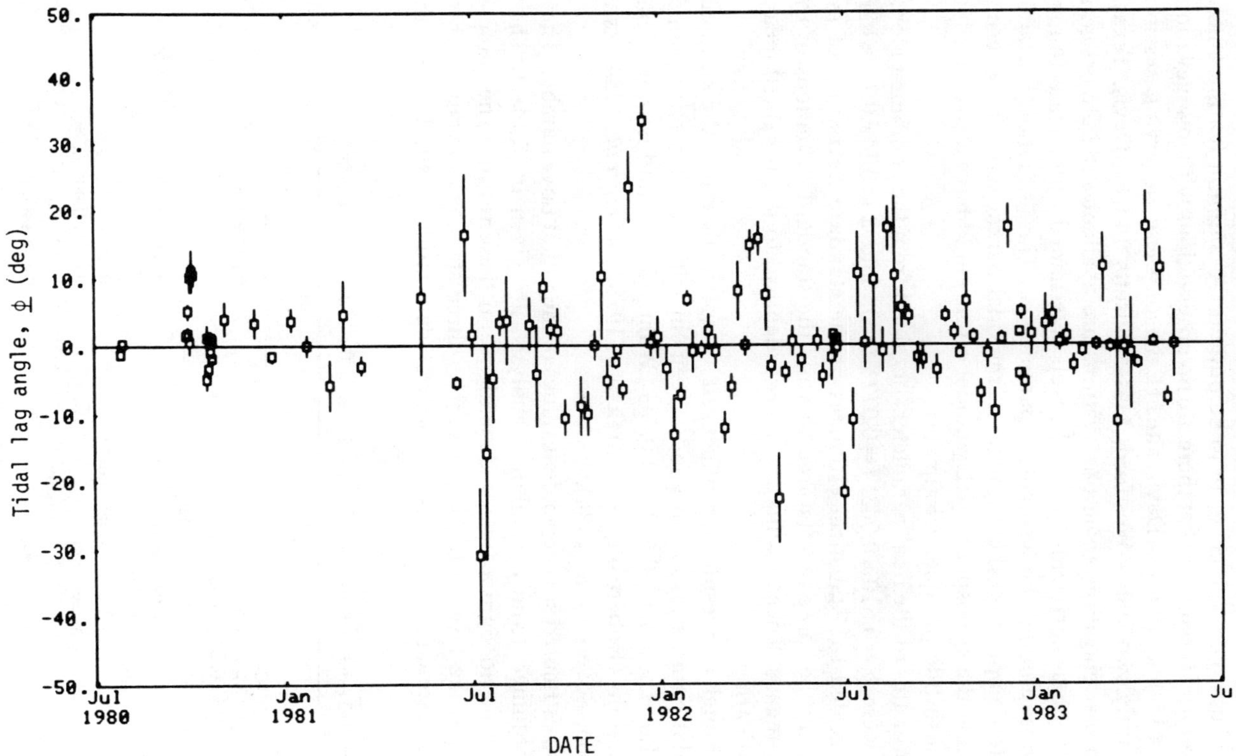

Fig. 5. Estimates of the tidal lag angle ϕ from 123 24-hr VLBI observing sessions. The wrms and nrms scatters of the estimates about the mean are 3.8° and 2.9°, respectively (see Fig. 1).

In each of the plots of the estimates of the tidal parameters, there are a number of estimates which seem to be outliers. A common feature among these observing sessions is that there are no observations with low elevations angles ($<15°$) at the Ft. Davis site. (Each of these observing sessions involved only two sites—Westford, Massachusetts and Ft. Davis, Texas.) When no low elevation angle observations are available, the atmospheric zenith delay bias correction cannot be well determined which leads to high correlations among the estimated parameters. This situation certainly appears to compromise the estimates of the tidal parameters. These results indicate that the atmospheric delay model errors could have a sizable effect on the Earth tide parameter estimates.

The estimates of the tidal parameters obtained from the combined global solution were $h = 0.610 \pm 0.003$, $l = 0.072 \pm 0.008$ and $\phi = 0.1° \pm 0.3°$, where we have scaled the statistical estimates of the standard deviations of the estimates by the same factors used to scale the standard deviations of the weighted means. These results are almost identical to the weighted means presented earlier.

Even though our results are influenced by systematic errors the estimates of the tidal parameters are in good agreement with the theoretical values of these Love numbers, 0.609 and 0.085 (Wahr, 1981b); and with previous determinations (Robertson, 1975; Melchior, 1978; Carter *et al.*, 1984; Ryan *et al.*, 1986; Sovers *et al.*, 1984).

We also estimated site-dependent values of the radial Love number. These results, obtained from a combined analysis, are given in Table I. There appears to be no significant site dependence in these results. The value at Effelsberg is the largest, but it is less than three standard deviations from the global estimate of h. It is the poorest determined value because the

TABLE I. Site-dependent estimates of the radial Love number h

Location	h^a
Haystack, Massachusetts	0.612 ± 0.004
Westford, Massachusetts	0.618 ± 0.003
Maryland Point, Maryland	0.601 ± 0.010
Green Bank, West Virginia	0.606 ± 0.006
Ft. Davis, Texas	0.610 ± 0.003
Owens Valley, California	0.616 ± 0.004
Onsala, Sweden	0.617 ± 0.006
Effelsberg, F.R. Germany	0.642 ± 0.012
Chilbolton, England	0.606 ± 0.009

[a] These values were computed from the combined analysis of the data set described in Section II.

antenna at Effelsberg was only used on five occasions (in July and September, 1980).

The VLBI measurements of the tidal parameters of the Earth confirm that the theoretical values of these parameters seem to be correct within the VLBI uncertainties. At this time, however, the tidal parameters from VLBI measurements seem to be more useful for studying the effects of systematic errors in the VLBI models. The geophysically interesting aspects of Earth tide measurements such as the effects of the core–mantle relative motions and the ellipticity of the Earth (Wahr, 1981b) seem to be below the level where they can be reliably studied by using the VLBI. The effects of core–mantle relative motions are expected to produce only 1-cm radial site displacements at the K_1 tidal frequency. The effects of ellipticity are expected to produce latitude-dependent radial and horizontal tidal displacements of 0.1 and 1 cm (Wahr, 1981b). Improvements in the VLBI models could allow such small displacements to be measured in the future.

V. The Nutations of the Earth

The most recently computed nutation series is based on a model of the forced nutations of an elliptical, rotating, elastic and oceanless earth (Wahr, 1981a). This theory has now been adopted by the International Astronomical Union (IAU) as the standard nutation series to be used in the reduction of astrometric observations. We discuss here the errors in the Wahr nutation series obtained from the analysis of VLBI observations, and the possible origins of these errors.

The technique we have adopted for estimating corrections to the nutation series is to estimate for each 24-hr VLBI observing session the corrections to the nutations in obliquity ($\Delta\varepsilon$) and longitude ($\Delta\psi$). We therefore produce a time series of the errors in the nutation series. In a post-processing operation, we obtain corrections to the nutation series coefficients for $\Delta\varepsilon$ and $\Delta\psi$ by estimating the amplitudes of signatures with the frequency and phase of the individual constituents in the nutation series. This technique allows us to estimate the scatter of the values from the individual 24-hr observing sessions about the curves generated by the estimated errors in the nutation series. The scaling of the statistical standard deviations of the estimates of the nutation series coefficients by the ratio of the observed scatter of the nutation angle estimates to the expected scatter allows us to obtain more reliable estimates of the uncertainties of the corrections to the nutation series coefficients.

We have estimated corrections to the nutation angles $\Delta\varepsilon$ and $\Delta\psi$ for each 24-hr VLBI observing session from a data set which spans 3.5 years. In

Figs. 1(a), (b) we show the estimates of $\Delta\varepsilon$ and $\Delta\psi \sin \varepsilon_0$ from these 123 observing sessions. We have estimated from these time series of $\Delta\varepsilon$ and $\Delta\psi$, corrections to the six largest coefficients of the nutations series with periods less than or equal to one year. The model we have adopted for the errors in the series is

$$\Delta\varepsilon(t) = \sum_{j=1}^{6} \delta\varepsilon_{rj} \cos(\theta_j(t)) + \sum_{j=1}^{6} \delta\varepsilon_{ij} \sin(\theta_j(t)) + C_\varepsilon \qquad (1)$$

$$\Delta\psi(t) = \sum_{j=1}^{6} \delta\psi_{rj} \sin(\theta_j(t)) + \sum_{j=1}^{6} \delta\psi_{ij} \cos(\theta_j(t)) + C_\psi \qquad (2)$$

where $\delta\varepsilon_{rj}$ and $\delta\psi_{rj}$ are the corrections to the coefficients for the nutations in obliquity and longitude with argument θ_j. These terms represent corrections to coefficients which are currently in the nutation series and are referred to as the "in-phase" ("real") coefficients; $\delta\varepsilon_{ij}$ and $\delta\psi_{ij}$ are coefficients for terms which are 90° out of phase with nutations series. These terms are called the "out-of-phase" ("*i*maginary") coefficients and their existence would indicate the presence of dissipation of energy in the Earth (see more detailed discussion in Section IV); θ_j are the arguments of the nutation series and are of the form $\theta_j = a1 + b1' + cF + dD + e\Omega$, where a, b, c, d and e are integer coefficients which arise from the Fourier series expansion of the orbital motions of the Earth and the moon; 1,1$'$, F, D and Ω are Brown's fundamental arguments (Eckert *et al.*, 1966); C_ε and C_ψ are constant offsets to the nutation angles which are necessary because VLBI observations of extragalactic radio sources (as used for this analysis) are insensitive to the location of the ecliptic. Corrections to the coefficients of the 18.6- and 9.3-yr terms could not be reliably estimated from our short span of data. The corrections to the Wahr nutation series are given in Table II. The quoted standard deviations given are twice the statistical estimates of uncertainties. The factor of 2 was chosen because the scatter of the individual nutation angle estimates about the curve produced by the coefficients given in Table II was twice that to be expected from the statistical estimates of uncertainties of the individual measurements. We discuss later in this section the possible reasons for this additional scatter.

In Table II, the largest corrections to the nutation series are for the annual and semiannual nutations. All other corrections are less than three standard deviations.

In order to study the possible origins of the corrections to the nutation series, we need to consider in more detail the construction of the nutation series. Here $\Delta\varepsilon$ and $\Delta\psi$ are convenient angles to use for the rotation of coordinate systems. However, for the interpretation of the errors in the nutation series, the corrections to the nutation amplitudes are more con-

TABLE II. Estimates of the corrections to the coefficients in the Wahr nutation series

	$\Delta\varepsilon$		$\Delta\psi \sin \varepsilon_0$	
Period (days)	In phase (mas)	Out of phase (mas)	In phase (mas)	Out of phase (mas)
182.6	-0.5 ± 0.2[a]	-0.8 ± 0.2[a]	0.7 ± 0.2[a]	-0.3 ± 0.2[a]
365.2	1.9 ± 0.2	-0.7 ± 0.2	1.7 ± 0.2	0.9 ± 0.2
13.7	0.0 ± 0.2	-0.1 ± 0.2	0.4 ± 0.2	0.2 ± 0.2
27.6	0.0 ± 0.2	0.1 ± 0.2	0.2 ± 0.2	-0.2 ± 0.2
121.7	0.6 ± 0.2	0.1 ± 0.2	-0.1 ± 0.2	-0.3 ± 0.2
9.1	0.0 ± 0.2	0.4 ± 0.2	-0.3 ± 0.2	-0.0 ± 0.2

[a] The quoted standard deviations are the statistical estimates multiplied by the ratio of the scatter of the nutation angle estimates to their expected scatter. The multiplicative factor was 2.0 for both $\Delta\varepsilon$ and $\Delta\psi \sin \varepsilon_0$.

venient quantities to use. Each nutation constituent is the superposition of two motions of the body axis with frequencies (when viewed on the rotating Earth) which are symmetric about one sidereal day (Melchior, 1971). These terms are called the retrograde and prograde amplitudes depending upon whether their frequency is greater than or less than one sidereal day, respectively.The relationship between the nutation amplitudes and $\delta\varepsilon$ and $\delta\psi \sin \varepsilon_0$ are given in Wahr (1981) for the in-phase components. The relationship for the out-of-phase components can be obtained in the same manner as used by Wahr for the in-phase components with the substitution of $\delta a = \delta a_r - i\delta a_i$. [We have adopted a sign convention for the imaginary component consistent with Wahr and Sasao (1981).] The equations relating the nutation amplitudes and the nutations in obliquity and longitude are

$$\delta a_{rj}^{+} = (\delta\varepsilon_{rj} + \delta\psi_{rj} \sin \varepsilon_0)/2 \tag{3}$$

$$\delta a_{rj}^{-} = -(\delta\varepsilon_{rj} - \delta\psi_{rj} \sin \varepsilon_0)/2 \tag{4}$$

$$\delta a_{ij}^{+} = -(\delta\varepsilon_{ij} - \delta\psi_{ij} \sin \varepsilon_0)/2 \tag{5}$$

$$\delta a_{ij}^{-} = (\delta\varepsilon_{ij} + \delta\psi_{ij} \sin \varepsilon_0)/2 \tag{6}$$

where the δa_{rj}^{+} and δa_{ij}^{+} are the real and imaginary amplitudes for the tidal constituent with frequency $\Omega + f_j$; δa_{rj}^{-} and δa_{ij}^{-} are the real and imaginary amplitudes for the tidal constituent with frequency $\Omega - f_j$; Ω is the frequency of the Earth's rotation (one cycle per sidereal day) and f_j is the nutation frequency (viewed from inertial space).

In Table III we give the corrections to the nutation amplitudes computed from Eqs.(3–6), and the values given in Table II. From this table we see that the errors in annual nutations in longitude and obliquity are due solely

TABLE III. Estimates of the corrections to the amplitudes of the circular nutations in the Wahr nutation series

Period (days)	Direction	Amplitude[a] (mas)	Corrections: In phase, δa_r (mas)	Corrections: Out of phase, δa_i (mas)
182.6	Prograde	−549.1	0.6 ± 0.2	−0.6 ± 0.2
182.6	Retrograde	−24.5	−0.2 ± 0.2	0.3 ± 0.2
365.3	Prograde	25.7	−0.1 ± 0.2	0.1 ± 0.2
365.2	Retrograde	−31.1	−1.9 ± 0.2	0.8 ± 0.2
13.7	Prograde	−94.1	−0.2 ± 0.2	0.1 ± 0.2
13.7	Retrograde	−3.6	0.2 ± 0.2	0.2 ± 0.2
27.6	Prograde	14.5	0.1 ± 0.2	−0.0 ± 0.2
27.6	Retrograde	−13.8	−0.1 ± 0.2	−0.1 ± 0.2
121.7	Prograde	−21.5	−0.3 ± 0.2	0.0 ± 0.2
121.7	Retrograde	−0.9	−0.3 ± 0.2	−0.3 ± 0.2
9.1	Prograde	−12.5	−0.1 ± 0.2	0.2 ± 0.2
9.1	Retrograde	−0.5	0.1 ± 0.2	−0.2 ± 0.2

[a] Data derived from Wahr (1981a).

to an error in the retrograde annual nutation amplitude. Similarly, the errors in the semiannual nutations are due to an error in the prograde semiannual amplitude.

We discuss two possible origins for the errors in the nutation series given in Table III—errors in the VLBI measurements and errors in the geophysical models used by Wahr to construct the nutation series. We commence by discussing the possible errors in the VLBI measurements.

The nutation angle error signature for the VLBI observations can be viewed as either an orientation change of the source coordinate frame or as an oscillation of the site coordinate frame with a period of approximately one sidereal day. (If the oscillation is coherent for the duration of our data set (~3.5 years), then the nutation amplitude errors would have the frequency of the oscillation minus one cycle per sidereal day.) Any error source in the VLBI models which could produce a diurnal oscillation of the site coordinate frame is likely to cause apparent errors in the nutation series results. One error source which could have a signature of this form is the delay due to the troposphere. (The delay due to the ionosphere is well modeled using dual-frequency observations.) The inadequately modeled tropospheric delay could easily contain almost-diurnal signatures. We believe that this error source probably contributes mostly to the scatter in our nutation results which cannot be accounted for by the statistical estimates of nutation angle standard deviations. Could this error source account for the error in the retrograde annual nutation amplitude? This possibility is unlikely because the error in the prograde annual nutation amplitude (the term which remains in-phase with the sun and hence most likely to be affected by the atmospheric delay model errors) is zero to within our quoted uncertainties, implying that the atmospheric delay errors do not remain coherent for periods of years.

A second unmodeled error which could produce errors in the nutation angle estimates is the loading effects of ocean tides. We do not think it has a large effect on our retrograde annual nutation because the tidal constituent for this nutation, ψ_1, is very small. (The tidal amplitude of ψ_1 is about 10% of the amplitude of P_1; see later discussion for the expected magnitude of the effect of P_1.) The most complete set of ocean tide models does not include this constituent because of its small size (Schwiderski, 1983). However, the prograde semiannual term could be influenced by the loading effects due to the P_1 tide. But even for this term, currently available ocean-loading models would indicate that the errors in the nutations induced by neglecting this effect would be ~0.2 mean atomic seconds (mas).

The ocean tides not only affect the VLBI measurements (through the neglect of the ocean-loading corrections), but they will also directly affect the nutations (Wahr and Sasao, 1981). We summarize in Table IV the expected contribution of ocean tides to the nutations of the Earth (compiled from Wahr and Sasao, 1981). From this table we see that our corrections to the semiannual nutations are consistent, within our uncertainties, with the expected effects of ocean tides. However, since we have neglected the ocean-loading effects in our analysis, our semiannual nutation results may be biased. Also the significance of the detection of an error in the semiannual nutation amplitude is marginal (when the effects of systematic errors on the estimate are considered). Consequently, we cannot claim that our results confirm the ocean tide effects computed by Wahr and Sasao.

The effects of the ocean tides on the nutations can not explain our estimated correction to the annual nutations. The frequency of the retrograde

TABLE IV. Effects of the ocean tides on the nutation amplitudes[a,b]

Period (days)	Direction	In phase, δa_r (mas)	Out of phase, δa_i (mas)
6789.4 (18.6 yr)	Prograde	0.1	−0.1
6789.4 (18.6 yr)	Retrograde	−1.1	1.1
182.6	Prograde	0.6	−0.6
365.3	Retrograde	−0.2	0.2

[a] Results compiled from Wahr and Sasao (1981) for the Schwiderski tide models.
[b] All other amplitudes are affected by <0.1 mas.

annual nutation is very close to the resonance frequency of the differential rotation of the fluid core and mantle of the Earth (Rochester *et al.*, 1974). The eigenmode associated with this differential rotation is called the "free core nutation" (FCN) or the "nearly diurnal free nutation" (Sasao and Wahr, 1981). The frequency of this resonance is thought to be $\sim -1/460$ cycle/sidereal day (Sasao *et al.*, 1980; Sasao and Wahr, 1981). This value is computed by assuming only pressure coupling between the core and the mantle. The proximity of the retrograde annual nutation to this resonance frequency makes the amplitude of this nutation very sensitive to the resonance frequency. Our estimated correction to this amplitude would imply that the resonance frequency used in computing the nutation series was incorrect. The existence of the imaginary component in our correction implies that the differential rotation of the core and the mantle involves a dissipative process. We modeled the frequency of the FCN resonance as $\eta = \eta_r + i\eta_i$, where η_r and η_i are the real and imaginary components of the frequency, respectively, and computed the values of η_r and η_i from the VLBI estimates of corrections to the retrograde annual nutation. The estimates of the FCN resonance frequency are given in Table V along with some theoretical estimates of this frequency. Our results differ significantly from the theoretical values. There are several possible reasons for this disagreement. Our results imply that the core is more strongly coupled to the mantle than would be predicted from pressure coupling only. The possible origins of this additional coupling are now being investigated.

The imaginary component of the estimated resonance frequency can be interpreted as the effects of dissipation on the core–mantle boundary. Bounds on the magnitude of the imaginary component are $5\times10^{-4}/\mathrm{yr} < \eta_i < 1/\mathrm{yr}$, in which the upper bound is provided by nutational analysis of optical observations (Sasao *et al.*, 1977), and the lower bound is provided by the analysis of Rochester *et al.* (1974) of the electromagnetic coupling of the core and mantle. Our estimate of η_i (in the units given above) is

TABLE V. Theoretical and estimated values of the FCN resonance frequency

Earth model	Real component (cycles/sidereal day)	Imaginary component (cycles/sidereal day)
1066A Gilbert and Dziewonski[a]	−1/460.5	0
Wang (1972)[b]	−1/466.6	0
Bullen and Haddon (1967)[b]	−1/472.7	0
Guttenberg and Bullen A (1961)[b]	−1/451.9	0
VLBI estimate	−1/(433 ± 3)	−1/(23,000 ± 4000)

[a] From Sasao and Wahr (1981).
[b] Compiled from Sasao *et al.* (1980).

TABLE VI. Estimated corrections to the nutation amplitudes due to the change in the FCN resonance frequency

Period (days)	Direction	In phase, δa_r (mas)	Out of phase, δa_i (mas)
6789.4 (18.6 yr)	Prograde	0.2	−0.1
6789.4 (18.6 yr)	Retrograde	−1.9	0.5
182.6	Prograde	0.3	−0.1
182.6	Retrograde	<0.1	<0.1
365.3	Prograde	<0.1	<0.1
365.3	Retrograde	−1.9	0.8

$\eta_i = 0.10 \pm 0.02$/yr (see Table V). A coupling mechanism which could generate the observed value of η_i is also being investigated.

The estimate of the correction to the FCN resonance frequency allows us to make a prediction of the expected correction to the 18.6-yr nutation amplitudes. In Table VI, we summarize the expected corrections to the nutation amplitudes due to the change in the FCN resonance frequency. The combined effects of the change in the FCN resonance frequency and the ocean tides, lead to an expected correction of $-3.0 - 1.7i$ mas in the amplitude of the retrograde 18.6-yr nutation and amplitude. In Figs. 8 and 9, we show the nutation angle residuals after removing the estimated correction to the annual and semiannual nutations, and the *predicted* correction to the 18.6-yr nutations. An encouraging, but certainly not definitive, sign that the predicted corrections to the 18.6-yr nutations are of the correct magnitude is that the slope of the residual nutation in obliquity is zero within our uncertainties (±0.02 arc sec/cycle). This result would imply that the slope of the nutation in obliquity residuals in Fig. 6 is due to an error in the 18.6-yr nutation terms. The expected secular decrease in the obliquity, not accounted for by planetary perturbations of the Earth's orbit is <0.0004 arc sec/cycle (Rochester, 1976).

The $\Delta\psi \sin \varepsilon_0$ residuals have a slope of -0.11 ± 0.02 arc sec/cycle (the quoted uncertainty is the scaled statistical standard deviation and does not take into account errors in the predicted values of the 18.6 yr nutation terms), indicating that the Lieske *et al.* (1977) precession constant may be in error by this amount divided by $\sin \varepsilon_0$, -0.28 ± 0.05 arc sec/cycle. However, our use of a predicted correction to the 18.6 yr nutation terms, places a high uncertainty on this correction to the precession constant. (Our estimate of the correction to the precession constant is most sensitive to the out-of-phase component of the 18.6 yr nutation in longitude. The sensitivity of the precession constant estimate to changes in the magnitude of this nutation component is ~ -0.03 (arc sec/cycle)/mas.) We can state that, unless the

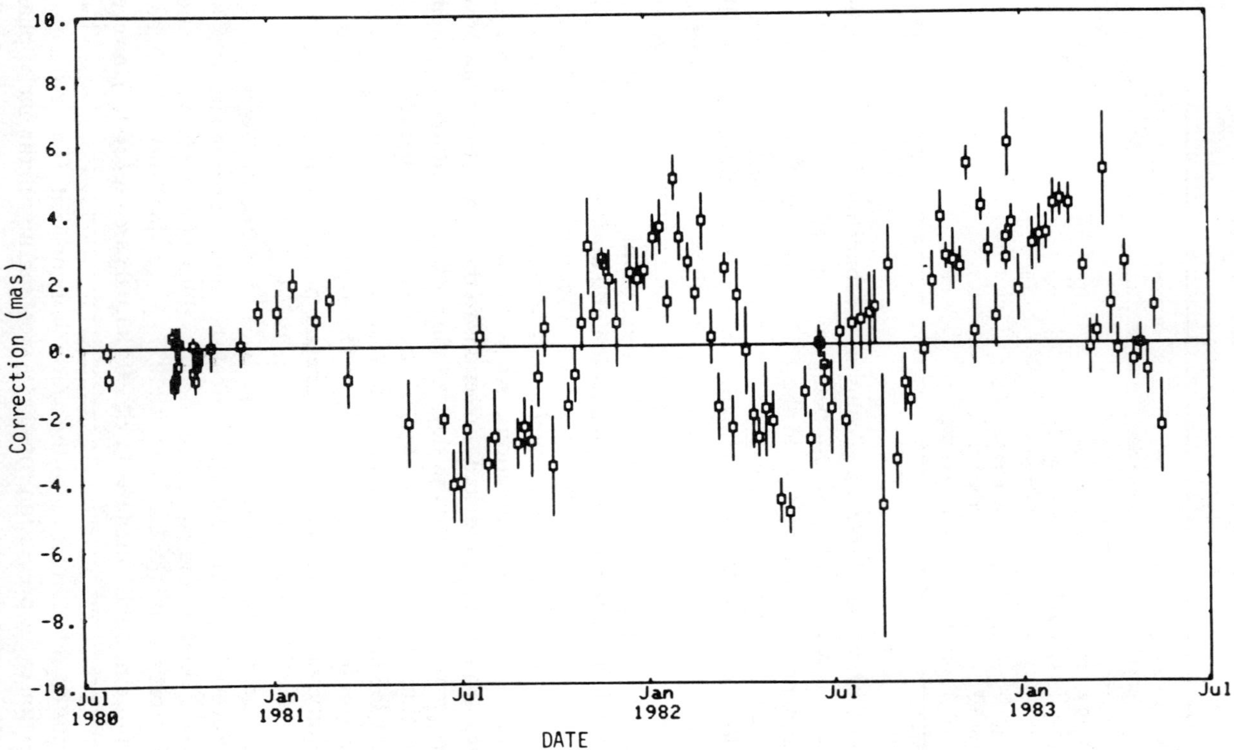

Fig. 6. Estimates of the corrections to the Wahr nutation in obliquity $\Delta\varepsilon$.

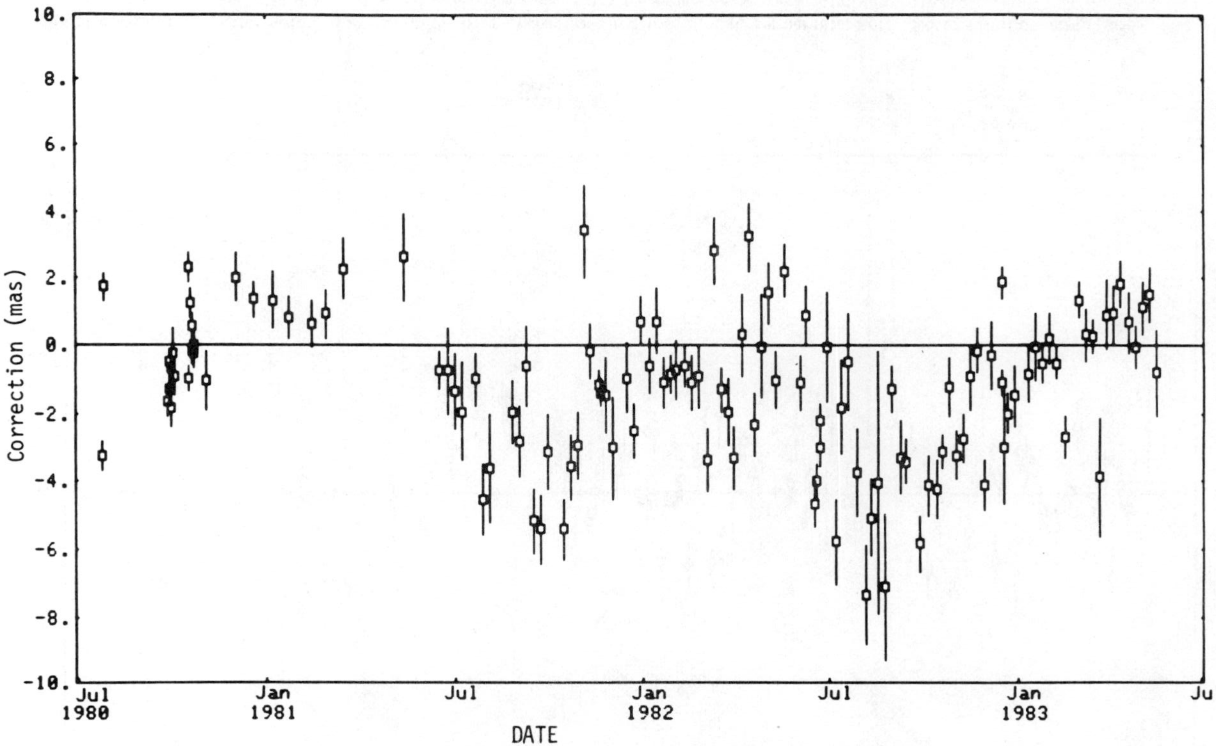

Fig. 7. Estimates of the corrections to the Wahr nutation in longitude $\Delta\psi \sin \varepsilon_0$.

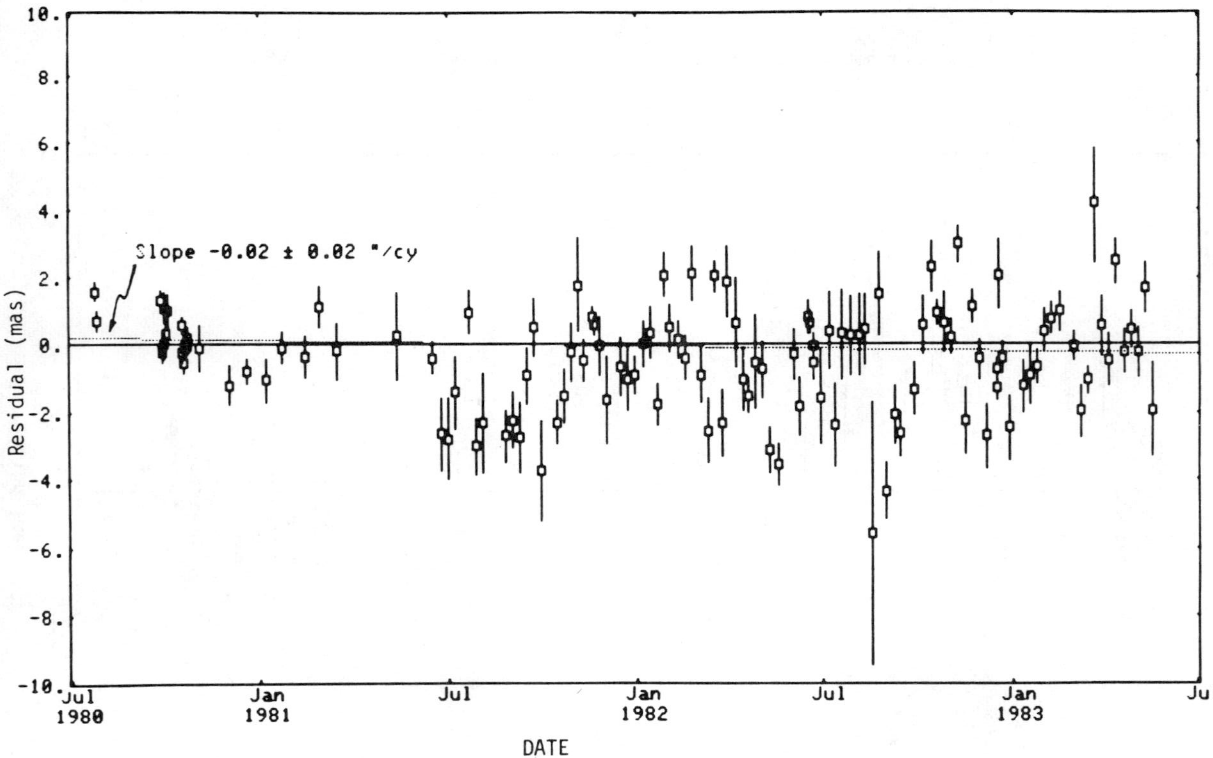

Fig. 8. Residual corrections to the nutation in obliquity after corrections for the annual, semiannual and 18.6-yr nutation amplitudes. The wrms and nrms scatters of the estimates about the mean value are 1.1 mas and 2.1 mas, respectively (see Fig. 1). The uncertainty of the slope estimate is one statistical standard deviation multiplied by the nrms scatter.

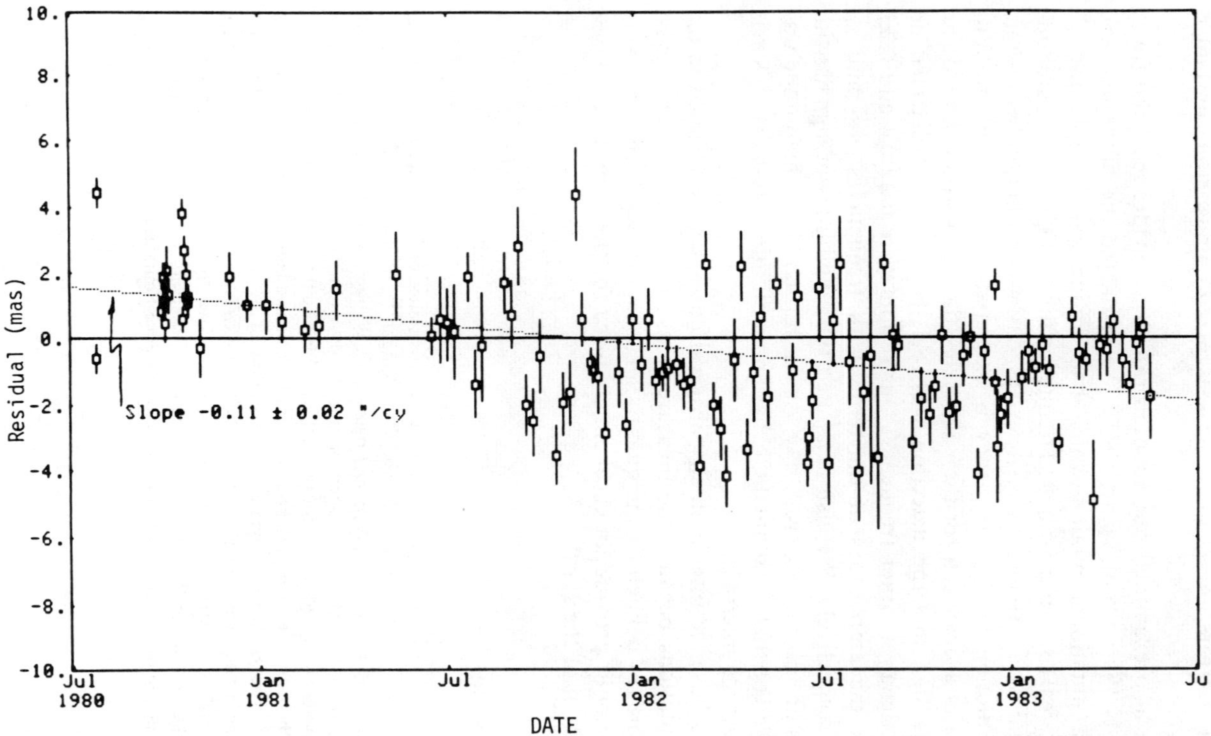

Fig. 9. Residual corrections to the nutation in longitude after corrections for the annual, semi-annual and 18.6-yr nutation amplitudes. The wrms and nrms scatters of the estimates about the mean value are 1.8 mas and 2.7, respectively; about the best-fit sloped line they are 1.3 mas and 2.1, respectively (see Fig. 1). The uncertainty of the slope is one statistical standard deviation multiplied by the nrms scatter of the residual from the best-fit slope line.

imaginary component of the 18.6-hr nutation terms are much larger than our predicted errors, the error in the Lieske *et al.* precession constant is small, <0.5 arc sec/cycle.

We have also examined our nutation results for evidence of the freely excited FCN (Sasao and Wahr, 1981). We estimated the amplitude and phase of a nutational motion which had frequencies of $-\frac{1}{433}$ and $\frac{1}{433}$ cycles/(sidereal) day (our estimated frequency for the FCN). The amplitude of any signal with a frequency of $\frac{1}{433}$ cycles/day was estimated because no nutational motion at this frequency is expected—a null result for this frequency and a detection for the $-\frac{1}{433}$-cycles/day term would provide an encouraging sign that the detection was real. However, we detected no significant signal at either frequency. Our results were 0.2 ± 0.4 and 0.2 ± 0.4 mas for the amplitudes of the terms with frequencies $-\frac{1}{433}$ and $\frac{1}{433}$ cycle/day, respectively. We also computed the amplitudes of signals with frequencies of $-\frac{1}{460}$ and $\frac{1}{460}$ cycle/day. Again, no significant amplitude was detected. We conclude that any FCN has an amplitude of less than 0.8 mas (95% confidence interval).

Very long baseline interferometry observations are ideally suited to the study of nutations because the measurements are referred to an almost inertial frame (as defined by the quasars). The continuing VLBI observing programs should provide an excellent resource for the continued studies of the nutation of the Earth.

Acknowledgments

This work would not have been possible without the assistance of I. I. Shapiro, T. A. Clark, C. Ma, J. W. Ryan, J. C. Pigg, B. R. Schupler, C. A. Knight, D. B. Shaffer, N. R. Vandenberg, R. J. Cappallo, H. F. Hinteregger, A. E. E. Rogers, J. C. Webber, A. R. Whitney, G. Elgered, G. Lundqvist, B. E. Corey, D. S. Robertson and J. M. Wahr. We thank the staffs of the participating observatories for their indispensable aid. We also thank R. J. Coates, E. A. Flinn, and T. L. Fischetti for their support, and NGS for making available the Polaris data used in these analyses. The HCO experimenters were supported by the United States Air Force Geophysics Laboratory, contract F19628-83-K-0031; NASA contract NAS5-27571 and NSF grant EAR-83-02221. The Haystack Observatory is operated by a grant from the National Science Foundation, and the National Radio Astronomy Observatory is operated by Associated Universities, Inc., under contract with the National Science Foundation.

References

Allenby, R. J. (1979). *Tectonophysics* **60**, T27–T35

Aoki, S., Guinot, B., Kaplan, G. H., Kinoshita, H., McCarthy, D. D. and Seidelman, P. K. (1982). *Astron. Astrophys.* **105**, 359–361.

Atkinson, R. d'E. (1973). *Astron. J.* **78**, 147-151.

Atkinson, R. d'E. (1975). *Mon. Not. R. Astron. Soc.* **71**, 381-386.

Cannon, W. H., Langley, R. B., Petrachenko, W. T. and Kouba, J. (1979). *J. Geophys. Res.* **84**, 229-236.

Carter, W. E., Rogers, A. E. E., Counselman, C. C. and Shapiro, I. I. (1980). *J. Geophys. Res.* **85**, 2685-2687.

Carter, W. E., Robertson, D. S., Mader, G. L. and MacKay, J. R. (1984). *J. Geophys. Res.* (to be published).

Doser, D. I. and Smith, R. B. (1982). *Bull. Seismol. Soc. Am.* **72**, 525.

Eckert, W. J., Walker, M. J. and Eckert, D. (1966). *Astron. J.* **71**, 314-332.

Hamilton, W. and Myers, W. B. (1966). *Rev. Geophys.* **4**, 509.

Herring, T. A., Corey, B. E., Counselman III, C. C., Shapiro, I. I., Ronnang, B. O., Rydbeck, O. E. H., Clark, T. A., Coates, R. J., Ma, C., Ryan, J. W., Vandenberg, N. R., Hinteregger, H. F., Knight, C. A., Rogers, A. E. E., Whitney, A. R., Robertson, D. S. and Schupler, B. S. (1981). *J. Geophys. Res.* **86**, 1647-1651.

Herring, T. A., Corey, B. E., Shapiro, I. I., Rogers, A. E. E., Whitney, A. R., Clark, T. A., Knight, C. A., Ma, C., Ryan, J. W., Schupler, N. R., Vandenberg, N. R., Elgered, G., Lundqvist, G., Ronnang, B. O., Campbell, J. and Richards, P. (1983). *Proc. Int. Symp. Earth tides 9th*, (J. Kuo, ed.), pp. 205-211. Schweizerbart'sche Verlagsbuchhandlung, Stuttgart.

Hinteregger, H. F., Shapiro, I. I., Robertson, D. S., Knight, C. A., Ergas, R. A., Whitney, A. R., Rogers, A. E. E. , Moran, J. M., Clark, T. A. and Burke, B. F. (1972), *Science* **178**, 396-398.

Kaula, W. M. (1966). *Theory of Satellite Geodesy.* Ginn (Blaisdell), Boston, Massachusetts.

Kinoshita, H., Nakajima, K., Kubo, Y., Nakagawa, I., Sasao, T. and Yokoyama, K. (1979). *Publ. Int. Latitude Observatory* **XII**, 71-107.

Lieske, J. H., Lederle, T., Fricke, W. and Morando, B. (1977). *Astron. Astrophys.* **58**, 1-16.

Marini, J. W. (1974). *Intern. NASA Mem.* Feb.

Melchior, P. J. (1978). *The Tides of the Planet Earth.* Pergamon, Oxford.

Minster, J. B. and Jordan, T. H. (1978). *J. Geophys. Res.* **83**, 5331-5354.

Morrison, N. (1969). *Introduction to Sequential Smoothing and Prediction.* McGraw-Hill, New York.

Moyer, T. D. (1981). *Cel. Mech.* **23**, 33-68.

Niell, A. E., Ong, K. M., MacDoran, P. F., Resch, G. M., Morabito, D. D., Claflin, E. S. and Dracup, J. F. (1979). *Tectonophysics* **52**, 49-58.

Ong, K. M., MacDoran, P. F., Thomas, J. B., Fliegel, H. F., Skjerve, L. J., Spitzmesser, D. J., Batelaan, P. D. and Paine, S. R. (1976). *J. Geophys. Res.* **81**, 3587-3593.

Robertson, D. S. (1975). Ph.D. thesis Massachusetts Institute of Technology, Cambridge, Massachusetts.

Robertson, D. S. and Carter, W. E. (1982). *IAU Colloq. no. 63: High Precision Earth rotation and Earth-Moon Dynamics* (O. Calame, ed.). Reidel Publ., Boston, Massachusetts.

Robertson, D. S., Carter, W. E., Corey, B. E., Cotton, W. D., Counselman, C. C., Shapiro, I. I., Whittels, J. J., Hinteregger, H. F., Knight, C. A., Rogers, A. E. E., Whitney, A. R., Ryan, J. W., Clark, T. A., Coates, R. J., Ma, C., Moran, J. M. (1979). In *Time and the Earth's Rotation* (D. D. McCarthy and J. D. Pilkington, eds), pp 217-224. Reidel Publ. Boston, Massachusetts.

Rochester, M. G. (1976). *Geophys. J. R. Astron. Soc.* **46**, 109-126.

Rochester, M. G., Jensen, O. G. and Smylie, D. E. (1974). *Geophys. J. R. Astron. Soc.* **38**, 349-363.

Rogers, A. E. E., Knight, C. A., Hinteregger, H. F., Whitney, A. R., Counselman, C. C., Shapiro, I. I., Gourevitch, S. A. and Clark, T. A. (1978). *J. Geophys. Res.* **83**, 325-334.

Rogers, A. E. E., Cappallo, R. H., Hinteregger, H. F., Levine, J. R., Nesman, E. F., Webber, J. C., Whitney, A. R., Clark, T. A., Ma, C., Ryan, J., Corey, B. E., Counselman, C. C., Herring, T. A., Shapiro, I. I., Knight, C. A., Shaffer, D. B., Vandenberg, N. R., Lacasse, R., Mauzy, R., Rayhrer, B., Schupler, B. and Pigg, J. C. (1983). *Science* **219**, 51-54.

Ryan, J. W., Clark, T. A., Coates, R. J., Ma, C., Wildes, W. T., Herring, T. A., Shapiro, I. I., Corey, B. E., Counselman III, C. C., Hinteregger, H. F., Rogers, A. E. E., Whitney, A. R., Knight, C. A., Vandenberg, N. R., Pigg, J. C., Schupler, B. R. and Ronnang, B. O. (1986). *J. Geophys. Res.* **91**, 1935-1946.

Sassao, T., and Wahr, J. M. (1981). *Geophys. J. R. Astron. Soc.* **64**, 729-746.

Sasao, T., Okamota, I. and Sakai, S. (1977). *Publ. Astron. Soc. Jpn* **29**, 83-105.

Sasao, T., Okubo, S. and Saito, M. (1980). *Proc. IAU Symp., no 78* (E. P. Fedorov, M. L. Smith and P. L. Bender, eds), pp. 165-183. Reidel Pub. Boston, Massachusetts.

Schwiderski, E. W. (1983). *Proc. Int. Symp. Earth tides 9th* (J. Kuo, ed.), pp. 419-428 Schweizerbart'sche Verlagsbuchhandlung, Stuttgart.

Shapiro, I. I. (1967). *Science* **157**, 806-808.

Shapiro I. I. (1976). In *Methods of Experimental Physics*, Vol. 12, Part C, (M. L. Meeks, ed.), pp. 261-276. Academic Press, New York.

Shapiro, I. I., Robertson, D. S., Knight, C. A., Counselman III, C. C., Rogers, A. E. E., Hinteregger, H. F., Lippincott, S., Whitney, A. R., Clark, T. A., Niell, A. E. and Spitzmesser, D. J. (1974). *Science* **186**, 920-922; ibid, **191**, 451, 1976.

Sovers, O. J., Thomas, J. B., Fanselow, J. L., Cohen, E. J., Purcell, Jr., G. H., Rogstad, D. H., Skjerve, L. J. and Spitzmesser, D. J. (1984). *J. Geophys. Res.* **89**, 7597-7607.

Stewart, J. H. (1977). *Geol. Soc. Am. Bull.* **82**, 1019-1044.

Thomas, J. B., Fanselow, J. L., MacDoran, P. F., Spitzmesser, D. J. and Skjerve, L. J. (1976). *J. Geophys. Res.* **81**, 995-1005.

Wahr, J. M. (1981a). *Geophys. J. R. Astron. Soc.* **64**, 705-727.

Wahr, J. M. (1981b). *Geophys. J. R. Astron. Soc.* **64**, 677-704.

Wahr, J. M., and Sasao, T. (1981). *Geophys. J. R. Astron. Soc.* **64**, 747-766.

Whitney, A. R., Rogers, A. E. E., Hinteregger, H. F., Knight, C. A., Levine, J. I., Lippincott, S., Clark, T. A., Shapiro, I. I. and Robertson, D. S. (1976). Radio Sci. **11**, 421-432.

3.3 GPS-Based Geodetic Measurement Systems

C. L. Thornton, J. L. Fanselow and N. A. Renzetti

Jet Propulsion Laboratory, California Institute of Technology, Pasadena, California, USA

I. Introduction

During the past decade, various U.S. agencies, such as the National Aeronautical and Space Administration (NASA), the National Oceanic and Atmospheric Administration (NOAA) and the Defense Mapping Agency (DMA), have developed a number of space techniques for precise measurement of geodetic baselines, Earth orientation and various other geophysical phenomena. Each of these methods relies upon extra-terrestrial reference sources, such as the distant radio stars used for very long baseline interferometry (VLBI) and the Earth orbiters (e.g. LAGEOS) used for satellite laser ranging (SLR). Both VLBI and SLR have achieved significant improvements over conventional approaches for measuring vector baselines. Subdecimeter accuracies are routinely obtained over distances of several thousand kilometers, and 1-3 cm performance on intercontinental scales is expected by the end of the decade (Sovers *et al.*, 1984; Davidson and Trask, 1985; Herring *et al.*, 1981; Robertson and Carter, 1984).

The application of these space systems for regional geodesy is limited, however, by the high cost of instrumentation and operation (Eller *et al.*, 1980). Operations costs are particularly high for SLR systems due to their susceptibility to inclement weather. Site occupations of several weeks duration are typically required. Mobile VLBI systems are less sensitive to poor weather, but involve substantial implementation and operations overhead due to the extremely low signal strength of the distant radio sources. The weak signals necessitate high-gain antennas, low-noise receivers and multimegabit recording systems, as well as expensive special purpose central site cross-correlation systems. Moreover, observing time required to achieve subdecimeter baseline accuracy is of the order of 24 hr. For geodetic applications requiring multiple site occupations within a few days, more compact, efficient systems are needed.

The global positioning system (GPS) under development by the Department of Defense offers an attractive opportunity for low-cost worldwide geodesy. The system will become fully operational in 1991 with the placement of 18 operating NAVSTAR satellites into equally-spaced orbits around the Earth. The orbits are spatially distributed so that at any moment from virtually any location on Earth a properly equipped user can receive the broadcast signals from at least four satellites. Each received signal provides a measure of pseudorange (range plus satellite/user clock offset) to the observed satellite as well as information on GPS ephemerides and time of day. Four such signals received simultaneously at two ground locations permit the instantaneous measurement of the vector baseline and the clock offset between the receiving stations.

Studies have shown that regional GPS-based geodetic networks can achieve baseline accuracies of 1-3 cm over distances of several thousand kilometers with no more than 4 hours of observing time (Larden and Bender, 1982; Thornton and Beckman, 1984). This level of accuracy requires high-performance GPS receivers and reliable calibration of both propagation and receiver delays. Receiver systems and calibration assemblies currently under development by NASA and several commercial ventures show promise of achieving the required performance. Moreover, these terminals will be sufficiently compact and portable that two people can set up, operate and disassemble them in less than 1 day. It is expected that the cost of a typical station, including the GPS receiver and related calibration equipment, will be less than $300,000.00, approximately an order of magnitude less than mobile VLBI or SLR systems of comparable performance.

The low-cost, high-performance features of potential GPS geodetic systems have sparked an interest within the NASA Geodynamics Program. System studies have been undertaken, and promising geodetic applications in the Caribbean and elsewhere have been identified (Thornton *et al.*, 1983; Dixon, 1984; Dixon *et al.*, 1985). This chapter describes the major elements of an envisioned NASA-sponsored GPS geodetic system. In the next section, the GPS program characteristics, including the expected modes of operation and implementation schedules, are described. The NASA geodetic system design objectives and related subsystem requirements are discussed in Sections III and IV. A description of potential GPS geodetic system performance and major error sources are given in Section V, while Section VI addresses the development status of key subsystems and related technologies.

II. The Global Positioning System

A. Program Description

The GPS is a multiagency program managed by the United States Air Force. The program is divided into three segments: space, control and user. The space segment includes the satellites. The control segment includes the ground monitor, command and control functions. This segment performs the tracking, orbit determination and time synchronization functions. It also periodically uplinks the computed ephemerides and timing information to the individual satellites. The user segment includes GPS receivers, calibration assemblies and navigation data-processing functions.

When fully operational, the space segment will involve a total of 18 operating satellites plus 6 on-orbit spares. These satellites will be placed in six different 12-hr orbits having nodal separations of 60° and inclinations of 55°. Each orbit will support three equally-spaced operating satellites plus an on-orbit spare.

There are currently six prototype satellites in operation being used for testing and proof-of-concept demonstrations. The Block II series of operational satellites are scheduled for deployment beginning in 1986. The system is expected to be operating with 12 satellites by the end of 1987 and with the full constellation of 18 in place by the end of 1988. The reader may refer to Fig. 1 for an illustration of the planned operating system and to Milliken and Zoller (1980) and Spilker (1980) for more details on system design and operation.

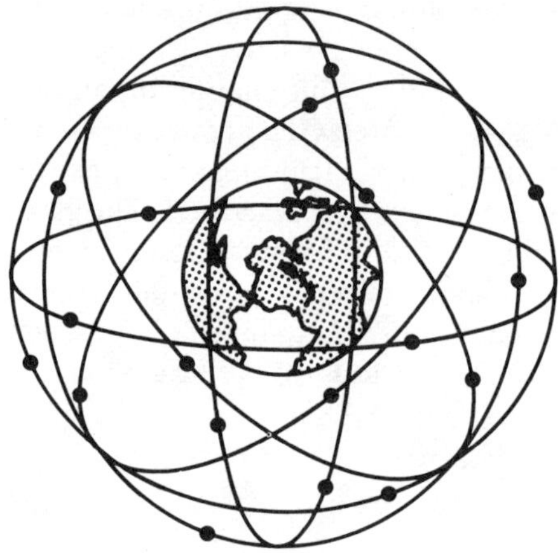

Fig. 1. GPS satellite configuration.

B. Technical Description

The GPS satellites each broadcast on two L-band carriers at 1575 and 1228 MHz. These frequencies are derived from the basic clock rate of 10.23 MHz by applying multipliers of 154 and 120, respectively. The L1 and L2 carriers are modulated by pseudorandom spread spectrum codes. The coarse acquisition (C/A) code has a chip rate of 1.023 MHz and appears on the L1 channel only, while the precision (P) code operates at 10.23 MHz and is available on both L1 and L2. Both carriers also support a 50 bits per second (bps) data message which contains satellite ephemeris and timing information. The power spectral density of an example signal is shown in Fig. 2. Further details on signal structure can be found in Spilker (1980).

In the nominal mode of operation, a user obtains a measure of one-way range to each visible satellite by aligning the ranging code in the receiver with the received code embedded in the spread spectrum signal. This

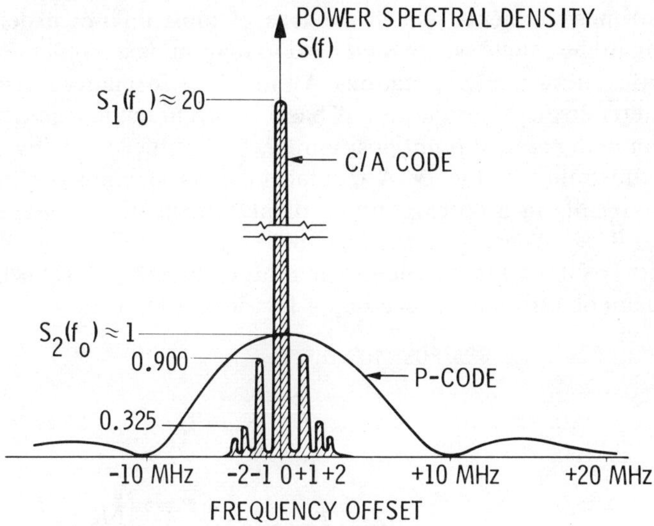

Fig. 2. Sample GPS signal spectra.

measurement, referred to as pseudorange, is the sum of true range, propagation delays and timing and phase delays in both the satellite and the receiver. It is also possible to obtain pseudorange without knowledge of the codes through nonlinear detection techniques that yield the tones associated with the P and C/A chip rates (Buennagel *et al.*, 1984; Crow *et al.*, 1984; Melbourne, 1984).

Ionospheric delays in the P-code pseudorange are essentially removed by dual-frequency calibration, using the L1 and L2 channels. Theoretically, the timing errors originating in the satellites are small since each is equipped with an atomic clock which is periodically reset to GPS-system time. The principal content of the calibrated pseudorange measurement, then, is true range plus user clock offset relative to GPS-system time. Therefore, four such measurements from the multiple satellites in view enable an instantaneous determination of user position and time relative to the GPS system. Position accuracies using the P-code are limited primarily by errors in the broadcast ephemerides. Satellite-position errors map nearly one-for-one into user position. The most optimistic projection of broadcast ephemeris accuracy is on the order of 1–2 m. Therefore, a static user operating in the conventional mode can expect position accuracies to no better than a meter.

However, improved accuracy can be obtained by using differential positioning techniques. This approach involves the simultaneous reception of signals at multiple locations. These observations are combined at a central processing site where they are differenced pairwise between terminals. These differential measurements, which are free of most unknown delay errors originating in the satellites, are then used to determine the relative positions of the participating receiving stations. Although performance is still affected by ephemeris errors, degradation of baseline accuracy is significantly less severe than in the case of point positioning. This reduced sensitivity to orbit errors is illustrated in Fig. 3. A displacement in satellite position by an amount α results in a corresponding displacement of the baseline by an amount $\alpha(B/D)$ where B is the baseline length and D is the distance to the satellite (~20,000 km). Hence, over short distances (<200 km), sensitivity to ephemeris errors is reduced by a factor of 100 or more.

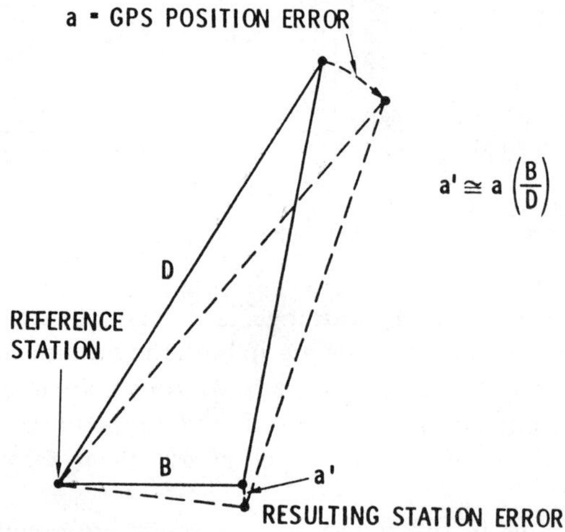

Fig. 3. Reduced sensitivity to GPS orbit errors via differential positioning.

The analysis in Section V shows, however, that for centimeter geodesy over distances greater than a few hundred kilometers, orbit errors still limit system performance. In order to achieve centimeter baseline accuracy over these longer distances, it is necessary to operate the geodetic stations in conjunction with three or more fiducial sites whose relative locations are known to a few centimeters or better. The known fiducial sites define an earth-fixed coordinate frame relative to which the satellite orbits and geodetic baselines can be simultaneously determined.

Precision geodesy also requires the measurement of carrier phase as well as the P- and C/A-code pseudorange measurements. Typical pseudorange precision is about 10–20 cm over a few minutes averaging time. The additional use of the carrier phase is significant since these measurements can be expected to have millimeter precision (1%). When corrected for the ionosphere, this data type provides a measure of range change with subcentimeter precision. In order to achieve comparable metric accuracy, however, it is necessary to carefully calibrate for troposphere delays and phase errors introduced by the receiver.

The GPS data types used for precision geodesy are the differential, calibrated carrier phases and P-code pseudoranges. The carrier phase observable provides a measure of differential range, modulo $2n\pi$, since the integer number of carrier cycles in the differential arrival time of the wavefront at the two stations is not directly measured. If cycle ambiguity resolution can be achieved by using sufficiently precise P-code differential range, or by other means, then the fully calibrated carrier phase yields a differential range measurement having centimeter precision. Removing the cycle ambiguity is no simple task, except over the shortest baselines. A carrier wavelength is approximately 20 cm, and hence an independent measure of differential range accurate to substantially better than 10 cm is required.

Even without cycle ambiguity resolution, the carrier phase observables provide valuable information in the form of an integrated Doppler data type. In this application, carrier phase is tracked continuously, without loss of cycles, during the observation interval. The $2n\pi$-cycle ambiguity then introduces an unknown bias which is constant during the tracking interval. This bias can be included as an estimated parameter which, in effect, allows the representation of carrier phase as a precise measure of range change. In this application of carrier phase the P-code pseudorange is a valuable additional data type for determining the unknown phase bias.

III. GPS Geodetic System Design

A. System Performance Objectives

The objectives of the NASA-sponsored GPS geodetic system development are to demonstrate the ability to measure vector baselines over distances of at least 3000 km with an accuracy of 1–3 cm and to apply this capability to the study of plate tectonics and crustal deformation. In order to fully utilize available time in the field, it is desirable to have highly portable, compact stations which can be easily moved, set up and operated by two people to obtain a baseline measurement in considerably less than 1 day. This goal maps into a need for field stations which can achieve the 1–3-cm baseline accuracy goal with no more than 4 hr of observing time. A related objective is to develop field stations which can be reproduced for less than $300,000.00 including the GPS receiver and all calibration assemblies.

B. System Description

The general system objectives lead to a system design which includes three major components: mobile GPS field stations, a network of well-known fiducial locations and a central processing facility. The GPS data acquired simultaneously at the mobile and fiducial sites are brought to a central processing facility where the differential observables are calculated, calibrated and analyzed to jointly determine the GPS orbits and the geodetic baselines.

IV. Subsystem Requirements

The broad objective of measuring geodetic baselines to an accuracy of 1–3 cm during a 1-day site occupation places requirements on each of the major subsystems. These requirements are delineated in the following three subsections.

A. Fiducial Network

The need for centimeter-level baseline accuracy in three dimensions over distances of several thousand kilometers translates into a requirement for a regional network of fiducial sites whose relative positions are known to approximately 1 cm (Thornton and Beckman, 1984). Sensitivity to fiducial baseline accuracy is proportional to the length of the geodetic baseline to be measured, and hence the 1-cm requirement can be relaxed to a few centimeters for some applications. Error analysis has shown that regional geodesy requires a minimum of three fiducial sites so positioned as to provide substantial (~3000 km) orthogonal baselines, e.g. north–south and east-west. The network must be sufficiently close to the geodetic region being surveyed so that adequate simultaneous viewing of the GPS satellites by the mobile and fiducial stations is assured. It is not necessary, however, for the mobile sites to lie within the fiducial net. Mobile sites can be located as far away from the center of the network as the size of the fiducial network itself without suffering a noticeable degradation in baseline accuracy. Given this latitude in placement of the fiducial sites, it is possible to establish a global network to support centimeter geodesy worldwide with no more than nine or ten fiducial stations.

In order to establish and maintain a 1-cm fiducial network, it will be necessary to place the fiducial sites in readily accessible locations. Site selection criteria must include considerations for timely maintenance of

GPS receivers and the necessity for periodic measurement of site locations via some independent system such as VLBI or SLR. Potential degradation of the network by unmodeled ocean loading suggests that fiducial sites be placed at least 50 km inland.

B. GPS Field Station

A system requirement for centimeter baseline accuracy from a few hours of observations maps into a requirement for 1-cm differenced carrier phase measurements every few minutes. Simultaneous, or near simultaneous, measurements from all satellites in view is necessary in order to remove clock errors. The need for simultaneous acquisition of multiple satellites dictates the use of omnidirectional antennas or an appropriately designed phased array. The 1-cm measurement errors are assumed to be zero mean and random, composed of the contributions from system noise, multipathing, uncalibrated media effects and instrumental phase errors. The receiver to support this task must simultaneously acquire the L1 and L2 carriers and preferably measure the P1 and P2 pseudoranges. In order to permit carrier-cycle ambiguity resolution, it is necessary that each single station P-code pseudorange be measured to a precision (including system noise, uncalibrated multipath and instrumental delays) of about 2 cm with a few minutes integration time.

The measurement of differential carrier phase to an accuracy of 1 cm depends on the careful calibration of the troposphere. A water vapor radiometer (WVR) is required at each site to correct for the water vapor in the troposphere. Allowable errors in this calibration are 0.5 cm random errors along the line-of-sight delay to each satellite and no more than 0.5 cm zenith delay error at each site. The zenith delay error introduces an elevation-dependent systematic error which tends to limit accuracy in the local vertical component of the baseline. Systematic errors such as these must be minimized, or at least be adequately characterized, in order to achieve better than 3-cm accuracy in baseline vertical measurements. Calibration of the dry portion of the troposphere requires the measurement of barometric pressure at each site to an absolute accuracy of 0.4 mbar or better.

C. Central Processing Facility

The central processing facility must be equipped to receive and process all data taken by the receivers in the field, solve for GPS orbits and determine geodetic baselines accurate to a centimeter. To achieve this goal, the facility must also provide the capability for establishing and maintaining a 1-cm fiducial network.

V. System Performance Analysis

A. Nominal System Configuration and Error Models

A number of system configurations have been evaluated for determination of geodetic baselines, both in the Caribbean region and in the South Pacific near Australia and New Zealand (Thornton *et al.*, 1983; Thornton and Beckman, 1984). By varying the number and geometry of the assumed fiducial stations it was quickly determined that regional geodesy at the centimeter level over distances of several thousand kilometers requires at least three fiducial sites near the region of interest. Moreover, three sites provide optimum system performance if they are positioned so that interfiducial baselines provide substantial (less than 3000 km) north–south and east–west projections. This arrangement assures adequate geometry to precisely determine the GPS orbits. Lack of adequate geometry is manifested in reduced accuracy over longer baselines where sensitivities to orbit errors are more substantial. Example fiducial networks are illustrated in Figs. 4 and 5. Figure 4 represents an ideal arrangement for regional geodesy in the Caribbean, while Fig. 5 presents a comparable network for the South Pacific.

Fig. 4. Example of fiducial network (●) and mobile sites (○) for Caribbean study.

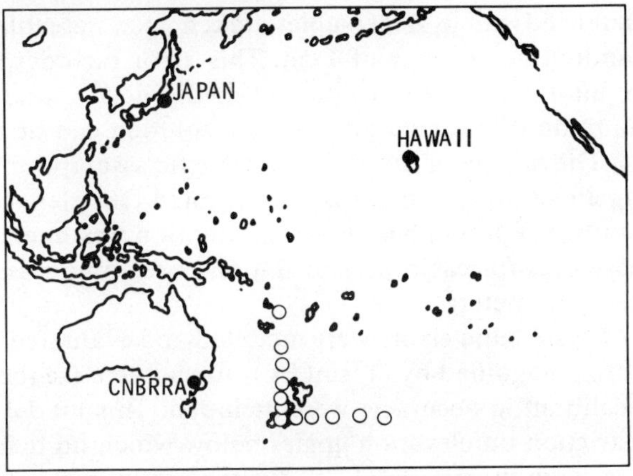

Fig. 5. Example fiducial network (●) and mobile sites (○) for South Pacific study.

Aside from the general requirements for good fiducial network geometry and reasonable proximity to the survey region, there is considerable latitude involved in the specific location of individual fiducial sites. For example, the Caribbean study (Thornton *et al.*, 1983) considered not only the Texas–Massachusetts–Columbia network shown in Fig. 4, but also arrangements such as Texas–Massachusetts–Brazil and California–Massachusetts–Ecuador. No appreciable differences in system performance were identified.

In each case studied, it was assumed that baselines between regional fiducial sites have been previously measured by some independent system such as VLBI or SLR to an accuracy of 1 cm. It was further assumed that each fiducial and mobile site was equipped with a field station meeting the requirements outlined in Section IV.B. Simultaneous, fully calibrated GPS measurements made at any two of the receiving sites were assumed to be explicitly differenced in order to eliminate clock errors and phase delays originating in the satellites. This operation leaves embedded in the data an error due to unknown clock offsets between the two ground stations. These errors could be removed by a second differencing of measurements taken simultaneously from two different satellites, or, if the ground clocks are sufficiently well-behaved, they could be modeled as stochastic processes in the estimator. The results given here are based on a linear model for the clocks over a full 4-hr observation period, and could therefore be considered somewhat optimistic for systems using quartz clocks. However, the most conservative approach, by performing a double differencing of observables, typically produces less than a 20% degradation in performance relative to the results given here (Davidson *et al.*, 1984).

Singly differenced range observables, taken at a rate of 12/hr, were assigned a random noise error of 1 cm. This error includes the effects of system noise, multipath and uncalibrated media delays, and is consistent with an assumption that carrier ambiquity resolution can be achieved (c.f. Section II.B.). The impact of this rather optimistic assumption was assessed by performing an additional analysis of integrated Doppler systems. In this companion study, the 1-cm phase noise assumption was unchanged, but an unknown phase bias for each satellite-station observable was added to the list of adjusted parameters.

A number of systematic errors were modeled and evaluated. Zenith delay errors of 7.5 mm, magnified by $(1/\sin(EL))$ to characterize the line-of-sight troposphere calibration accuracy, were included. This model was coupled with a 20° restriction on elevation angles below which no data were taken. Also evaluated were the effects of unadjusted errors in the gravity model. Error contributions due to imperfectly modeled GPS solar-radiation pressure and random nongravitational forces have been found [c.f. Larden and Bender (1982) and Davidson *et al.* (1984)] to have a negligible effect on geodetic system performance and are, therefore, omitted from consideration. Errors in universal time (UT) and polar motion are also omitted since the system described here assumes a crust-fixed reference frame defined by the three fiducial stations. Any errors in Earth orientation during the relatively brief observing time (<4 hr) would essentially be absorbed by the satellite orbits. The position of the geocenter in the fiducial-based reference frame would be known to about 2–3 m if the fiducial network were established by VLBI. This error was included as an adjusted parameter along with the mobile station locations, satellite epoch states and clock parameters. A list of errors and associated uncertainties is given in Table I.

TABLE I. Error models

Parameter	Type	*A priori* sigma
GPS initial position	Adjusted	10 m
GPS initial velocity	Adjusted	1.0 mm/s
Mobile station x, y, z	Adjusted	10 m
Clock epoch each sta pair	Adjusted	4 μs
Clock rate each sta pair	Adjusted	1000 sec/sec
Geocenter, x, y, z	Adjusted	10 m
Fiducial baseline x, y, z	Considered	1.0 cm
Mapped trop at all stations	Considered	0.75 cm (Mapped 1/sin(el)
GM (Earth)	Considered	1.0e-8 Fractional error
Gravity model error	Considered	0.03° (GEM6-APL 50)
Singly-differenced phase error	Random noise	1.0 cm

B. Covariance Analysis Results

Typical baseline accuracies achievable with the assumed system models are shown in Figs 6 and 7. Each of these figures displays one-sigma accuracy in the local vertical and horizontal baseline components as a function of baseline length. Only one horizontal component is shown since results for the two directions were virtually indistinguishable. Although the performance described here is derived from the South Pacific configuration depicted in Fig. 5, baseline accuracies are nearly identical for the Caribbean scenario of Fig. 4. The results in Fig. 6 pertain to baselines with a north–south orientation, while Fig. 7 illustrates the east–west baseline performance. In each case the vertical component is the most poorly determined with an uncertainty of about 3 cm regardless of baseline length. The dominant source of error in the vertical-component determination is the troposphere.

The horizontal-component accuracies, on the other hand, are dominated by data noise. Each of the horizontal components can be determined to about 1 cm or better, although accuracy tends to degrade over longer distances (>1000 km). This can be explained by the observation that, over longer distances, the notion of a local baseline vertical loses physical

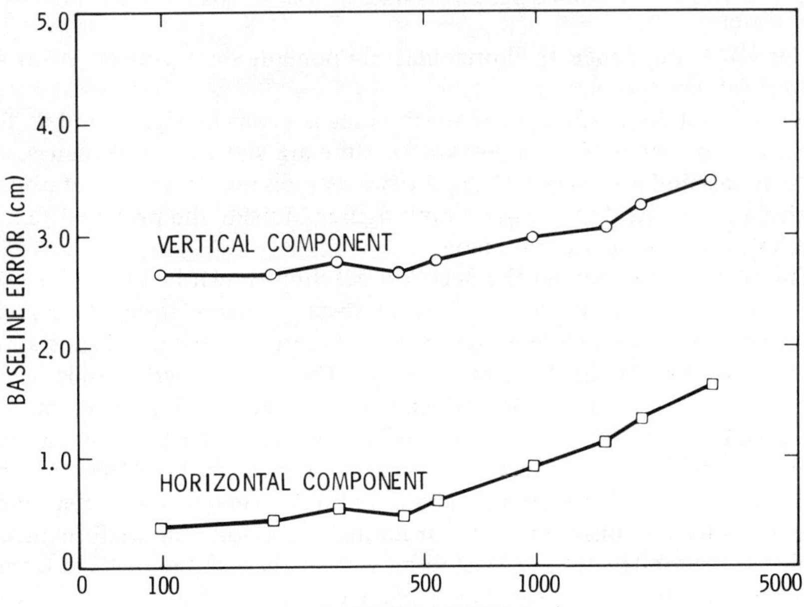

Fig. 6. Typical north–south baseline accuracies in South Pacific study (4-hr data arc; 1-cm range).

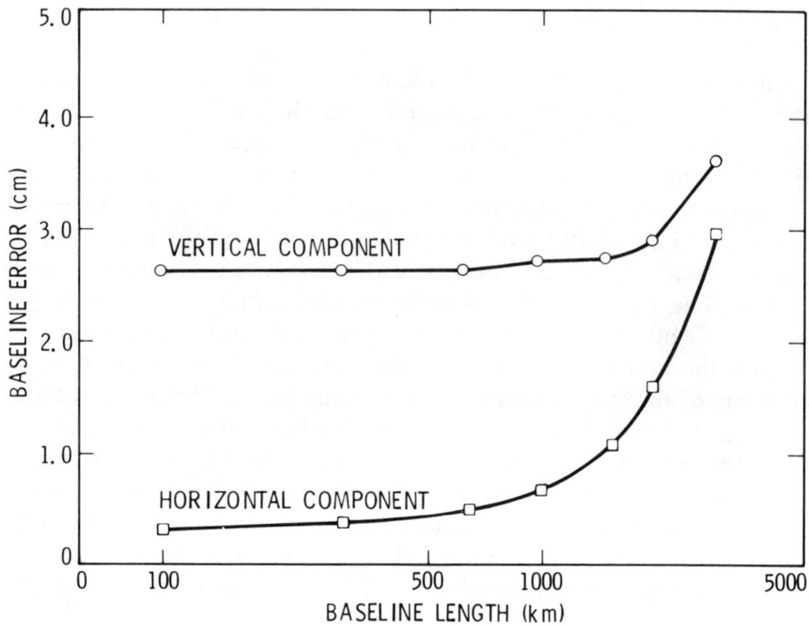

Fig. 7. Typical east–west baseline accuracies in South Pacific study (4-hr data arc; 1-cm range).

significance, and hence, the horizontal components show greater sensitivity to troposphere errors.

An illustration of these error sensitivities is given in Figs. 8 and 9. The one-sigma dispersions for a 300-km baseline are shown in two planes, the vertical-east and north–east. Figure 8 demonstrates the dominance of troposphere in the vertical error, while both figures indicate the principal role of data noise in the horizontal plane.

The increased errors on the 3000-km baseline shown in Fig. 7 illustrate the sensitivity of baseline accuracy to distance away from the fiducial network. In this case, mobile sites were deliberately selected at a substantial distance to the east of the central region. The most easterly site is nearly 6000 km from the center of the fiducial network, and from previous studies (Davidson *et al.*, 1984) we would expect the longest east–west baseline associated with that site to be less accurately measured. The comparable 3000-km north–south baseline, as shown in Fig. 6, is much better determined. These results are consistent with our findings that optimum performance is achieved if mobile sites are located no further than about 5000 km from a fiducial network of this size. Allowable distance from the fiducial network would be somewhat less, however, for a system using doubly-differenced data due to the increased emphasis on mutual satellite visibility.

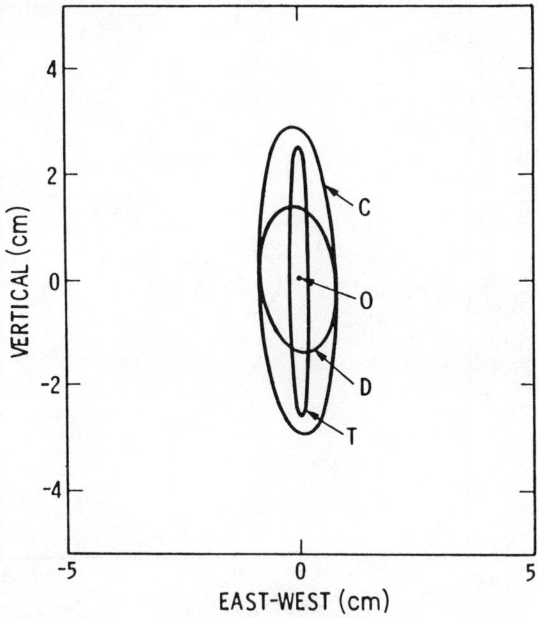

Fig. 8. Error contributions to the vertical-east one-sigma dispersion ellipse (300-km baseline). c, total error; D, data; T, mapped trop; O, other.

Measurement of shorter (<300 km) baselines may be possible without the need for a fiducial network. Of course performance depends on the accuracy of the *a priori* satellite ephemerides. Sensitivities to GPS orbits are illustrated in Fig. 10. From these results it appears that over distances of no more than 400 km orbit errors of the order of 1–2 m cause an insignificant degradation in accuracy. However, degradation is directly proportional to baseline length and, over longer distances, 1-m orbits are inadequate for centimeter geodesy. Even on this 370-km baseline, typical current orbit errors of 10 m would introduce serious degradations in each baseline component.

The results in Figs 6–10 reflect the performance of a system using singly-differenced range. In the eventuality that carrier-cycle ambiguity is not achieved, the observable would be integrated Doppler having 1-cm phase noise and an unknown initial value. An indication of integrated Doppler system performance is given in Fig. 11. In this case, Doppler and range baseline accuracies are compared for baseline lengths of approximately 400 and 2700 km. The geodetic sites for this analysis are in the Caribbean, and fiducial stations are located in California, Massachusetts and Ecuador. Notice that regardless of baseline length, the most serious degradation in

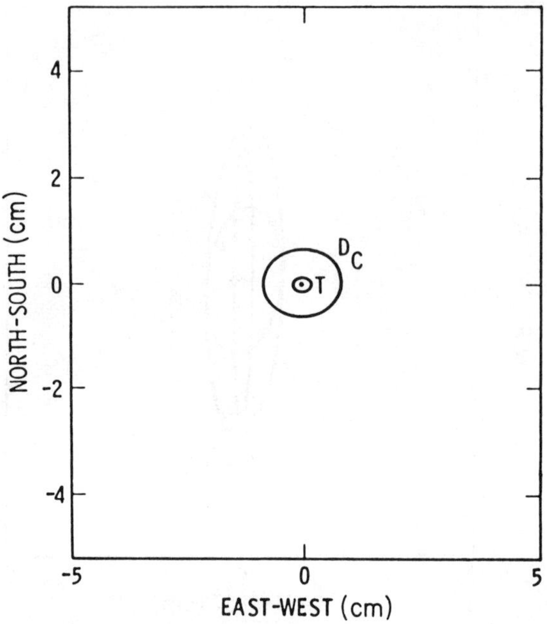

Fig. 9. Error contributions to the north–east one-sigma dispersion ellipse (300-km baseline) C, total error; D, data; T, mapped trop; O, other.

accuracy is incurred by the east component in the local horizontal plane. The estimation of initial biases increases the data noise contribution to errors in the baseline components and augments the east horizontal uncertainties by as much as a factor of 5. The local vertical component is so highly affected by troposphere that accuracy is degraded by no more than 40% due to the use of integrated Doppler.

VI. Subsystem Development Status

A. GPS Receivers

Several precision GPS receivers are under development by government agencies and private industry. Two designs which operate without knowledge of the precison code have been built and tested at the Jet Propulsion Laboratory under the sponsorship of the NASA Office of Space Science and Applications. The first such design, called Series (Satellite Emission Range Inferred Earth Survey) was initiated by P. M. MacDoran and D. J.

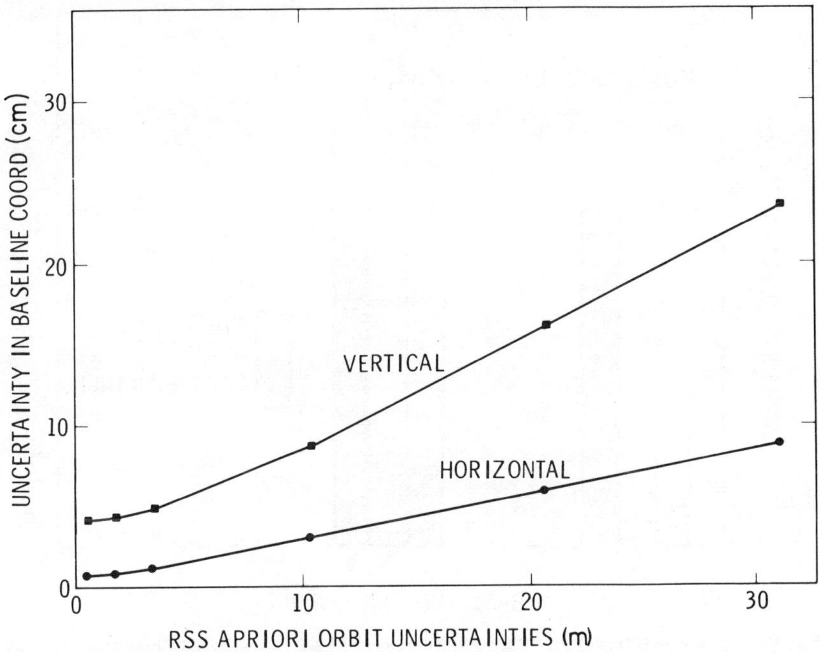

Fig. 10. Sensitivity to unadjusted GPS orbit errors on a 370-km baseline.

Spitzmesser (MacDoran *et al.* 1982) and was later demonstrated by L. E. Young, Spitzmesser, Buennagel and others for the NASA Geodynamics Program (Buennagel, *et al.*, 1984). These proof-of-concept tests confirmed the potential for centimeter geodesy by measuring baselines to a few centimeters over distances up to 22 km.

A related effort, under the sponsorship of the NASA Oceanic Processes Program, expanded upon the SERIES concept and resulted in an improved code-free design referred to as SERIES-X. Both designs utilize a delay-and-multiply technique for extracting the tones associated with the carrier and the C/A and P-code clock rates. By measuring the phases of these tones, they obtain precise pseudoranges without knowledge of either the C/A or P-codes. A shortcoming of the SERIES design for measuring geodetic baselines is the use of a directional antenna, which has definite advantages in improved SNR, but limits reception to one satellite at a time. The SERIES-X approach using two omnidirectional antennas, one at L1 and one at L2, enables the differencing of simultaneous measurements and the removal of common errors without regard for ultra-stable receiver clocks. The SERIES-X receivers also provide measurements of carrier phase with millimeter precision. Tests have been successfully conducted over varying

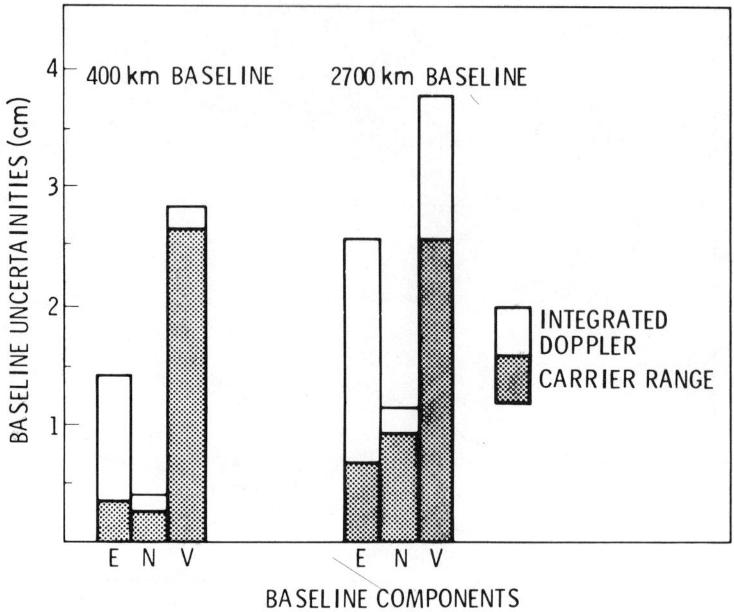

Fig. 11. Comparison of carrier range and integrated Doppler baseline estimation accuracies.

baselines from 15 m to 171 km. The results have demonstrated vector baseline repeatability to a few millimeters over distances where GPS ephemeris errors do not limit performance. On the 171-km baseline, repeatability was consistent with the expected 10-m GPS orbit errors, i.e. 10 cm (Melbourne, 1984; Crow *et al.*, 1984). The principle observable employed was differential carrier phase, calibrated for ionosphere, receiver phase errors and troposphere.

Similar performance is expected from a dual-frequency code-free receiver, termed the Macrometer and built by Macrometrics, Inc. of Woburn, Massachusetts (Bock *et al.*, 1983). The Macrometer provides dual-frequency carrier phase measurements, but does not measure pseudorange directly. Carrier phase precision is believed to be comparable to other non-code designs. Tests over baselines up to several thousand kilometers in length have shown promise of vector baseline accuracies of 0.1 parts per million, or 1 cm over 100 km.

Another commercially available geodetic quality receiver called the TI-4100 has been built by Texas Instruments under a tri-agency sponsorship including NOAA, DMA and USGS (Ward, 1980). The TI-4100 is a dual-frequency design which uses the P-code to measure precise pseudorange and also extracts carrier phase. Analysis by the Applied Research Laboratory of the University of Texas, where the development is managed, indicates that fully calibrated, differential pseudorange would be accurate to about 10-20 cm with an integration time of 100 s. No reports of baseline tests are available at this writing, but system specifications would lead one to expect performance, by using calibrated carrier phase, to be comparable to the SERIES-X and Macrometer systems.

In summary, several geodetic quality GPS receivers have been developed and tested. At least three designs (dual-frequency Macrometer, TI-4100 and SERIES-X) will support integrated Doppler-based geodetic systems, presumably at the performance level assumed here. Additional high-performance receivers are under development by Magnavox, Rockwell-Collins and others. None of these receiver systems is capable of achieving cycle-ambiguity resolution over long baselines. This achievement will most likely require a combination of approaches, including the development of new, multidimensional algorithms and the measurement of P-code pseudorange with ten times the precision of current models. At least one receiver is under development which shows promise of measuring pseudorange to the required precision (Yunck, 1985). It will be mid 1987, however, before this capability is fully demonstrated.

B. Fiducial Networks

In order to meet fiducial network requirements, it is necessary to collocate GPS terminals with VLBI, SLR or other facilities adequate to establish and maintain centimeter baselines over several thousand kilometers. A number of VLBI sites are expected to be instrumented in the near future with high-performance GPS receivers. The National Geodetic Survey (NGS) has already implemented its POLARIS sites in Massachusetts, Texas and Florida with Macrometers. Likewise, the NASA Deep Space Network (DSN) plans to place GPS receivers at the Deep Space Communication Complexes in Spain, Australia and California by early 1989, initially for the purpose of ionosphere calibration and time synchronization, and later to support precision tracking of low Earth orbiters (Yunck, 1985). The DMA will place receivers (most likely the TI-4100) at numerous sites around the world and is considering collocating with VLBI sites in Japan and South America. Each of the NGS, DSN and DMA locations would be equipped with water-vapor radiometers and other necessary calibration assemblies. If this implementation occurs, there will be at least eight worldwide high-performance GPS terminals collocated with VLBI facilities. The three DSN sites plus the proposed DMA locations in Japan, South America and elsewhere would provide an even distribution of fiducial locations around the globe.

Such a network would enable worldwide subdecimeter geodesy, as well as precise determination of low-altitude satellite orbits (Yunck, 1985). For centimeter geodetic baselines, a greater densification of fiducial sites near the region of interest is required. Regional networks, such as the type described in Section IVA, could be established by the addition of two or three appropriately located fiducial stations to the set of proposed sites.

Collocation of GPS receivers with VLBI facilities would enable the establishment of fiducial baselines with near-centimeter accuracy. The VLBI determination of DSN vector baselines is expected to reach the 1–3 cm level for each component by 1990, following the installation of wideband-data-acquisition systems, instrumental phase calibrators, and new state-of-the-art water vapor radiometers. These geodetic measurements will be performed on a routine basis in support of spacecraft tracking.

Performance projections for the DSN VLBI systems are based on current accuracies (Sovers, 1984) coupled with recent technology demonstrations using mobile VLBI systems equipped with comparable phase calibrators, dual-frequency wideband receivers and WVRs. The mobile VLBI systems, developed at the Jet Propulsion Laboratory for the NASA Crustal Dynamics Project, have achieved baseline vector accuracies of 1–3 cm in the horizontal plane and 7–10 cm in the local vertical over distances of 1000 km (Davidson and Trask, 1985). Troposphere calibration errors, which currently limit the vertical baseline accuracies to about 7 cm, are expected to improve substantially with the use of a new three-frequency WVR under development at JPL for the Crustal Dynamics Project (Janssen, 1985). The new, highly-digital WVR is expected to achieve average delay accuracies at zenith of 5 mm, a factor of four improvement over present-day radiometers.

The VLBI systems at the NGS POLARIS sites are equipped with wideband Mark III recording systems. Recent baseline measurements, in conjunction with the Wettzell Observatory of the Federal Republic of Germany, indicate the capability for 2-cm repeatability in baseline length over distances up to 6000 km (Robertson and Carter, 1984). Thus the technology to support the determination of fiducial positions with an accuracy approaching 1 cm is essentially in hand and demonstrated.

C. Data Processing Systems

The analysis of GPS geodetic data will require large-scale data-processing and archiving systems analogous to those developed for VLBI. It is anticipated that a large inheritance can be obtained from the VLBI estimation program MASTERFIT (Fanselow, 1983). The estimation of satellite orbits will require modification to MASTERFIT, incorporating major modules from the NASA Orbit Determination Program which has successfully supported numerous deep-space and near-earth missions (Moyer, 1971). A first-generation GPS data analysis system is expected to be operational at JPL in 1986. Similar systems will likely be developed by NGS and others over the next several years.

VII. Summary

The GPS offers a cost effective means of measuring geodetic baselines to an accuracy of 1–3 cm over distances of several thousand kilometers. This level of performance will require high-precision receivers and calibration assemblies, regional fiducial networks with baselines independently determined to a few centimeters and data-processing programs having state-of-the-art Earth models and filtering algorithms. The GPS receiver technology promises acceptable performance within the next few years. Calibration of troposphere is likely to limit the ultimate performance of both the GPS geodetic measurement systems and the VLBI systems which will establish the fiducial baselines. New calibration equipment may enable the requisite measurement of wet troposphere delay. The advancing development of VLBI and related technologies provides encouragement that centimeter fiducial baselines can be established by the end of the decade. The planned GPS tracking sites, supported by NASA, the NGS and the DMA, would provide the framework for a global fiducial network to support centimeter geodesy in the 1990s.

REFERENCES

Bock, Y. *et al.* (1983). *Proc. Gen. Assem. Int. Union Geod. Geophys, IAG Symp. VIII, Hamburg, Federal Republic, Germany.*

Buennagel, L. A., *et al.* (1984). Satellite emission range inferred earth survey (SERIES) project. Final Report on Research and Development Phase, 1979–1983, pp. 84–116. Jet Propulsion Lab., Calif. Inst. Tech., Pasadena, California.

Crow, R. B., *et al.* (1984). SERIES-X final engineering report, D-1476. Jet Propulsion Lab., Calif. Inst. Tech., Pasadena, California.

Davidson, J. M. and Trask, D. W. (1985). *IEEE Trans. Geosc. Rem. Sens.* Special Issue on Geodynamics.

Davidson, J. M. *et al.* (1984). Covariance analysis for geodetic measurements utilizing the global positioning system satellites, pp. 84–94. Jet Propulsion Lab., Calif. Inst. Tech., Pasadena, California.

Dixon, T. H. (1984). Caribbean geodetic measurements with GPS receivers D-1485. Jet Propulsion Lab., Calif. Inst. Tech., Pasadena, California.

Dixon, T. H., Golombek, M. E. and Thornton, C. L. (1985). *IEEE Trans. Geosc. Rem. Sens.* Special Issue on Geodynamics.

Eller, E. Roman, N. and Wellen, A. (1980). T. R. 2335, Operations Research Inc. (ORI). Silver Spring, Maryland.

Fanselow, J. L. (1983). Observation model and parameter partials for JPL VLBI parameter estimation software 'MASTERFIT'-VI.0, pp. 83–89. Jet Propulsion Lab., Calif. Inst. Tech., Pasadena, California.

Herring, T. A. *et al.* (1981). *J. Geophys. Res.* **86**, 1647–1651.

Janssen, M. A., (1985). A new instrument for the determination of radio path delay variations due to atmospheric water vapor. *IEEE Trans. Geosc. Rem. Sens.*

Larden, D. R. and Bender, P. L. (1982). *Proc. Int. Geod. Symp. Satellite Doppler Positioning, 3rd* **2**, 1247–1252.

MacDoran, P. F., Spitzmesser, D. J. and Buennagel, L. A. (1982). *Proc. Int. Geod. Symp. Satellite Positioning, 3rd, Las Cruces,* New Mexico.

Melbourne, W. G. (1984). *Proc. SPIE Tech., Symp. East 1984, Arlington,* Virginia.

Milliken, R. J. and Zoller, C. J. (1980). in *Global Positioning System,* pp. 3–14, Institute of Navigation, Washington, D.C.

Moyer, T. D. (1971). Mathematical formulation of the double precision orbit determination program (DPODP), TR 32-1527. Jet Propulsion Laboratory, Calif. Inst. Tech., Pasadena, California.

Robertson, D. S. and Carter, W. E. (1984). *Proc. Int. Symp. Space Tech. Geod. Soporon, Hungary.*

Sovers, O. J. *et al.* (1984). *J. Geophs. Res.*

Spilker, J. J. (1980). In *Global Positioning System,* pp. 29–54, Institute of Navigation, Washington, D.C.

Thornton, C. L. and Beckman, B. C. (1984). *Proc. Symp. Recent Crustal Movements Pacific Region, Wellington, New Zealand.*

Thornton, C. L. *et al.* (1983). System study for global positioning satellite (GPS) geodesy in the Caribbean, D-941. Jet Propulsion Lab., Calif. Inst. Tech., Pasadena, California.

Ward, P. (1980). *Proc. IEEE Plans, 1980, Position Location Navigation Symp., Atlantic City, New Jersey.*

Yunck, T. P. (1985). *IEEE Trans. Inst. & Meas.*

Part 4

Earth Rotation and Polar Motion

4.1 The Application of Space Geodesy to Earth Orientation Studies

J. O. Dickey and T. M. Eubanks

Jet Propulsion Laboratory, California Institute of Technology, Pasadena, California, USA

I. Overview

New technologies are beginning to have a dramatic impact on the study of changes in Earth rotation and polar motion. The development of space geodesy has greatly increased the accuracy and precision of Earth orientation measurements, while the analysis of global weather data for operational weather forecasting now routinely provides high-quality estimates of the atmospheric excitation of Earth rotation (UT1-UTC) and the polar motion (PM). The combination of these data types has also increased the understanding of the nonatmospheric excitations of Earth orientation changes and of the Earth's angular momentum balance in general. Study of the

SPACE GEODESY AND GEODYNAMICS
ISBN 0-12-057840-9

transfer of angular momentum between the solid Earth, atmosphere, oceans and the core is thus emerging as a problem of great scientific interest. This research is interdisciplinary by its very nature, with real or potential contributions to many areas of geophysics, such as atmospheric science, oceanography, geodesy and geodynamics.

In this chapter, the measurements of Earth rotation and polar motion (collectively referred to as Earth orientation) are described together with the combination, smoothing, and comparison of results from the various techniques. The calculation of atmospheric angular momentum (AAM) excitation functions and recent comparisons of these functions with variations in the length of day (LOD) and polar motion are discussed. This chapter also presents some of the advances in geophysics made possible by the new techniques, and concludes by describing some of the anticipated advances and prospects for the future.

Several other chapters in this book complement the discussion in this section. Three chapters provide a more complete description of the various techniques of space geodesy [Anderle, Chapter 3.1; Carter and Robertson, Chapter 4.2 and Herring, Chapter 3.2]. The contribution by Wahr (Chapter 4.3) and that of (Chapter 2.3) deal with the implications of Earth orientation results.

II. Introduction

In the absence of interaction with other astronomical bodies and internal energy sources, the Earth would move as a rigid body with its various parts (crust, mantle, inner and outer core, atmosphere and oceans) rotating together at a constant rate about a fixed axis. In such a world, the diurnal rotation of the Earth could serve as the basis for a unit of time, and the rotation axis of the Earth could be used to define an irrotational coordinate system. In addition, the rotation axis would always maintain a fixed orientation with respect to the Earth's surface, and points on the surface would remain fixed with respect to each other. In this case, both the interval between successive transits of stars across the meridian of an observatory (the sidereal day) and the elevation angle of the stars at each meridian passage would be constant. There would thus be no variation in the astronomical latitude or longitude of any observatory (see, for example, Munk and MacDonald, 1960).

The real world is considerably more complicated. The Earth's rotation rate is not constant, but exhibits a number of fluctuations with amplitudes on the order of 1 part in 10^9 or more over many time scales. The rotation axis moves with respect both to the fixed stars and the Earth's surface. In addition, the existence of plate tectonics has become widely accepted, and this implies slow relative motions of points on the Earth's surface. The study of the Earth's orientation is in large part an attempt to understand the nature and excitations of these motions and, if possible, to predict them. Variations in astronomical latitudes were first conclusively detected in the 1890s through a comparison of latitude results from Berlin and Honolulu. In 1891 and 1892, Chandler showed that polar motion is dominated by oscillations with periods of 1 year and about 433 days. The polar wobble (Figs. 1 and 2) has a total amplitude of a few meters. Shortly thereafter Newcomb showed that the longer period component (the Chandler wobble) was the resonant oscillation predicted by Euler, by refining Euler's calculations to include the elasticity of the Earth. These discoveries led to the creation of the International Latitude Service (ILS) in late 1899 to observe polar motion through optical astrometry (Munk and MacDonald, 1960). Although these data form the raw material of most polar motion studies since that time, they are about an order of magnitude less precise than the polar motion data available from the new techniques. The long-time constants involved with polar motion imply that it will be many years before these observations are completely superseded by the modern data. One

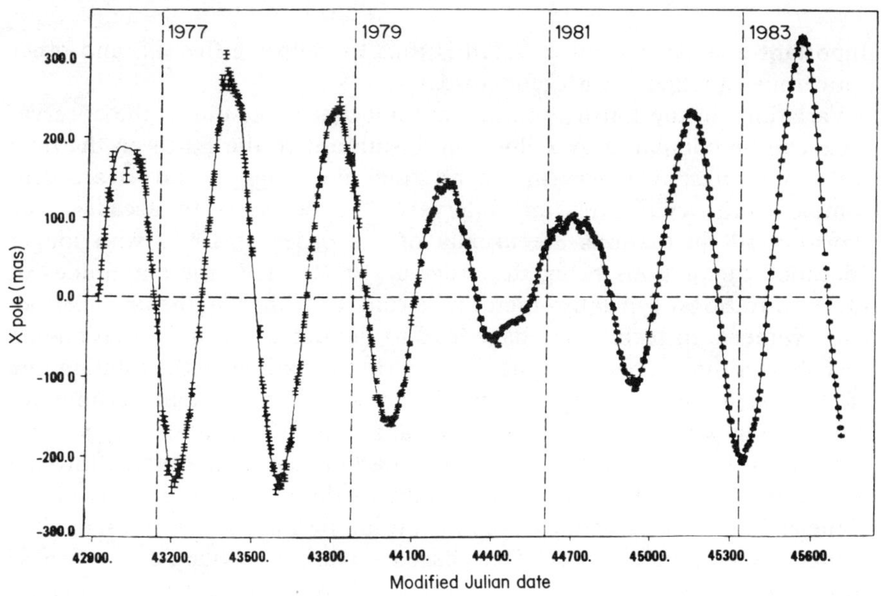

Fig. 1. X pole as obtained from LAGEOS long-arc solution 8402 for the period May 1976–January 1984 (Tapley *et al.,* 1985).

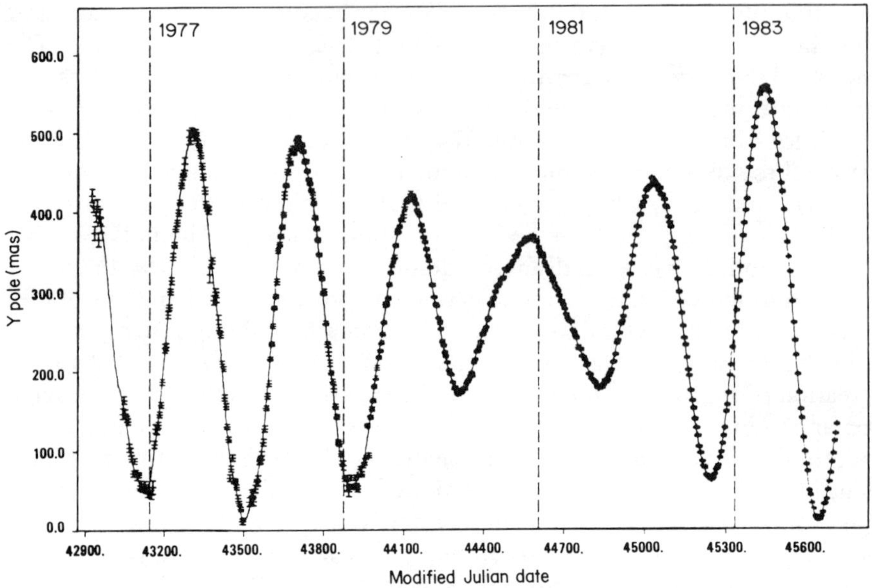

Fig. 2. Y pole as obtained from LAGEOS long-arc solution 8402 for the period May 1976–January 1984 (Tapley *et al.*, 1985).

important goal of current research is thus to calibrate the ILS and other optical measurements with modern data.

Variations in the Earth's rotation are not detectable unless the observations can be compared to a clock more stable than the Earth on the time scale of interest. Changes in the rotation rate (Figs. 3 and 4) are very complex and occur over time intervals of a few days to decades. The short-period fluctuations are usually of the order of 10^{-3} s with longer "decade" fluctuations being the order of 10^{-2} s. Since the amplitudes of LOD variations typically increase strongly with lengthening period, improvements in technology have lead to the detection of fluctuations on ever shorter time scales. Irregular long-period variations in the rotation rate of the Earth were first hypothesized in the 19th century to account for otherwise inexplicable variations in the apparent motion of the Moon. This hypothesis was confirmed in the early part of the 20th century through observations of corresponding variations in the apparent motions of the planets. This work essentially used the orbital periods of the Moon and the planets as reference clocks. Time based on orbital dynamics, known as Ephemeris Time (ET), is the only time scale with long-term stability for the period before 1955. With continuing improvements in positional astronomy and pendulum clocks, it became evident in the 1930s that there

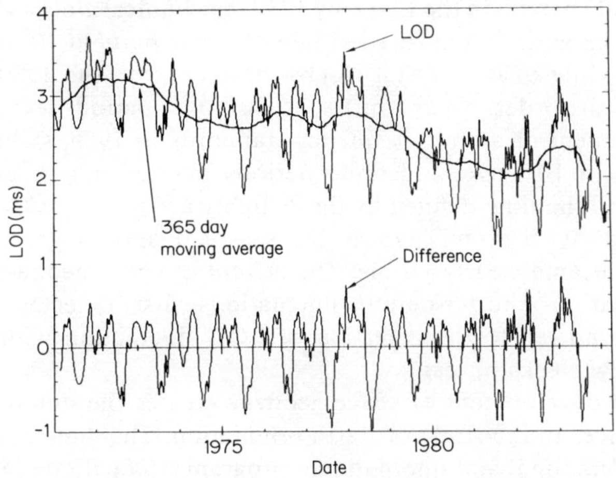

Fig. 3. The LOD curve shows changes in LOD from a base state of 86,400 s for the period from January 1971–March 1984. All tidal terms with amplitudes greater than 0.01 and periods of 18.6 yr or less have been removed. The heavy curve is a 365-day moving average of this time series. The difference curve shows the difference between the original series and its moving average (Eubanks *et al.*, 1985c).

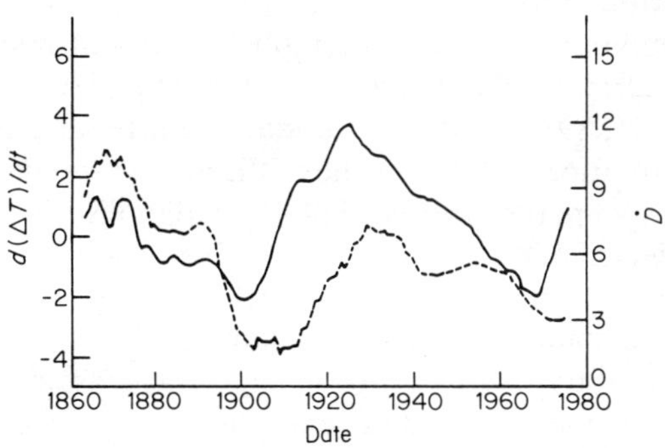

Fig. 4. The secular variation of declination, D (solid curve and right-hand scale, in arc minutes per year), and the excess LOD $d(\Delta T)/dt$ (dashed curve and left–hand scale, in milliseconds) for the period 1861–1978. Rotation data are from Morrison (1979) and D data are from compilation of Paris observatory data. The relative changes in the Earth's rotation rate $\Delta\Omega/\Omega$ are obtained from the $d(\Delta T)/dt$ values through multiplication by -1.157×10^{-8} (after Le Mouel *et al.*, 1981).

were seasonal changes in the LOD. By 1930, mechanical clocks were capable of detecting changes in the rotation rate of a few parts in 10^8 over periods of months, while quartz crystal clocks, developed in the 1930s, obtained stabilities of the order of parts in 10^{11} to 10^{12} over periods less than a year. The development of atomic frequency standards in 1955, stable to within a few parts in 10^{13} over indefinite periods marked a new era in "time keeping", and the time defined by these highly accurate clocks has become the standard. With atomic clocks the seasonal and longer-period LOD fluctuation became clearly visible. The advent of space geodesy confirmed the existence of shorter-period fluctuations first detected by optical astrometry, and led to the detection of LOD fluctuation with periods as short as a few weeks or less.

The rapid development of space geodesy creates the potential for profound advances in the study of Earth orientation. This potential has led to a number of national and international programs to facilitate the collection and analysis of data from all techniques. Monitor Earth Rotation and Intercomparison of Techniques (MERIT) of observation and analysis, an international effort to evaluate the new techniques sponsored by the International Union of Geodesy and Geophysics (IUGG) and the International Astronomical Union (IAU) (Wilkins, 1980), has obtained the most accurate Earth rotation data ever obtained during its main campaign (September 1983–October 1984). One result of the new techniques should be the establishment of an accurate and consistent set of global coordinates for participating observatories (a terrestrial reference system). A joint IUGG/IAU Working Group on the Establishment and Maintenance of a Conventional Terrestrial reference System (COTES) is concentrating on the connection between the various terrestrial and inertial reference frames used during the MERIT campaign and on the establishment and maintenance of the new conventional terrestrial reference system (Feissel and Wilson, 1983). Daily (Eanes *et al.*, 1984) and even subdaily (Carter and Robertson, 1985) values of Earth rotation well below the millisecond level have been reported during the intensive part of the MERIT campaign (April–June 1984); daily polar motion determinations were also obtained during this period (Eanes *et al.*, 1984). This effort is providing a rich data set which no doubt will increase our understanding of Earth dynamics.

The excitations of Earth orientation changes have fascinated and intrigued geophysicists for some time and have applications in a multitude of areas (geodesy, meteorology, astrometry and solid Earth geophysics among others). The first complete book length review of this area was written by Munk and MacDonald (1960); Lambeck (1980) reviewed and discussed advances in this multi-faceted area. Hide (1985) reviewed the short-term variations in Earth rotation and polar motion; Wahr (1985) emphasized the implications of the Earth's rotation measurements. A shorter overview article by the authors (Dickey and Eubanks, 1985) highlighted the topics discussed here with a less mathematical treatment.

The availability of global atmospheric angular momentum measurements along with highly accurate geodetic Earth orientation results permits the study of the angular momentum budget of the Earth. Understanding the transfer of angular momentum between the atmosphere, solid Earth and oceans is rapidly emerging as a problem of great scientific importance. For example, research has demonstrated that the angular momentum transfer between the atmosphere and the solid Earth is an important source seasonal and short-term variations in the length of day. The correlations between these fluctuations and El Niño in 1982 and 1983 are of interest to the meteorological community, and the LOD may serve as a useful proxy index for global wind changes. The IUGG/IAU Special Study Group 5.98 strives for the improvement and promotion of further collaborative work in these areas and serves as an advocate for improved measurement techniques (Dickey, 1985a, b).

III. Definitions and Measurements

A. Earth

Here, the "Earth" is defined to be the core, mantle, crust and hydrosphere, i.e. the entire Earth minus the atmosphere, while the term "solid Earth" includes only the crust and mantle. Earth orientation refers collectively to the rotations of the solid Earth. Universal time (UT1) is the rotation angle of Earth: the direction of the Greenwich meridian in an equinox-based inertial frame. The UT1 measurements are usually given in units of time (1 msec is 46.05 cm at the equator). The actual observable is UT1-T, where T is some independent time series such as ET (ephemeris time) or IAT (atomic time). The LOD can be estimated from the time rate of change of UT1-T. The motion of the pole can be divided into two classes, nutational motion (which includes precession) and the polar motion (PM).

Nutations are motions of the spin axis of the Earth with respect to an inertial reference frame; while polar motion or wobble is the motion of the solid Earth with respect to the spin axis of the Earth. Polar motion changes are easily distinguished from nutations observationally. On any clear night the stars can be seen to rotate about a spot on the sky called the celestial pole (currently near Polaris in northern skies and near the Southern Cross in the south). Polar motion changes the angle between the celestial pole and the local horizon (the observer's latitude), while nutation changes the angle between Polaris (or any star) and the celestial pole. From observations at one location over the course of the night; it is thus possible to estimate both components of nutation but only one component of the polar motion (the variation of latitude). Although nutations and polar motion both change the instantaneous angle between a celestial object and the local zenith, they can be clearly separated if observations are available over a substantial fraction of a day. In reality, one type of motion of the pole cannot occur without some accompanying motion of the others; however, one can characterize the Earth's response as either predominately wobble or nutation.

Earth orientation measurements in current use can be classified into two subsets: classical (optical astrometry) and modern space techniques. Optical astrometry is based on measurements of the apparent angular positions of selected bright stars. The new techniques are all based on measurements of electromagnetic signal delay or its time derivative (generally expressed as a Doppler shift). Delay and delay rate measurements are both more precise and somewhat less sensitive to systematic error than angular measurements. The modern measurement types include: satellite Doppler tracking, lunar laser ranging (LLR), satellite laser ranging (SLR) and very long baseline interferometry (VLBI). In each technique, changes in Earth orientation are monitored by observations of extraterrestrial objects from the surface of the Earth. The observed objects are used to approximate a nonrotating reference frame, either directly in the case of slow moving objects, or from dynamical theories of their motion, in the case of planetary and satellite observations. In each case, Earth orientation is estimated from the apparent motion of the Earth with respect to this frame. The accuracy of each technique is thus limited by unmodeled reference frame rotations, as well as by intrinsic measurement error. In addition, observations are of necessity made from specific locations on the Earth and are thus corrupted by any unmodeled motions of the observatory.

These measurement types (both classical and modern) are briefly described below, with references made to more complete treatments.

1. Classical Astrometry

Optical astrometry (e.g. Guinot, 1978; Feissel, 1980) estimates the Earth rotation and polar motion from observations of selected bright stars. Both the latitude and the longitude of the observatory can be determined from observations of the time and elevation of meridian passage. The measurements are thus generally determinations of the angle between a star and the local vertical, together with the time of the observation. Measurements from a single location can determine only the latitude and longitude (UT0) of the observatory since they are insensitive to rotations about the local vertical. With a global network of more than 50 observatories, several days of measurements are still needed to determine Earth rotation at the millisecond level and the polar motion at the 10-mas (milliarcsec) level.

2. Doppler Satellite

The measurement and analysis of the Doppler shift of radio signals received from artificial Earth satellites have proved to be a multidisciplinary tool useful in many areas of geodesy (Kouba, 1983). Polar motion (PM) results are obtained as by products of the satellite orbit computations; PM values, computed by the Defense Mapping Agency (DMA), have been provided to the International Polar Motion Service (IPMS) and the Bureau International de l'Heure (BIH), for over 15 years. Precise ephemerides are computed by numerical integration, using orbit elements estimated by a least-squares fit to Doppler observations made by a global network of stations over contiguous 2-day time intervals. In theory, UT1 can also be estimated from these data; however, hard to model orbital perturbations cause the orbital plane to precess in an unpredictable manner. Such motion cannot be distinguished from UT1 changes. In fact, in the operational ephemeris, UT1 is extrapolated and used in the determination of the orbital plane for every 48-hr span. Hence, for Earth orientation studies, the strength of the Doppler techniques lies in its determination of PM; other techniques are better suited for Earth rotation studies.

3. Lunar Laser Ranging

Lunar laser ranging (LLR) data consist of estimates of the round trip time of flight of laser pulses from terrestrial observatories to retroreflectors placed on the moon's surface by the Apollo astronauts and by unmanned probes from the Soviet Union. Lunar laser ranging can provide the monitoring of submillisecond changes in Earth rotation with a temporal resolution of the order of a day or less. Lunar laser ranging has vastly improved our knowledge of lunar motion and has provided important observational constraints on generalized gravitational theories (e.g. general relativity). The ability to model accurately the lunar orbit since the beginning observations in 1969 allows long-term studies of variations in the Earth's rotation as well as highly accurate determination of many parameters of the Earth-Moon system (Langley *et al.*, 1981a; Dickey *et al.*, 1984a, b; 1985). Most of the LLR observations acquired prior to 1984 were obtained by a single observatory (McDonald Observatory in Texas). Lunar laser ranging from a single site is sensitive to two components of Earth orientation: apparent observatory longitude (UT0) and variation of latitude. Universal time (UT1) can be estimated from these data using independent estimates of the polar motion. The advent of multi-station LLR, beginning in 1984, permits the determination of both components of polar motion, along with UT1.

4. Satellite Laser Ranging (SLR)

In this technique, very short laser-generated pulses of light are transmitted to retroreflectors mounted on the surface of artificial satellites; the round-trip travel times of these pulses (and hence the path lengths) are measured. The return power in laser ranging is proportional to distance to the negative fourth power, and this, plus the relative nearness of these satellites, means that SLR is technologically much less demanding than LLR. A series of SLR measurements from a global network of stations permits the determination of orbital parameters, a satellite ephemeris and relative station locations. High-quality Earth rotation data, as well as other quantities of geophysical interest (e.g. plate motion and gravity field studies), can be generated by SLR only if the satellite orbit can be calculated with high precision. The best geodetic data comes from laser ranging to the LAGEOS (*La*ser *Geo*dynamics *S*atellite) target, a dense sphere placed in a high (6000 km) orbit. The shape and density of the satellite and the high orbital altitude serve to minimize the effects of orbital perturbations from atmospheric drag, high-frequency gravity field and radiation pressure from sunlight reflected by the Earth, all of which are difficult to model. Even so, the orbit of this satellite can be used as a nonrotating reference system only for periods short compared to a year. Since polar motion induces a diurnal residual motion of an observatory (as seen from a nearly inertial reference frame), the stability of the LAGEOS orbit is sufficient to provide highly accurate and stable polar motion estimates, while LAGEOS UT1 estimates diverge from the true UT1 after a few months (see e.g. Goddard Space Flight Center, 1983; Tapley *et al.*, 1985; Smith *et al.*, 1985). The LAGEOS data can be used, however, in studies of high-frequency changes in the UT1 or LOD (Carter *et al.*, 1984). Nutational motions are slow from an inertial reference frame, and thus are not well determined by LAGEOS.

5. Very Long Baseline Interferometry (VLBI)

Radio interferometry is now used to make highly-accurate measurements of changes in UT1 and polar motion with observing times of from a few hours to a day. The VLBI measures the interferometric group delay: the difference in the time of arrival of a radio signal at two or more radio telescopes. The delay rate (the time derivative of the interferometric phase delay) is frequently measured as well. The interferometric delay between two telescopes is a direct estimate of the projection of the baseline vector (the vector between the telescopes) in the direction of the radio source. Observations from one baseline are thus not sensitive to rotation about that baseline. For single-baseline results (two stations), two components of Earth orientation are determined: Earth rotation parallel to a change in longitude (UT0) and the variation of latitude. Multibaseline VLBI can measure all three components of the Earth orientation if some of the baselines are nonparallel. The VLBI analyses also provide valuable estimates of tectonic plate motion over distances ranging from the small regional measurements (e.g. the southern California network) to long intercontinental baselines (e.g. Europe to America).

Regular independent VLBI estimates of the Earth orientation are now produced routinely under projects POLARIS/IRIS (Carter and Robertson, 1984) and by the TEMPO program of the Deep Space Network (DSN) at the Jet Propulsion Laboratory (JPL) (Eubanks *et al.*, 1982). Earth orientation estimates are also provided by geodetic and astrometric observations conducted on a less regular basis both by the DSN (Sovers *et al.*, 1984) and by the NASA Crustal Dynamics Project (Ryan and Ma, 1983). The DSN measurements are Europe to America and Australia to America, while the IRIS/POLARIS measurements are Europe to America (IRIS) and within America (POLARIS). A number of other countries (including China and Japan) are in the process of building VLBI systems and have or will take part in international collaborations.

B. Atmosphere

Changes in atmospheric winds and the distribution of surface pressure cause variations in the atmospheric angular momentum (AAM). Estimates of the AAM, a three-component vector, are currently available based on the reduction and analysis of meteorological data for weather forecasting at three centers: European Centre for Medium-Range Weather Forecasts (ECMWF), Japanese Meteorological Agency (JMA) and the U.S. National Meteorological Center (NMC). A variety of atmospheric variables, including the local wind velocity vector, are estimated at each model grid point. The grid-point data are updated several times daily by using a so-called forecast/assimilation, a weighted average of the forecast from the previous update and any new meteorological data near the grid point (see e.g. McPherson *et al.*, 1979). The total angular momentum is estimated at 12-hr or 24-hr intervals (depending on the service used) from the appropriate integral of the grid-point wind velocity and surface pressure estimates from the current update. Accurate estimation of the AAM would be impossible without the efforts of the weather forecasters. At present, for example, the ECMWF system has 273,630 grid points and receives over 10 million measurements daily; Kerr (1983) provides a nontechnical description of operational weather forecasting. These atmospheric models thus incorporate vast amounts of meteorological data and conveniently provide grid-point estimates over the entire globe. The model's predictive power is used to estimate grid-point values over regions with sparse data coverage; climatological averages may be used as well if the model becomes unstable in some region due to absent or conflicting data. This reliance on model forecasts in regions (such as the South Pacific) with continually sparse coverage could introduce systematic errors in the resulting excitation estimates.

Wahr (1982, 1983) and Barnes *et al.* (1983) re-examined and improved some aspects of the underlying theory of nonrigid body rotational dynamics and angular momentum exchange between the atmosphere and solid Earth. Barnes *et al.* concluded that the excitational functions, $(\psi_1, \psi_2) = \psi$, used previously in polar motion-related studies are proportional to the equatorial component of the total frictional and pressure torque on the solid Earth and hence cannot be computed reliably from currently available meteorological data. The new atmospheric effective angular momentum function (EAMF), χ, a three-dimensional vector, includes Love number corrections for rotational and surface-loading deformation of the Earth and can be evaluated accurately from meteorological data.

The AAM changes can be divided into two categories: the changes in the net atmospheric rotation rate (the wind term), and those in the net atmospheric moment of inertia (the pressure term). These terms can be decoupled to first order and are generally calculated separately. Allowance for the response of the oceans and the solid Earth to atmospheric surface pressure changes can be made via the "inverted barometer" and "Love number" corrections (see Wahr, Chapter 4.3). Results are generally computed separately for the wind and pressure terms and for the Northern and Southern Hemispheres. Daily values of the EAMF have been calculated from the "initialized analysis global data base" of the ECMWF starting in December 1979 (Barnes *et al.*, 1983). Rosen and Salstein (1983 and private communication) have calculated twice-daily values of the EAMF from the U.S. National Meteorological Center (NMC) wind and surface pressure analyses following the formulation of Barnes *et al.* The time series component that corresponds to Earth rotation (χ_3) begins in January 1976; the two components (χ_1 and χ_2) that relate to polar motion begin in July 1976. These functions are now calculated operationally (every 12 hr) at the NMC and are accessed and archived by the National Geodetic Survey (NGS). Naito and Yokoyama (1984) have computed the atmospheric angular momenta due to wind and mass by using JMA global analysis data beginning in September 1983; these computations use the formulation presented by Wahr (1983).

IV. Excitations of Earth Rotation and Polar Motion

Changes in Earth orientation are caused by deformations of the solid Earth and by exchanges of angular momentum between the solid and fluid parts of the Earth, as well as by exchanges of angular momentum with extraterrestrial objects. Since the angular momentum changes are small compared to the total angular momentum of the Earth, UT-PM changes can be regarded as a linear function of a three-dimensional excitation vector with three components. Meteorological excitation estimates are generally expressed in terms of the χ vectors of Barnes *et al.* (1983). In this scheme the excitation of polar motion is described by χ_1 and χ_2, and that of earth rotation by χ_3. Although polar motion is generally described by a left-handed coordinate system, the χ vectors follow a right-handed system, which at times has led to confusion. The relation between the χ vector and the UT-PM vector **U** is given by the linearized Liouville equations. Let

$$\mathbf{u} = \begin{bmatrix} \text{PM–X} \\ \text{PM–Y} \\ \text{UT1} - \text{IAT} \end{bmatrix}$$

then

$$\mathbf{du}/\mathbf{dt} = \mathbf{Fu} + \mathbf{G}\chi \tag{1}$$

where

$$F = \frac{2\pi}{T_c}\begin{bmatrix} -\frac{1}{2Q} & 1 & 0 \\ -1 & -\frac{1}{2Q} & 0 \\ 1 & 0 & 0 \end{bmatrix}$$

and

$$G = \left[\begin{array}{ccc} \frac{2\pi}{T_c}\begin{bmatrix} -\frac{1}{2Q} \\ 1 \end{bmatrix} & \begin{matrix} 1 \\ -\frac{1}{2Q} \end{matrix} & \begin{matrix} 0 \\ 0 \end{matrix} \\ 0 & 0 & -1 \end{array}\right]$$

In this equation, T_c is the Chandler wobble period and Q is the Chandler wobble quality factor. Note that the difference in handedness between the χ and $\mathbf{u}$ vectors is accounted for in the G matrix. Equation (1) is a first-order constant coefficient differential equation, and the homogeneous solution can be given in terms of the propagation matrix, defined by the matrix exponential of F:

$$\phi(\Delta t) = I + F\,\Delta t + \frac{1}{2!}F^2\,\Delta t^2 + \cdots = e^{F\,\Delta t}$$

The analytic expression for ϕ is

$$\phi(\Delta t) = \begin{bmatrix} \exp(-\pi\Delta t/QT_c)\begin{bmatrix} \cos(2\pi f_c\Delta t) & \sin(2\pi f_c\Delta t) & 0 \\ -\sin(2\pi f_c\Delta t) & \cos(2\pi f_c t) & 0 \end{bmatrix} \\ \begin{matrix} 0 & 0 & 1 \end{matrix} \end{bmatrix}$$

where f_c is the Chandler wobble frequency. Given an expression for ϕ, the UT-PM can be found from excitation estimates and an initial condition by:

$$\mathbf{u}(t) = \phi(t-t_0)\mathbf{u}(t_0) + \int_{t_0}^{t} \phi(t-t')G\chi(t')\,dt' \qquad (2)$$

Notice that the excitation of the Chandler wobble and Earth rotation can easily be decoupled. When done, the polar motion parts of $\mathbf{u}$ and χ are frequently expressed as complex numbers, which requires reversing the handedness of either χ or $\mathbf{u}$. While the UT1 − IAT is a simple integration of χ_3, the changes in polar motion depend not only upon the excitation but also on the initial condition. In the absence of any excitation, the LOD would be constant and the rotation pole would rotate at the Chandler frequency about a fixed point on the Earth, the excitation pole (the axis of maximum moment of inertia). The quality factor Q has a finite positive value since the real Earth has some dissipation; thus, the rotation pole will slowly spiral into the excitation pole. In the presence of excitation, the pole undergoes a damped spiral about a moving excitation pole. It can be shown that the coordinates of the excitation pole are given by χ_1 and $-\chi_2$.

Comparisons of UT-PM results with atmospheric or other excitation estimates can take two basic forms. Equation (2) can be used to estimate the UT-PM from excitation estimates (the direct approach) or UT-PM data can be used to estimate the excitation via a differenced form of Eq. (2), or by some other means (the deconvolution approach). Deconvolution is preferred, especially for the polar motion (Wilson and Haubrich, 1976). The deconvolution approach automatically prewhitens the data by removing most of the power near the Chandler wobble frequency (for polar motion) and at very low frequencies (for Earth rotation).

A number of different filters have been developed for polar motion deconvolution. These include direct numerical differentiation (Wilson and Haubrich, 1976), numerical differentiation followed by smoothing (Wilson and Vincente, 1980), Wiener deconvolution filters (Smylie *et al.*, 1973), Backus–Gilbert inverse approach of Gross and Chao (1984) and the Kalman filter approach developed by the authors and others at JPL, which can be viewed as a time domain version of the Wiener filter approach.

In the frequency domain, any convolution can be expressed as a multiplication

$$M(f)=H(f)C(f)+N(f)$$

where M is the Fourier transform of the (measured) polar motion, H is the excitation transfer function (generally assumed known) and C and N are the transforms of excitation and noise, respectively. In general, M, C and N are vectors and H is a matrix, but use of the complex rotation for the polar motion reduces this to a complex scalar equation. In this case any linear deconvolution filter can be described in terms of multiplication by a scalar transfer function G in the frequency domain, with the excitation estimate being:

$$C(f)=G(f)M(f)$$

The Wiener filter is designed to minimize the mean square difference between the excitation estimate and the true excitation (Smylie *et al.*, 1973). The acausal Wiener filter (Wiener smoother) transfer function is thus

$$G(f)=\frac{1}{H(f)}\frac{H^2\langle C^2\rangle}{\langle N^2\rangle+H^2\langle C^2\rangle}$$

Notice that the Wiener filter consists of a direct inverse deconvolution coupled with a bandpass filter that depends on the ratio of the signal power and the signal plus noise power. The Weiner filter is thus a direct inverse at frequencies where the signal-to-noise ratio (SNR) is high and attenuates where the SNR is low. The Wiener filter requires knowledge of both the excitation and noise power spectra to estimate the SNR. Generally these spectra are assumed to follow some particular parametric form (such as a white noise or random walk) and the model parameters are estimated from the data. Most linear deconvolution filters are a direct inverse coupled with some sort of bandpass or low-pass filter, and many of them [such as the approach of Gross and Chao (1984)] are equivalent to the Wiener filter for some particular choice of spectral models.

V. Earth Orientation Estimates and Analysis

The past decade has seen the development of several promising new techniques capable of monitoring changes in Earth orientation. An understanding of the true errors in these new measurements is essential to realizing the full geophysical potential of the new data. The best method to validate the new results is through comparisons of independent measurement series. Fortunately, changes in orientation are dominated by rotations of the Earth as a whole, and it is thus possible to compare Earth orientation results from widely separated locations. Comparisons of these results should be of general interest to the geodetic community, since virtually the only other way to independently verify the accuracy of the new techniques is the expensive (and thus rare) long-term collocation of independent measurement systems at the same observatories.

Orientation measurements from different techniques are generally not acquired simultaneously and so any comparison requires averaging or interpolating the data in some fashion. Earth rotation and polar motion form a three-component vector (defined here as UT-PM). Many measurement techniques (e.g. single-baseline VLBI or single-station LLR) do not determine all three components of this vector; they determine linear combinations of Earth rotation and polar motion. Hence, most comparisons require smoothing over different linear combinations of the UT-PM vector to estimate the full vector at any one time. To get an accurate and long-term UT-PM series, data from several techniques must be included in the smoothing. This section discusses the smoothing and combination of UT-PM time series and the results from recent comparisons.

A number of different smoothing techniques have been applied to Earth orientation data recently: Gaussian filtering, Fourier decomposition, Kalman smoothing (e.g. Dickey *et al.*, 1984c) and Vondrak smoothing (Feissel and Lawandowski, 1984). All of these filters are linear low-pass filters, and thus they can all be discussed in terms of their transfer function or their impulse response.

The Gaussian filter, which is probably the most common, is really a Gaussian weight moving average. The value at any epoch t is a weighted average over all data in which the weights decay according to

$$\exp[-(t_i - t)^2/2\tau^2]$$

where t_i is the measurement time and τ is the filter-time constant. Generally the points are also weighted by the inverse of their variance. The filter-time constant can be determined by an examination of filter residuals. Note that for equally spaced data with equal errors, the transfer function of the Gaussian filter is a Gaussian function of the frequency. It can be shown that of all linear low-pass filters with a given frequency cutoff, the Gaussian filter has the smallest possible combination of bandwidth and impulse response. For a given cutoff frequency, filters with smaller impulse responses have a slower frequency cutoff, and pass more power at high frequencies, while filters with a sharper frequency cutoff have a wider impulse response and are more subject to edge effects.

Fourier smoothing involves the modeling of a time series by a Fourier series which is then truncated to achieve a low-pass filter. If the truncation is a step function, then the filter impulse response will be very long and the edge effects could be severe. In addition, this introduces ripples into the filtered data. Vondrak smoothing, used mostly by the Bureau International de l'Heure (BIH), is a time-domain polynomial filter designed to minimize the filter bandwidth for a given frequency cutoff. Both Fourier and Vondrak smoothing can introduce ripples into the filtered time series.

The Kalman filter approach offers a number of advantages in filtering and smoothing of multiple-dimensional time series such as UT1 and Polar Motion. A UT-PM Kalman filter implemented at the JPL is designed to provide the DSN with accurate estimates of Earth orientation changes during periods of spacecraft navigation (Eubanks *et al.*, 1985a). With this filter it is possible to produce both predictions and smoothings using raw measurements at irregular times and with widely varying accuracies. The Kalman filter can accept measurements of any linear combinations of UT1 and polar motion and can also use the full covariance matrix of any multi-dimensional measurements. The output of the filter includes not only estimates of the UT-PM vector but also the formal error covariance matrix for each estimate. These attributes are very useful both in research applications and in providing the best possible estimates for spacecraft navigation.

The design of a Kalman filter is relatively straightforward, given linear models of the signal and the measurement noise (Gelb, 1974; Jazwinski, 1970). It is not practical to model the physical causes of Earth orientation changes from first principles, since this would involve modeling the fluid dynamics of the atmosphere, oceans and core. Instead, this filter uses empirical time-series models for the excitation of UT1 and polar motion. These models were derived mainly from studies of the meteorological excitation data (only the power information is included) and from International Latitude Service (ILS) polar motion data. To simplify the filter design, the models were chosen to be algebraic sums of white noises, integrals of white noises (random walks) and second-order linear oscillators driven by white noise.

Dickey *et al.* (1984c) present a comparison of Gaussian, Fourier and Kalman smoothing, while Feissel and Lawandowski (1983) describe a comparison of Vondrak and Gaussian smoothing. Figure 5 shows UT1 values from raw LLR data and from five different smoothings of the same data. The BIH Circular-D smoothing (Feissel, 1984) and the UT1 tidal oscillations (Yoder *et al.*, 1981) have been subtracted from each time series. The BIH is an international service whose charter includes providing Earth rotation and polar motion to the general community in a timely fashion. Their Circular-D values are widely used, and other results are generally differenced with them for comparison purposes. The tidal oscillations are periodic variations in Earth rotation caused by the attraction of the Moon and Sun. Gravitational attraction of the Moon and Sun deforms the Earth and changes the Earth's polar moment of inertia which in turn, causes periodic variations in Earth rotation (see Wahr, Chapter 4.3). The amplitude of these variations is proportional to the response of the Earth to periodic changes in the gravitational potential. Since this contribution can vary as much as 1–2 ms from day to day and because it is predicatable, these oscillations are removed before most comparisons. The Gaussian filter is shown for several smoothing time constants. If the data quality varies widely, it would be useful to increase the amount of smoothing during periods of

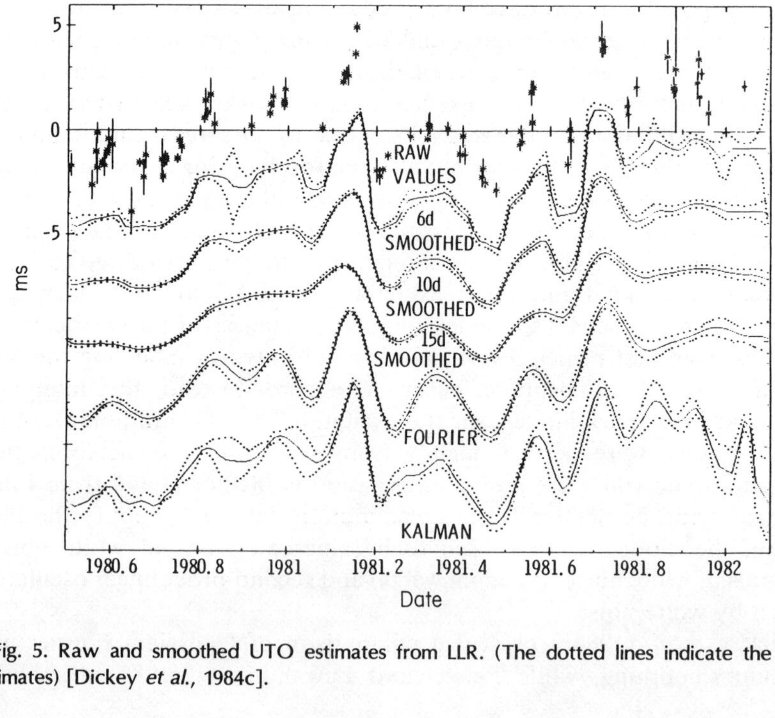

Fig. 5. Raw and smoothed UT0 estimates from LLR. (The dotted lines indicate the error estimates) [Dickey *et al.*, 1984c].

poor data. With a Gaussian filter, this would have to be done in an *ad hoc* fashion; the Kalman filter varies the amount of smoothing automatically in response to changes in the quality or density of the data. However, the accuracy of the Kalman filter depends on the accuracy of the excitation models embedded in the filter. The Gaussian filter would be more robust if the excitation was not properly modeled. There is still room for diversity and research into the problem of filtering changes in Earth orientation.

When the new techniques were first introduced with their improved accuracies, it was unclear whether any of the new features observed in universal time were real changes. In an early comparison, Feissel and Gambis (1980) (Fig. 6) showed that the UT1 results from four different techniques were indeed consistent and discovered rapid "50-day" fluctuations in UT1 as well. In a later comparison of SLR, LLR, VLBI and optical astrometry data, Robertson *et al.* (1983) showed that SLR and VLBI polar motion results agree to within 6-mas or equivalently 18 cm at the equator for the period September 1980 to December 1981 (Fig. 7a). Polar motion comparisons are somewhat easier, given the slowness of the motion and the density of the SLR data. The UT1 comparisons are considerably more difficult. The corresponding UT1 intercomparison is shown in Fig. 7b; the VLBI, LLR and BIH astrometric determinations of UT1 are plotted as differences from heavily smoothed astrometric determinations with the mean differences between the various sets removed. Common fluctuations are seen in all three data sets. The LLR and VLBI results had a root-mean-square (rms) scatter of 0.9 ms of time, equivalent to about 40 cm at the equator.

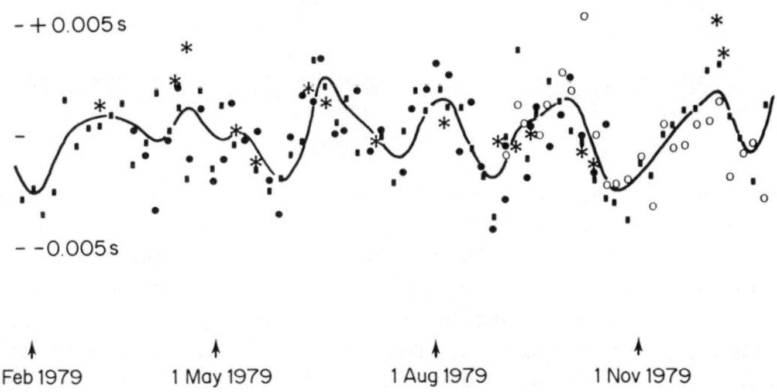

Fig. 6. UT1–TAI corrected for long-term variations. Measurement techniques: ●, line of sight velocity from artificial satellite; ○, interferometry of radio sources; □, optical astronomy; ★, lunar laser ranging (Feissel and Gambis, 1980).

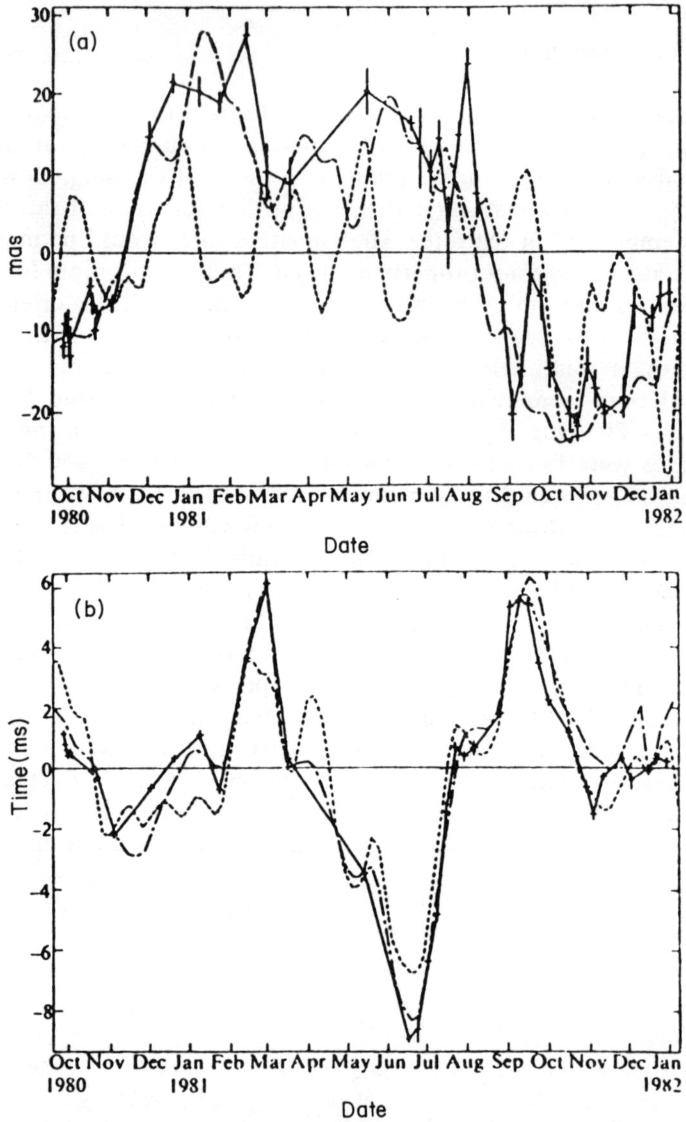

Fig. 7. (a) VLBI (——), SLR (–––) and BIH (----) astrometric determinations of *X*, shown as differences from the values determined from heavily smoothed astrometric measurements, after removal of the mean differences between the various sets of values (1 mas is equal to a displacement of about 3 cm at one Earth radius). The vertical bar on each VLBI point indicates the statistical standard error of the estimate (no allowance for possible systematic errors). The SLR and BIH astrometric values have been convolved with a Gaussian function with FWHM of 10 days (b) VLBI, LLR and BIH astrometric determinations of UT1 plotted as differences from heavily smoothed astrometric determinations after removal of the mean differences between the various sets of values (1 ms of time is equivalent to a displacement of nearly 50 cm at the equator). The error bars on the VLBI data indicate the statistical standard errors (no allowance for possible systematic errors), and the lines connecting the VLBI points are intended only to guide the eye (Robertson *et al.*, 1983).

The corresponding rms differences between the VLBI and LLR sets and the classical astrometric set are both 1.2 ms of time (or 55 cm at the equator). Carter *et al.* (1984) used SLR data in an analysis of high-frequency changes in the UT1; Fig. 8 illustrates their results. POLARIS/IRIS VLBI measures the UT1 directly. The SLR UT1 estimates are corrupted by orbital perturbations, and the difference between the two (Part C) was smoothed and then subtracted. In effect, the SLR data were put through a high-pass filter, and the remainder agree quite well, with an rms scatter of 0.7 ms. Comparisons are extremely useful in estimating the accuracy of various data sets. For example, the estimated accuracy of the LAGEOS–Center for Space Research (University of Texas) series (Tapley *et al.*, 1985) is 2–4 mas in polar motion and 0.2 ms in UT1 based on VLBI comparisons. Studies by Colquitt *et al.* (1984) indicate that the precision of the combined Doppler 5-day solution is at the 20–30 cm level for the period of day 90, 1983 through day 90, 1984. Doppler results are expected to improve with the addition of drag-compensated navigation satellites.

The results from comparisons between different techniques are quite impressive. The progress is such that several groups are currently forming combined Earth rotation parameter series. The BIH (e.g. Feissel, 1984) and the U.S.Naval Observatory (e.g. Babcock and McCarthy, 1983) have been involved for some time in the gathering of the relevant data and providing Earth rotation and polar motion values. King *et al.* (1984) combined various results (both UT1 and PM) to form a combination series using a Gaussian filter. Eubanks *et al.* (1985a) have applied the Kalman filter technique to space geodetic data to form a combination series. This technique, described in some detail earlier in this section, has several advantages including the treatment of UT-PM as a three vector.

A combination solution (excluding the results being tested) compared with values from a single technique provides valuable tests and an indication of the technique's accuracy. Modern VLBI measurements through such comparisons have a demonstrated ability to monitor Earth orientation changes to within a few milliarcseconds over a period of years. Figure 9 shows multi-baseline IRIS/POLARIS UT1 determinations after subtracting a Kalman smoothing of LLR determinations of the McDonald UT0, SLR polar motion estimates and DSN VLBI results (Steppe *et al.*, 1984). The data here had a root-mean-square value of 5.37 mas (or 0.36 ms) and a χ^2 of 93, statistically consistent with the 89 degrees of freedom. The rms scatter of the IRIS/POLARIS polar motion data is 3 mas in both components, consistent with the residual formal errors.

In summary, the new techniques are providing a data set of unprecedented accuracy. These data sets, coupled together with the combined series, which are designed to optimize the results of each technique, should provide a

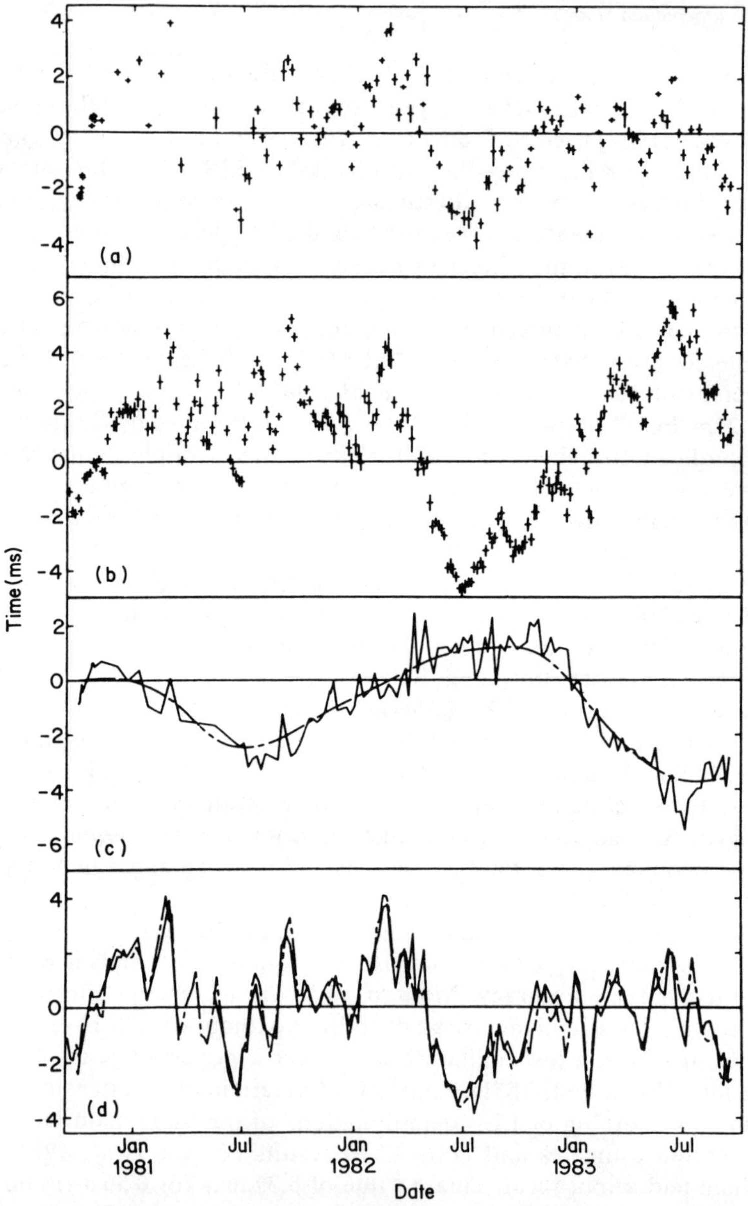

Fig. 8. (a) The UT1 determinations from the POLARIS observations. (b) The UT1 determinations from the SLR observations. Both series are shown as differences from the BIH Circular-D series: the vertical bars represent the formal standard errors of the determinations. (c) Differences between the SLR and POLARIS determinations and the curve generated as a result of smoothing the differences. (d) the POLARIS (solid line) and SLR (dashed line) series after removal of the smoothed differences shown in (c). The POLARIS line is discontinuous during April and May 1981, when equipment failures caused a gap in the VLBI observations (Carter *et al.*, 1984).

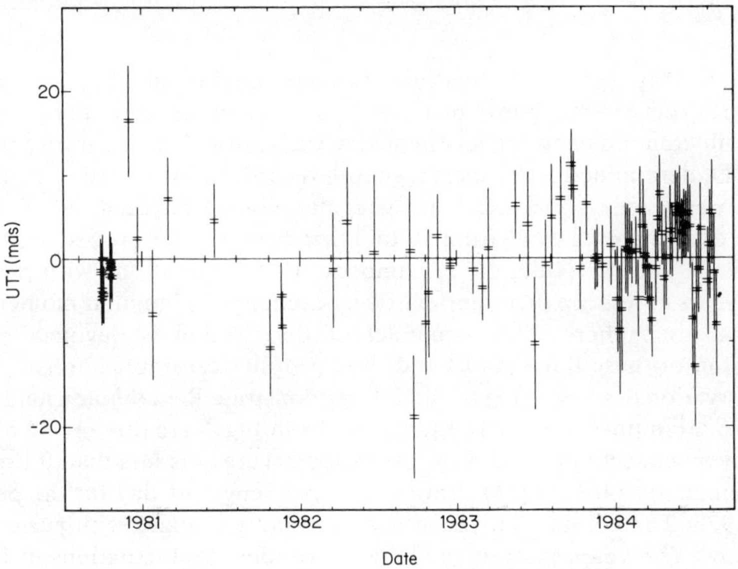

Fig. 9. IRIS/POLARIS multi-baseline VLBI UT1 estimates minus a combined smoothing of LLR, SLR and DSN VLBI as described in the text. One-sigma error bars are presented with each data point (Steppe *et al.*, 1984).

strong foundation to study the geophysical implications of Earth rotation and polar motion.

VI. Intercomparisons and Geophysical Implications

A. Introduction

The excitations of variations in Earth rotation and polar motion can be grouped into three areas: changes in the total angular momentum or the moment of inertia due to the application of external forces (e.g. lunar and solar tides); changes in the inertia tensor due to internal effects (such as earthquakes); and exchange of angular momentum between the solid Earth and the oceans, atmosphere and fluid core. The external effects are now largely predictable and can be removed from geodetic or astronomical measurements allowing the study of the latter two variations (Hide, 1985). The comparisons and the resulting geophysical implications will be discussed separately for each motion type: Earth rotation, polar wobble, precession and nutation.

B. Earth Rotation

Length of day (LOD) fluctuations can be roughly divided into three categories: the overall linear increase from tidal dissipation, the irregular decade fluctuations and higher-frequency variations. The linear increase in the LOD, determined from ancient eclipse records, is estimated to be about 1–2 ms per century. The LOD has irregular decade fluctuations of about 4–5 ms over time scales of from 20 to 30 yr; core–mantle processes are the most likely source (Hide, 1977; Lambeck, 1980). Variations with periods of five years or less are driven primarily by exchanges of angular momentum with the atmosphere. The remainder of this section is devoted to the discussion of these fluctuations and their geophysical implications.

Observational results (Figs. 3 and 4) demonstrate the complex nature of Earth rotation phenomena. The LOD results in Fig. 3 are dominated by the modern space techniques and the estimated accuracy is less than 0.1 ms. A complementary plot (Fig. 4) shows the excess length of day for the period 1861–1978. These data were smoothed to remove changes with periods of less than a few years to show the longer-term decade fluctuations in Earth rotation. Even 400 years of optical astrometric data are not enough to separate a linear trend, expected from tidal dissipation, from the irregular decade fluctuations. The tidal slowing of Earth rotation is best studied using ancient astronomical records. Historical records of the times of the ancient lunar eclipses differ by several hours from the time predicted by a uniform terrestrial rotation rate. From this one can infer the amount the Earth has slowed down in the last few thousand years (Brosche and Sunderman, 1978; Lambeck, 1980). Recent studies indicate that the Earth's rotation is not uniform, even on time scales of thousands of years (Newton, 1985).

Changes in Earth rotation with periods of a decade or larger are generally assumed to be caused by the angular momentum transfer between the Earth's solid mantle and the fluid core (Lambeck, 1980). Since all Earth rotation measurements are performed on the Earth's surface, they indicate the mantle rotation alone. The secular decrease in the LOD since 1970 (Fig. 4) represents an angular acceleration of the Earth and thus cannot possibly be due to tidal dissipation. The decrease after 1976 was not shared by the AAM data available during that period. The LOD has changed by several milliseconds over the past century (Morrison, 1979), an effect too large to be accounted for by changes in atmospheric circulation.

Several mechanisms have been forwarded for the core–mantle interaction; they include drag from bumps on the core–mantle interface (similar in nature to the atmospheric pressure torques) and a viscous drag in the fluid core (the counterpoint of the atmospheric wind stress torque). There is renewed interest in the core–mantle bumps with the recent geodynamic results based on seismic tomography (e.g., Hager *et al.*, 1985); more activity is expected in this area in the near future (see final section). Another mechanism, electromagnetic coupling between the conducting fluid core and a partially conducting lower mantle, has no corresponding atmospheric torque. Electromagnetic coupling is the most generally accepted mechanism for the decade fluctuation. The Earth's magnetic field is thought to arise from dynamos driven by motions in the fluid outer core (Hide, 1966). Long-period variations in the surface magnetic field are thought to be related to changes in the dynamos motions. Since the lower mantle is thought to be a moderate electrical conductor, Lorentz forces from the core's magnetic field could result in significant motions of the mantle.

The nature of the electromagnetic coupling suggests a correlation between LOD changes and temporal variations in the Earth's magnetic field. A general correlation has not been conclusively demonstrated; however, this could be caused by the attenuation of high-frequency magnetic variations within the mantle. In addition, the general acceptance of the electromagnetic coupling relies on the fact that the observed LOD decade fluctuations could result from magnetic field variations and from mantle conductivity values that are consistent with those estimated directly from magnetic data.

A number of investigations have examined the relationship between decade fluctuations and the geomagnetic field; many have focused on the 1969 geomagnetic jerk event. Data from a global network of magnetic observatories indicates that a declination jerk (step-change in the second time derivative) occurred over an interval of less than 2 years around 1970. This was first reported by Courtillot *et al* (1978) and has been studied by many others (e.g. Achache *et al.*, 1980; Mizuno 1980; Alldredge 1979, 1984; Courtillot *et al.*, 1979; Ducruix, *et al.*, 1980; Le Mouel *et al.*, 1981; Chan *et al.*, 1981; Malin and Hodder, 1982). The effect is easily seen in Fig. 10; Fig. 10a shows a plot of 3-yr running means of declination observed at Eskdalemuir, Scotland, while Fig. 10b gives the first derivative of these data obtained by differencing successive means. The secular acceleration is the slope of curve b. The resulting slope is approximately constant before 1970 but changes abruptly around this time and assumes a new value for the remainder of the time considered. Figure 4 (Le Mouel and Courtillot, 1982) shows the low-pass filtering of LOD changes and the magnetic declination at Paris, France for the period 1861–1978. Although the LOD seems to lag the magnetic field changes, this may be an artifact of the heavy smoothing used. Indeed, Fig. 3 shows that the angular acceleration of the Earth

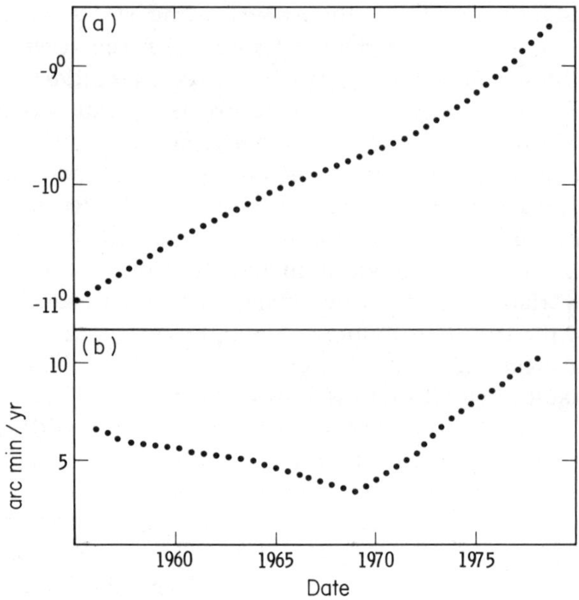

Fig. 10. Declination observed at Eskdalemuir, Scotland: (a) mean values and (b) secular variations obtained from first differences of the mean values. Note the sudden change of slope around 1970. [From Malin and Hodder (1982).]

"predicted" for the late 1970s by Le Mouel *et al.* (1981) on the basis of the 1969 magnetic jerk had actually been occurring since the early 1970s.

The reality and short-time duration of the 1969 magnetic jerk has been established from worldwide measurements; however, its origin is still under active investigation (see for example, Runcorn and Cain, 1984). Many authors believe this effect to be of internal origin (e.g. Courtillot *et al.*, 1979; Malin and Hodder, 1982, while others (e.g. Alldredge, 1979, 1980, 1984) believe that all such short-period variations are of external origin. Alldredge (1984) gives a critical assessment of these changes and concludes that abrupt changes in the geomagnetic field probably arise in the aurora and magnetosphere, not the core. Observations of the attenuation of a short-lived phenomenon originating in the Earth's core, after passing through the mantle to the surface, provide constraints on the electrical conductivity of the mantle. Malin and Hodder (1982) applied the method of spherical harmonic analysis (originally developed by Gauss in 1829 to determine whether the field is of internal or external origin) to the change of secular acceleration that occurred around 1970. They conclude that internal sources can give rise to changes in secular variation on time scales as short as one or two years and these were indeed the major factor in the 1969 geomagnetic jerk. Their study suggests that the jerk originates deep within the Earth rather than near the surface. This work implies a much lower conductivity for the lower mantle than had been previously believed.

The most outstanding characteristic of the general circulation of the Earth's atmosphere is its average "super-rotation" west to east relative to the solid Earth (at about 10 m/s) with concentration of motion in jet streams (Hide, 1985). The variance in the zonal angular momentum as a function of latitude is shown in Fig. 11 from the NMC zonal wind estimates (Rosen and Salstein, 1983). Changes in the atmospheric angular momentum (Fig. 12) contain a large seasonal cycle, dominated by annual and semiannual harmonics. The seasonal cycle of angular momentum causes seasonal LOD changes with an amplitude of about 0.5 ms, and UT1 changes with an amplitude of about 30 ms. This figure shows zonal angular momentum

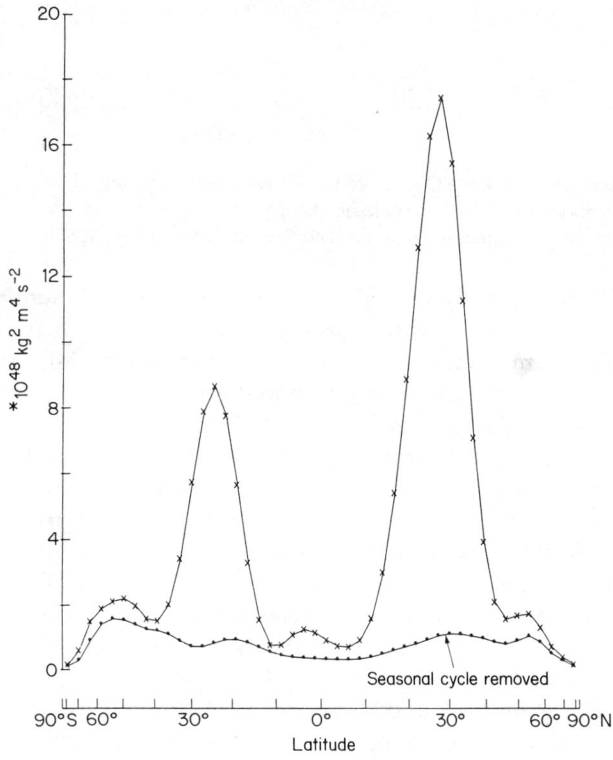

Fig. 11. The variance in zonal angular momentum as a function of latitude for 46 equal area belts. The upper curve shows the total variance for the period 1976–1980, while the lower curve is from the same data after removal of the seasonal cycle from each belt (Rosen and Salstein, 1983).

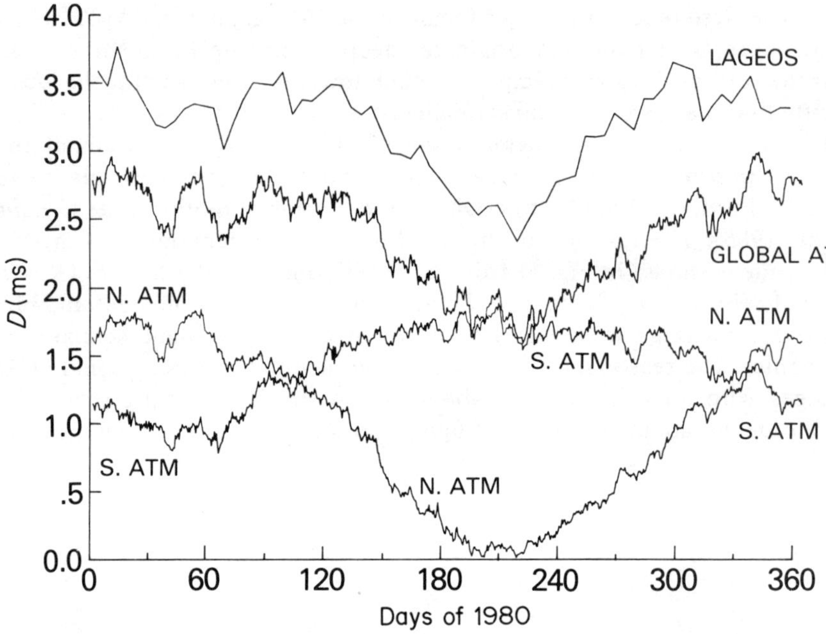

Fig. 12. Comparison of LAGEOS derived LOD for 1980 with the Northern and Southern Hemispheric components of the atmospheric changes, and total atmospheric change in LOD as calculated from the NMC analysis by Rosen and Salstein (Smith *et al.*, 1985).

estimates for the atmosphere above the Northern and Southern Hemispheres, as well as for the entire globe for 1980. In most years, the annual angular momentum cycles in the two hemispheres are 180° out of phase, and the northern cycle has a larger annual amplitude. The seasonal cycle is mostly due to changes in the mid-latitude westerlies, including the subtropical jet streams. Rosen and Salstein (1983) show that most of this variance comes from the high-velocity winds at or near the 250-bar level. Superimposed on the seasonal cycles in Fig. 12 is the irregular 50-day oscillation, which varies in period roughly from 40 to 60 days. Most of the nonseasonal changes occur in regions from 10–25° S and 20–35° N.

Comparison of astronomical measurements with axial atmospheric angular momentum indicates the significance of the atmospheric contribution to Earth rotation. Various studies (e.g. Rudloff, 1973; Lambeck and Cazenave, 1977; Lambeck and Hopgood, 1981; Eubanks *et al.*, 1985b) have related LOD variation to changes in the atmospheric angular momentum on time scales ranging from months to a few years by using mean monthly or longer period atmospheric data. Hide *et al.* (1980) evaluated the zonal component of angular momentum at 12-hr intervals by using wind and pressure data collected by the Global Atmospheric Research Program (GARP) during the special observing periods (two-8-week periods in 1979) of the First GARP Global Experiment (FGGE) and compared these with LOD data from the Bureau International de l'Heure (BIH). The comparison (Fig. 13), especially for the second period, showed generally good agreement and indicated that on these short time scales, the angular momentum transfer between Earth and the atmosphere probably accounts for most of the observed variation. This implies that core–mantle interactions are probably not significant on these time scales. Recent atmospheric and Earth rotation LOD results are shown in Figs. 12 and 14. A spectral analysis of the Earth's angular momentum budget (Eubanks *et al.*, 1985b) performed by using Earth rotation and atmospheric angular momentum data shows that significant imbalances in the angular momentum budget exist at the annual and semiannual periods (8% and 15%, respectively). Most of the annual imbalance can be attributed to the neglect of the upper-atmosphere winds (Rosen and Salstein, 1985), but changes in the oceanic circulation may contribute as much as 0.1–0.2 ms to the unexplained seasonal changes in

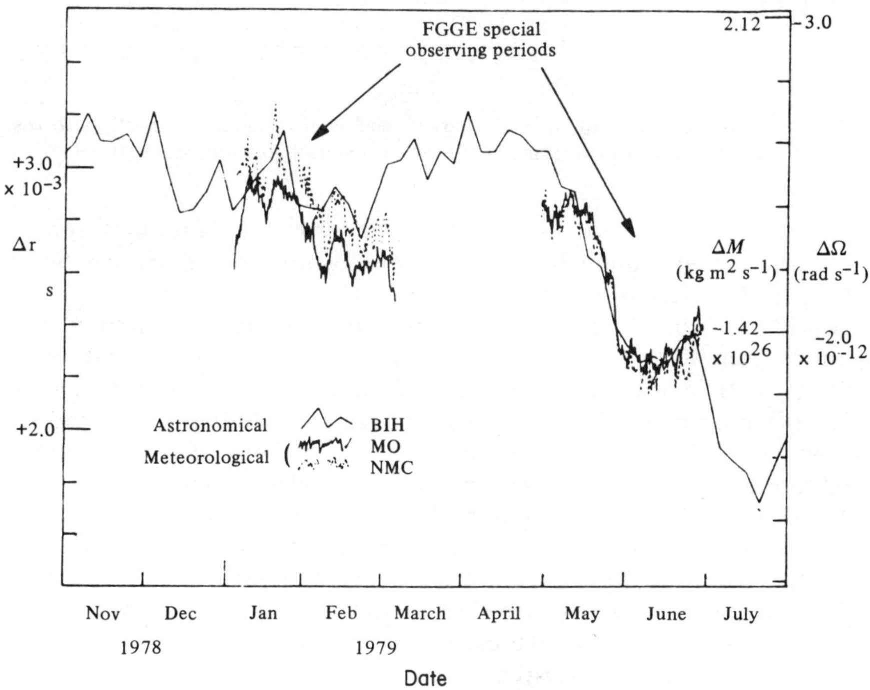

Fig. 13. Comparison of the expected changes in LOD derived from the meteorological results alongside observed changes from astronomical measurements (Hide *et al.*, 1980).

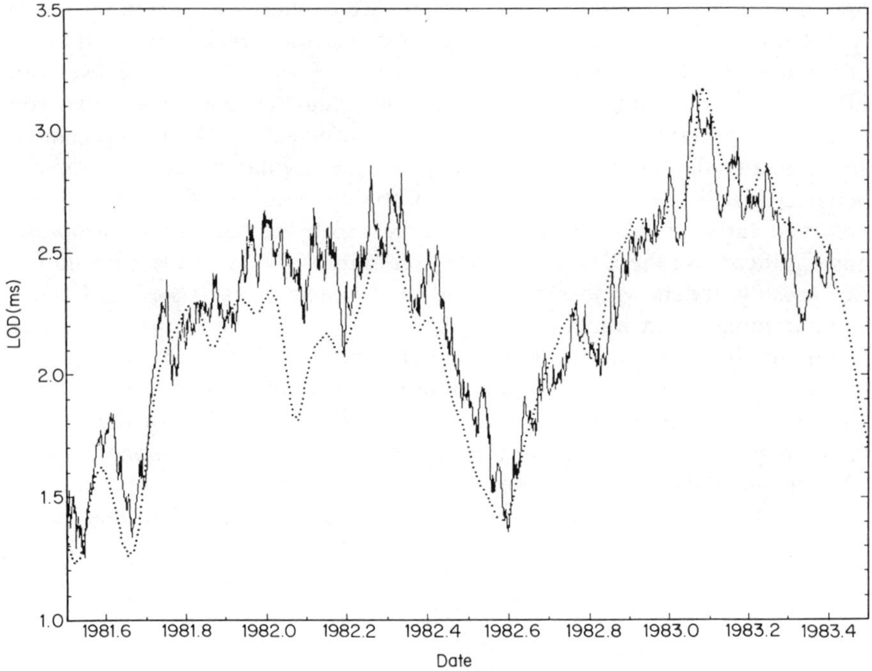

Fig. 14. AAM changes during the 1982–1983 El Niño as estimated by the NMC (solid line) together with Kalman smoothed estimates of LOD (dotted line) (Rosen *et al.*, 1984).

the LOD (Eubanks *et al.*, 1984, 1985b). The correlation of the atmospheric data with modern results has been reported by several authors (e.g. Smith *et al.*, 1985; Carter *et al.*, 1984; Langley *et al.*, 1981b).

Oscillations in zonal winds and other meteorological quantities with periods from 40 to 60 days were first discovered by Madden and Julian (1971, 1972) in wind data from the equatorial Pacific. Since that time, higher-frequency oscillations have also been seen (Miller, 1974), and the 40–60 day oscillation has been observed on a global scale (Krishnamurti and Subrahmanyam, 1982; Yasunari, 1981; Anderson and Rosen, 1983). The corresponding Earth rotation changes were first detected by Feissel and Gambis (1980) in four independent data types (shown in Fig. 6); Langley *et al.* (1981b) later reported this effect in both the AAM and the LLR LOD, as is shown in Fig. 15. The peak amplitude seems to be about 0.25 ms of LOD, although the oscillation exhibits changes in both period and amplitude on a year-to-year basis.

The atmosphere presumably excites some of the interannual fluctuations in the LOD (Lambeck, 1980; Stephanic, 1983). The Southern Oscillation

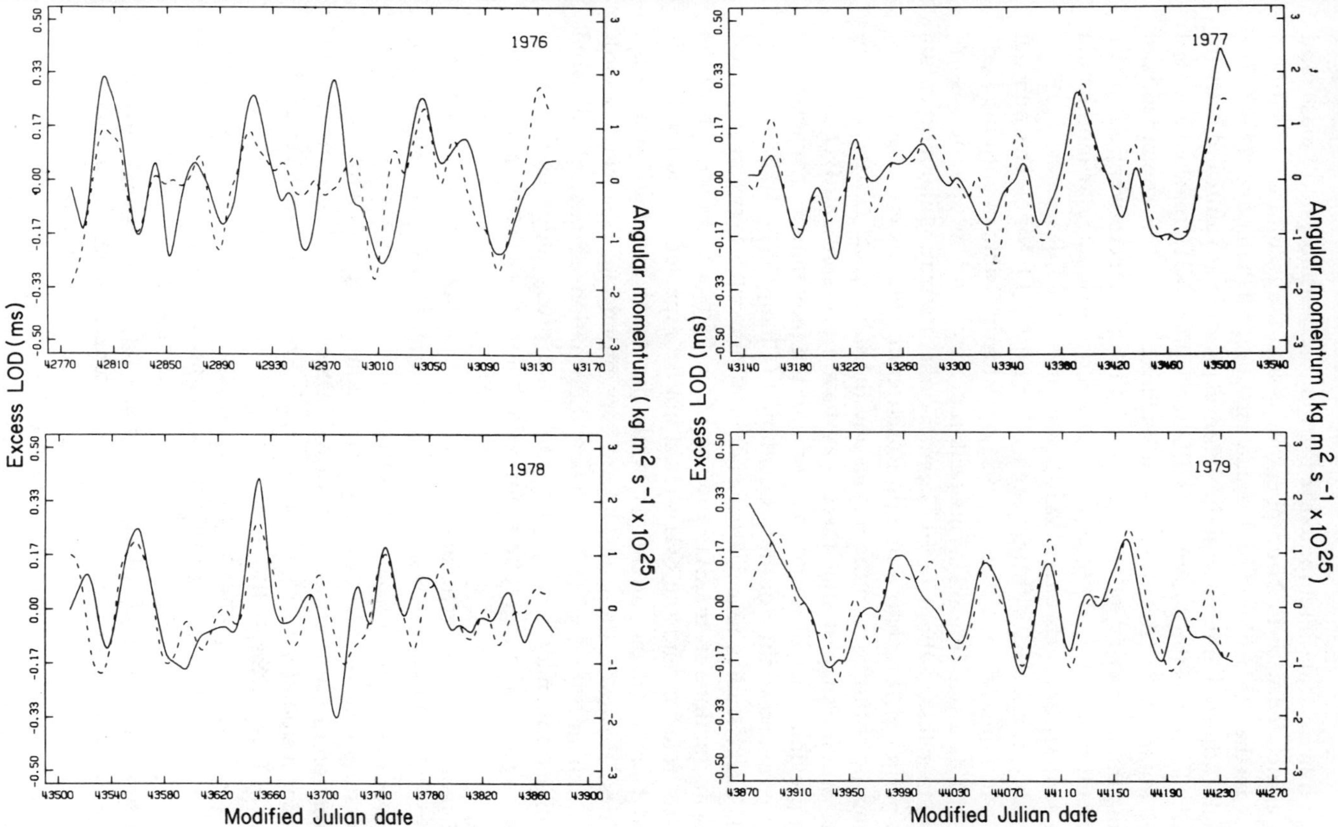

Fig. 15. Variation in LOD determined from LLR observations (solid line) and from the angular momentum of the atmosphere (dashed line) for the years 1976–1979. The annual and semiannual variations and mean levels of both sets of LOD values, as well as the fortnightly and monthly tidal terms in the lunar-ranging values have been removed (Langley *et al.*, 1981b).

involves a large-scale interannual redistribution of atmospheric mass, and is also associated with El Niño phenomena and the resulting wide-spread changes in the atmospheric and ocean circulation (Philander, 1983). The seasonally adjusted difference of the sea-level pressure at Tahiti and Darwin, Australia, known as the Southern Oscillation Index (SOI), reached a record low in January, 1983 (Rasmusson and Wallace, 1983). This coincided with the most intense period of the 1982–1983 El Niño. There were unusually large changes in the LOD and the axial atmospheric angular momentum during January and February, 1983 (Carter *et al.*, 1984; Rosen *et al.*, 1984), presumably connected with the intense 1982–1983 El Niño. The unusual maxima in January and February, 1983, are clearly visible in both the LOD and AAM data sets (Fig. 14). This peak represents an atmospheric absorption and then loss of about 3×10^{25} kg m^2 s^{-1} of eastward angular momentum over a period of less than a month. Rosen *et al.* (1984) show that most of the anomalous torque is in the tropics and the subtropical Northern Hemisphere. Studies suggest that there may have been similar LOD changes during previous El Niños and that some of the interannual changes in the LOD are related to the Southern Oscillation (Chao, 1984a; Eubanks *et al.*, 1985c). These studies reveal a correlation of about 0.5 between the interannual fluctuations in the SOI and the LOD. Perfect correlation between the Earth rotation and the SOI should not be expected, since the SOI is derived from local measurements, while the AAM and LOD are global quantities. However, these results are very promising. Historical records of LOD changes undoubtedly contain information on past global wind fluctuations and may serve as a global wind proxy index of interest to meteorologists, oceanographers and climatologists (Salstein and Rosen, 1985).

The modern space techniques are in many senses complementary; the quantities not measured well in one technique are often determined by another. Comparisons of data types frequently allow for additional insight not possible with data from a single technique. An example of this is the combination of Earth rotation with LAGEOS results to determine changes in the Earth's oblateness (Yoder *et al.*, 1983; Rubincam, 1984). An analysis of $5\frac{1}{2}$ yr of LAGEOS satellite range data together with LLR UT1 determinations reveals three significant residual signatures in the LAGEOS orbital node: an acceleration, an annual period and a semiannual period (Yoder *et al.*, 1983). These orbital node changes apparently reflect variations in the gravitational harmonic J_2 (the second zonal term in a spherical harmonic expansion), which corresponds to the oblateness of the Earth and hence the polar moment of inertia. The observed acceleration implies a decrease in J_2, requiring a motion of mass towards the rotation axis of the Earth and a reduction of the difference between the equatorial and polar moments of inertia. The implied change of J_2 is consistent with historical astronomical observations of the nontidal acceleration of the Earth's rotation and models of viscous rebound of the solid Earth from the decrease in load due to the last deglaciation. Recently, LAGEOS and the astronomical measurements have been analyzed to help further constrain models of the Earth's viscosity (Peltier, 1983). The determination of the secular J_2 change is a significant milestone in that it is the first clear-cut demonstration of changes in the Earth's global gravity field. Tidal oscillations in the gravity field have, of course, long been measured by gravity meters, but the establishment of long-term global trends from surface measurements has been plagued both by instrumental difficulties and by local effects of lesser interest, such as groundwater motion.

C. Polar Motion

The polar motion consists mainly of oscillations at periods of one year (the annual cycle) and about 433 days (the Chandler wobble), with amplitudes of about 100 and 200 mas, respectively (Lambeck, 1980), together with a long-period drift of a few milliarcseconds a year. There is also an apparent long-period oscillation along about 60° East longitude, having an amplitude of about 20 mas and a period of about 24 years, known as the Markowitz wobble. The reality of this feature is still controversial (e.g., Dickman, 1981). The yearly oscillation is driven mostly by the atmosphere (Merriam, 1982; Wahr, 1983), while the Chandler wobble is a free oscillation of the Earth (Smith and Dahlen, 1981; Wilson and Vincente, 1980). The wobble is frequently treated as a linearly damped oscillator with a resonance at the Chandler period and with dissipation described by a quality factor Q. While the period is known to within a few days (Lambeck, 1980; Wilson and Vincente, 1980), estimates of Q range over an order of magnitude or more about the current consensus of about 100–200 (Wilson and Haubrick, 1976). Note that these Q estimates are derived from about 80 years of International Latitude Service (ILS) data, and that the consensus values represent $1/e$ dissipation times of about 35–70 yr. These Q estimates are not reliable with data spans of only 1–2 dissipation times, and it is not yet possible to rule out considerably larger values for the Q of the Chandler wobble. Seismic studies provide information about the Q at seismic frequencies (Lambeck, 1980). Since the Earth is generally more rigid at high frequencies, the Q at the much lower Chandler wobble frequency is thus constrained to be less than 2000. The dissipation time thus has a rough upper bound of about 800 years.

Consider the complex form of the wobble excitation equations

$$\frac{dm}{dt}=\frac{2\pi}{T_c}\left[i-\frac{1}{2Q}\right][m-\chi]$$

where

$$m = PM_x - iPM_y \text{ and } \chi = \chi_1 + i\chi_2$$

This can be simplified still further through use of the complex Chandler frequency

$$\sigma=\frac{2\pi}{T_c}\left[1+\frac{i}{2Q}\right]$$

thus

$$\dot{m} = i\sigma(m-\chi)$$

This complex scalar equation can be Fourier transformed to yield

$$m(f)=\frac{\sigma}{\sigma-2\pi f}\chi(f)$$

Note that the use of the complex notation requires the use of the Fourier transform at positive and negative frequencies for prograde and retrograde components of circular polarization, respectively.

Clearly, the transfer function is

$$H=\frac{\sigma}{(\sigma-2\pi f)}$$

and the power spectra are related by

$$|m|^2=|H|^2|\chi|^2$$

$$=\frac{1+1/4Q^2}{(1-T_cf)^2+1/4Q^2}|\chi|^2$$

The quality factor Q thus controls the spectral bandwidth and the peak amplification of the resonance. If, as seems likely, $Q \gg 1$, then only a narrow spectral band centered on the (prograde) Chandler wobble frequency is sensitive to changes in the Q. The half power full width of this region is

$$\Delta f=\frac{1}{QT_c}=\frac{1}{\pi\tau}$$

where τ is the $1/e$ dissipation time of the wobble. Estimates of the wobble quality factor generally depend in some fashion on the apparent width of the Chandler wobble resonance. If the $Q \gtrsim 70$, then a reliable estimate of the Q requires a spectral resolution less than 1/80 yr.

Since only 80 years of ILS data exist, and since the Q probably is greater than 70, a reliable Q estimate requires a spectral resolution that would violate the uncertainty principle of classical spectral analysis. Many authors have made use of auto regressive (AR), autoregressive integrated moving average (ARIMA) or maximum entropy (ME) spectral analyses (Okubo, 1982; Wilson, 1979; Wilson and Haubrich, 1976), which can achieve higher resolution at the cost of decreased reliability. These methods assume that the spectrum can be approximated by a ratio of polynomials of the frequency; they estimate the spectrum by fitting to these polynomials in either the time or the frequency domain. Priestley (1981, Chapter 7) discusses spectral analysis in practice and Section 7.9 provides examples of spectra obtained with the different techniques. The ME technique works well if the chosen functional form can adequately represent the true spectrum, but can be very misleading if this is not the case. The uncertainty in the true nature of the Chandler wobble and its excitation, together with the evidence for wobble nonstationarity (see, e.g. Lambeck (1980), Fig. 5.14) indicates that caution should be used in interpreting the results from ME spectral analysis of the Chandler wobble.

The source of the excitation of the Chandler wobble is uncertain; the two major candidates are the atmosphere and seismic deformation. Since polar motion is assumed to be a linear response to geophysical excitation, the problem is to find an excitation with enough power at periods near the Chandler period of about 433 days. Note that the uncertainty in Q translates into a large uncertainty in the power needed to maintain the Chandler wobble, complicating the interpretation of excitations by either source.

Wahr (1983) and Wilson and Gabay (1981) used 80 years of rereduced ILS data in an attempt to understand the excitation of polar motion. These studies essentially take two complementary approaches. The power approach assumes some value for Q and compares power in the observed excitation with that required by the observed polar motion. The coherence approach assesses the significance of the coherence (in the frequency domain) or the correlation (in the time domain) between the observed excitation and that required from the observed polar motion. One advantage of the coherence approach is that it does not require knowledge of the Chandler wobble Q.

While Wilson and Gabay (1981) find essentially no coherence between the polar motion and an earthquake excitation function, they do find some correlation at the times of the large Chilean (1960) and Alaskan (1964) earthquakes. Both Wilson and Gabay (1981) and Wahr (1983) find a formally significant square coherence of about 0.45 between the observed polar motion and various atmospheric excitation data sets at the Chandler wobble frequency.

Wahr, however, using the power approach found that his atmospheric data had, at the Chandler wobble frequency, only 22% of the power in the excitation spectrum computed from the observed excitation, mostly due to the effects of atmospheric pressure and mountain torques. The effects of friction stress, currents and upwelling are not a major contribution. The coherence and the power results do not appear to be totally consistent. Suppose that the true excitation consisted of an atmospheric component together with some independent nonatmospheric component:

$$\chi = \chi_{\text{atm}} + \chi_{\text{other}}$$

The maximum square coherence (in the absence of any measurement noise) would thus be

$$C^2 = \frac{1}{1 + (P_{\text{other}}/P_{\text{atm}})} = \frac{P_{\text{atm}}}{P_{\text{atm}} + P_{\text{other}}}$$

where P_{atm} and P_{other} are the spectral power in χ_{atm} and χ_{other} near the Chandler wobble frequency. A coherence of about 0.45 thus implies that the atmosphere contributes about 45% of the total excitation power. The presence of measurement noise makes this a lower bound on the total atmospheric contribution. The discrepancy between this value and the 22% indicated by the power argument implies that either the assumed value for the Q is too low, the atmospheric data used systematically underreports the true excitation, or there is actually some correlation between χ_{atm} and χ_{other}.

The atmosphere and oceans, together with the effect of variable ground water storage, are clearly exciting the annual wobble. Hide (1985) (see Fig. 16) evaluated the atmospheric equatorial effective angular momentum functions for the period 1 December 1979–15 February 1984 from the ECMWF analyses. By comparing these results with polar motion estimates from the Bureau International de l'Heure (BIH), he concluded that the atmospheric excitation alone was sufficient to account for nearly all of the observed polar motion over this interval (3.6 Chandlerian periods or 0.5 beat periods) and that substantial excitation either by the fluid core of the Earth or by movements in the mantle associated with earthquakes was not required. It should be noted that there were several earthquakes exceeding 7.2 in magnitude during the interval of this study, but none exceeding 7.9. Analyses (Dickey, *et al.*, 1984d; Eubanks *et al.*, 1986) of the NMC data set (Rosen and Salstein, 1983 and private communication) compared (see Fig. 17) with

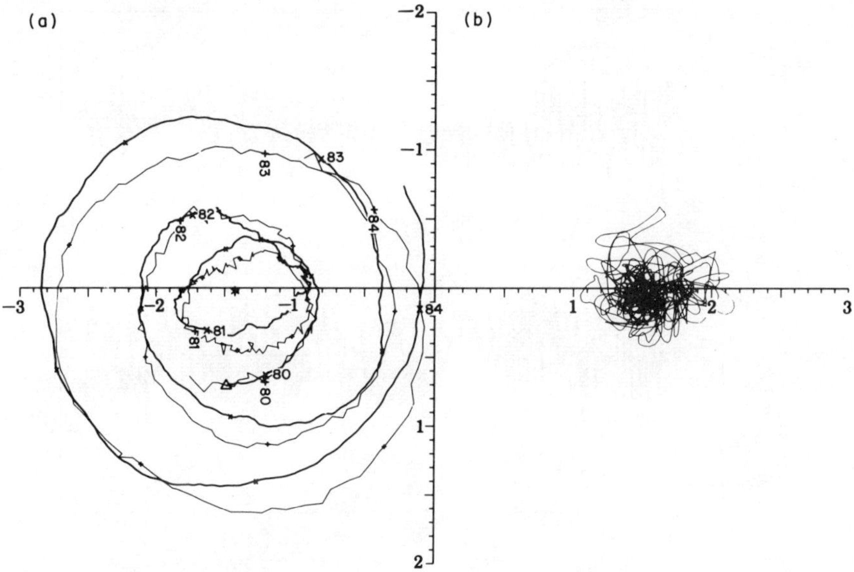

Fig. 16. Demonstration that the observed polar motion (1 December 1979–15 February 1984) may be accounted for in terms of angular momentum exchange between the atmosphere and solid Earth, without having to involve significant non-meteorological excitation. (a) Observed polar motion *m* [BIH] (—+—), as determined by BIH compared with the expected polar motion *m* [χ] (—×—) calculated from the time-series of χ_1 and χ_2. The units of both axes are μ radians: 10^{-6} rad is equivalent to a displacement of about 6.4 m over the Earth's surface. (b) Polar plot of smoothed (for convenience of display) values of χ. Time marks × meteorological; + astronomical are given at intervals of three months, with those corresponding to 1 January 1980, 1981, 1982, 1983 and 1984 labelled 80, 81, 82, 83, 84, respectively. (Hide, 1985).

a Kalman smoothed polar motion series for the period July 1976–August 1984 indicate that the atmosphere may be able to explain non-nutational polar motion with periods less than one year. The overall coherence between the χ_2 data sets is statistically significant for periods greater than 100 days, although significant coherence is not seen between any of the χ_1 data sets. There are restricted time intervals with apparent high-frequency coherence in both the χ_1 and χ_2 comparisons. A study of the ILS estimates and the Southern Oscillation Index (SOI) (Chao, 1984b) found a statistically significant coherence between χ_1 data and the SOI, at near zero phase (Chao, private communication). There was no evidence of any χ_2–SOI coherence.

D. Nutations and Precession

Nutations are driven mostly by tidal-induced exchanges of angular momentum between the Earth, Moon, Sun and planets; thus, they are largely

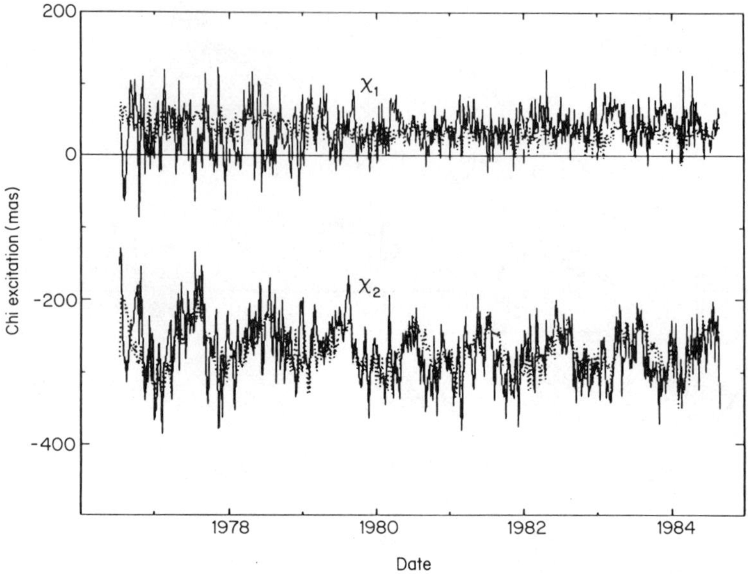

Fig. 17. Atmospheric equatorial angular momentum functions χ_1 and χ_2 derived from a Kalman-filtered combination polar motion series (——) and from the National Meteorological Center Analysis pressure terms (· · · ·) with inverted barometer model (Dickey *et al.*, 1984a).

predictable. Tidal-driven nutations are called "forced" nutations, while "free" nutations are those from any internal cause (particularly exchanges of angular momentum with the core or atmosphere). The currently recommended nutation model, the IAU 1980 Theory of Nutation (Seidelmann, 1982), is based on perturbation analysis of the effects of elasticity on a rigid Earth model (Wahr, 1981). The old IAU nutation theory used a rigid Earth model and an Earth-fixed coordinate system defined by the principal axes of the Earth's moment of inertia tensor. The Earth is, however, deformable, and this system does not have the accuracy required by modern space geodesy. The current theory instead uses a coordinate system defined by the location of the ILS observatories. One advantage of new theory is that the predictable diurnal polar motions of the old theory are now accounted for in the nutation theory.

From an inertial coordinate system, the Earth's nutation is dominated by changes with periods much longer than a day. The largest such motion is the long-period precession, a rotation of the spin axis with a period of about 26,000 years. The precession is caused by the gravitational torque on the Earth's equatorial bulge from the Sun and the planets and thus has an angular amplitude of 23.5°, the angle between the equator and the plane of the Earth's orbit (the ecliptic). The inclination of the Moon's orbit with respect to the Earth's equator induces a lunar nutation on the Earth. The Sun perturbs the lunar orbit and causes it to precess (rotate in the plane of the ecliptic) with a period of 18.6 years, the period of the induced nutation. The amplitude of this nutation is about 9 s of arc. There are smaller nutations with periods of one solar year, one lunar month and at the harmonics of the dominant frequencies. All of these nutations are called forced. The new geodetic techniques will be best able to cleanly separate errors in the precession constant and the lunar nutation after one or more 18.6-yr cycles have passed. This will occur in 1988 for LLR and about 1995 for VLBI. The accuracy of the precession constant estimates from the new techniques will thus improve steadily throughout the rest of this century.

Exchanges of equatorial angular momentum (EAM) between the Earth and the atmosphere will in general excite both a free nutation and polar motion, although one or the other will dominate, depending on the frequency involved. The Chandler wobble effectively acts as an amplifier of EAM changes that occur near the Chandler wobble frequency. At a nearly diurnal frequency the wobble transfer function is less than the resonance peak by about 5 orders of magnitude. Any power in atmospheric χ_1 and χ_2 functions at periods near one sidereal day will thus predominantly excite nutations, not polar motions. It may be possible to provide meteorological estimates at 6-hr intervals (R. D. Rosen, personal communication), which might be sufficient time resolution to study the atmospheric excitation of the nutation. The high-frequency atmospheric excitation is probably dominated by the thermally driven atmospheric diurnal and semidiurnal pressure tides, with total amplitudes of about $\frac{1}{2}$ and 1 mbar, respectively (Chapman and Lindzen, 1970). A spherical harmonic analysis of these tides reveals, however, that less than 10% of this total amplitude actually contributes to the atmospheric excitation. In the absence of a strong resonance at tidal period, these tides are several orders of magnitude below detectability, even with the most modern equipment. For example, the χ_3 component of these tides would give rise to diurnal and semidurnal UT1 oscillations with amplitudes on the order of 1 s of UT1 ($\sim 1/2$ μmm at the equator).

The free core nutation (FCN) provides a possible mechanism for amplifying the effects of diurnal EAM changes through resonance. This nutation is a free mode of oscillation of a liquid core contained within a rotating cavity, and has a retrograde period of about 460 days with respect to an inertial frame (Toomre, 1974; Shen and Mansinha, 1976). The FCN is easy to account for heuristically. Suppose that, in the absence of any excitation, the core and mantle were left to reach hydrostatic equilibrium. In that case, the core would reside within a rotating oblate spheroid container, and the only forces at the core–mantle boundary would be normal to that surface. If the mantle were then deformed or perturbed in some fashion, then the core and its container would no longer be in hydrostatic equilibrium, the core–mantle boundary would no longer be an equipotential surface, and the resulting pressure differences at the boundary would give rise to a torque and thus change the spin axis. (Note that this mechanism would not operate if the core–mantle boundary were a perfect sphere.)

Wahr and Sasao (1981) and Sasao and Wahr (1981) considered the effects of diurnal atmospheric and oceanic loading on the FCN. Atmospheric and oceanic loading cause compensating deformations of the surface of the Earth, and presumably of the core–mantle boundary as well. If the load had an Earth-fixed period near 0.99511 solar days, then the FCN resonance could be excited. Unfortunately the P_2^1 spherical harmonic of the diurnal pressure tide is predominantly prograde from an Earth-fixed viewpoint with an inertial period of about 0.5 solar days (Chapman and Lindzen, 1970). The hypothetical atmospheric tidal spectrum of Sasao and Wahr (1981, Section 5) would thus not have sufficient bandwidth to excite a FCN nutation with a dissipation time much greater than 1 year, as they assume. A semiannual modulation of the diurnal atmospheric pressure tide would produce spectral lines with Earth-fixed periods of about 0.9945 and 1.0055 solar days; the former period is close enough to the FCN period to play a possible role in its excitation. As in the case of the Chandler wobble, these arguments require some knowledge of the FCN quality factor or dissipation time, which is essentially unknown. An analysis of VLBI nutation estimates indicates that the FCN amplitude since 1980 is certainly less than 1 mas, and probably less than $\frac{1}{2}$ mas (T. A. Herring, personal communication). Herring and his collaborators claim to have detected the FCN indirectly, by attributing observed annual and semiannual retrograde nutation errors to the FCN resonance (see Herring, Chapter 3.2). If the liquid core contains density stratification, there might be several core nutations (see Toomre, 1974 (Section 4) and Shen and Mansinha, 1976). In addition, the core might be subject to a variety of internal oscillations, which might have observable effects on the Earth orientation or other geodetic parameters (Crossley, 1984). The solid inner core could undergo its own nutations and polar motions and these could cause observable effects if there is enough coupling with the mantle (Kakuba *et al.*, 1975; Szeto and Smylie, 1984). So little is known about the inner and outer core that detection of any of these possible core effects would make a significant contribution to core research.

VII. Summary and Prospects for the Future

Fluctuations in Earth orientation over time scales of a year or less are dominated by atmospheric effects. Rotation and polar motion measurements provided by modern techniques, and studies of the Earth's rotation and atmospheric dynamics have become closely associated. The continual improvement in the accuracy and density of data from the new techniques will allow the study of the Earth's exchange of angular momentum over even shorter time scales. The new techniques are producing data of unprecedented accuracy and are providing a rich data set which no doubt will increase our knowledge of Earth dynamics. Intercomparisons of Earth orientation measurements provide confidence in the reliability of the other geodetic results from the new techniques. The comparison of Earth orientation and meteorological data is leading to a better understanding of the coupling between the atmosphere and the solid Earth and is facilitating the study of other excitations of the UT-PM. New insights are expected with the merging of results from seismic tomography, core–mantle interactions, and Earth rotation. Historical records of LOD changes may serve as a global wind proxy index useful to meteorologists, oceanographers and climatologists as well.

Tectonic motions, thought from geological evidence to amount to a few centimeters per year (Minster and Jordan, 1978), will need to be accounted for in the very near future in the combined Earth orientation smoothings. Modern geodetic techniques can measure changes in the Earth orientation to within a few milliarcseconds or better. These accuracies are equivalent to knowing the location of the observatories to within a few centimeters. A number of effects can cause "local" motions of an observatory and modeling these motions will be important to understanding the results from the new techniques. If the current level of monitoring UT-PM changes can be maintained through the rest of the decade, then the effects of global plate motions should be detectable as a slow divergence in the UT-PM results.

The conceptual difference between large-scale tectonic motions and Earth orientation changes becomes somewhat arbitrary on a deformable planet monitored from a limited number of locations. The best operational procedure is to define Earth orientation as the global rotation of the network that minimizes the sum of the χ^2 for the residual motion of each station (or baseline). Any true residual motion is then ascribed to tectonic or local motion of the observatory (Cannon, 1978, 1981). The combination of nonsimultaneous data from different locations will thus require a statistical or deterministic model for the local station motion. Given such a model, the Kalman-filter approach can easily be extended to estimate local station motions together with the total Earth orientation. Eventually it will be useful to include distance information, such as VLBI baseline length estimates, together with geometrical constraints, in this procedure. Both plate motion and Earth orientation are of intense geophysical interest; these studies will become increasingly intertwined with the increasing accuracy and length of the data sets.

Acknowledgment

The work of the authors was carried out by the Jet Propulsion Laboratory, California Institute of Technology, under contract with the National Aeronautics and Space Administration.

References

Achache, J., Courtillot, V., Ducruix, J. and Le Mouel, J.-L. (1980). *Phys. Earth Planet. Int.* **23**, 72–75.

Alldredge, L. R. (1975). *J. Geophys. Res.* **80**, 1571–1578.

Alldredge, L. R. (1979). *C.R. Hebd. Seanc. Acad. Sci.*, B **289**, 169–171.

Alldredge, L. R. (1980). *J. Geomagn. Geolectr.* **32**, 421–425.

Alldredge, L. R. (1984). A discussion of impulses and jerks in the geomagnetic field. *J. Geophys. Res.* **89**, 4403–4412.

Anderson, J. A. and Rosen, R. D. (1983). The latitude-height structure of 40–50 day variations in atmospheric angular momentum. *J. Atmos. Sci.* **40**, 1584–1591.

Babcock, A. K. and McCarthy, D. D. (1983). Predicting Earth Orientation—The Ideal System *EOS, Trans., Am. Geophys. Union* **64**, 675.

Barnes, R. T. H., Hide, R., White, A. A. and Wilson, C. A. (1983). Atmospheric angular momentum fluctuations, length of day changes and polar motion. *Proc. R. Soc. London,* **387**, 31–73.

Brosche, P. and Sundermann, J. (1978). "Tidal Friction and the Earth's Rotation". Springer-Verlag, Berlin and New York.

Cannon, W. H. (1978). The measurement of Earth rotation on a deformable earth. *NASA Conf. Publ. CP-2115.*

Cannon, W. H. (1981). The definition of the terrestrial coordinate frame by long baseline interferometry. *In* "Reference Coordinate Systems for Earth Dynamics" (E. M. Gaposckin and B. Kolaczek, eds.) pp. 111–118.

Carter, W. E. and Robertson, D. S. (1984). IRIS Earth rotation and polar motion results. *Proc. Int. Symp. Space Tech. Geodyn., Sopron, Hungary.*

Carter, W. E. and Robertson, D. S. (1985). High frequency variations in the rotation of the Earth. *IEEE Trans. Geosci. Remote Sensing* (Special Issue on Geodynamics) *GE-23,* 369–372.

Carter, W. E., Robertson, D. S., Pettey, J. E., Tapley, B. D., Schutz, B. E., Eanes, R. J. and Lufeng, M. (1984). Variations in the rotation of the Earth. *Science* **224**, 957–961.

Chan, G. D., Ducruix, J. and J.-L. Le Mouel, *C.R. Hebd. Seanc. Acad. Sci.*, B **293**, 157–160.

Chao, B. F. (1984). Interannual length-of-day variations with relation to the Southern Oscillation/El Niño. *Geophys. Res. Lett.* **11**, 541–544.

Chao, B. F. (1984). Excitation of Chandler wobble by the Southern Oscillation based on ILS Data. *EOS, Trans. Am. Geophys. Union* **65**, 45, 859.

Chapman, S. and Lindzen, R. S. (1970). *Atmospheric Tides.* Reidel Publ. Boston, Massachusetts.

Colquitt, E. S., Stein, W. L. and Anderle, R. J. (1984). Doppler polar motion measurements during project MERIT. *EOS, Trans. Am. Geophys. Union* **65**, 188.

Courtillot, V., Ducruix, J. and Le Mouel, J.-L. (1978). *C.R. Hebd. Seanc. Acad. Sci.*, D **287**, 1095–1098.

Courtillot, V., Ducruix, J. and Le Mouel, J.-L. (1979). *C.R. Hebd. Seanc. Acad. Sci.*, B **289**, 173–175.

Courtillot, V., Le Mouel, J., Ducruix, J. and Cazenave, A. (1982). Magnetic secular variations as a precursor of climatic change. *Nature* **297**, 386–7.

Crossley, D. (1984). Oscillatory flow in the liquid core. *Phys. Earth Planet. Inter.* **36**, 1–16.

Dickey, J. O. (1985a) Activities and goals of the IUGG/IAG special study group 5-98. Atmospheric excitation of the Earth's rotation. *Proc. Int. Symp. Space Tech. Geodyn., Sopron, Hungary.*

Dickey, J. O. (1985b). Activity Report of IUGG/IAG special study group 5-98. Atmospheric excitation of the Earth's rotation. (*Int. Coord. Space Tech. Geod. Geodyn. Bull*). Dept. of Geodetic Science and Surveying, Ohio State Univ., Columbus, Ohio.

Dickey, J. O. and Eubanks, T. M. (1985). Earth rotation and polar motion: measurements and implications, (invited article). *IEEE Trans. Geosci. Remote Sensing*, (special issue on geodynamics) *GE-23*, 373–384.

Dickey, J. O., Williams, J. G., Newhall, X. X. and Yoder, C. F. (1984). Geophysical application of lunar laser ranging. *Proc. Int. Assoc. Geod.* (*IAG*) *Symp. Int. Union Geod. Geophys.* (*IUGG*) *XVIIIth, General Assembly, Hamburg, FRG.* 1983, Ohio State Univ., Dept. of Geodetic Science and Surveying, Columbus, Ohio **2**, 509–521.

Dickey, J. O., Newhall, X. X. and Williams, J. G. (1984b). Earth orientation from lunar laser ranging. *BIH Ann. Rep. 1983*, D31–D34.

Dickey, J. O., Williams, J. G., and Eubanks, T. M. (1984c). Earth rotation and polar motion: results from lunar laser ranging analysis and an intercomparison study. *Proc. Int. Assoc. Geod. Symp. Int. Union Geod. Geophys., IUGG XVIIIth General Assembly, Hamburg, FRG 1983* **2**, 12–27.

Dickey, J. O., Eubanks, T. M. and Steppe, J. A. (1984d). The atmospheric excitation of the polar motion. *EOS Trans. Am. Geophys. Union* **65**, 859.

Dickey, J. O., Newhall, X. X. and Williams, J. G. (1985). Earth orientation from lunar laser ranging and an error analysis of polar motion services (Special LAGEOS issue). *J. Geophys. Res.* **90**, 9353–9363.

Dickman, S. R. (1981). Investigation of controversial polar motion features using homogeneous ILS data. *J. Geophys. Res.* **86**, 4904.

Ducruix, J., Courtillot, V. and Le Mouel, J.-L. (1980). *Geophys. J. R. Astron. Soc.* **61**, 73–94.

Eanes, R. J., Schutz, B. E. and Tapley, B. D. (1984). Earth rotation from LAGEOS: The 1984 CSR System. *EOS, Trans. Am. Geophys. Union* **65**, 16, 187–188.

Eubanks, T. M., Roth, M. G., Esposito, P. B., Steppe, J. A. and Callahan, P. S. (1982). An analysis of JPL TEMPO Earth orientation results. *Proc. Symp. No. 5: Geod. Appl. Radio Interferometry, NOAA Techn. Rep. NOS 95 NGS* **24**, 63–70.

Eubanks, T. M., Dickey, J. O. and Steppe, J. A. (1984). "The geophysical significance of systematic errors in the Earth's angular momentum budget". *Proc. Int. Assoc. Geod. Symp. Int. Union Geod. Geophys. XVIIIth* General Assembly, *Hamburg, FRG*; *1983* **2**, 122–143.

Eubanks, T. M., Spieth, M. A. and Steppe, J. A. (1985a) The accuracy of radio interferometric estimates of Earth rotation *JPL/TDA Prog. Rep.* **42–80**, 229–235.

Eubanks, T. M., Steppe, J. A., Dickey, J. O. and Callahan, P. S. (1985b). A spectral analysis of the Earth's angular momentum budget. *J. Geophys. Res.* **90**, 5385–5404.

Eubanks, T. M., Dickey, J. O. and Steppe, J. A. (1985c) The 1982–83 El Niño, the Southern Oscillation, and the Length of Day. *Trop. Ocean Atm. Newslett.* **29**, 21–22.

Eubanks, T. M., Steppe, J. A., Dickey, J. O., Rosen, R. D. and Salstein, D. A. (1986). (In preparation.)

Feissel, M. (1980). Determination of the Earth rotation parameters by the BIH 1962–1979. *Bull. Geod.* **54**, 81–102.

Feissel, M. (1984). *BIH Ann. Rep. 1983.*

Feissel, M. and Gambis, D. (1980). La mise en evidence de variations rapides de la duree du jour. *C.R. Hebd. Sceanc. Acad. Sci.* B **291**, 271–273.

Feissel, M. and Lawandowski, W. (1984). A comparative analysis of Vondrak's and Gaussian smoothing techniques. *Bull. Geod.* **58**, 464.

Feissel, M. and Wilson, P. (1983). MERIT campaign: connection of reference frames, implementation plan. MERIT/COTES Joint Working Groups, Bureau International de l'Heure, Paris.

Gelb, A., ed. (1974). "Applied Optimal Estimation". MIT Press, Cambridge, Massachusetts.

Goddard Space Flight Center (1983). Earth rotation from laser ranging to LAGEOS. *BIH Ann. Rep. 1982*, D31–D55.

Gross, R. S. and Chao, B. F. (1984). The observed excitation of the LAGEOS derived Chandler wobble. *EOS, Trans. Am. Geophys. Union* **65**, 187.

Guinot, B. (1978). Rotation of the Earth and polar motion services In *Proc. GEOP Conf. Int. Symp Appl. Geod. Geodyn. 9th* pp. 13–18, Dept. of Geodetic Science, Rep. no. 280, Ohio State Univ.

Hager, B. H., Clayton, R. W., Dziewonski, A., Richards, M. and Comer, T. (1985). Lower mantle heterogeneity, dynamic topography and the geoid. *Nature* **313**, 541–545.

Hide, R. (1966). Free hydromagnetic oscillations of the Earth's core and the theory of the geomagnetic secular variation. *Philos. Trans. R. Soc.*, A **259**, 615–647.

Hide, R. (1977). *Philos. Trans. R. Soc. London, A* **284**, 547–554.

Hide, R., Birch, N. T., Morrison, L. V., Shea, D. J. and White, A. A. (1980). Atmospheric angular momentum fluctuations and changes in the length of the day. *Nature* **286**, 114–117.

Hide, R. (1985). Rotation of the atmospheres of the Earth and planets. *Philos. Trans. R. Soc. London, A.* **313**, 107–121.

Jazwinski, A. H. (1970). "Stochastic Processes and Filtering Theory" Academic Press, New York.

Kakuba, C., Okamoto, I. and Sasao, T. (1975). Is the nutation of the solid inner core responsible for the 24-year libration of the pole? *Publ. Astron. Soc. Jpn* **27**, 357–365.

Kerr, Richard A. (1983). The race to predict next week's weather. *Res. Science* April 1, 39–41.

King, R. W., Kolaczek, B. A. and Shapiro, I. I. (1984). Accuracies of recent observations of the Earth's rotation. *EOS, Trans. Am. Geophys. Union* **65**, 187.

Kouba, J. (1983). A review of geodetic and geodynamic satellite Doppler positioning. *Rev. Geophys. Space Phys.* **21**, 27–40.

Krishnamurti, T. N. and Subrahmanyam, D. (1982). The 30–50 day mode at 850 mb during MONEX. *J. Atmos. Sci.* **39**, 2088–2095.

Lambeck, K. and Cazenave, A. (1977). *Philos. Trans. R. Soc. London A* **284**, 495–506.

Lambeck, K. (1980). "The Earth's Variable Rotation." Cambridge Univ. Press, London and New York.

Lambeck, K. and Hopgood, P. (1981). The Earth's rotation and atmospheric circulation from 1963 to 1973. *Geophys. J. R. Astron. Soc.* **71**, 67–84.

Langley, R. B., King, R. W., and Shapiro, I. I. (1981a). Earth rotation from lunar laser ranging. *J. Geophys. Res.* **86**, 11,913–11,918.

Langley, R. B., King, R. W., Shapiro, I. I., Rosen, R. D. and Salstein, D. A. (1981b). Atmospheric angular momentum and the length of the day: a common fluctuation with a period near 50 days. *Nature* **294**, 730–733.

Le Mouel, J. L. and Courtillot, V. (1981). Core motions, electromagnetic core-mantle coupling and variations in the Earth rotation—new constraints from geomagnetic secular variation impulses. *Phys. Earth Planet. Int.* **24**, 236–241.

Le Mouel, J. L. and Courtillot, V. (1982). On the outer layers of the core and geomagnetic secular variation. *J. Geophys. Res.* **87**, B5, 4103–4108.

Le Mouel, J. L., Madden, T. R., Ducruix, J. and Courtillot, V. (1981). Decade fluctuations in geomagnetic westward drift and Earth rotation. *Nature* **290**, 763–765.

Madden, R. A. and Julian, P. R. (1971). Detection of a 40–50 day oscillation in the zonal wind in the tropical Pacific. *J. Atmos. Sci.* **28**, 702–708.

Madden, R. A. and Julian, P. R. (1972). Description of global-scale circulation cells in the tropics with a 40–50 day period. *J. Atmos. Sci.* **29**, 1109–1123.

McPherson, R. D., Bergman, K. H., Kistler, R. E., Rash, G. E. and Gordon, D. S. (1979). The NMC operational global data assimilation system. *Mon. Weather Rev.* **107**, 1445–1461, November 1979.

Malin, S. R. C. and Hodder, B. M. (1982). Was the 1970 geomagnetic jerk of internal or external origin? *Nature* **296**, 726–728.

Merriam, J. B. (1982). Meteorological excitation of the annual polar motion. *Geophys. J. R. Astron. Soc.* **70**, 41–56.

Miller, A. J. (1974). Periodic variation of atmospheric circulation at 14–16 days. *J. Atmos. Sci.* **31**, 720–726.

Minster, J. B. and Jordan, T. H. (1978). Present day plate motions. *J. Geophys. Res.*, **83**, 5331–5354.

Mizuno, H. *J. Geomagn. Geolect* **32**, 413–420.

Morrison, L. V. (1979). Re-determination of the decade fluctuations in the rotation of the Earth in the period 1861–1978. *Geophys. J. R. Astron. Soc.* **58**, 349–360.

Munk, W. H. and MacDonald, G. J. F. (1960). "The Rotation of the Earth, A Geophysical Discussion". Cambridge Univ. Press, London.

Naito, I. and Yokoyama, K. (1984). A computation of atmospheric angular momenta based on JMA global analysis data. *Int. Symp. Space Tech. Geodyn. Sopron, Hungary.*

Newton, R. R. (1985). The secular acceleration of the Earth's spin. *Geophys. J. R. Astron. Soc.* **80**, 313–328.

Okubo, S. (1982). Theoretical and observed Q of the Chandler wobble. *Geophys. J. RAS* **71**, 647–657.

Peltier, W. R. (1983). Constraint on deep mantle viscosity from LAGEOS acceleration data. *Nature* **304**, 434–436.

Philander, S. G. H. (1983). El Niño Southern Oscillation phenomena. *Nature* **302**, 295–301.

Priestley, M. B. (1981). "Spectral Analysis and Time Series." Academic Press, New York.

Rasmusson, E. U. and Wallace, J. M. (1983). Meteorological aspects of the El Niño/Southern Oscillation. *Science* **222**, 1195–1202.

Robertson, D. S., Carter, W. E., Eanes, R. J., Schutz, B. E., Tapley, B. D., King, R. W., Langley, R. B., Morgan, P. J. and Shapiro, I. I. (1983). Comparison of Earth rotation as inferred from radio interferometric, laser ranging and astrometric observations. *Nature* **302**, 509–511.

Rosen, R. D. and Salstein, D. A. (1983). Variations in atmospheric angular momentum on global and regional scales and the length of day. *J. Geophys. Res.* **88**, 5451–5470.

Rosen, R. D. and Salstein, D. A. (1985). Contribution of stratospheric winds to annual and semi-annual fluctuations in atmospheric angular momentum and the length of day. *J. Geophys. Res.* **90**, 8033–8041.

Rosen, R. D., Salstein, D. A., Eubanks, T. M., Dickey, J. O. and Steppe, J. A. (1984). An El Niño signal in atmospheric angular momentum and Earth rotation. *Science* **225**, 411–414.

Rubincam, D. (1984). Postglacial rebound observed by LAGEOS and the effective viscosity of the lower mantle. *J. Geophys. Res.* **89**, 1077–1087.

Rudloff, W. (1973). *Ann. Meteorol. Neue Folge* **6**, 221–223.

Runcorn, S. K. (1982). The role of the core in irregular fluctuations of the Earth's rotation and the excitation of the Chandler wobble. *Philos. Trans. R. Soc. London, A* **306**, 261–270.

Runcorn, S. K. and Cain, J. C. (1984). Chaired Special Session entitled Irregularities in the secular variations and geodynamic implications. *EOS, Trans. Am. Geophys. Union* **65**, 203.

Ryan, R. C. and Ma, C. (1983). Results from the NASA Crustal Dynamics Project. *BIH Ann. Rep. 1982.*

Salstein, D. A. and Rosen, R. D. (1985). Earth rotation data as a proxy index of global wind fluctuations. *Conf. Climate Variations Symp. Contemp. Climate, 3rd, 1985, Los Angeles,* 1850–2100.

Sasao, T. and Wahr, J. M. (1981). An excitation mechanism for the free 'core nutation'. *Geophys. J. R. Astron. Soc.* **64**, 729–746.

Seidelmann, P. (1982). *Celestial Mech.* **27**, 79.

Shen, P. and Mansinha, L. (1976). Oscillation, nutation and wobble of an elliptical rotating Earth with liquid outer core. *Geophys. J. R. Astron. Soc.* **46**, 467–496.

Smith, M. L. (1977). Wobble and nutation of the Earth. *Geophys. J. R. Astron. Soc.* **50**, 103–140.

Smith, M. L. and Dahlen, F. A. (1981). The period and *Q* of the Chandler wobble, *Geophys. J. R. Astron. Soc.* **64**, 223–281.

Smith, D. E., Christodoulidis, D. C., Wyatt, G. H., Dunn, P., Klosko, S. and Torrence, M. (1985). Private communication.

Smylie, D. E., Clarke, G. K. C. and Ulrych, T. J. (1973). Analysis of irregularities in the Earth's rotation. In "Methods in Computational Physics" Vol. 13, pp. 391–430.

Sovers, O. J., Thomas, J. B., Fanselow, J. L., Cohen, E. J., Purcell Jr., G. H., Rogstad, D. H., Skjerve, L. J. and Spitzmesser, D. J. (1984). Radio interferometric determination of intercontinental baselines and Earth orientation utilizing deep space network antennas: 1971 to 1980. *J. Geophys. Res.* **89**, B9, 7597–7607.

Stephanic, M. (1983). Interannual atmospheric angular momentum variability 1963–1973 and the Southern oscillation. *J. Geophys. Res.* **87**, 428–432.

Steppe, J. A., Eubanks, T. M. and Spieth, M. A. (1984). Systematic error in "short" baseline VLBI estimates of the Earth orientation. *EOS, Trans. Am. Geophys. Union* **65**, 45, 854.

Szeto, A. M. K. and Smylie, D. E. (1984). Coupled motions of the inner core and possible geomagnetic implications. *Phys. Earth Planet. Inter.* **36**, 27–42.

Tapley, B. D., Schutz, B. E. and Eanes, R. J. (1985). Station coordinates, baselines and Earth rotation from LAGEOS laser ranging: 1976–1984. (LAGEOS special issue.) *J. Geophys. Res.* **90**, 9234–9248

Toomre, A. (1974). On the "nearly diurnal wobble" of the Earth. *Geophys. J. R. Astron. Soc.* **38**, 335–348.

Wahr, J. M. (1981). The forced nutations of an elliptical, rotating, elastic and oceanless Earth. *Geophys. J. R. Astron. Soc.* **64**, 705–727.

Wahr, J. M. (1982). The effects of the atmosphere and oceans on the Earth's wobble and on the seasonal variations in the length of day—I. Theory. *Geophys. J. R. Astron. Soc.* **70**, 349–372.

Wahr, J. M. (1983). The effects of the atmosphere and oceans on the Earth's wobble and on the seasonal variations in the length of day—II. Results. *Geophys. J. R. Astron. Soc.* **74**, 451–487.

Wahr, J. M. (1985). The Earth's rotation rate. *Am. Sci.* **73**, 41–46.

Wahr, J. M. and Sasao, T. (1981). A diurnal resonance in the ocean tide and in the Earth's load response due to the resonant free 'core nutation'. *Geophys. J. R. Astron. Soc.* **64**, 747–765.

Wilkins, G. A., ed. (1980). PROJECT MERIT, Joint working group on the rotation of the Earth. IAU/IUGG.

Wilson, C. R. (1979). Estimation of the parameters of the Earth's polar motion, Time and the Earth's rotation, IAU, 307-312.

Wilson, C. R., and Gabay, R. (1981). Excitation of the Earth's polar motion: A reassessment with new data. *GRL* **8**, 745-748.

Wilson, C. R. and Haubrich, R. A. (1976). Meteorological excitation of the Earth's wobble. *Geophys. J. R. Astron. Soc.* **46**, 707-743.

Wilson, C. R. and Vincente, R. O. (1980). An analysis of the homogeneous ILS polar motion series. *Geophys. J. R. Astron. Soc.* **62**, 605-616.

Yasunari, T. (1981). Structure of an indian summer monsoon system with around 40-day period. *J. Met. Soc. Jpn* **59**, 336-354.

Yoder, C. F., Williams, J. G. and Parke, M. E. (1981). Tidal variations of Earth rotation. *J. Geophys. Res.* **86**, B2, 881-891.

Yoder, C. F., Williams, J. G., Dickey, J. O., Schutz, B. E., Eanes, R. J. and Tapley, B. D. (1983). Secular variation of Earth's gravitational harmonic J_2 from LAGEOS and the nontidal acceleration of Earth rotation. *Nature* **303**, 757-762.

4.2 Projects POLARIS and IRIS: Monitoring Polar Motion and UT1 by Very Long Baseline Interferometry

W. E. Carter and D. S. Robertson

National Geodetic Survey, Charting and Geodetic Services, National Ocean Service, NOAA, Rockville, Maryland, USA

I. Introduction

In striving to improve the spatial and temporal resolutions of geodetic surveying to the levels required for monitoring geodynamics, i.e. spatial resolutions of parts in 10^8 in observing periods of one day or less, the National Geodetic Survey (NGS) in the late 1970s concluded that completely new methods of observation based on fundamentally more powerful technologies must be implemented. Considerations of cost, achievable accuracy and operability led to the selection of radio interferometry as the primary technology, and specifically to a combination of very long baseline interferometry (VLBI), using natural radio sources, and phase difference observations of the radio signals broadcast by the satellites of the global positioning system (GPS).

The vanguard project in the NGS effort to realize these advanced methods of surveying was POLar-motion Analysis by Radio Interferometric Surveying (POLARIS), initiated in 1977. The primary goal of POLARIS was the development of an advanced polar motion and Earth rotation (UT1) monitoring service, capable of determining both the x and y components of polar motion to ±3 mas (milliarcseconds) and UT1 to 0.1 ms, in observing periods of 1 day (Carter and Strange, 1979; Carter *et al.*, 1979). This was an ambitious goal considering that VLBI was then still widely viewed as a developmental technique with formidable data handling and processing problems—indeed some geodesists questioned if the technique could ever be made practical for operational applications. But a consortium of scientists, engineers and technicians at the Massachusetts Institute of Technology, Haystack Radio Observatory, NASA Goddard Space Flight Center and the National Radio Astronomy Observatory were already building the third-generation Mark III VLBI system (Rogers *et al.*, 1983). The wideband recording capacity and computer automation built into the Mark III were the keys to transforming VLBI into a viable geodetic surveying system. Progress has been rapid, and the original goals of POLARIS have already been met and surpassed. Today the International Radio Interferometric Surveying (IRIS) network, which includes the POLARIS stations, is routinely monitoring, at 5-day intervals, changes in x, y and UT1 with accuracies of 1–2 mas and 0.05–0.10 ms, respectively. Also a 90-day series of experiments with a single North America–Europe interferometer was part of the activities organized by IRIS to support the MERIT intensive campaign (Wilkins, 1984), which obtained daily values of UT1 accurate to about 0.1 ms in observing periods of only about 1 hr.

The time series produced by the POLARIS and IRIS programs have spatial and temporal resolutions of at least an order of magnitude better than the classical optical astrometric monitoring systems, and it would be surprising if they did not clarify certain past ambiguities and reveal previously undetected patterns in the variations in polar motion and UT1.

II. Review of Results

The first advanced technology applied to monitoring polar motion used Doppler observations of the Navy navigation satellites (Oesterwinter, 1979). The system was introduced by the U.S. Department of Defense in 1964 and began producing regularly spaced time series in 1969. Systematic differences between the U.S. Department of Defense and Bureau International de l'Heure (BIH) time series soon became obvious, and in 1973 the BIH began incorporating the Doppler values in computing the annual and semiannual "corrections" applied to the optical observations. However, the weighting given to the Doppler values ranged from only 13–24% during the years 1973–1977 (Guinot, 1979). As a result the two series continued to display significant systematic differences, which could not be attributed with certainty to deficiencies in one or the other. Robertson *et al.* (1978) reported the results of an analysis of 13 geodetic VLBI observing sessions with the

Haystack–Owens Valley Radio Observatory interferometer during the period September 1976–May 1978, which produced estimates of the x component of polar motion and UT1. Figure 1 clearly shows that the discrepancy between the Doppler and the BIH polar motion series was predominantly caused by errors in the BIH reductions. The VLBI UT1 values had an rms scatter of 1.6 ms relative to the BIH values, but the sparsity of the VLBI series made it impossible to detect any systematic components. Subsequent comparison of the VLBI and BIH UT1 estimates with a third series derived from lunar laser ranging (LLR) observations produced Fig. 2. The rms difference between the VLBI and LLR values is only 0.8 ms, indicating that the difference between the VLBI and BIH series was largely due to errors in the BIH values.

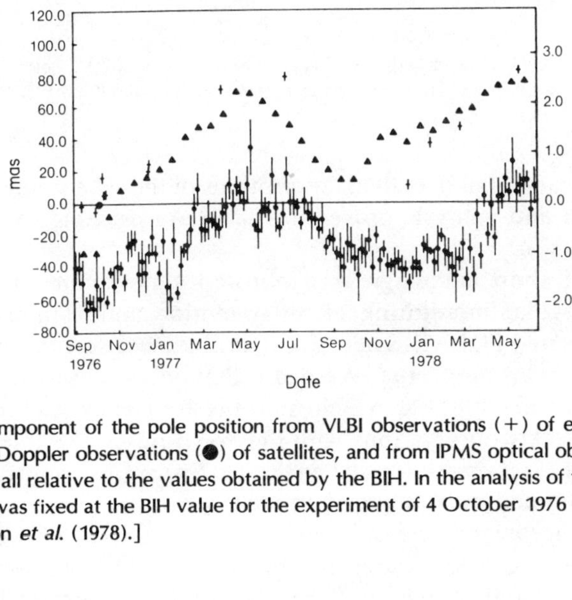

Fig. 1. x-component of the pole position from VLBI observations (+) of extragalactic radio sources, from Doppler observations (●) of satellites, and from IPMS optical observations (▲) of Galactic stars, all relative to the values obtained by the BIH. In the analysis of the VLBI data, the pole position was fixed at the BIH value for the experiment of 4 October 1976 (●). [Reproduced from Robertson *et al.* (1978).]

The next important step in the application of VLBI to monitoring polar motion and UT1 was organized around the MERIT short campaign (Wilkins, 1980). During September and October 1980, two 7-day series of observations were conducted by an international network of stations in North America and Europe. The polar motion and UT1 estimates obtained were almost certainly the most accurate ever achieved, but perhaps even more important was that the sessions served as the first full-scale operational application of the Mark III VLBI system. Many of the operating procedures which have since become standard were developed or proven during the

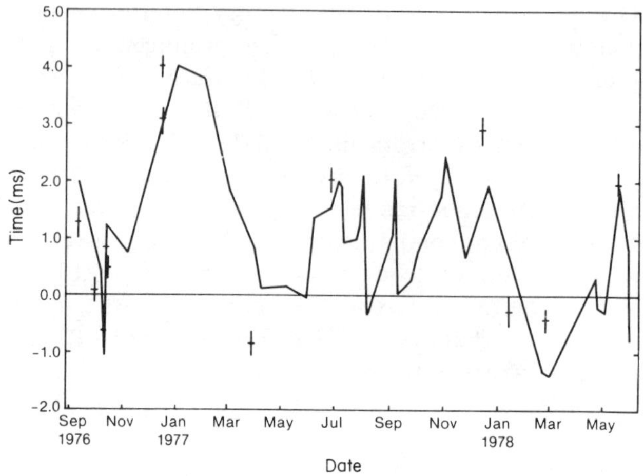

Fig. 2. VLBI (crosses) and lunar laser (solid line) determinations of UT1, shown as differences from the BIH Circular-D values. [Reproduced from Robertson *et al.* (1979).]

MERIT short campaign. Detailed descriptions of the scheduling, observing, data reduction and analysis procedures used are presented in Robertson and Carter (1982).

The MERIT short campaign was followed almost immediately by the start of the regular monitoring of polar motion and UT1 by using the partially completed POLARIS network. The observing sessions were spaced at 2-week intervals until the Westford POLARIS observatory became operational in June 1981, at which time the frequency was increased to weekly. The weekly observations with the Westford-GRAS interferometer, and occasional 3 station sessions when Onsala Space Observatory in Sweden joined in, continued until September 1983, when the density was further increased to once every 5 days. Even the relatively sparse 2-week interval single-baseline series of measurements quickly obtained unexpected results. During February and March 1981 the POLARIS estimates of UT1 veered away from the BIH Circular-D values by more than 4 ms, and then in a period of only 2 weeks reversed direction and plunged 5 ms. Fortunately this event was also captured by the LLR UT1 time series as shown in Fig. 3 (see also Robertson *et al.*, 1983). The detection of such a large variation of UT1 in such a short time period made it clear that the degree of smoothing applied by the BIH in calculating their Circular-D values was too strong. Real variations in the orientation of the Earth were being obscured by the excessive smoothing. The BIH has since reduced the degree of smoothing in generating their Circular-D values. The detection of such an event also

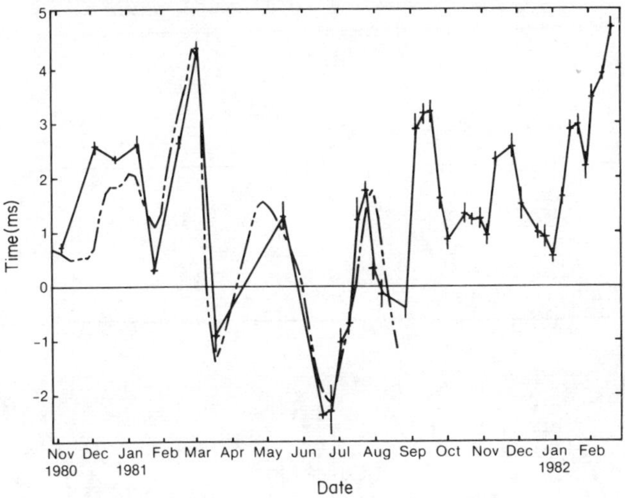

Fig. 3. A plot of the differences between the POLARIS estimates UT1 and BIH Circular-D values (solid line). For purposes of comparison, smoothed estimates from LLR (dashed line) are plotted for the period November 1980 until August 1981. [Reproduced from Carter and Robertson (1982).]

focused attention on the question of rapid variations in UT1, i.e. is there significant power during periods as short as a few days, or even hours?

As the POLARIS UT1 time series continued to grow in length and the frequency of the observations was increased to weekly, we learned that variations of several milliseconds in periods of a few weeks are not uncommon. Unfortunately, the valuable comparisons with the lunar laser values were no longer possible after a combination of managerial decisions and technical problems resulted in an extended interruption of the LLR UT1 time series, beginning in January 1982. The only other UT1 time series potentially accurate enough to verify the POLARIS results was one derived from the LAGEOS satellite laser ranging (SLR) observations by researchers at the University of Texas (Shutz *et al.*, 1983). The SLR series was expected to contain long-period systematic errors associated with tidal terms, but we found that after high-pass filtering the series could be used to check the VLBI results over periods up to a few months. Figure 4 shows the excellent agreement (rms of 0.7 ms) between the two series at periods less than 90 days.

Most of the results discussed thus far were based on single-interferometer observations. It was not until January 1984, when the Wettzell Observatory, Federal Republic of Germany, and the Richmond POLARIS Observatory, Florida, began operations, that a minimum configuration VLBI network

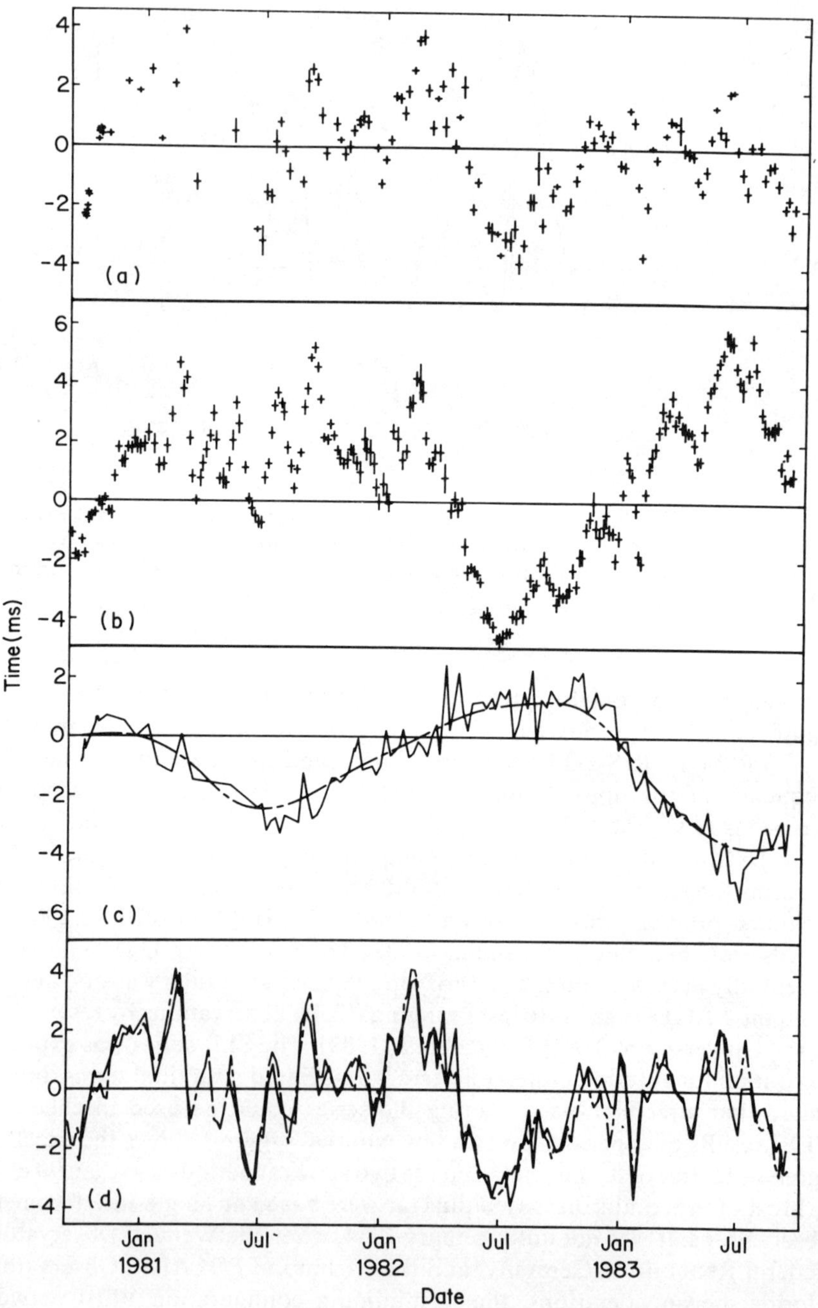

(a)
(b)
(c)
(d)
Time (ms)
Jan 1981
Jul
Jan 1982
Jul
Jan 1983
Jul
Date

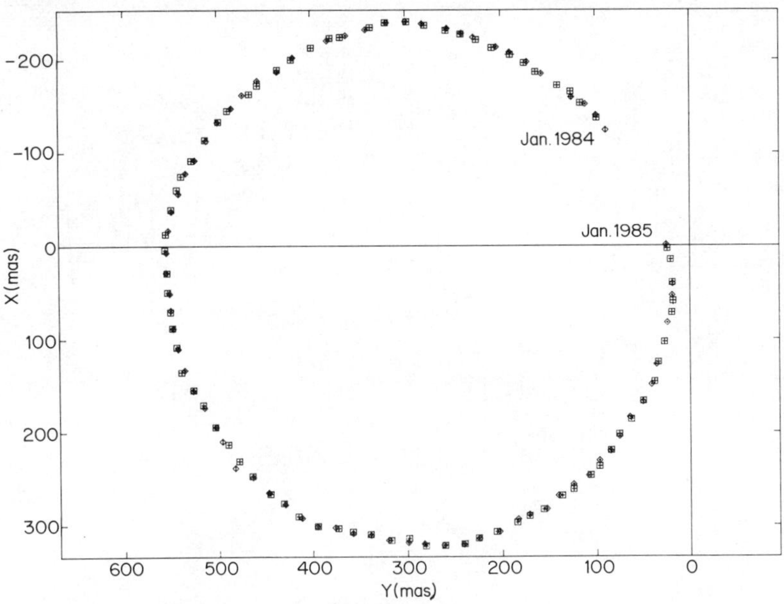

Fig. 5. IRIS VLBI (diamonds) and LAGEOS SLR (squares) determinations of the position of the pole for the time period from January 1984–January 1985.

dedicated to monitoring both components of polar motion and UT1 became a reality. This international network forms the IRIS Earth orientation monitoring system (Carter and Robertson, 1984). A comparison of the IRIS and SLR polar motion time series for the period January 1984–January 1985 is presented in Fig. 5. The rms difference over this time span is only 2 mas in both x and y, corresponding to approximately 6 cm in the location of the pole at the surface of the Earth. In contrast to the rapid variations in UT1 discussed earlier, the track of the pole is very smooth on time scales of several days, and the current 5-day monitoring interval may be adequate to track the pole at the few-centimeter level (Robertson *et al.*, 1985a).

Fig. 4. (a) The UT1 determinations from the POLARIS observations. (b) The UT1 determinations from the SLR observations. Both series are shown as differences from the BIH Circular-D series; the vertical bars represent the formal standard errors of the determinations. (c) Differences between the SLR and POLARIS determinations and the curve generated as a result of smoothing the differences. (d) The POLARIS and SLR series after removal of the smoothed differences shown in (c). The POLARIS line is discontinuous during April and May 1981, when equipment failures caused a gap in the VLBI observations. [Reproduced from Carter *et al.* (1984).]

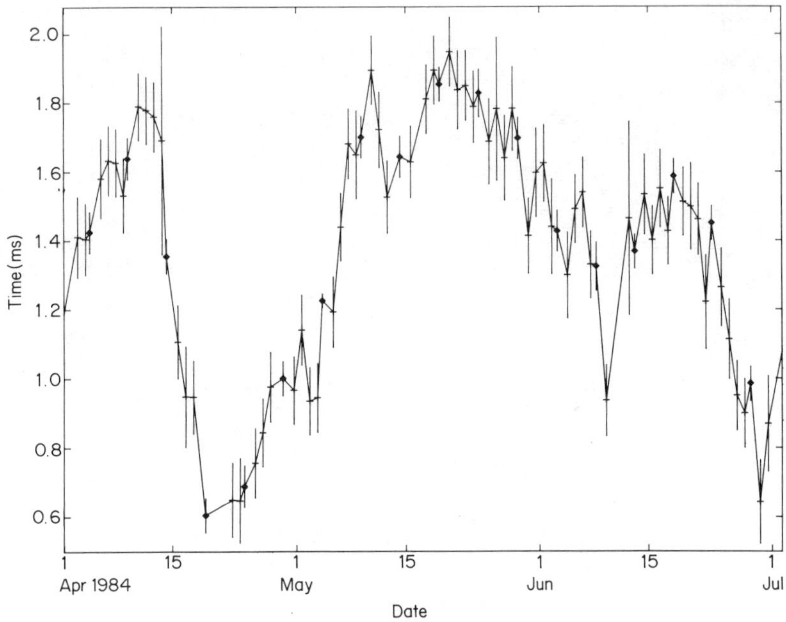

Fig. 6. UT1 determinations from the IRIS intensive observation campaign (crosses) and the regular IRIS 24-hr observing sessions (diamonds), all shown as differences from the BIH Circular-D values.

The completion of the Westford–Wettzell IRIS interferometer opened a new opportunity to explore the short-period variations in UT1. For the first time we had available a nearly equatorial (5° declination) 6000-km long interferometer formed by two fast-slewing altitude–azimuth telescopes dedicated to geodetic VLBI. During April–June 1984, this interferometer was used to make daily observations optimized for monitoring UT1. The observing sessions were only 74 minutes in duration, consisted of eight observations on four sources, and consumed two tapes per station—a relatively inexpensive operation. The daily UT1 time series obtained (plotted in Fig. 6) indicated that variations in UT1 as large as 0.3 ms can occur in periods as short as 1 day (Robertson *et al.*, 1985b).

III. Concluding Remarks

The observing campaigns organized under projects MERIT, POLARIS and IRIS have amassed more than 50,000 delay and delay rate observations during the past 4 years. We have focused here on the polar motion and UT1 information extracted from the data, but the same data can be, and have been, mined for interesting new results on a variety of subjects (Carter *et al.*, 1985; Herring, Chapter 3.4).

The lengths of the baselines between the observatories can be estimated for each observing session with uncertainties of only a few centimeters. The resulting baseline-length time series will, assuming that the results are not badly contaminated by systematic error, reflect the relative motions between the observatories associated with crustal deformations and plate motions. The Westford–Onsala baseline-length estimates currently display an increase of 1.4 ± 0.4 cm/yr, while those of the Westford–GRAS baseline show a decrease of 0.8 ± 0.1 cm/yr. If confirmed by continuing observations, these may turn out to be the first detections of the predicted motion between the North American and European tectonic plates, and an unpredicted deformation of the North American plate (Robertson and Carter, 1984a).

The apparent positions and changes in positions of the radio sources can be determined with uncertainties of a few milliseconds of arc. Temporal variations in the apparent positions of sources near the ecliptic have enabled us to measure the deflection of the electromagnetic radiation in the gravity field of the Sun. The observed deflections are consistent with Einstein's general theory of relativity to within a few tenths of a per cent (Robertson and Carter, 1985).

The approximately semidiurnal deformations of the Earth by the luni-solar gravity field result in motions of observatories of a few decimeters, which can be estimated from the VLBI observations. The observations agree with the best theoretical values to within a fraction of a centimeter in equivalent displacement (Carter *et al.*, 1985).

The variations in the luni–solar gravitational field also cause the Earth to nutate. Estimates of the components of nutation, ψ and ε, based on the VLBI observations suggest that certain coefficients in the adopted theoretical models may be in error by several milliseconds of arc (Robertson, 1984; Herring, Chapter 3.2).

The VLBI observations even contain useful information about the performance of the VLBI instrumentation itself. Most notably, the observations can be used to map the variations in the frequencies of the hydrogen maser frequency standards. From multi-station observing sessions it is possible to identify clocks which are performing below the norm (Carter *et al.*, 1985).

We have by no means exhausted the list of phenomena that can be studied using the geodetic VLBI observations amassed under projects MERIT, POLARIS and IRIS. Other subjects include studies of the variations in the ionosphere and troposphere which result in refraction of the radio signals, variations in the source brightness and structures, proper motions of the sources and the precession constant. The data are available to the entire community and we invite researchers to make us of it. It will continue to grow and improve as the IRIS network develops in the years ahead, and based on the results already obtained, we can look forward to the future results with a great deal of optimism.

References

Carter, W. E. and Strange, W. E. (1979). The National Geodetic Survey Project "POLARIS". *Tectonophysics* **52**, 39–46.

Carter, W. E., Robertson, D. S. and Abell, M. D. (1979). An improved polar motion and Earth rotation monitoring service using radio interferometry. *In* "Time and the Earth's Rotation" (D. D. McCarthy and J. D. H. Pilkington, eds.), pp. 191–198. Reidel Publ., Boston, Massachusetts.

Carter, W. E. and Robertson, D. S. (1982). Geodynamic measurements from the HRAS-Westford POLARIS interferometer. *Proc. IAG Gen. Meet.*, Tokyo, Japan, 1982.

Carter, W. E. and Robertson, D. S. (1984). IRIS Earth rotation and polar motion results. *Proc. Int. Symp. Space Tech. Geodyn., Sopron, Hungary.*

Carter, W. E., Robertson, D. S., Pettey, J. E., Tapley, B. D., Schutz, B. E., Eanes, R. J. and Miao Lufeng (1984). Variations in the rotation of the Earth. *Science* **224**, 957–961.

Carter, W. E., Robertson, D. S. and Mackay J. R. (1985). Geodetic radio interferometric surveying: applications and results, *J. Geophys. Res.* **90**, **B6**, 4577–4587.

Guinot, B. (1979). Irregularities of the polar motion. *In* "Time and the Earth's Rotation" (D. D. McCarthy and J. D. H. Pilkington, eds.), pp. 279–286. Reidel Publ., Boston, Massachusetts.

Herring, T. A., Corey, B. E., Shapiro, I. I., Rogers, A. E. E., Whitney, A. R., Clark, T. A., Knight, C. A., Ma, C., Ryan, J. W., Shupler, B. R., Vandenberg, N. R., Elgered, G., Lundqvist, G., Ronnang, B. O., Campbell, J. and Richards, P. (1983). Determination of tidal parameters from VLBI observations. *In* "Proceedings of the Ninth International Symposium on Earth Tides" (J. Kuo, ed.), pp. 205–211. E. Schweizerbart'sche Verlagsbuchhandlung, Stuttgart.

Oesterwinter, C. (1979). Polar motion through 1977 from Doppler satellite observations. *In* "Time and the Earth's Rotation" (D. D. McCarthy and J. D. H. Pilkington, eds.), pp. 263–278. Reidel Publ., Boston, Massachusetts.

Robertson, D. S. (1984). The astrometric possibilities of VLBI. *In* "Proceedings of the IAU Symposium 109" (H. Eichhorn, ed.). Reidel Publ., Boston, Massachusetts (in press).

Robertson, D. S. and Carter, W. E. (1982). Earth rotation information derived from MERIT and POLARIS VLBI observations. *In* "High Precision Earth Rotation and Earth–Moon Dynamics" (O. Calame, ed.), pp. 97–122. Reidel Publ., Boston, Massachusetts.

Robertson, D. S. and Carter, W. E. (1984a). Continental scale baselines determined by VLBI, *Proc. Int. Symp. Space Tech. Geodyn., Sopron, Hungary.*

Robertson, D. S. and Carter, W. E. (1984b). Relativistic deflection of radio signals in the solar gravitational field measured with very long baseline interferometry. *Nature* **310**, 572–574.

Robertson, D. S., Carter, W. E., Corey, B. E., Cotton, W. D., Counselman III, C. C., Shapiro, I. I., Wittels, J. J., Hinteregger, H. F., Knight, C. A., Rogers, A. E. E., Whitney, A. R., Ryan, J. W., Clark, T. A., Coates, R. J., Ma, C. and Moran, J. M. (1979). Recent results of radio interferometric determinations of a transcontinental baseline, polar motion, and Earth rotation. *In* "Time and the Earth's Rotation" (D. D. McCarthy and J. D. H. Pilkington, eds.), pp. 217–224. Reidel, Dordrecht.

Robertson, D. S., Carter, W. E., Eanes, R. J., Schutz, B. E., Tapley, B. D., King, R. W., Langley, R. B., Morgan P. J. and Shapiro, I. I. (1983). Comparison of Earth rotation as inferred from radio interferometric, laser ranging and astrometric observations. *Nature* **302**, 509–511.

Robertson, D. S., Carter, W. E., Tapley, B. D., Schutz, B. E. and Eanes, R. J. (1985a). Polar motion measurements: sub-decimeter accuracy verified by intercomparison. *Science* **229**, 1259–1261.

Robertson, D. S., Carter, W. E., Campbell, J. A. and Schuh, H. (1985b). Daily UT1 determinations from IRIS very long baseline interferometry. *Nature* **316**, 424–427.

Rogers, A. E. E., Cappallo, R. J., Hinteregger, H. F., Levine, J. I., Nesman, E. F., Webber, J. C., Whitney, A. R., Clark, T. A., Ma, C., Ryan, J. W., Corey, B. E., Counselman III, C. C., Herring, T. A., Shapiro, I. I., Knight, C. A., Shaffer, D. B., Vandenberg, N. R., Lacasse, R., Mauzy, R., Rayhrer, B., Schupler, B. R. and Pigg, J. C. (1983). Very long baseline radio interferometry: the Mark III system for geodesy, astrometry, and aperture synthesis. *Science* **219**, 51–54.

Schutz, B. E., Tapley, B. D. and Eanes, R. J. (1983). UT1 from LAGEOS laser ranging. *Ann. Rep. 1982*, D25–D28. Bureau International de l'Heure, Paris.

Wilkins, G. A., ed. (1980). "Project MERIT: A review of the techniques to be used during project MERIT to monitor the rotation of the Earth". Royal Greenwich Observatory, Herstmonceux.

Wilkins, G. A., ed. (1984). "Project MERIT: a report on the second MERIT workshop and on other activities in 1983". Royal Greenwich Observatory, Herstmonceux.

4.3 Geophysical Aspects of Polar Motion, Variations in the Length of Day, and the Luni–Solar Nutations

John M. Wahr

Department of Physics and Cooperative Institute for Research in Environmental Sciences, University of Colorado, Boulder, Colorado, USA

I. Introduction

There is abundant observational evidence that the rotation of the Earth is not constant with time. In fact, both the rotation rate and the position of the rotation axis are observed to be variable. Variations of the rotation rate cause apparent perturbations of the length of day, and so of universal time (UT)—the time determined by the meridian passage of stars and other extraterrestrial objects. To understand the relationship between UT and the length of day (LOD), note that the latter can be defined as the atomic time between successive transits of the same star. If the length of day was constant at 86,400 atomic seconds, then atomic time and UT would be equal: each star would pass overhead at the same time every day. For a variable length

SPACE GEODESY AND GEODYNAMICS
ISBN 0-12-057840-9

of day, however, there is a discrepancy between atomic time and UT, usually denoted as ΔUT. The time derivative of ΔUT is proportional to the variability of the length of day. This variability is caused by a number of different kinds of forces, as we shall see, both internal and external in origin.

Observed variations of the position of the rotation axis include changes in the orientation of that axis both with respect to inertial space and with respect to fixed geographical locations on the Earth's surface. Inertial space motion is called nutation, and geographical motion is called wobble or, sometimes, polar motion. Strictly speaking, wobble cannot occur without some accompanying nutational motion, and vice versa. In practice, though, it is usually possible to characterize the Earth's response to a given excitation mechanism as either predominately wobble or predominately nutation. That is because wobble is most effectively induced by forces with time scales much longer than 1 day as seen by an observer on the diurnally rotating Earth; while nutation is best induced by forces which vary nearly diurnally. For the real Earth, almost all of the observed nutational motion is caused by the gravitational force from the Sun and Moon. The observed wobble is induced by a number of different mechanisms, though all are apparently internal in origin.

A thorough understanding of all aspects of the Earth's variable rotation requires some familiarity with topics in meteorology, oceanography, astronomy, hydrology, geodesy and, particularly, geophysics. Excellent summaries of the subject are Munk and MacDonald (1975) and Lambeck (1980). The goal of this review is to present an overall description of the observed variability and, especially, its geophysical implications. Considerable emphasis is placed on the derivation of the equations used to relate an arbitrary excitation source to the variability (Section III). Specific excitation sources are discussed in Section IV. Some of these excitation sources and their effects on rotation are presented in more detail elsewhere in this volume.

Our first order of business, though, is to offer a brief review of Earth tides. Tides are of some interest in their own right as an observable geophysical phenomenon. But, it is also necessary to have some understanding of tidal deformation in order to fully understand the theoretical description of the Earth's rotation. For a more complete discussion of tides, see Melchior (1983).

II. Tides

The tidal deformation of the Earth and oceans is caused by the gravitational force from the Sun and Moon. The primary effect of that force is, of course, to produce a net acceleration of the Earth's center of mass, i.e. orbital motion. But the gravitational accelerations at points within the Earth are not exactly the same everywhere; some points are closer to the Sun or Moon than others. Consequently, we can separate the luni-solar gravitational force into a spatial constant, responsible for the Earth's orbital motion, and a relatively small spatially variable remainder which imparts no net acceleration to the Earth's center of mass. It is this remainder, the "tidal force", which is responsible for the tidal deformation. The tidal force also produces a torque on the Earth that is responsible for the luni-solar precession and nutations.

The tidal force can be represented as the gradient of a "tidal potential". The total gravitational potential V (the negative of the potential energy per unit mass) is a real function which can be expanded within the Earth in terms of complex spherical harmonics $Y_l^m(\theta, \lambda)$ as

$$V(r, \theta, \lambda, t) = \mathrm{Re}\left[\sum_{l=1}^{\infty} \sum_{m=0}^{l} c_l^m(t) \frac{r^l}{a^l} Y_l^m(\theta, \lambda)\right] \tag{1}$$

where θ and λ are the co-latitude and eastward longitude, a is the radius of the Earth, $c_l^m(t)$ are coefficients which depend on the motion of the Sun and Moon relative to the Earth and Re denotes the real part. No $l=0$ term is included in (1), because that term would describe a spatially uniform potential and so would be physically meaningless. The $l=1$ terms in (1) represent a uniform force. These terms are responsible for the Earth's orbital motion, and so they are removed from (1) in order to define the tidal potential:

$$V_T(r, \theta, \lambda, t) = \sum_{l=2}^{\infty} \sum_{m=0}^{l} c_l^m(t) \frac{r^l}{a^l} Y_l^m(\theta, \lambda) \tag{2}$$

where the real part of the right-hand side is implicitly assumed.

The $c_l^m(t)$ decrease with increasing l as $(a/R)^l$, where R is the distance to either the Moon or the Sun depending on which body is being considered. Since $a/R \approx \frac{1}{60}$ for the Moon, and less than $\frac{1}{20{,}000}$ for the Sun, it is sufficient for most purposes to keep only the $l=2$ terms in (2):

$$V_T \approx \frac{r^2}{a^2} \sum_{m=0}^{2} c_2^m(t) Y_2^m(\theta, \lambda) \tag{3}$$

Each of the three $c_2^m(t)$ in (3) can be well represented by a finite sum of periodic terms

$$c_2^m(t) \approx \sum_{\omega} H_2^m(\omega)\, e^{i\omega t} \tag{4}$$

where each term in the sum has a frequency close to m cycles/day. Thus, the $m=2,1,0$ terms in (3) are referred to as semidiurnal, diurnal and long-period tides, respectively. (The long-period tides have periods from about 10 days to 18.6 years.)

The tidal potential induces deformation of both the solid Earth and the oceans. Of these, the ocean tides have proven to be much the more difficult to model, due to the asymmetry of the ocean basins, the importance of the awkward Coriolis force, and the unknown characteristics of oceanic dissipation. The most accurate ocean tide models solve an idealized set of dynamical equations by using tide-gauge data as external constraints on the solution (see, e.g., Schwiderski, 1980). These models have proven to be useful in removing the effects of ocean tides from various types of geodetic observations.

Models of the solid Earth tide exploit the near spherical symmetry of the Earth. For a nonrotating, spherically stratified Earth, spherical harmonics exactly separate the equations of motion. In that case, a term $r^l Y_l^m(\theta, \lambda)$ in the tidal potential induces a displacement field within the Earth which can be characterized with vector spherical harmonics with the same l, m indices. For example, if s_r, s_θ and s_λ are the displacements in the radial, southward and eastward directions, respectively, as induced by a $H_l^m(\omega)(r^l/a^l)Y_l^m(\theta, \lambda)$ term in the tidal potential (2), then at the outer surface $r=a$:

$$
\begin{aligned}
s_r(a, \theta, \lambda) &= \frac{h_l}{g} H_l^m(\omega) Y_l^m(\theta, \lambda) \\
s_\theta(a, \theta, \lambda) &= \frac{l_l}{g} H_l^m(\omega)\, \partial_\theta Y_l^m(\theta, \lambda) \\
s_\lambda(a, \theta, \lambda) &= \frac{l_l}{g \sin\theta} H_l^m(\omega)\, \partial_\lambda Y_l^m(\theta, \lambda)
\end{aligned}
\tag{5}
$$

And, if δV is the perturbation in the Earth's gravitational potential caused by the tidal deformation of the Earth's density field, then at $r=a$:

$$
\delta V(a, \theta, \lambda) = k_l H_l^m(\omega) Y_l^m(\theta, \lambda) \tag{6}
$$

Here, h_l, l_l and k_l are dimensionless tidal "Love numbers", and g is the gravitational acceleration at the Earth's surface.

The Love numbers in (5) and (6) depend on l, but they are exactly independent of m and effectively independent of the frequency ω. This means that the response of a spherical Earth to the dominant $l=2$ tides,

(3) can be completely characterized by three dimensionless numbers, h_2, l_2 and k_2, usually written simply as h, l and k (the subscript 2 is omitted).

For a rotating, elliptical Earth the situation is more complicated. Spherical harmonics do not perfectly separate the equations, and the tidal response to a Y_l^m potential also includes small $Y_{l'}^m$ terms, where $l' \neq l$ (Wahr, 1981a). However, these terms are only about 1% of the dominant response which is still described by (5) and (6).

A more important effect of rotation and ellipticity is that h, l and k become dependent both on m and on frequency. Most notable is a predicted resonance in the diurnal ($m=1$) Love numbers at frequencies near (1+1/460) cycles/day. This resonance is associated with the nearly diurnal free nutation, a normal mode of the Earth discussed later in more detail. Although results from tidal observations are consistent with models of this resonance (see, e.g., Warburton and Goodkind, 1978; Levine, 1978; Lecolazet, 1979), the data have not, so far, proven accurate enough to help constrain models of the resonance.

In fact, tidal observations, in general, have not proven particularly useful in learning about global properties of the Earth. One of the problems has traditionally been the uncertainties associated with the ocean tides which load, and so deform, the Earth. That deformation cannot be observationally separated from the deformation induced directly by the tidal force. Instead, it must be modeled.

The usual procedure for modeling the load deformation is to convolve an ocean tide model with a load Green's function, which describes the response of the solid Earth to a point-mass load on its surface (Farrell, 1972). Green's functions are constructed from load Love numbers, which will be used in later sections and which are defined as follows. The forces on the Earth associated with a mass load include surface pressure plus an incremental gravitational force. Both the pressure and the gravitational force depend linearly on the mass distribution, and so are directly related to each other. In particular, suppose the induced gravitational potential and the surface pressure are expanded in terms of spherical harmonics. Then the coefficient of a Y_l^m term in the surface pressure turns out to be proportional to the coefficient of the same Y_l^m term in the gravitational potential, evaluated at $r=a$. And both terms are proportional to the same Y_l^m term in an expansion of the mass-load distribution. The proportionality constants depend, in general, on l.

So, since spherical harmonics separate the equations of motion for a spherically symmetric Earth, the response of such an Earth to a Y_l^m load can be described by the same Y_l^m. Suppose, for example, that the Y_l^m mass load produces a gravitational potential $H_l^m(r^l/a^l)Y_l^m(\theta,\lambda)$. Then, dimensionless load Love numbers, h_l', l_l' and k_l' can be defined so that at

$r=a$:

$$
\begin{aligned}
s_r(a,\theta,\lambda)&=\frac{h_l'}{g}H_l^m Y_l^m(\theta,\lambda)\\
s_\theta(a,\theta,\lambda)&=\frac{l_l'}{g}H_l^m\,\partial_\theta Y_l^m(\theta,\lambda)\\
s_\lambda(\alpha,\theta,\lambda)&=\frac{l_l'}{g\sin\theta}H_l^m\,\partial_\lambda Y_l^m(\theta,\lambda)\\
\delta V(a,\theta,\lambda)&=k_l' H_l^m Y_l^m(\theta,\lambda)
\end{aligned}
\tag{7}
$$

The load Love numbers represent the Earth's response to the combined gravitational and pressure forces. For a spherically symmetric Earth these numbers are exactly independent of *m*, and essentially independent of frequency.

III. Rotational Transfer Functions

Models of the observed variability of the Earth's rotation require an identification of the internal or external mechanisms driving the variability, plus knowledge of the appropriate rotational transfer functions for the Earth: functions which describe the rotational response of the Earth to applied disturbances. Inferences about either the source mechanism or the transfer function can help extend our knowledge of the dynamical behavior of the Earth and its environment. In this section, we will review the theoretical description of the transfer function. In the next section we discuss particular excitation sources.

The question we face here is: if you do something to the Earth, how can you compute the induced rotation? The most general effect of doing something to the Earth is to induce both rotation and deformation. (Plus rigid body translations. Translations are equivalent to changes in the orbital motion, and will occur only if we apply a net force to the Earth. To make our job here easier, we will assume there is no net force, and so we will ignore rigid translations.) As we shall see, rotational motion cannot usually be modeled without some simultaneous modeling of deformation. The equations used to describe the deformation are similar to the equations used by seismologists to model free oscillations. We will discuss these equations, or at least how they apply to our problem, later.

First, though, we will discuss the equations which describe conservation of the Earth's angular momentum. Suppose we are in a coordinate system which is attached to the Earth in some as yet unspecified way and is rotating with angular velocity ω. Let **H** be the angular momentum of the Earth as seen in that system, and let **L** be the external torque on the Earth. Then, conservation of angular momentum in the rotating frame is described by

$$\frac{d}{dt}\mathbf{H}=(\partial_t+\boldsymbol{\omega}\times)\mathbf{H}=\mathbf{L} \tag{8a}$$

H is given by

$$\mathbf{H}=\mathbf{I}\cdot\boldsymbol{\omega}+\mathbf{h} \tag{8b}$$

where $\mathbf{I}$ is the Earth's instantaneous inertia tensor as seen in our frame, $\mathbf{h}=\int_{earth}\rho\mathbf{r}\times\mathbf{v}\,d^3\mathbf{r}$ is the relative angular momentum, and ρ and $\mathbf{v}$ are the density and particle velocity, with respect to our coordinate system, at points within the Earth. If we were to choose our coordinate system to rotate with the mean rotation of the Earth, then **h** would vanish. In fact, the mean rotation of the Earth is *defined* as that choice of $\boldsymbol{\omega}$ where $\mathbf{H}=\mathbf{I}\cdot\boldsymbol{\omega}$. In general, though, we prefer to use a coordinate system which rotates with observatories and instruments on the Earth's surface, rather than with the entire Earth. To this end, we now define our system so that its rotation coincides with the mean rotation of the mantle. In other words, we assume that in our system there is no contribution to **h** from the mantle. Since the mantle need not rotate with the rest of the Earth (e.g. with the core), **h** in (8b) may be non-zero.

Our intent is to describe induced departures of the Earth from its equilibrium state. Equilibrium is characterized by uniform rotation at one cycle per day about a fixed axis. We define our coordinate system so that when the Earth is at equilibrium, the $\hat{\mathbf{z}}$ axis is along the rotation axis and the $\hat{\mathbf{x}}$ axis is along the Greenwich meridian. Then, for an Earth at equilibrium

$$\boldsymbol{\omega}=\boldsymbol{\Omega}=\Omega\begin{bmatrix}0\\0\\1\end{bmatrix}$$

If we assume, as we shall, that at equilibrium, the Earth is in hydrostatic balance with the internal gravitational and centrifugal forces, then the equilibrium Earth is an ellipsoid of revolution and so has a diagonal inertia tensor of the form:

$$\mathbf{I}=\mathbf{I}_0=\begin{bmatrix}A&0&0\\0&A&0\\0&0&C\end{bmatrix}$$

For the real Earth, C and A differ by about 1 part in 300, which reflects the fact that the Earth is slightly elliptical rather than spherical. Of course, the real Earth has an equilibrium inertia tensor which is considerably more complicated than $\mathbf{I}_0$. It has non-zero off-diagonal elements, and all three diagonal elements are unequal. In general, though, these modifications are less than 1% of $C-A$, and it turns out to be sufficient for our purposes to use $\mathbf{I}_0$ for the equilibrium inertia tensor. The other equilibrium characteristics are $\mathbf{h}=0$ (since $\mathbf{v}=0$) and $\mathbf{L}=0$.

Now, suppose we do something to the Earth. In general, this will result in small perturbations of $\mathbf{I}$ and $\boldsymbol{\omega}$, and small, non-zero values of $\mathbf{L}$ and $\mathbf{h}$:

$$\boldsymbol{\omega}=\boldsymbol{\Omega}+\Omega\begin{bmatrix} m_1 \\ m_2 \\ m_3 \end{bmatrix}=\boldsymbol{\Omega}+\boldsymbol{\delta\omega} \tag{9a}$$

$$\mathbf{I}=\mathbf{I}_0+\begin{bmatrix} c_{11} & c_{12} & c_{31} \\ c_{21} & c_{22} & c_{32} \\ c_{31} & c_{32} & c_{33} \end{bmatrix}=\mathbf{I}_0+\mathbf{c} \tag{9b}$$

$$\mathbf{h}=\begin{bmatrix} h_1 \\ h_2 \\ h_3 \end{bmatrix} \tag{9c}$$

$$\mathbf{L}=\begin{bmatrix} L_1 \\ L_2 \\ L_3 \end{bmatrix} \tag{9d}$$

Using these expressions in (8a) and (8b) and dropping second-order terms, such as $(m_i c_{ji})$ and $(m_i m_j)$ gives

$$\Omega A\partial_t m_1+\Omega^2(C-A)m_2+\partial_t(h_1+\Omega c_{13})-\Omega(h_2+\Omega c_{23})=L_1 \tag{10a}$$

$$\Omega A\partial_t m_2-\Omega^2(C-A)m_1+\partial_t(h_2+\Omega c_{23})+\Omega(h_1+\Omega c_{13})=L_2 \tag{10b}$$

$$\Omega C\partial_t m_3+\partial_t(h_3+\Omega c_{33})=L_3 \tag{10c}$$

Equations (10) are called the scalar Liouville equations. The parameters m_1 and m_2 describe departures of the rotation axis from its equilibrium position, and so represent wobble and/or nutation. The parameter m_3 reflects a change in the rotation rate, and so describes a variation in the length of day. Note that m_1 and m_2 appear to be decoupled from m_3, suggesting that changes in the length of day can be modeled separately from wobble and nutation. Although this does turn out to be true in practice, it cannot be inferred directly from (10), since, as we shall see, the c_{i3} and h_i in (10) depend on the m_j.

Equations (10a) and (10b) can be represented in a more compact form by using complex variables. Define $m_+ = m_1 + im_2$, $c_+ = c_{13} + ic_{23}$, $h_+ = h_1 + ih_2$ and $L_+ = L_1 + iL_2$. Then, taking (10a) + $i \times$ (10b) gives the complex equation

$$\Omega A \partial_t m_+ - i\Omega^2(C - A)m_+ + \partial_t(h_+ + \Omega c_+) + i\Omega(h_+ + \Omega c_+) = L_+ \quad (11)$$

The real and imaginary parts of (11) are equivalent to (10a) and (10b), respectively.

The implication of (10c) and (11) is that rotational motion can be produced by external torques (L_i), by a redistribution of mass within the Earth (c_{i3}), and by motion of material within the Earth (h_i). Suppose an internal or external force is applied to the Earth. Estimating the torque (L_i) from knowledge of the forcing mechanism is, in principle, straightforward. But, how would we compute the induced c_{i3} and h_i for use in (10c) and (11)? This is where the deformation equations are needed. Those equations are a pointwise representation of the equations describing the conservation of linear momentum $\mathbf{F} = m\mathbf{a}$ in our coordinate system; $\mathbf{F}$ includes things like internal stresses and gravitational restoring forces, as well as the applied force. The acceleration $\mathbf{a}$ represents the sum of inertial and frame dependent (Coriolis and centripetal) accelerations.

The frame dependent accelerations are functions of the m_i. For example, define the Lagrangian displacement field $\mathbf{s}(\mathbf{r}, t)$, relative to our variably rotating frame, so that a particle with an equilibrium position at $\mathbf{r}$ gets displaced to $\mathbf{r} + \mathbf{s}(\mathbf{r}, t)$. Then, the total apparent acceleration vector in the rotating frame is $\mathbf{a} = (\partial_t + \omega \times)(\partial_t + \omega \times)(\mathbf{r} + \mathbf{s}(\mathbf{r}, t))$. If we use (9a) for ω and ignore terms of order $m_i s_j$ and $m_i m_j$, then the m_i dependent terms in $\mathbf{a}$ have the form

$$\begin{aligned} &m_i - \text{dependent accelerations} \\ &\quad = (\partial_t \boldsymbol{\delta\omega}) \times \mathbf{r} + \boldsymbol{\delta}\omega(\boldsymbol{\Omega} \cdot \mathbf{r}) + \boldsymbol{\Omega}(\boldsymbol{\delta}\omega \cdot \mathbf{r}) - 2\mathbf{r}\delta\omega \cdot \boldsymbol{\Omega} \end{aligned} \quad (12)$$

This is the only dependence of the linear momentum equation on the m_i. It is convenient to think of (12) (or, more precisely, its negative) as an additional applied force per unit mass. In this case, we could describe the deformational displacement field $\mathbf{s}(\mathbf{r}, t)$ as being induced partially by the internal or external driving force and partially by the m_i-dependent terms in this fictitious force. Since, for small perturbations, the linear momentum equation can be linearized in $\mathbf{s}$, we can separate $\mathbf{s}$ into terms linear in the m_i and terms induced directly by the applied force. Since the c_{i3} and h_i are obtained by direct integration of the displacement field $\mathbf{s}$ (or the velocity field $\mathbf{v}$) a similar linear separation is also possible for these variables. We represent this separation, in its most general form, by

$$\begin{aligned} c_+ &= p_{++}m_+ + p_{+-}m_- + p_{+3}m_3 + \bar{c}_+ \\ c_{33} &= p_{3+}m_+ + p_{3-}m_- + p_{33}m_3 + \bar{c}_{33} \\ h_+ &= \Omega q_{++}m_+ + \Omega q_{+-}m_- + \Omega q_{+3}m_3 + \bar{h}_+ \\ h_3 &= \Omega q_{3+}m_+ + \Omega q_{3-}m_- + \Omega q_{33}m_3 + \bar{h}_3 \end{aligned} \tag{13}$$

where $m_- = m_1 - im_2$ is the complex conjugate of m_+. If we transform to the frequency domain, where $\partial_t = i\sigma$, then the p_{ij} and q_{ij} are multiplicative constants (possibly frequency dependent) which reflect the dynamical response of the Earth to the m_i-dependent fictitious forces. The $\bar{c}_{ij}$ and $\bar{h}_i$ in (13) represent that part of the deformation induced directly by the internal or external applied force. In some cases that force is known well enough to solve the linear momentum equation directly. In other cases, though, the force is not well understood, and the $\bar{c}_{ij}$ and $\bar{h}_i$ are regarded as the external perturbation.

If (13) for c_{i3} and h_i are used in (10c) and (11), then, in the frequency domain, those equations reduce to

$$\begin{aligned} &[\sigma A + \Omega(A - C) + (\sigma + \Omega)(q_{++} + p_{++})]m_+ + (\sigma + \Omega)(q_{+3} + p_{+3})m_3 \\ &\quad + (\sigma + \Omega)(q_{+-} + p_{+-})m_- + (\sigma + \Omega)\left(\frac{\bar{h}_+}{\Omega} + \bar{c}_+\right) = \frac{-iL_+}{\Omega} \end{aligned} \tag{14a}$$

$$(C + q_{33} + p_{33})m_3 + (q_{3+} + p_{3+})m_+ + (q_{3-} + p_{3-})m_- + \frac{\bar{h}_3}{\Omega} + \bar{c}_{33} = \frac{-iL_3}{\Omega\sigma}. \tag{14b}$$

Note that, in general, there might be coupling between m_3, m_+ and m_-. However, for an ellipsoidal Earth, such as we are assuming here,

$$p_{+3} = p_{3+} = q_{+3} = q_{3+} = p_{+-} = q_{+-} = p_{3-} = q_{3-} = 0 \tag{15}$$

To understand, physically, why $p_{+3} = q_{+3} = 0$, note that a change in m_3 represents a corresponding change in the Earth's rotation rate. This causes a perturbation in the centrifugal force which is everywhere away from the rotation (the $\hat{\mathbf{z}}$) axis. Consequently, material is forced outwards from the $\hat{\mathbf{z}}$ axis, thus changing the principal moments of inertia, but, by symmetry, not changing the orientation of the principal axes. So, the induced c_{+3} is zero, which implies that p_{+3} must vanish. It is also evident, by symmetry, that any induced flow would have to be about the $\hat{\mathbf{z}}$ axis, and so the induced h_+ must also be zero, which implies that $q_{+3} = 0$. Similar hand waving arguments can be invoked to demonstrate that p_{3+}, q_{3+} *and* p_{i-}, q_{i-} should also vanish.

The mathematical explanation for these simplifications of the p_{ij} and q_{ij} is as follows. Suppose we expand the displacement $\mathbf{s}$ in terms of vector spherical harmonics of angular position within the Earth, where the coefficient of each spherical harmonic is an unknown function of radius. In this way, the vector partial differential equations reduce to an infinite set of ordinary differential equations for the radially dependent scalars. For a spherically symmetric, nonrotating Earth, the scalar equations are perfectly decoupled from one another. In other words, spherical harmonics separate these equations, just as they separate many other equations with spherical symmetry. The symmetry, in this case, is a consequence of the fact that a nonrotating, spherical Earth looks the same from every direction. As a result of this mathematical separation, a force with angular order l and azimuthal number m can only induce a displacement field with the same l, m values.

For a rotating, elliptical Earth the equations separate in the azimuthal number m, but not in the angular order l. So, an input l, m force can induce displacements with different angular orders l', but still only with the same m value. Coupling between scalars with different l values occurs because rotation and ellipticity define a preferred direction in space. The separation in the azimuthal number m is due to the fact that a rotating, elliptical Earth still looks the same from all longitudes.

The m_i-dependent acceleration (12) can be separated into terms proportional to each of m_+, m_- and m_3. An expansion of these terms into vector spherical harmonics shows that the m_+, m_- and m_3 terms are proportional to spherical harmonics with, respectively, $l=1$ and 2, $m=-1$; $l=1$ and 2, $m=+1$; and $l=0$, 1 and 2, $m=0$. The implication is that for an elliptical Earth m_3 may induce displacements with possibly all angular orders l' (although mostly $l=0$, 1 and 2) but only with azimuthal number $m=0$, and that m_+ and m_- induce displacements with all angular orders l' (mostly $l=1$ and 2) but only with $m=-1$ and $m=+1$, respectively.

It is straightforward to show that a non-zero c_+ is directly related to the $l=2$, $m=-1$ terms in **s**. And, c_3 depends only on the $l=2$, $m=0$ terms. Similarly, h_+ and h_3 depend only on $l=1$, $m=-1$ and $l=1$, $m=0$ terms, respectively. The implication, then, is that for a rotating, elliptical Earth, where a force with azimuthal number m only induces displacements which have the same value of m, (15) is valid.

The real Earth is, of course, not exactly elliptical, and one consequence is that none of the scalars in (15) actually vanish identically. The most important departure from ellipticity is due to the irregular distribution of oceans. Dahlen (1976) considered the effects of the oceans on these scalars and concluded that for most purposes the approximation (15) is adequate. We will here assume that (15) is valid.

Equation (14) with the substitution (15) then shows that since the L_i, $\bar{c}_{ij}$, and $\bar{h}_i$ do not depend on the m_i, the equations for m_3 and m_+ are, indeed, perfectly decoupled. Furthermore, there is no dependence on m_-. Solving

(14) for m_+ and m_3 gives

$$m_+ = \frac{-[iL_+/\Omega + (\sigma+\Omega)(\bar{c}_+ + \bar{h}_+/\Omega)]}{[\sigma A + \Omega(A-C) + (\sigma+\Omega)(q_{++} + p_{++})]} \tag{16a}$$

$$m_3 = -\frac{[i(L_3/\Omega\sigma) + \bar{c}_{33} + \bar{h}_3/\Omega]}{[C + q_{33} + p_{33}]} \tag{16b}$$

An excitation process is characterized by L_i, $\bar{c}_{ij}$ and $\bar{h}_i$. The transfer function for the Earth is described by (16), together with results for the p_{ii} and q_{ii}. In general, the p_{ii} and q_{ii} are reasonably well understood theoretically. Although there are some cases where rotational observations have been used to infer information about the p_{ii} and q_{ii} and so help to further constrain models of the dynamical behavior of the Earth, in most cases uncertainties either in the excitation process or in the observations are large enough that the p_{ii} and q_{ii} can be assumed to be known.

A careful perusal of the development of (16) will reveal an apparent ambiguity. These equations reflect angular momentum conservation for a physical system denoted here as "the Earth". What, precisely, should we include in that system? We defined the rotation of the coordinate frame so that it coincided with the mean rotation of the mantle. This was done so that the rotational parameters m_3 and m_+ correspond directly to observations of the rotation. Consequently, it is probably reasonable (although it is not necessary) to at least include the Earth's mantle in our physical system. Should we also include the atmosphere and oceans, or, for that matter, the Earth's core? We can, of course, include anything we like, so long as we are careful in modeling the $\bar{c}$, $\bar{h}$, L, p and q.

In practice, for a given excitation source we use a definition of "the Earth" which depends on the type of data we might have access to, and also on how easy it is to model the behavior of different regions of the Earth. It is useful here to illustrate this situation with a particular important example. Suppose we wish to compute the effects of the atmosphere on the rotation. If we choose the atmosphere to be part of our system, then $\bar{c}$ and $\bar{h}$ include contributions from both the atmosphere and the solid Earth (atmospheric effects may cause deformation of the solid Earth), but the L_i vanish (the external energy source in this case is solar radiation which provides no external torque). Modeling the rotational variability using this system is commonly referred to as the angular momentum approach. If we choose the atmosphere to be external to "the Earth", then the $\bar{c}$ and $\bar{h}$ include contributions only from the solid Earth, but there may now be external atmospheric torques L_i. This is the torque approach. The approach we choose to use depends on the quality and availability of the pertinent atmospheric data.

In principle, values for the p_{ii}, q_{ii}, C and A in (16) are different in the angular momentum and torque approaches. In this example, though, the extremely low density of the atmosphere compared to the density of the solid Earth implies that in practice the values are effectively the same. For example, the rotation-induced forces, which are reflected in the p_{ii} and q_{ii}, are proportional to the density and so are negligible in the atmosphere. Similarly, A and C directly reflect the unperturbed internal density distribution. This does not mean that in the angular momentum approach, for example, the atmospheric contributions to the $\bar{c}$ and $\bar{h}$ are much smaller than the solid Earth contributions. The contribution from the solid Earth reflects only those perturbations induced by the atmosphere. As a result, those contributions are of the same order as (and in some cases are much smaller than) the corresponding atmospheric contributions.

As another example, the decision about whether or not to include the core in "the Earth" is, in most cases, reasonably clear cut. When modeling variations in the rotation rate, it is usually best to regard the core as external to the system. This is because for a non- or weakly dissipative, elliptical Earth, variations in m_3 (the rotation rate of the mantle) cannot induce displacements within the core. As a result, p_{33} and q_{33} in (16b) are the same whether the core is included in the system or not. Since there is thus no reason to model the possibly complicated displacement field induced in the fluid core by the applied force, the core is best left out of the system. For wobble and nutation, however, there can be induced elastic pressure torques between the core and mantle due to the elliptical shape of the core–mantle boundary. In this case, the non-zero m_1 and m_2 imply that there is some incremental rotation about an equatorial axis. Since the equator is perpendicular to the axis of symmetry of the ellipse, then any relative rotational motion between the core and mantle causes a restoring torque. So, p_{++} and q_{++} in (16a) do depend on the dynamical behavior of the core, and it is best in this case to include the core in the system.

For a complete description of the transfer functions, (16), we still must model the coefficients p_{ii} and q_{ii}. For wobble and nutation (16a) we need p_{++} and q_{++}. For variations in the length of day (16b) we need p_{33} and q_{33}. We now describe models for these scalars, first for wobble, second for nutations, and finally for variations in the length of day.

A. Wobble

Wobble is characterized by long-period ($|\sigma/\Omega| \ll 1$) disturbances of m_+. The coefficients p_{++} and q_{++} describe the Earth's response to the m_+-dependent part of (12). To find p_{++}, note that at long periods the $\partial_t \delta\boldsymbol{\omega}$ term in (12) is much smaller than the other terms in that equation, and so the c_+ it induces is correspondingly negligible. (The $\partial_t \delta\boldsymbol{\omega}$ term is important in modeling q_{++}.) If we ignore this term, then the negative of (12) (the centrifugal force per unit mass) is

$$\begin{aligned} -(\Omega^2 m_+/2)[z(\hat{\mathbf{x}} - i\hat{\mathbf{y}}) + \hat{\mathbf{z}}(x - iy)] &= -\nabla[(\Omega^2 m_+/2) z(x - iy)] \\ &= -\nabla[m_+ \Omega^2 \sqrt{(2\pi/15)}\, r^2 Y_2^{-1}(\theta, \lambda)] \end{aligned} \tag{17}$$

where we have included only the m_+-dependent terms. Here, θ and λ are co-latitude and eastward longitude, respectively, and the spherical harmonics Y_l^m are assumed to be normalized so that $\int_{earth} |Y_l^m|^2 \sin\theta\, d\theta\, d\lambda = 1$.

The centrifugal potential, $-m_+\Omega^2 \sqrt{(2\pi/15)}\, r^2 Y_2^{-1}(\theta, \lambda)$ has exactly the same spatial dependence ($r^2 Y_2^{-1}$) as does the diurnal tidal potential. Consequently, we can use the Love number formalism, described by (5), to model the induced deformation of the solid Earth. For example, there is an induced perturbation of the Earth's gravitational potential measured at the Earth's surface given by

$$\delta V(a, \theta, \lambda) = -k m_+ \Omega^2 \sqrt{(2\pi/15)}\, a^2 Y_2^{-1}(\theta, \lambda) \tag{18}$$

where $k \approx 0.301$ is the tidal potential Love number, and a is the radius of the Earth. $\delta V(a, \theta, \lambda)$ is a useful function in our problem because it can be related directly to p_{++}, as follows.

Let $\delta\rho(r_0, \theta_0, \lambda_0)$ be the perturbation in the Earth's density distribution at a spatial point $\mathbf{r}_0 = (r_0, \theta_0, \lambda_0)$ (so $\delta\rho$ is the incremental Eulerian density). Then, the gravitational potential at a surface point $\mathbf{r} = (a, \theta, \lambda)$ is

$$\delta V(a, \theta, \lambda) = G \int_{earth} \frac{\delta\rho(\mathbf{r}_0)}{|\mathbf{r} - \mathbf{r}_0|}\, d^3\mathbf{r}_0 \tag{19}$$

Expanding $\delta\rho$ and $1/|\mathbf{r} - \mathbf{r}_0|$ in terms of spherical harmonics (assuming $a > r_0$) gives

$$\begin{aligned} \delta\rho(\mathbf{r}_0) &= \sum_{l,m} \delta\rho_l^m(r_0)\, Y_l^m(\theta_0, \lambda_0) \\ \frac{1}{|\mathbf{r} - \mathbf{r}_0|} &= \sum_{l,m} \frac{4\pi}{2l+1} \frac{r_0^l}{a^{l+1}}\, Y_l^m(\theta, \lambda)\, \bar{Y}_l^m(\theta_0, \lambda_0) \end{aligned} \tag{20}$$

where the bar denotes complex conjugation. Using these expansions in (19), together with the orthogonality of spherical harmonics gives:

$$\delta V(a, \theta, \lambda) = \sum_{l,m} \frac{4\pi G}{a^{l+1}(2l+1)}\, Y_l^m(\theta, \lambda) \int_0^a \delta\rho_l^m(r_0)\, r_0^{l+2}\, dr_0 \tag{21}$$

A comparison with (18) shows that

$$\int_0^a \delta\rho_l^m(r_0) r_0^4 \, dr_0 = \begin{cases} -km_+ \dfrac{\Omega^2 a^5}{G} \sqrt{(5/24\pi)} & \text{for } \; l=2,\ m=-1 \\ 0 & \textit{otherwise} \end{cases} \tag{22}$$

Next, we can relate c_+ to the integral of $\delta\rho_2^{-1}$ in (22). Since, by definition of the inertia tensor,

$$c_{ij} = \int_{earth} \delta\rho(\delta_{ij} \sum_k x_k^2 - x_i x_j) \, d^3\mathbf{x}$$

it is easy to obtain

$$c_+ = -\sqrt{(8/15\pi)} \int_{earth} \delta\rho(\mathbf{r}_0) r_0^2 \bar{Y}_2^{-1}(\theta_0, \lambda_0) \, d^3\mathbf{r}_0 \tag{23}$$

Using the expansion (20) for $\delta\rho$ and the orthogonality of the spherical harmonics gives:

$$c_+ = -\sqrt{(8/15\pi)} \int_0^a \delta\rho_2^{-1}(r_0) r_0^4 \, dr_0 \tag{24}$$

Comparing (22) with (24) gives

$$c_+ = km_+(\Omega^2 a^5/3G) \tag{25}$$

So, for a zonally symmetric Earth

$$p_{++} = (c_+/m_+) = k(\Omega^2 a^5/3G) \tag{26}$$

The oceans modify this result. The centrifugal force (17) induces displacements within the ocean which produce further perturbations in δV. These perturbations have a more complicated angular dependence than the result, (11), for an oceanless Earth, due to the high degree of asymmetry of the ocean-continent distribution. In general, we can expand the total (Earth plus ocean) induced potential at $r = a$ as

$$\delta V(a, \theta, \lambda) = \sum_{l,m} \delta V_l^m Y_l^m(\theta, \lambda) \tag{27}$$

where the δV_l^m are constants. Then, (21) and (24) are still valid, and they, together with (27), imply

$$c_+ = -\sqrt{(5/6\pi)} \frac{a^3}{G} \delta V_2^{-1} \tag{28}$$

Writing

$$\delta V_2^{-1} = -\kappa m_+ \Omega^2 \sqrt{(2/15\pi)}\, a^2$$

(this just defines the constant κ) gives, finally

$$p_{++} = c_+/m_+ = \kappa(\Omega^2 a^5/3G) \tag{29}$$

For an oceanless Earth, $\kappa = k \approx 0.301$, as we have seen. Dahlen (1976) found that for oceans which remain in a continual state of equilibrium adjustment to the incremental centrifugal potential, κ increases by about 15%, to 0.3464. The equilibrium assumption is consistent with the oceanographic equations of motion for a barotropic (uniform density) ocean (Carton and Wahr, 1986; O'Connor and Starr, 1983; Dickman, 1985), although observations of the induced oceanic displacement (the "pole tide") are inconclusive.

Models for q_{++} are harder, and we will here simply review the results. By definition of our coordinate frame, there is no contribution to h_+ from the mantle. And, at long periods, the barotropic oceanic equations of motion imply that the induced oceanic currents are negligible (the solution is equilibrium again). Consequently, the only portion of the Earth which contributes to h_+ is the core.

The core contribution to h_+ is induced almost entirely by the seemingly small $\partial_t \delta\boldsymbol{\omega}$ term in (12). The results of Hough (1895) (for a summary, see Smith, 1977) imply that if the ellipticity of the core–mantle boundary ε_b is much less than 1, then for a homogeneous, incompressible core:

$$q_{++} \approx \frac{-\sigma A_c}{\sigma + (1+\varepsilon_b)\Omega} \tag{30}$$

where A_c is a principal moment of inertia of the core. At long periods ($|\sigma/\Omega| \ll 1$), and to lowest order in small quantities

$$q_{++} \cong -A_c(\sigma/\Omega) \tag{31}$$

What happens is that the core rotates approximately rigidly with respect to the coordinate system and so with respect to the mantle. If m_+^c describes this relative rotation, then the contribution to h_+ from the core is

$$h_+ = A_c \Omega m_+^c \tag{32}$$

Using (31) and (32) in (13) for h_+ gives

$$m_+^c = -(\sigma/\Omega) m_+$$

which implies that the total inertial space rotation vector of the core, as seen from the mantle-fixed coordinate system, is

$$\boldsymbol{\omega}_c = \boldsymbol{\Omega} + \boldsymbol{\delta\omega}\left(1 - \frac{\sigma}{\Omega}\right)$$

with $\boldsymbol{\delta\omega}$ given by (9a). As a result, the total time rate of change of $\boldsymbol{\omega}_c$ with respect to inertial space is

$$\frac{d\boldsymbol{\omega}_c}{dt} = \frac{\partial \boldsymbol{\omega}_c}{\partial t} + (\boldsymbol{\Omega} + \boldsymbol{\delta\omega}) \times \boldsymbol{\omega}_c \cong 0$$

to first order in σ/Ω. Thus, to this order, the core rotation axis is constant in inertial space; or, as it is often put, the core does not participate in the wobble.

We now use these results for p_{++} (29) and q_{++} (31) in (16a). If we replace $\sigma + \Omega$ by Ω in both the numerator and the denominator, as is appropriate at long periods, (16a) reduces to

$$m_+ = -\frac{1}{\Omega A_m}\left[\frac{iL_+ + \Omega^2 \bar{c}_+ + \Omega \bar{h}_+}{\sigma - \sigma_{CW}}\right] \tag{33}$$

where A_m is a principal moment of inertia of the mantle, and

$$\sigma_{CW} = \Omega\left[\frac{C - A - \kappa\Omega^2 a^5/3G}{A_m}\right] \tag{34}$$

The fact that A_m and not A is in the numerator in both (33) and (34), reflects the non-participation of the core in long-period wobble.

The result (33) predicts a resonance in m_+ at a frequency $\sigma = \sigma_{CW}$, which is indeed evident in the observed wobble spectrum. This resonance is properly described as a normal mode of the Earth, and is called the Chandler wobble, after S. C. Chandler who first observed the oscillation in 1891. The observed period of the mode is about 435 days, and its decays with a Q of about 100 (Wilson and Haubrich, 1976), corresponding to a damping time of several decades. For an elastic Earth with a fluid core and equilibrium oceans, the theoretical period, as predicted by (34), is 427 days (Smith and Dahlen, 1981), and the Q is infinite. Smith and Dahlen interpreted the observed finite Q and the 8-day discrepancy between the observed and computed periods as due to mantle anelasticity. They, in essence, included the effects of anelasticity on κ (κ becomes complex in this case) by using an adjustable parameter to scale a starting seismic anelastic model, and then adjusted this parameter to best fit the data. With this single parameter they were able to simultaneously fit both the observed Q and the observed period to within the observational limits. The principal goal of this sort of enterprise is to infer information about the variation of the Earth's anelastic behavior with frequency. For example, the adjustable scaling parameter is sensitive to any differences between the Earth's anelasticity at the Chandler wobble period and its anelasticity as inferred from the much shorter period (<1 hr) seismological data.

B. Nutations

Nutations are rotational perturbations which, like wobble, are characterized by $m_+ \neq 0$ and $m_3 = 0$. But, for nutations $|(\sigma/\Omega)+1| \ll 1$, and so the motion is nearly diurnal as seen from the diurnally rotating Earth. Nutations are driven by the external gravitational force from the Sun and Moon, so that $L_+ \neq 0$ in (16a). In general, $\bar{c}_+$ and $\bar{h}_+$ in (16a) are also non-zero.

At nearly diurnal periods, p_{++} and q_{++} are much more difficult to model than at the long periods which characterize wobble, and we will not attempt a full derivation here. One special case we can consider, though, is an Earth with a rigid mantle and a homogeneous, incompressible fluid core. In this case, the Earth's instantaneous inertia tensor in the mantle-fixed frame is always equal to the equilibrium result $\mathbf{I}_0$ (see 9b), and so both p_{++} and $\bar{c}_+$ are zero. Furthermore, for this relatively simple Earth model the relative angular momentum $\bar{h}_+$ induced directly by the applied force is zero. Hough's result (30) for q_{++} is still approximately valid.

By using these results in (16a), and ignoring terms quadratic in small quantities (such as $(C-A)/A$, $(\sigma/\Omega)+1$, and ε_b) gives

$$m_+ \approx \frac{iL_+(\lambda - \varepsilon_b\Omega)}{\Omega^2 A_m[\lambda - \varepsilon_b\Omega(A/A_m)]} \tag{35}$$

where $\lambda = -(\sigma+\Omega)$ is the frequency of the motion as seen from nonrotating, inertial space.

The variable m_+ describes the rotation vector of the mantle, relative to mantle-fixed axes. Nutation is more familiarly characterized by the same vector as seen in inertial space. If we write the inertial space representation of this rotation vector as $\mathbf{\Omega}+\Omega\mathbf{n}$, then $n_+ = n_1 + in_2$ is given by

$$n_+ = \frac{\sigma}{\sigma+\Omega} m_+ \approx \frac{iL_+(\lambda - \Omega\varepsilon_b)}{\lambda\Omega A_m[\lambda - \varepsilon_b\Omega(A/A_m)]} \tag{36}$$

From (36) we infer that for this Earth model the nutations are resonant at $\lambda = 0$ and at $\lambda = \varepsilon_b\Omega A/A_m$.

The effects of the resonance at $\lambda = 0$ are clearly evident in the observed nutation amplitudes. It is because of this resonance, for example, that the 18.6-yr nutations ($\lambda/\Omega = \pm 1/18.6$ yr) are so large. However, this resonance is not particularly interesting, since its frequency is not affected by any dynamical process within the Earth. The resonance merely reflects the fact that the longer a torque acts on a body, the more the angular velocity vector will change.

On the other hand, the resonance at $\lambda = \varepsilon_b \Omega A / A_m$ (corresponding to an inertial space period of about 348 days for this simple model) is of interest. This resonance is associated with a normal mode of the Earth, the nearly diurnal free wobble or free core nutation, and is characterized by a relative rotation between the core and the mantle. In this case, unlike in the case of the Chandler wobble, the induced rotations of both the core and the mantle are non-zero as seen in inertial space.

Theoretical results for the nutations of an Earth with an elastic mantle and inhomogeneous, compressible core, have the same general characteristics as (36). The most important differences are that the numerator of (36) also includes terms proportional to $\bar{c}_+$ and $\bar{h}_+$, and the inertial space period of the nearly diurnal free wobble resonance increases from 348 to ~460 days (see, for example, Smith, 1977; Wahr, 1981b). The resonance at $\lambda = 0$ is unaffected.

The effects of the 460-day resonance on the forced nutation amplitudes are evident in the astrometric data (see, for example, Herring, Chapter 3.2). In fact, as described above, the resonance is also evident in Earth tide observations (the luni-solar forcing terms responsible for the torque in (36) also causes the tides). So far, however, few of these indirect observations of the resonance have proven particularly useful at constraining our knowledge of the Earth; the observational error bars are too large (although, again, see Herring, Chapter 3.2). Furthermore, the nearly diurnal free wobble—the expected 460-day oscillation itself—has never been observed conclusively.

C. Variations in the Length of Day

For m_3 the transfer function (16b) is much simpler than for wobble and nutation (16b). As already discussed, in the absence of dissipative torques across the core–mantle boundary it is easiest to define the fluid core to be external to "the Earth". Then C in (16b) is replaced by C_m, a principal moment of inertia of the mantle. The parameter q_{33} in (16b) represents the relative angular momentum h_3, induced by the m_3 terms in (12). Since there are no contributions to h_3 from the mantle (by definition of the coordinate system), and assuming there is no appreciable induced flow in the atmosphere (low density) or the oceans (e.g. assuming an equilibrium response at long periods), then $q_{33} = 0$.

The steps leading to the result (29) for p_{++} can be repeated, with minor modifications, to deduce a similar result for p_{33} (here, the core is ignored when computing p_{33}). But, it turns out that numerically $p_{33} \ll C_m$. Equation (16b) then reduces approximately to:

$$m_3 = \frac{-1}{\Omega C_m}\left[\Omega \bar{c}_{33} + \bar{h}_3 + \frac{iL_3}{\sigma}\right] \tag{37}$$

In this case, $\bar{c}_{33}$ and $\bar{h}_3$ do not include the effects of deformation within the core, but L_3 may include torques on the mantle from the core.

This approximate transfer function is uninteresting, in the sense that it does not depend on any adjustable parameters descriptive of the Earth's dynamical behavior. An important assumption in deriving (37), though, is that the core–mantle dissipative coupling is weak enough so that relative core–mantle rotation does not induce notable restraining torques on the mantle. Independent estimates of the dissipation suggest that this assumption is probably valid at periods of a few years or less (Wahr *et al.*, 1981). In fact, there have so far been no rotational observations at any period which indicate that (37) is incomplete in any way.

IV. Excitation Sources

From the transfer functions (33) and (37), we can estimate amplitudes for wobble and for variations in the length of day as induced by either internal or external forcing. In some cases, the applied forces and the Earth's deformational response to those forces are well enough understood that the variations $\bar{c}_+$, $\bar{c}_3$, $\bar{h}_+$ and $\bar{h}_3$ can be modeled directly in terms of those forces. In other cases, the $\bar{c}$ and $\bar{h}$ variables are not easily modeled, but can be estimated reasonably well from existing geophysical data. Finally, there are important forces for which results for the $\bar{c}$ and $\bar{h}$ variables can neither be modeled nor reliably inferred from data. In these cases, you simply do the best you can.

We will discuss excitation sources and their effects, first for wobble and then for variations in the length of day. Nutations are not discussed. The only observed nutations (characterized by a nearly diurnal m_+) are those caused by the luni–solar gravitational force, as discussed previously.

A. Wobble

Figure 1 shows the observed wobble frequency spectrum ($|m_+|^2$) as inferred from de-trended ILS astrometric data over the last 70–80 years. The power

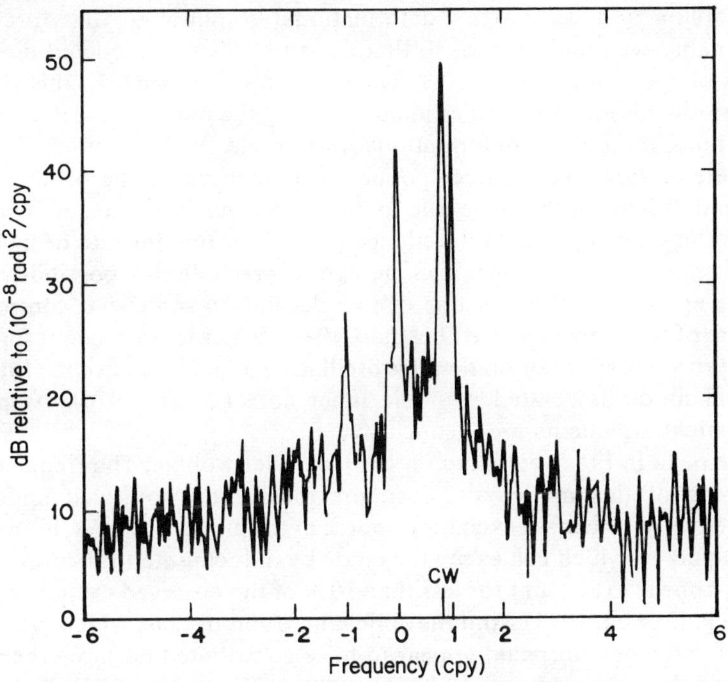

Fig. 1. The observed power spectrum for the wobble parameter m_+ computed from detrended 1900–1973 monthly ILS data. The spectrum has been smoothed using a five-weight triangular filter. Prominent peaks are evident at periods of 14 mon (the Chandler wobble), ±1 yr (the annual wobble), and several decades. The retrograde annual wobble (−1 cpy) is smaller than the prograde annual wobble (1 cpy), because the retrograde forcing frequency is further from the Chandler wobble eigenfrequency.

is concentrated in three peaks, centered respectively at 1 yr, 14 mon, and 10–20 yr. There is also some indication in the original data of a secular drift of the pole over the last century (see e.g. Dickman, 1981).

The observed drift has been hypothesized as due to post-glacial rebound. The Earth's surface near Hudson Bay and Scandinavia is presently uplifting due to the Earth's viscous response to the removal, several thousand years ago, of the ice loads in those areas (see e.g. Wu and Peltier, 1983). This uplift presumably causes the product of inertia $\bar{c}_+$ to vary linearly with time, and this induces linear drift in m_+. (The variation is actually exponential, but the time constant is long enough so that over periods as short as a century, the drift appears to be linear.) The observed drift has been compared with theoretical results from post-glacial rebound models to help constrain the viscosity of the mantle. This is an example of an excitation source where the forcing and the Earth's deformational response to that forcing are reasonably well understood, so that $\bar{c}_+$ in (33) can be modeled directly in terms of the applied force (the removal of the ice load). In this case, the end product is an improved estimate of one of the parameters (the viscosity) describing the Earth's deformational response.

There is, however, another possible explanation for the observed drift. It could at least partially be due to linear changes in $\bar{c}_+$ caused by melting ice in the Greenland or Antarctic ice fields. The present rate of melting in those regions is not known well enough to preclude this possibility.

The apparent wobble at one or two decades, a somewhat controversial feature of the spectrum (see Dickman, 1981), is harder to account for. There has been some speculation that this oscillation might be an excited rotational normal mode associated with the inner core (Busse, 1970). At present, theoretical arguments are inconclusive.

The peak in Fig. 1 at 14 mon is the Chandler wobble. The frequency and damping of this mode, discussed previously, are now well understood theoretically. The main excitation source of the mode, however, is unknown. Earthquakes, which can excite the mode by inducing step-function changes in $\bar{c}_+$, appear to account for less than 10% of the observed excitation power (see e.g. Dahlen, 1973). And the atmosphere and oceans, which *are* responsible for the annual peak, appear to have contributed no more than about 25% of the observed power during 1900–1973 (Wahr, 1983). Torques on the mantle from the fluid core are probably too small by several orders of magnitude (Rochester and Smylie, 1965). An identification of the excitation source is important, not only for its own sake, but also because it would help to improve estimates of the Q of the mode: the width of the 14-mon peak, which is used to determine Q, could also be partially due to unknown spectral characteristics of the excitation source.

The annual wobble (the peak at 1 year in Fig. 1) is well understood as due to seasonal processes in the atmosphere, the oceans and the distribution of groundwater (see e.g. Wilson and Haubrich, 1976; Merriam, 1982b; Wahr, 1983). Most important is the annual variation in $\bar{c}_+$ which accompanies the seasonal redistribution of mass in the atmosphere. This variation in $\bar{c}_+$ is mostly due to the increased amount of mass over central Asia during the winter. The mass distribution can be inferred from global observations of surface pressure, since the pressure at a surface point is linearly proportional to the atmospheric mass directly above that point.

Also important in the excitation is the global response of the oceans to the annual variations in atmospheric pressure. That response is probably close to an "inverted barometer" response (i.e. if the atmospheric pressure increases, the ocean surface depresses so that there is no net change in pressure at the sea floor), and it contributes to the annual wobble excitation by modifying $\bar{c}_+$ (Merriam, 1982b; Wahr, 1982). The effects on the annual wobble of atmospheric winds and ocean currents [non-zero $\bar{h}_+$ in (33)], and of wind-driven oceanic upwelling (non-zero $\bar{c}_+$) are apparently small (Wahr, 1983).

The annual variability of global water storage (including snow and ice) over land is apparently another important contributing excitation source for the annual wobble. Estimates of the water storage excitation (Van Hylckama, 1970) are difficult, but the indication is that the annual contributions to $\bar{c}_+$ from the variable water storage are about $\frac{1}{4}$ as large as the contributions from the redistribution of atmospheric mass.

In each of these cases, the total induced $\bar{c}_+$ includes contributions not only from the atmosphere, oceans and groundwater, but also from deformation within the solid Earth caused by the atmospheric-, oceanic- and groundwater-induced pressures. This deformational contribution can be derived in a manner similar to the derivation of (25), except that in this case the centrifugal potential is replaced by a term proportional to the Y_2^{-1} component of the pressure, and k in (25) is replaced by the $l=2$ load Love number.

This discussion of the annual wobble excitation implicitly assumes the use of the angular momentum approach, where the atmosphere and oceans are defined to be part of the Earth. In this case, only the observed change in angular momentum of the atmosphere and oceans is pertinent. The mechanisms responsible for the transfer of that angular momentum are irrelevant. It is of interest, however, to also consider those mechanisms and their effects on the annual wobble. This is accomplished, quantitatively, by using the torque approach, where the atmosphere and oceans are defined to be external to the Earth.

Consider, first, the atmosphere. There are two possible mechanisms by which angular momentum can be exchanged between the atmosphere and the underlying solid Earth and oceans. One, the friction torque, is due to frictional drag in the atmospheric boundary layer as winds blow over the Earth's surface. The other mechanism is the torque due to pressure on one side of a topographic feature being larger than on the other. This torque is usually called the "mountain torque", even though any sort of topographic feature is relevant. In fact, the most extreme topographic feature on the Earth is the elliptical bulge: the Earth's equatorial radius is about 20 km longer than its polar radius. It is useful, here, to define the mountain torque as the torque due to pressure against any topographic feature *except* the elliptical bulge. The pressure torque against the elliptical bulge is considered separately.

A little reflection will reveal that because the elliptical bulge is symmetric about the Earth's polar radius vector, the torque against that bulge cannot have a component along the polar direction ($L_3 = 0$). As a result, pressure against the elliptical bulge cannot change the Earth's rotation rate. This torque is, however, the primary excitation source for the annual wobble, as described with the torque approach. In fact, if the elliptical bulge torque is expressed as a global integral of surface pressure, the result is identical to the pressure integral describing wobble excitation from the redistribution of atmospheric mass, as computed with the angular momentum approach (Wahr, 1982). One implication is that the effects of winds on wobble are exactly described, within the torque approach, by the combined effects of the friction and mountain torques. The indications are that the friction and mountain torques are about equally important in exciting wobble (Wahr, 1983). In any event, the net effect of winds on the annual wobble is small, as noted above.

The transfer of angular momentum between the oceans and the solid Earth is accomplished by oceanic equivalents of the friction, mountain, and elliptical bulge torques. As with the atmosphere, the dominant exchange of oceanic angular momentum in the case of wobble is accomplished through the torque against the elliptical bulge.

B. Variations in the Length of Day

The observed variability of the length of day can be conveniently separated into variations with periods less than and periods greater than, say, 5 years. This separation is motivated, primarily, by the different excitation mechanisms operative in the two regimes, but also by the different observational techniques presently used to infer the short-period and long-period variability.

The short-period variability is reasonably well understood. It is caused partly by the long-period tidal deformation of the Earth, and partly by atmospheric perturbations. The tidal potential induces perturbations in the Earth's internal density distribution, and this causes corresponding perturbations of the Earth's inertia tensor. The long-period tidal potential, $V_0(r^2/a^2)Y_2^0(\theta, \lambda)$, causes $\bar{c}_{33}$ variations which, in turn, lead to variations in the length of day as described by (37). There are no significant tidal-induced variations of $\bar{h}_3$ (the core is assumed to be external to the Earth in this case, and so does not contribute to $\bar{h}_3$).

These tidal-induced variations in m_3 occur, of course, at tidal periods, particularly at the fortnightly (13.7 days), monthly (27.6 days), semiannual and annual periods, where the long-period tidal forcing is particularly large. The fortnightly and monthly variations in m_3 are clearly visible in the astrometric data. The semiannual and annual contributions are large enough to be seen, but there are significant contributions from the atmosphere at these seasonal periods (see below) which cannot be confidently separated from the tidal effects.

The induced $\bar{c}_{33}$ can be computed directly from the tidal potential with an expression similar to (25), except that, here, the centrifugal potential in the derivation of (25) is replaced with the tidal potential. The result is

$$\bar{c}_{33} = -\kappa_0(a^3/3G)\sqrt{5\pi}^{-1}\, V_0 \tag{38}$$

If there were no oceans, and if $\bar{c}_{33}$ in (38) was meant to describe the moment of inertia of the entire Earth, then $\kappa_0 = k = 0.301$. Because $\bar{c}_{33}$ in (38) should only reflect the change in inertia of the mantle alone, κ_0 is reduced by about 20%. On the other hand, the Earth's oceans increase κ_0 by about 20% (Agnew and Farrell, 1978). The best theoretical estimate for κ_0, including the effects of the core and oceans but ignoring anelasticity, is $\kappa_0 \approx 0.315$ (Merriam, 1982a). This result can be further modified for mantle anelasticity. In fact, as length of day data continues to improve, observations of the fortnightly and monthly amplitudes can be used to infer the effects of anelasticity on κ_0 at these periods. This can, in turn, provide additional constraints on the anelasticity in this frequency range. This procedure has been applied by Merriam (1985) in an analysis of about 5 years of BIH (Bureau International de L'Heure) rotation rate data.

All other observed variations in m_3 at periods shorter than a few years are apparently due to the atmosphere and oceans. These variations include periodicities at 50 day, 6 mon, 1 yr and 2 yr, superimposed on a background continuum spectrum (see Dickey and Eubanks, Chapter 4.1). Most of the variable polar angular momentum (the component of angular momentum in the $\hat{\mathbf{z}}$ direction) in the atmosphere is due to variations in the zonal (east–west) winds. The effects of these winds on m_3 can be estimated by using wind data to find $\bar{h}_3$, and then using that result in (37). The effects of the redistribution of atmospheric mass on m_3 are observationally significant, but are only about 10% of the effects of winds (Wahr, 1983).

This difference from the corresponding situation for the atmospheric excitation of the annual wobble, where the effects of winds are much smaller than the effects of the seasonal redistribution of atmospheric mass, can be understood from the torque approach. The redistribution of mass is dominant in the case of wobble, because its effects correspond to the effects of the pressure torque against the Earth's enormous elliptical bulge. Since that torque cannot have an $\hat{\mathbf{x}}_3$ component and so cannot excite variations in m_3, it is not surprising that the effects of mass redistribution are much less important here. Instead, all angular momentum exchange in this case occurs through friction and mountain torques which, in turn, are primarily related to the wind field (the mountain torque is often described as the torque caused by winds blowing against topographic features).

Processes in the oceans are much less important than winds in exciting variations in the length of day. The circumpolar current around Antarctica is apparently responsible [through a non-zero $\bar{h}_3$ in (37)] for about 5% or less of the observed annual and semiannual variations in m_3 (Ray Peterson, personal communication). The effects of other current systems, of wind-induced oceanic upwelling, and of variations in groundwater storage are probably negligible at seasonal periods (Lambeck and Hopgood, 1981; Wahr, 1983). And, there is no reason to believe that any of these variables are relatively more important at nonseasonal periods.

The good short-period agreement between the atmospheric wind data and observations of the length of day (less the effects of tides) has been strikingly demonstrated in a number of recent papers (see e.g. Barnes *et al.*, 1983; Rosen and Salstein, 1983). Even the catastrophic 1982–1983 El Niño event over the southern Pacific Ocean correlates well with recent rotation data (Rosen *et al.*, 1984). Probably the most significant aspect of comparisons such as these, is the inferred indication that both the meteorological and rotational data sets are of high quality. For example, the good agreement at 50 days (Langley *et al.*, 1981) demonstrates that that periodicity in the atmospheric data is indeed real and deserves attention from the meteorological community.

The observed long-period variability of the Earth's rotation rate is less well understood. There are apparently irregular variations, the decade fluctuations, with typical periods of 10 years or longer. These are evident in observational results from conventional astrometric techniques over the last century (see Lambeck, 1980, for a summary). There is also evidence for a linear increase in the length of day over at least the last few thousand years. This linear increase cannot be cleanly separated from the decade fluctuations over periods as short as 100 years. Instead, the strongest evidence for the linear term comes from the recorded times of eclipses and planetary occultations and conjunctions from ancient observational records. The inferred results are probably fairly uncertain, but they indicate an increase of about 2 ms/100 yr.

Most of the linear increase in the length of day is due to tidal dissipation. Consider, for example, the lunar tidal force. That force induces a tidal bulge in the Earth which, for a dissipationless Earth, would have an axis of symmetry exactly along the Earth–Moon vector. Because of dissipation, however, there is a small time lag after the tidal force is maximum at a point before that point reaches its maximum deformed state. As a result, the tidal bulge is slightly ahead of the Earth–Moon vector, as indicated by δ in Fig. 2. The Moon can then exert a gravitational torque on this bulge

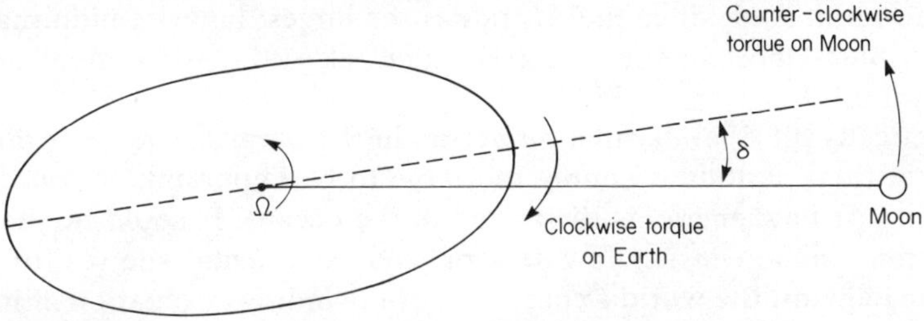

Fig. 2. The effects of tidal dissipation on the Earth's rotation and on the lunar orbit. Because of dissipation, the Earth's tidal bulge (greatly exaggerated) leads the Earth–Moon vector by the angle δ. The result is a clockwise gravitational torque on the Earth tending to decrease the Earth's rotation rate Ω. There is also a counter-clockwise gravitational torque on the Moon causing the Moon to recede from the Earth.

in the direction which tends to decrease the rotation rate. Since the torque is time independent, the resulting increase in the length of day is linear in time (for a more complete description see Zschau, 1978).

This effect on m_3 cannot be described within the context of (37). Instead, when L_3 is time independent, (10c) implies that in the time domain

$$C\Omega m_3 + h_3 + \Omega c_{33} = L_3(t - t_0) \tag{39}$$

where the left-hand side of (39) is assumed to vanish at $t = t_0$. When the period of a disturbance is infinite, as it is for a constant torque, the core and mantle are probably perfectly coupled through dissipative forces acting across the core–mantle boundary, and the core and mantle rotate together. In this case, unlike at shorter periods, it is best to define "the Earth" so that it does include the core. It is also useful to include the oceans in the system. Then, L_3 in (39) is the total torque on the Earth and oceans. Furthermore, since the core, mantle and oceans (assuming the latter are equilibrium) all rotate together, h_3 in (39) is zero. If we ignore the induced c_{33} compared with C, as we did in deriving (37), then (39) reduces to

$$m_3 = \frac{L_3}{C\Omega}(t - t_0) \tag{40}$$

The expression for the tidal torque L_3 reduces to a sum of terms, each proportional to the square of the amplitude of a different semidiurnal term in the tidal potential. Every semidiurnal term contributes to the sum. The coefficient of each term depends on the dissipation within the Earth and oceans at that frequency. The coefficient can be conveniently expressed as proportional to the phase lag between the Y_2^2 term in the tidal-induced gravitational potential of the Earth and oceans, and the Y_2^2 term in the tidal potential. Since the lunar semidiurnal tides are roughly twice as large as the solar semidiurnal tides, about 80% of the dissipative torque is provided by the Moon. And, since the M_2 tide is the largest lunar semidiurnal tide, it contributes most to the sum (see Lambeck, 1977, for a more detailed description).

Practically all tidal dissipation occurs in the oceans. The Q of the solid Earth at tidal periods is simply too large to be important. It is not clear, though, just how energy is dissipated in the oceans. It could be the result of bottom friction in shallow seas or over continental shelves, of waves breaking against the world's coastlines, of the indirect excitation of internal tides or any of a host of other possibilities (see Lambeck, 1980, for a summary).

In spite of this uncertainty, the dissipative torque can still be estimated from independent data, namely tidal observations from artificial satellites. Satellite orbits are sensitive to the perturbation in the gravitational potential caused by the tidal-induced deformation of the solid Earth and oceans. Ranging data can thus be used to determine the phase lag between that induced potential and the tidal potential of the Moon at the different tidal frequencies. That phase lag reflects the combined effects of dissipation in the oceans and solid Earth. The observed semidiurnal phase lags can then be used directly as coefficients in the summation for L_3, discussed previously. These results predict that the length of day should be increasing linearly by slightly more than 2.5 ms/100 yr, about 30% larger than the observed rate (see e.g. Cazenave, 1982).

There is independent evidence that the results inferred from satellite observations are reasonably accurate. The linear decrease in the Earth's rotation rate implies that the Earth's rotational angular momentum is also decreasing linearly with time. But, the total angular momentum of a closed system must remain constant. This implies that the de-acceleration of the Earth's rotation due to lunar tidal dissipation ought to be accompanied by an increase in the orbital angular momentum of the Moon about the Earth. This transfer of angular momentum to the Moon can be anticipated from Fig. 2, by noting that the Earth's tidal bulge in that figure should produce a counter-clockwise torque on the Moon. There is a similar effect on the apparent solar orbit due to the solar tidal dissipation, but the observational consquences in that case are negligible.

The increase in lunar angular momentum causes the Moon to recede from the Earth and induces a linear decrease in its orbital angular velocity. The linear change in orbital angular velocity is now well determined from lunar laser ranging (Dickey *et al.*, 1982). The results are consistent, within a few percent, with the satellite tidal observations and with conservation of angular momentum of the Earth–Moon system.

One implication of these orbital effects is that the Moon must have been closer to the Earth in the geological past than it is at present. If the lunar orbit is extrapolated backwards in time using the present rate of tidal dissipation, the Moon is found to have been within a couple of Earth radii about $1\text{-}2 \times 10^9$ years ago. This inferred maximum age for the Moon conflicts with a number of terrestrial and lunar observations, all of which indicate that the Moon is much older. The inference is that the present dissipation rate must be anomalously large. The dependence of the oceanic dissipation rate on the shape of the global ocean basins and on the Earth's rotation rate are currently receiving attention from oceanographers in order to test this hypothesis (see, e.g. Brosche and Sundermann, 1982).

The 30% disagreement between the observed linear increase in the length of day and the increase as inferred from satellite and lunar laser ranging data is most likely due either to errors in the rotational results or to some non-tidal excitation source tending to decrease the length of day. Two possible mechanisms which could accelerate the Earth are changes in the volume of the Antarctic ice field, and post-glacial rebound. In both cases the excitation would be induced through apparently linear changes in c_{33}. In fact, a linear change in c_{33} (proportional to a linear change in J_2) has recently been observed in data from laser ranging to the artificial satellite LAGEOS (Yoder *et al.*, 1983; Rubincam, 1984). These results are numerically consistent with the non-tidal acceleration inferred from the ancient astronomical record. The implication is that the apparent non-tidal acceleration is probably real.

The variability of the Antarctic ice volume is not presently known well enough to confidently assess its effects on m_3. Instead, most interpretations of the non-tidal acceleration and of the linear term in c_{33} assume that they are the result of post-glacial rebound. With this assumption, these observations (particularly the better determined c_{33} result) provide tight constraints on the viscosity of the lower mantle (Peltier, 1983).

The decade fluctuations are believed to be caused by electromagnetic torques acting across the core–mantle boundary. Suppose the core is internally stirred up, so that its average rotation with respect to the mantle is non-zero and time variable. Since the Earth's magnetic field is generated within the core, we would expect that field to also rotate with respect to the mantle. In the mantle-fixed frame, then, there is a time-dependent magnetic field and therefore, by Faraday's law, an induced electric field. The mantle is an electrical conductor, albeit probably a weak one, and so the electric field induces currents in the mantle, particularly in the lower mantle. These currents interact with the Earth's primary magnetic field through the Lorentz force, producing a torque on the mantle (see, e.g. Rochester, 1960).

It is difficult to assess this model quantitatively. Variations in the magnetic field at the Earth's surface do not appear to be well correlated with variations in m_3 (although see the possible exception noted below). This is likely due to the apparent attentuation of magnetic field variations as they move outwards through the mantle. Other relevant parameters, such as the electrical conductivity and the strength of the primary magnetic field in the lower mantle, are not well known. The general acceptance of elecromagnetic coupling as the excitation source for the decade fluctuations is partly because torques of the right order of magnitude can be obtained with reasonable values for the magnetic field and its variability, and partly because no other mechanism seems to be important enough (see Lambeck, 1980, for a general review).

There is one intriguing case, however, where the effects of a specific magnetic field variation might, indeed, be visible in the length-of-day data. There is evidence that the magnetic field at the surface may have experienced a step change in its second time derivative during 1969 (see, e.g. Malin and Hodder, 1982). This "magnetic jerk" might conceivably cause a corresponding, but preceding, step change in the second time derivative of the length of day. There is, in fact, some indication of such a change in the length of day during the 1950s (Backus, 1983). This correlation, if real, could help constrain estimates of the electrical conductivity of the mantle.

V. Summary

We have reviewed the theoretical framework used to model variations of the Earth's rotation, and have discussed what are probably the most important geophysical applications of the theory. The rotational transfer functions for the Earth—i.e. mathematical descriptions of how the Earth's rotation responds to internal or external disturbances—are reasonably well understood. These functions depend on certain artificial parameters [such as κ in (29)] which, in turn, depend on the Earth's internal material properties. Although most of these properties are well determined from seismological data, there are good reasons to believe that values of, for example, the effective elastic and anelastic constants depend on the period of the disturbance, and so may not be adequately represented by the seismic values. As we have seen, rotational observations can help in constraining these parameters at sub-seismic frequencies.

In most cases, though, either the source of the disturbance is so poorly known, or the observational results are inaccurate enough, that seismically inferred Earth structure can be usefully applied to estimate the transfer functions. In fact, most theoretical work is directed towards identifying or quantifying the pertinent excitation mechanisms. The broad spectrum of possible mechanisms from so many different fields makes this task particularly intriguing. And, as we have seen, there is still a great deal to learn.

Acknowledgment

The support provided by NASA contract NAS5-27644 is gratefully acknowledged.

References

Agnew, D. C. and Farrell, W. E. (1978). Self-consistent equilibrium ocean tides. *Geophys. J. R. Astron. Soc.* **55**, 171–182.

Backus, G. E. (1983). Application of mantle filter theory to the magnetic jerk of 1969. *Geophys. J. R. Astron. Soc.* **74**, 713–746.

Barnes, R. T. H., Hide, R., White, A. A. and Wilson, C. A. (1983). Atmospheric angular momentum fluctuations correlated with length of day changes and polar motion. *Proc. R. Soc. London, A* **387**, 31–73.

Brosche, P. and Sundermann, J. (1982). "Tidal Friction and the Earth's Rotation II". Springer-Verlag, Berlin and New York.

Busse, F. H. (1970). The dynamical coupling between inner core and mantle of the Earth and the 24-year libration of the pole. *In* "Earthquake Displacement Fields and the Rotation of the Earth" (L. Mansinha, ed.), p. 88. Reidel Publ., Boston, Massachusetts.

Carton, J. A. and Wahr, J. M. (1986). Modelling the pole tide and its effects on the Earth's rotation. *Geophys. J. R. Astron. Soc.* **84**, 121–138.

Cazenave, A. (1982). Tidal friction parameters from satellite observations. *In* "Tidal Friction and the Earth's Rotation II" (P. Brosche and J. Sundermann, eds.), pp. 4–18. Springer-Verlag, Berlin and New York.

Dahlen, F. A. (1973). A correction to the excitation of the Chandler wobble by earthquakes. *Geophys. J. R. Astron. Soc.* **32**, 203–217.

Dahlen, F. A. (1976). The passive influence of the oceans upon the rotation of the Earth. *Geophys. J. R. Astron. Soc.* **46**, 363–406.

Dickey, J. O., Williams, J. G. and Yoder, C. F. (1982). Results from lunar laser ranging analysis. *In* "High-Precision Earth Rotation and Earth-Moon Dynamics" (O. Calame, ed.), pp. 209–216. Reidel Publ., Boston, Massachusetts.

Dickman, S. R. (1981). Investigation of controversial polar motion features using homogeneous International Latitude Service data. *J. Geophys. Res.* **86**, 4904–4912.

Dickman, S. R. (1985). The self-consistent dynamic pole tide in global oceans. *Geophys. J. R. Astron. Soc.* **81**, 157–174.

Farrell, W. E. (1972). Deformation of the Earth by surface loading. *Rev. Geophys. Space Phys.* **10**, 761–797.

Hough, S. S. (1895). The oscillations of a rotating ellipsoidal shell containing fluid. *Philos. Trans. R. Soc. London, A* **186**, 469–506.

Lambeck, K. (1977). Tidal dissipation in the oceans; astronomical, geophysical and oceanographic consequences. *Philos. Trans. R. Soc. London, A* **287**, 545–594.

Lambeck, K. (1980). "The Earth's Variable Rotation: Geophysical Causes and Consequences". Cambridge Univ. Press, London and New York.

Lambeck, K. and Hopgood, P. (1981). The Earth's rotation and atmospheric circulation from 1963 to 1973. *Geophys. J. R. Astron. Soc.* **64**, 67–89.

Langley, R. B., King, R. W., Shapiro, I. I., Rosen, R. D. and Salstein, D. A. (1981). Atmospheric angular momentum and the length of the day: a common fluctuation with a period near 50 days. *Nature* **294**, 730–733.

Lecolazet, R. (1979). Sur une relation etroite entre la vitesse de rotation de la Terre et l'amplitude des ondes diurnes ψ_1 et K_1 de la maree gravimetrique. *C. R. Acad. Sci. Paris* E **289**.

Levine, J. (1978). Strain-tide spectroscopy. *Geophys. J. R. Astron. Soc.* **54**, 27–41.

Malin, S. R. C. and Hodder, B. M. (1982). Was the 1970 geomagnetic jerk of internal or external origin? *Nature* **296**, 726–728.

Melchior, P. (1983). "The Tides of the Planet Earth". Pergamon, Oxford.

Merriam, J. B. (1982a). A comparison of recent theoretical results on the short-period terms in the length of day. *Geophys. J. R. Astron. Soc.* **69**, 837–840.

Merriam, J. B. (1982b). Meteorological excitation of the annual polar motion. *Geophys. J. R. Astron. Soc.* **70**, 41–56.

Merriam, J. B. (1985). LAGEOS and UT measurements of long period Earth tides and mantle Q. *J. Geophys. Res.* **190**, 9423–9430.

Munk, W. H. and MacDonald, G. J. F. (1975). "The Rotation of the Earth". Cambridge Univ. Press, London and New York.

O'Connor, W. P. and Starr, T. B. (1983). Approximate particular solutions for the pole tide in a global ocean. *Geophys. J. R. Astron. Soc.* **75**, 397–405.

Peltier, W. R. (1983). Constraint on deep mantle viscosity from LAGEOS acceleration data. *Nature* **304**, 434–436.

Rochester, M. G. (1960). Geomagnetic westward drift and irregularities in the Earth's rotation. *Philos. Trans. R. Soc. London, A* **252**, 531–555.

Rochester, M. G. and Smylie, D. E. (1965). Geomagnetic core-mantle coupling and the Chandler wobble. *Geophys. J. R. Astron. Soc.* **10**, 289–315.

Rosen, R. D. and Salstein, D. A. (1983). Variations in atmospheric angular momentum on global and regional scales and the length of day. *J. Geophys. Res.* **88**, 5451–5470.

Rosen, R. D., Salstein, D. A., Eubanks, T. M., Dickey, J. O. and Steppe, J. A. (1984). An El Niño signal in atmospheric angular momentum and Earth rotation. *Science* **225**, 411–414.

Rubincam, D. P. (1984). Postglacial rebound observed by LAGEOS and the effective viscosity of the lower mantle. *J. Geophys. Res.* **89**, 1077–1088.

Schwiderski, E. W. (1980). On charting global ocean tides. *Rev. Geophys. Space Phys.* **18**, 243–268.

Smith, M. L. (1977). Wobble and nutation of the Earth. *Geophys. J. R. Astron. Soc.* **50**, 103–140.

Smith, M. L. and Dahlen, F. A. (1981). The period and *Q* of the Chandler wobble. *Geophys. J. R. Astron. Soc.* **64**, 223–282.

Van Hylckama, T. E. A. (1970). Water balance and unbalance. *In* "International Association of Scientific Hydrology Proceedings of the Reading Symposium on World Water Balance". Publication 92, AIHS-UNESCO.

Wahr, J. M. (1981a). Body tides on an elliptical, rotating, elastic and oceanless Earth. *Geophys. J. R. Astron. Soc.* **64**, 677–704.

Wahr, J. M. (1981b). The forced nutations of an elliptical, rotating, elastic and oceanless Earth. *Geophys. J. R. Astron. Soc.* **64**, 705–728.

Wahr, J. M. (1982). The effects of the atmosphere and oceans on the Earth's wobble—I. Theory. *Geophys. J. R. Astron. Soc.* **70**, 349–372.

Wahr, J. M. (1983). The effects of the atmosphere and oceans on the Earth's wobble and on the seasonal variations in the length of day—II. Results. *Geophys. J. R. Astron. Soc.* **74**, 451–487.

Wahr, J. M., Sasao, T. and Smith, M. L. (1981). Effect of the fluid core on changes in the length of day due to long period tides. *Geophys. J. R. Astron. Soc.* **64**, 635–650.

Warburton, R. J. and Goodkind, J. M. (1978). Detailed gravity-tide spectrum between one and four cyles per day. *Geophys. J. R. Astron. Soc.* **52**, 117–136.

Wilson, C. R. and Haubrich, R. (1976). Meteorological excitation of the Earth's wobble. *Geophys. J. R. Astron. Soc.* **46**, 707–743.

Wu, P. and Peltier, W. R. (1983). Glacial isostatic adjustment and the free air gravity anomaly as a constraint on deep mantle viscosity. *Geophys. J. R. Astron. Soc.* **74**, 377–450.

Yoder, C. F., Williams, J. G., Dickey, J. O., Schutz, B. E., Eanes, R. J. and Tapley, B. D. (1983). Secular variations of Earth's gravitational harmonic J_2 coefficient from LAGEOS and the non-tidal acceleration of Earth rotation. *Nature* **303**, 757–762.

Zschau, J. (1978). Tidal friction in the solid Earth; loading tides versus body tides. *In* "Tidal Friction and Earth's Rotation" I, (P. Brosche and J. Sundermann, eds.). Springer-Verlag, New York.

4.4 Tidal Friction in the Solid Earth: Constraints from the Chandler Wobble Period

J. Zschau

Institute für Geophysik, Christian Albrechts Universität, Kiel, Federal Republic of Germany

I. Introduction

Astronomical evidence suggests that the Earth's rotation rate is steadily slowing down, resulting in a lengthening of the day by about 2 ms within 100 years. Simultaneously, there is a secular change $\dot{n} \approx 25$ arcsec/(cy)^2 of the Moon's mean motion, and the Earth–Moon distance is increasing by about 4 cm/yr. This is commonly attributed to tidal friction, i.e. to internal friction within and between the Earth's solid and fluid layers during tidal deformation. It results in an imperfectly elastic response of the

SPACE GEODESY AND GEODYNAMICS
ISBN 0-12-057840-9

solid Earth, as well as in an imperfectly fluid response of the Earth's oceans to tidal forces, and, hence, yields an effective tidal bulge which is not in equilibrium, but which is retarded by about 10 minutes with respect to the primarily acting tidal potential. The nonequilibrium tidal bulge is held by the gravitational forces of the Moon and acts as a brake block on the Earth which is rotating underneath. The Earth's rotation rate is, thus, slowing down, and conservation of angular momentum in the Earth-Moon system ensures that at the same time the Earth-Moon distance as well as the Moon's mean motion is changing.

A similar effect occurs due to the tidal forces of the Sun. However, the Sun is responsible for only 20% of the astronomically observed tidal friction, whereas the remaining 80% must be attributed to the influence of the Moon.

The process of tidal friction is accompanied by dissipation of tidal energy, i.e. by conversion of tidal energy into heat or into some other form of non-tidal energy. It adds up to the enormous energy flux of about 3×10^{19} erg/s, which is more than 10% of the total heat flux at the Earth's surface, about 10 to 100 times the total energy release due to earthquakes, and about equal to the present energy consumption of mankind.

It is commonly assumed that the oceans are the main consumers of the tidal energy and that dissipation in the solid Earth can only account for a few percent of the total tidal dissipation (Zschau, 1978). However, despite important progress in understanding the energy budget in the Earth-Moon system (Platzman, 1984; Schwiderski, 1983b), the exact share between both the oceanic dissipation and the dissipation in the solid Earth, is not known. On the one hand this is due to shortcomings in the various ocean tide models, particularly with regard to the inclusion of realistic dissipation mechanisms and on the other hand, it can be attributed to insufficiencies in the knowledge of the rheology in the Earth's interior, particularly to the unknown frequency dependence of the specific dissipation function Q^{-1} in the Earth's mantle.

Zschau (1978) and Platzman (1984) assumed that the Q structure at tidal periods is the same as the one valid in the seismological frequency band, and they came to the conclusion that tidal dissipation in the solid Earth accounts for only 1% of the total dissipation. Comparisons of Q values for seismic waves and free oscillations of the Earth with the Chandler wobble Q (Anderson and Minster, 1979; Minster, 1980; Anderson and Given, 1982; Okubo, 1982) as well as a detailed study of the Chandler wobble period (Smith and Dahlen, 1980) suggest, however, that the specific dissipation function in the mantle is slightly dependent on frequency. This is also confirmed by laboratory measurements of seismic attenuation in a forsterite sample at high temperatures and at loading periods of up to 1000 s (Berckhemer *et al.*, 1979, 1982).

For the description of these various observations the power law $Q \sim \omega^{\alpha}$ has become popular (Anderson and Minster, 1979; Müller, 1983). Despite many attempts to determine the coefficient α, it is still only badly known. Quoted values lie in the range between $\alpha = 0$ and $\alpha = 0.5$ with, however, a preference of values less than 0.3. From Fig. 1 it is seen that the rate of tidal dissipation in the solid Earth is crucially determined by this α value. For Q-model PREM (Dziewonski and Anderson, 1981), for instance, the solid Earth dissipation rate would be less than 2.5% of the total dissipation rate provided that $\alpha = 0$. For the same Q model, but for $\alpha = 0.3$, the solid

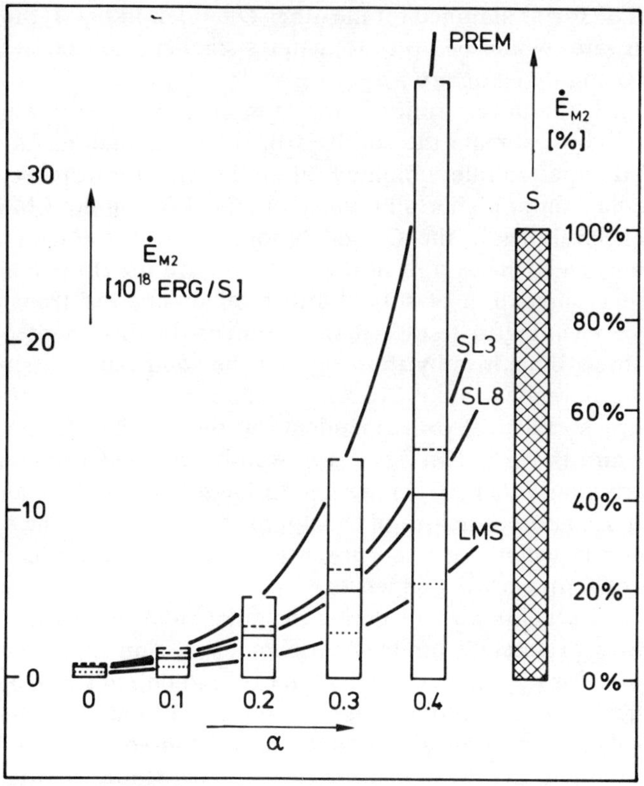

Fig. 1. Solid Earth M_2 tide dissipation rate $\dot{E}_{M2}$ for different Q models and different degrees of frequency dependence of Q. Q models SL3, SL8 and LMS are taken to be valid at a reference period of 30 s, whereas the reference period for Q model PREM is chosen to be 1 s; α: coefficient of frequency dependence in Q; S: total M_2 tide dissipation rate (solid Earth and ocean) as deduced from tidal perturbations of satellite orbits. For the method see Section IV.C.

Earth dissipation rate would amount to nearly 50% of the total rate, and for $\alpha = 0.4$ it would go up to the unrealistic value of more than 130%. Thus, the uncertainty in the knowledge of α seems to make it impossible to reliably estimate the solid Earth part of tidal dissipation.

A further uncertainty in estimating tidal dissipation within the solid Earth from Q models is the fact that these Q models are usually determined from observations in a broad frequency range. The frequencies or periods at which the given Q models such as SL3 (Anderson and Hart, 1978a), SL8 (Anderson and Hart, 1978b), LMS (Smith, 1972) or PREM (Dziewonski and Anderson, 1981) are valid, are, therefore, not exactly known. The results given in Fig. 1 for Q-model PREM, for instance, assume that these Q values are valid for a period of 1 s. If however, the Q models were related to a period of 30 s as assumed for the other Q models in Fig. 1, the estimated dissipation rates would be up to four times smaller, the exact value of the reduction being dependent on α.

Finally, it is the uncertainties in the Q models themselves which do not allow to reliably estimate the solid Earth tidal dissipation. As seen from Fig. 1, the dissipation rates obtained for model SL3, for instance, are about two and a half times higher than those obtained for model LMS.

These uncertainties in the Q models together with the uncertainties in the reference frequencies and in the α values are the three main reasons why a direct calculation of solid Earth tidal dissipation from Q models cannot give meaningful results at the moment. In this chapter we try to avoid all three difficulties by showing that the solid Earth dissipation rate can be linearly related to the Chandler wobble period, and that this linear relationship is practically independent of the Q model, the reference frequency and the α value. Thus, the well-determined Chandler wobble period, particularly its lengthening due to inelasticity in the mantle can be used to obtain good estimates of the apparent tidal Q and the solid Earth dissipation rate which are rather insensitive to the assumed Q model and its degree of frequency dependence.

In the next section we will derive the total tidal dissipation rate from perturbations of satellite orbits. A critical discussion of correcting these perturbations for the ocean tide effect in order to obtain a reliable estimate of the solid Earth dissipation rate, is given in Section III. The following chapter deals with theoretical foundations and consequences of the Earth's inelasticity, especially with respect to the lengthening of the Chandler wobble period and to several parameters of tidal friction. Quantitative results on tidal friction in the solid Earth as deduced from the lengthening of the wobble period are presented in Section V. The final Section VI gives a summary of the main conclusions.

II. Total M_2-tide Dissipation from Satellite Solutions

The total rate of gravitational working by tidal forces on tidal displacements can be expressed as

$$\dot{E} = -\frac{5}{8\pi^2}\frac{\omega}{GR}\oiint_S \oint_t \psi^S \frac{\partial \psi^P}{\partial t}\, dt\, dS \qquad (1)$$

with G the gravitational constant; R, Earth radius; ω, circular frequency of the tide; ψ^P primary tidal potential on a rigid Earth; ψ^S, secondary tidal potential due to tidal deformation of the solid Earth, oceans and atmosphere. The inner integral is taken over one complete tidal cycle (12.42 hr in the case of the M_2 tide), and the outer integral over the Earth's undeformed surface. Equation (1) was derived by Zschau [1978, Eq. (18)] for the solid Earth alone. Platzman [1984, Eq. (5)] showed that Eq. (1) is also applicable to the whole planetary tidal work, provided that the secondary potential does not only include tidal effects in the solid Earth, but also those in the oceans and atmosphere.

Only the second-degree harmonics of the tidal potential terms in (1) are important in the tidal dissipation problem (Lambeck, 1975, 1977; among various other authors). Thus, the primary potential can be written as

$$\psi^P = \frac{3}{4} C_0 \frac{Gm}{D^3} r^2 \cos^2 \varphi \cos(\omega t + 2\lambda) \qquad (2)$$

with m and D being the mass and the mean distance of the tide-producing body (Moon or Sun), respectively. Constant C_0 is dependent on the tidal wave, r the distance from the Earth's centre, φ the geocentric latitude and λ the longitude. The potential is defined in a reference frame that rotates with the Earth.

The secondary tidal potential consists of three terms, solid Earth, oceanic and atmospheric term. However, the atmospheric term can be neglected in the dissipation problem if the main tidal wave M_2 is considered (Kagan, 1981; Platzman 1984). Thus, we have

$$\psi^S \approx \psi^S_B$$

$$\psi^S_B = \frac{3}{4} C_0 k \frac{Gm}{D^3} R^2 \left(\frac{R}{r}\right)^3 \cos^2 \varphi \cos(\omega t + 2\lambda - \varepsilon) \qquad (3)$$

$$\psi^S_O = \frac{12}{5} \pi C^+_{22}(1+k')G\rho_W R \left(\frac{R}{r}\right)^3 \cos^2 \varphi \cos(\omega t + 2\lambda - \delta^+_{22} - \delta)$$

where ψ^S_B is the secondary body tide potential, and ψ^S_O is the secondary ocean tide potential comprising the gravitational effect of the varying water mass as well as the effect of loading deformations underneath the water load. Here k is one of the well-known Love numbers of second degree; k', the corresponding loading Love number; ρ_W, the density of the ocean water; C_{22}^+, the amplitude; and δ_{22}^+ the phase of the ocean tide. The additional phase delays ε and δ are due to the inelastic responses of the Earth's mantle to body tide forces and to the ocean tide bottom pressure, respectively.

From (3) it is clear that the total secondary potential can be expressed in the form

$$\psi^S = B\frac{\cos^2\varphi}{r^3}\cos(\omega t + 2\lambda - \beta) \tag{4}$$

where B and β are independent of φ and r, respectively. Inserting (2) and (4) into (1) yields the simple expression

$$\dot{E} = C_0\frac{m\omega}{D^3}B\sin\beta \tag{5}$$

for the total M_2-tide dissipation rate. Here B and β can be determined directly from perturbations of satellite orbits, and the other quantities in (5) are well known for the M_2-tide. A small problem arises from the fact that in geophysical literature the quantities B and β are usually not given explicitly, but C_{22}^+ and $\varepsilon_{22}^+ = \frac{5}{2}\pi - \delta_{22}^+$ are given instead. The latter quantities are usually derived from satellite observations after correcting for the body tide perturbations and by assuming that $k = 0.3$, $k' = -0.3$, and $\varepsilon = \delta = 0$. However, putting these quantities back into Eqs. (3) and (4) enables us to redetermine B and β from C_{22}^+ and ε_{22}^+. The total M_2-tide dissipation rate $\dot{E}$ then follows readily from applying B and β in Eq. (5).

This was carried out for various M_2-tide satellite solutions involving TRANSIT satellites, GEOS-3 and STARLETTE. For ω we have used the value $2\pi/12.42$ hr. For the remaining parameters in Eqs. (3) and (5) we have adopted the values given in Table III, Section V. The results are presented in Table I.

The weighted mean of different satellite solutions gives a total M_2-tide dissipation rate (ocean and solid earth) of

$$\dot{E} = [2.69 \pm 0.13] \times 10^{19} \text{ erg/s} \tag{6}$$

and a total phase delay of the secondary M_2-tide potential (ocean and solid Earth) with respect to the primary one of

$$\beta = 5.32° \pm 0.26° \tag{7}$$

These results apply only to the second-degree harmonics. But as mentioned before in the tidal dissipation problem these are, by far, the most important terms.

TABLE I. Results of various satellite solutions[a]

Satellites	Source	C_{22}^{+} [cm]	ε_{22}^{+} [deg]	$B \sin \beta$ [10^{30} cm^5/s^2]	$B \cos \beta$ [10^{30} cm^5/s^2]	β [deg]	$\dot{E}$ [10^{19} erg/s]
S+G3	Cazenave and Daillet (1981)	3.1±0.3	333±5	0.164±0.018	1.768±0.015	5.30±0.58	2.71±0.30
S+G3+T	Felsentregger *et al.* (1978 a, b)	3.4±0.24	325±4	0.165±0.014	1.735±0.014	5.43±0.46	2.72±0.23
G3+T	Goad and Douglas (1978)	3.2±0.25	331±6	0.166±0.012	1.759±0.022	5.39±0.39	2.74±0.20
G3+3T	Daillet (1978)	3.2±0.25	325±5	0.156±0.013	1.742±0.018	5.12±0.43	2.58±0.22
Weighted Mean		3.24±0.13	328±2.5	0.163±0.008	1.749±0.008	5.32±0.26	2.69±0.13

[a] C_{22}^{+}, ε_{22}^{+} are amplitude and phase, respectively, of the M_2-ocean tide (degree 2); $B \sin \beta$ and $B \cos \beta$ are the sin and cos term, respectively, of the secondary M_2 tide potential (solid Earth and ocean; degree 2); E is total M_2 tide dissipation rate (solid Earth and ocean; degree 2); S is STARLETTE; G3 is GEOS-3; T is TRANSIT. The weighted means were deduced from the weighted mean of $C_{22}^{+} \sin \varepsilon_{22}^{+}$ and $C_{22}^{+} \cos \varepsilon_{22}^{+}$, respectively: As weight, the inverse square of the error in the sin and cos term, respectively, was taken.

The quantities in (6) and (7) contain two different portions; one caused by dissipation in the ocean and the other by dissipation in the solid Earth. It will be checked in the next section whether both portions can be separated from each other by making use of existing ocean tide models.

III. Solid Earth Tide Dissipation from Combined Satellite–Ocean Tide Solutions

Cazenave and Daillet (1981) have tried to separate solid Earth tide dissipation from ocean tide dissipation by determining the Moon's mean motion $\dot{n}$ from the previously mentioned satellite solutions and subtracting the ocean contribution of recent ocean tide models. In this way they obtained the solid Earth phase delay [see Eq. (3)] and finally the tidal effective Q of the solid Earth. Inherent in their calculation was the assumption that the Love number k is equal to 0.3. This neglects the fact that satellite solutions as well as ocean tide models provide information on both phase *and* amplitude. Thus, not only ε but also the Love number k can be extracted from these data and there is no need to make any assumption on k.

Both quantities, ε and k, are not independent from each other because they are effected by inelastic processes in the mantle which have the property of causality. The relationship between both quantities is obtained after expressing them in terms of the real and imaginary parts of the complex Love number k^*. Due to causality, the real and imaginary parts must be Hilbert-transforms of each other. By determining both ε and k from the previously mentioned data provides an additional tool for judging the reliability of the results on ε, Q and the solid Earth tide dissipation rate $\dot{E}_{SE}$ ($\dot{E}_{SE} \sim k \sin \varepsilon$; see Zschau (1978) and Section IV): ε and k must be realistic pairs which do not violate the principle of causality.

By following Cazenave and Daillet (1981), we have redetermined the solid Earth phase shift ε and the Love number k from the previously mentioned satellite solutions and from the ocean tide models of Estes (1977), Parke and Hendershott (Parke, 1982), and Schwiderski (1983a). The difference from Cazenave and Daillet (1981) is that no assumption was made on k. For determining ε

$$\begin{aligned} k \sin \varepsilon &= C_1 B \sin \beta + C_2 (1+k') C_{22}^{+} \sin(\varepsilon_{22}^{+} - \tfrac{5}{2}\pi - \delta) \\ k \cos \varepsilon &= C_1 B \cos \beta - C_2 (1+k') C_{22}^{+} \cos(\varepsilon_{22}^{+} - \tfrac{5}{2}\pi - \delta) \end{aligned} \tag{8}$$

with

$$C_1 = \frac{4}{3} \frac{D^2}{C_0 G m R^5}, \qquad C = \frac{6}{5} \pi \frac{\rho_W D^3}{C_0 m R}$$

They follow directly from combining Eqs. (3) and (4). It is immediately verified that errors in δ have no significant influence on k or ε. Thus, it is justified to set δ equal to zero. Errors in k', on the other hand, may significantly effect the $k \sin \varepsilon$ term and thus ε. The error in ε, however, will be smaller than 0.1° if the error in k' is not greater than 6%. The latter requirement will definitely be fulfilled, if we assume k' to be (−0.3). Thus, Eq. (9) reads

$$\begin{aligned} k \sin \varepsilon &= C_1 B \sin \beta + 0.7 C_2\, C_{22}^{+} \sin(\varepsilon_{22}^{+} - \tfrac{5}{2}\pi) \\ k \cos \varepsilon &= C_1 B \cos \beta - 0.7 C_2 C_{22}^{+} \cos(\varepsilon_{22}^{+} - \tfrac{5}{2}\pi) \end{aligned} \tag{9}$$

With B and β from satellite solutions (see Table I) and C_{22}^{+} and ε_{22}^{+} from ocean tide models, Eq. (9) allows k and ε to be derived separately. The results are in Table II. For the case of the "best" satellite solution (weighted mean of the different satellite solutions) the results are shown in Fig. 2. The error bars in Fig. 2 reflect the uncertainties in the satellite solutions. They result in uncertainties which are typically ±0.25° for the solid Earth phase delay ε and ±0.0013 for the Love number k. The spread of the results due to different ocean tide models is much higher: about 1.8° in ε and 0.0075 in k. Considering that the solid Earth dissipation rate for the M_2-body tide can be expressed as

$$\dot{E}_{SE} = \tfrac{3}{4} G (C_0 m)^2 \omega \frac{R^5}{D^6} k \sin \varepsilon \tag{10}$$

TABLE II. Combined satellite–ocean tide solutions for the tidal Love number k and the solid Earth tide phase delay ε (M_2 tide only)[a]

		Estes		Parke and Hendershott		Schwiderski	
		C_{22}^{+} [cm]	ε_{22}^{+} [deg]	C_{22}^{+} [cm]	ε_{22}^{+} [deg]	C_{22}^{+} [cm]	ε_{22}^{+} [deg]
		3.31	333	3.38	313.1	2.95	311.6
S+G3	k	0.3010±0.0024		0.3103±0.0024		0.3079±0.0024	
	ε	−0.35±0.56		0.80±0.54		1.43±0.54	
S+G3+T	k	0.2957±0.0023		0.3050±0.0023		0.3025±0.0023	
	ε	−0.32±0.44		0.84±0.43		1.49±0.43	
G3+T	k	0.2996±0.0036		0.3089±0.0036		0.3064±0.0036	
	ε	−0.29±0.37		0.86±0.36		1.50±0.36	
G3+3T	k	0.2968±0.0029		0.3061±0.0029		0.3036±0.0029	
	ε	−0.60±0.41		0.57±0.39		1.21±0.40	
Weighted	k	0.2979±0.0013		0.3073±0.0013		0.3048±0.0013	
Mean	ε	−0.38±0.25		0.78±0.24		1.42±0.24	

[a] C_{22}^{+}, ε_{22}^{+}: amplitude and phase, respectively, of the M_2 ocean tide (degree 2). S, STARLETTE; G3, GEOS 3; T, TRANSIT. For the weighted mean see Table I.

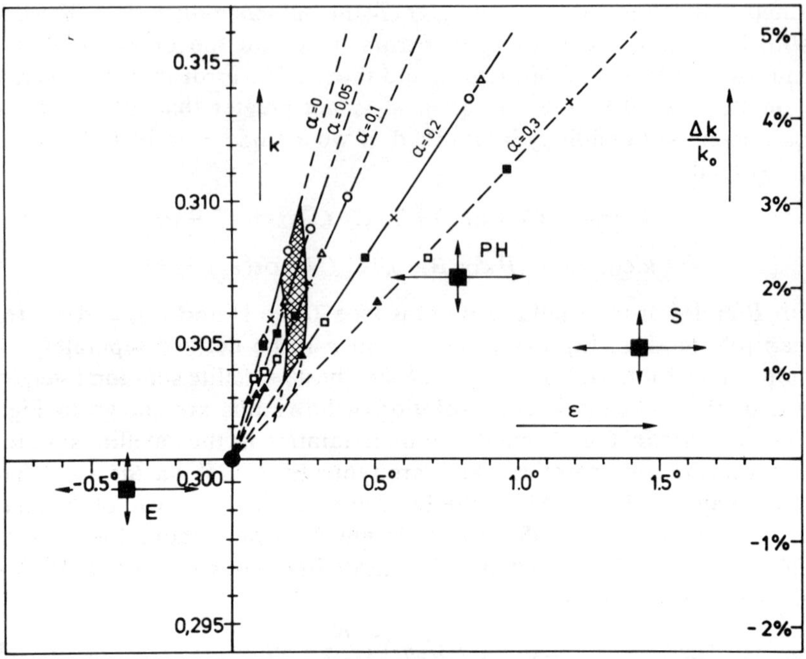

Fig. 2. Combined satellite–ocean tide solutions for the tidal Love number k and the solid Earth tide phase delay ε (M_2 tide only). The error bars reflect the uncertainties in the satellite solutions; E, PH and S are solutions for ocean tide model of Estes (1977), Parke and Hendershott (Parke, 1982), and Schwiderski (1983a) respectively; $\Delta k/k_0$ relative increase of the tidal Love number k with respect to the unrelaxed one k_0 on the elastic Earth. The lines represent results of model calculations based on different Q models for the mantle (▲ LMS, □ $Q=450=\text{const}$, ■ SL8, × SL3, ○ $Q=170=\text{const}$, △ PREM) and on different degrees α of frequency dependence of $Q(Q\sim\omega^{\alpha})$. Only the hatched area is in agreement with the observed Chandler wobble period.

[see Zschau, 1978, Eq. (25); see also Eq. (31) of this chapter], it follows that the uncertainty in $\dot{E}_{\mathrm{SE}}$ due to errors in the satellite solutions amounts to $\pm 1.3\times 10^{18}$ erg/s, and the spread due to different ocean tide models is roughly 10^{19} erg/s. Recalling that the total dissipation rate is only 2.7×10^{19} erg/s [Eq. (6)], these results clearly demonstrate that it is still mainly the problem of modeling the ocean tides which does not allow a reliable estimation of the solid Earth dissipation rate from satellite solutions.

Estimations of k, ε and the Chandler wobble period based on different Q models for the Earth's mantle support this statement. The theoretical basis for these estimations is described in Section IV [see especially Formulas (23), (28) and (30)] and the numerical results are presented in Section V. In order to judge the reliability of the combined satellite–ocean tide model solution, we have anticipated some of the results and plotted them in the k–ε diagram of Fig. 2. Only the hatched area is in agreement with the observed Chandler wobble period. Therefore, we expect ε to be near 0.2° and k to increase by about 2% with respect to the elastic case. None of the combined satellite-ocean tide model solutions shown in Fig. 2 is anywhere near this area. The uncertainties in these solutions are clearly much larger than those due to uncertainties in the knowledge of the Chandler wobble period. Thus, the observed wobble period seems to be a much better indicator for the k and ε values, and consequently for the solid Earth tidal dissipation rate than the satellite solutions corrected for ocean tide dissipation. How the wobble period can be used for this purpose is described in the next section.

IV. Internal Friction in the Mantle: Theory and Consequences

In this section we describe the theoretical fundamentals of the so-called absorption band which allows for the model of frequency dependent internal friction in the mantle with Q approximately proportional to ω^{α}. This model is used to determine the lengthening of the Chandler wobble period due to inelasticity as well as several parameters of tidal friction such as the tidal Love number k of the inelastic Earth, the apparent tidal $Q = Q_k^*$, the solid Earth phase delay ε and the solid Earth tide dissipation rate $\dot{E}_{SE}$. The goal is to provide the theoretical basis for relating the poorly known tidal friction parameters to the well-known Chandler wobble period.

A. The Absorption Band Model with $Q \sim \omega^{\alpha}$

Liu *et al.* (1976) as well as Anderson and Minster (1979) argued that absorption of strain energy in the Earth's mantle can best be described by assuming that within a certain frequency band—the so-called absorption band—the specific dissipation function Q^{-1} is a continuous and smooth function of ω. They claimed that the absorption band model is especially capable of describing the apparently flat or slightly inclined Q^{-1} spectrum observed in the band of seismic periods. Applications of the absorption band for seismic observations have been demonstrated by Lundquist and Cormier (1980), Anderson and Given (1982) and Doornbos (1983). In Zschau (1983) it was shown that the seismic absorption band is also likely to include the main tidal periods as well as the Chandler wobble. Therefore, the absorption band is appropriate to investigate relationships between certain tidal and wobble parameters, provided that Q^{-1} is a smooth function of ω between the wobble frequency and the main tidal frequencies.

The latter requirement will be fulfilled if absorption in the mantle is due to a distribution of strain retardation times or stress relaxation times, respectively, which is continuous and smooth within a certain band width. Liu *et al.* (1976) assumed a box-like distribution function for strain retardation times, and, thus, obtained a flat Q^{-1} spectrum within the absorption band. Anderson and Minster (1979) modified this box-like distribution in order to obtain a slightly inclined Q^{-1} spectrum with $Q \sim \omega^{\alpha}$. This was approximately achieved by the following distribution function:

$$D(\tau) = \begin{cases} \dfrac{\alpha}{\tau_2^{\alpha} - \tau_1^{\alpha}} \dfrac{1}{\tau^{1-\alpha}} & \text{for} \quad \tau_1 \ll \tau \ll \tau_2 \\ 0 & \text{otherwise} \end{cases} \tag{11}$$

where α is a constant between 0 and 1, and τ_1, τ_2 are the lower and upper cut-off periods of the absorption band (see Fig. 3). In (11) $D(\tau)$ defines the relative contribution of each single creep mechanism to the total creep

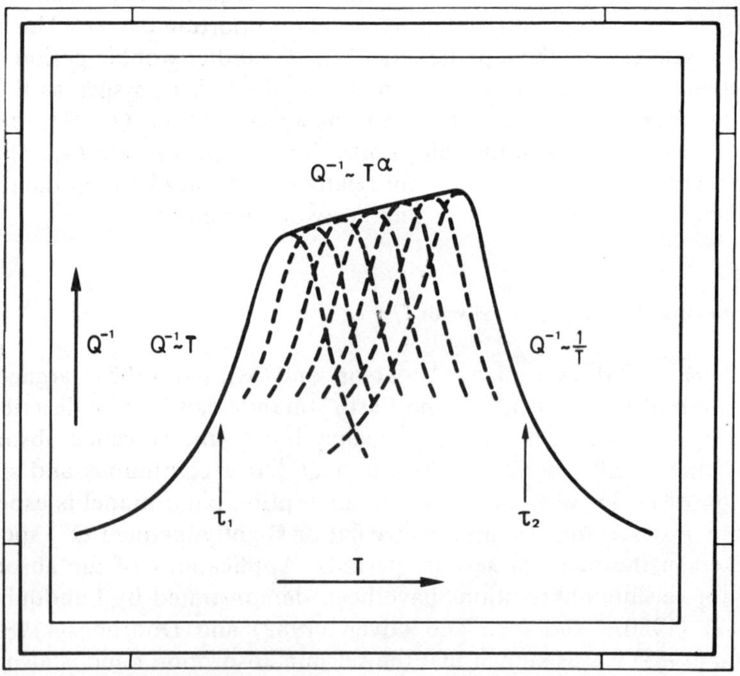

Fig. 3. The absorption band model for Q: Q^{-1}, specific dissipation function; T, period, τ_1 and τ_2 lower and upper cut-off period. It can be constructed from a continuous distribution of strain retardation times which corresponds to a continuous distribution of Debye peaks in the Q^{-1}–T diagram.

strength Δ. The results Anderson and Minster obtained turned out to be valid only if α is not too small ($\alpha \geqslant 0.2$). Therefore, their equations were rederived by Zschau (1981) for the same distribution of strain retardation times, but for arbitrary α values. There we gave the shear strain response* to harmonic shear loading in the following complex form:

$$\varepsilon(j\omega) = \frac{\delta(j\omega)}{2\mu_0}\left\{1 + \frac{\Delta}{1-(\tau_1/\tau_2)^\alpha}\frac{\pi\alpha/2}{(\tau_2\omega)^\alpha}\left[\frac{1}{\sin(\pi\alpha/2)} - \frac{(\tau_1\omega)^\alpha}{\pi\alpha/2} - \frac{j}{\cos(\pi\alpha/2)}\right]\right\} \tag{12}$$

$$\tau_1 \ll T = 2\pi/\omega \ll \tau_2$$

Here ε and δ denote strain and stress, respectively, $j = \sqrt{-1}$, ω circular frequency, and T period. The lower and upper cut-off periods τ_1 and τ_2, respectively, the creep strength Δ, the unrelaxed shear modulus μ_0 and the coefficient α are the 5 quantities which determine the absorption band. The coefficient α describes the frequency dependence of Q within the band; $\alpha = 0$ is identical with the case treated by Liu *et al.* (1976). With the complex shear modulus

$$\mu^* = \mu^R + j\mu^y = \tfrac{1}{2}\delta(j\omega)/\varepsilon(j\omega) \tag{13}$$

and the quality factor

$$Q = \mu^R/\mu^y \tag{14}$$

it follows from (12) that well within the absorption band ($\tau_1 \ll T \ll \tau_2$) we have

$$\mu^y(\omega) = \mu(\omega_0)\frac{[Q^2(\omega_0)+1]^{1/2}}{Q^2(\omega)+1}\left(\frac{\omega}{\omega_0}\right)^\alpha$$

$$\mu^R(\omega) = \mu^y(\omega)Q(\omega)$$

with

$$\mu(\omega_0) = |\mu^*(\omega_0)| \tag{15}$$

$$Q(\omega) = \left\{Q(\omega_1) - \cot\frac{\pi\alpha}{2}\right\}\left(\frac{\omega}{\omega_1}\right)^\alpha + \cot\frac{\pi\alpha}{2}$$

$$Q(\omega_0) = \left\{Q(\omega_1) - \cot\frac{\pi\alpha}{2}\right\}\left(\frac{\omega_0}{\omega_1}\right)^\alpha + \cot\frac{\pi\alpha}{2}$$

If α is not too small ($\alpha \geqslant 0.2$), the cot $\pi\alpha/2$ term can be neglected and we have $Q \sim \omega^\alpha$.

* Dissipation in bulk is commonly neglected for the mantle; therefore, the stress–strain relationship for pure compression is elastic.

It becomes clear from Eqs. (15) that, once knowing α, the complex modulus at frequency ω can be constructed from the modulus $\mu(\omega_0)$ at some reference frequency ω_0 and from the quality factor $Q(\omega_1)$ at some reference frequency ω_1. Furthermore, the dispersion relation valid for the absorption band is obtained;

$$\frac{\mu(\omega)}{\mu(\omega_0)}=\left[\frac{Q^2(\omega_0)+1}{Q^2(\omega)+1}\right]^{1/2}\left(\frac{\omega}{\omega_0}\right)^{\alpha} \tag{16}$$

with $Q(\omega)$ and $Q(\omega_0)$ as in (15).

In the limit $\alpha\to 0$ these equations reduce to

$$\begin{aligned}\mu^{y}(\omega)&=\mu(\omega_0)\frac{[Q^2(\omega_0)+1]^{1/2}}{Q^2(\omega)+1}\\ \mu^{R}(\omega)&=\mu^{y}(\omega)Q(\omega)\\ \frac{\mu(\omega)}{\mu(\omega_0)}&=\left\{\frac{Q^2(\omega_0)+1}{Q^2(\omega)+1}\right\}^{1/2}\end{aligned} \tag{17}$$

with

$$Q(\omega)=Q(\omega_1)-\frac{2}{\pi}\ln\left(\frac{\omega}{\omega_1}\right)$$

$$Q(\omega_0)=Q(\omega_1)-\frac{2}{\pi}\ln\left(\frac{\omega_0}{\omega_1}\right)$$

For many practical cases the logarithmic term in the Q expression is small, and Q may be taken as frequency-independent.

For Q being not too small, the dispersion relation in (17) may be approximated by

$$\frac{\mu(\omega)}{\mu(\omega_0)}\approx\frac{Q(\omega_0)}{Q(\omega)} \tag{18}$$

and with

$$Q(\omega_0)=Q(\omega)-\frac{2}{\pi}\ln\left(\frac{\omega_0}{\omega}\right) \tag{19}$$

we have

$$\frac{\mu(\omega)}{\mu(\omega_0)}\approx 1-\frac{2}{\pi Q(\omega)}\ln\left(\frac{\omega_0}{\omega}\right) \tag{20}$$

From this we can immediately verify the well-known and often-used dispersion relation for the shear wave velocity $v(\omega)$ (see for instance Liu *et al.*, 1976, Kjartansson, 1979)

$$\frac{v(\omega)}{v(\omega_0)}=\left\{\frac{\mu(\omega)}{\mu(\omega_0)}\right\}^{1/2}\approx 1-\frac{1}{\pi Q}\ln\left(\frac{\omega_0}{\omega}\right) \tag{21}$$

The real and imaginary part of the complex shear modulus in (15) and (17), respectively, can be used directly to calculate complex Love numbers for a radially stratified self-gravitating Earth (for the method see Zschau, 1977, 1979). As input the method needs a depth distribution of Q given at the reference frequency ω_1 and an elasticity structure given at another reference frequency ω_0. The complex Love numbers are then able to determine Chandler wobble deformations as well as tidal deformations for a viscoelastic Earth. Different from the complex Love numbers given in Zschau (1978, 1979) those derived as described above do allow for the modification of their real part due to dispersion.

The next section shows how the Chandler wobble period is related to the complex Love number k^*.

B. The Chandler Wobble Period

The Chandler wobble period is partly determined by the solid Earth's response to the varying centrifugal forces of the wobble motion. This is expressed in the dependence of the wobble period on the Earth's global Love number k. Due to inelasticity in the mantle, k is effected by dispersion (see last section). Thus, k is frequency-dependent and its value at the wobble period is different from the unrelaxed value in the completely elastic case. This results in a lengthening of the wobble period with respect to the elastic period.

For determining the wobble period we make use of the fact that the wobble can be regarded as a free mode. Thus, we may start from the complex wobble frequency

$$\omega^* = \omega^{\mathrm{R}} + j\omega^{\mathrm{I}} \tag{22}$$

with ω^{R} and ω^{I} being the real and imaginary part of the complex frequency, respectively. By following Smith and Dahlen (1981), ω^* can be written as

$$\omega^* = \frac{a - bk^*_{\mathrm{eff}}}{d + bk^*_{\mathrm{eff}}}\,\Omega \tag{23}$$

with

$$a = \frac{C - A}{A}, \qquad b = \frac{R^5\Omega^2}{3GA}, \qquad d = \frac{A_{\mathrm{M}} + \varepsilon_{\mathrm{b}} A_{\mathrm{C}}}{A}$$

Here R is the Earth's radius, Ω the Earth's angular speed of rotation, while A and C are its least and greatest principal moments of inertia, respectively;

A_M and A_C are the mantle's and the core's (inner and outer) equatorial moments of inertia, respectively; ε_b denotes the ellipticity of the core–mantle boundary and G the gravitational constant. The effective Love number k^*_{eff} consists of three terms:

$$k^*_{eff} = k^* + \Delta k_1^* + \Delta k_2^* \tag{24}$$

All three terms will be complex if the mantle exhibits inelasticity. Here k^* is the ordinary static Love number and Δk_1^* is a correction for the effect of the Earth's rotation and ellipticity on its response. For the elastic case Smith and Dahlen (1981) used the value 0.00070. This is a small quantity compared to k^*_{eff}. Thus, we do not correct this value for inelasticity, but we take this value for the real part of Δk_1^* and assume its imaginary part to be zero; Δk_2^* is a correction term that allows for the effect of the pole tide on the wobble frequency. We use the following expression from Smith and Dahlen's (1981) Eqs. (5.7), (8.22) and (8.24):

$$\Delta k_2^* = 0.708 \times \frac{0.04484}{0.0465} \times \frac{3}{5} \frac{\rho_W}{\bar{\rho}} \frac{(1+k^*-h^*)^2}{1-\frac{3}{5}(\rho_W/\bar{\rho})[1+k^*-h^*-(h')^*]} \tag{25}$$

The factor 0.708 is the fraction of the Earth's surface area covered by oceans and the factor 0.04484/0.0465 has been included to allow for the approximation in this method (see Smith and Dahlen, 1981, page 264). In (25), $(h')^*$, k^*, h^* are the load Love number and Love numbers of degree 2, respectively; ρ_W is the density of sea water and $\bar{\rho}$ is the mean density of the Earth. As before the asterisk indicates that in the viscoelastic case the corresponding quantities are complex. Equation (23) will be valid if the pole tide is at equilibrium. Dynamic effects are not taken into account.

From Eqs. (23), (24), (25) and from the explanations in the last section it is obvious that the Earth's Q structure enters into the wobble frequency through k^*, h^* and $(h')^*$. Having once determined these quantities as described in Chapter IV, the above equations can be used to determine the complex frequency ω^*, and the observable Chandler wobble period T can be readily obtained from

$$T = 2\pi/\omega^R \tag{26}$$

This allows for the elastic and inelastic yielding of the Earth due to the wobble's body forces and to the surface forces of a static pole tide.

In this section the Chandler wobble period was expressed in terms of complex Love numbers and loading Love numbers, respectively. In the next section we will do the same for several parameters of tidal friction in order to provide the tool for comparing these parameters with the wobble period.

C. Parameters of Tidal Friction

The parameters of tidal friction in which we are interested are the dispersion Δk of the tidal Love number k, the apparent tidal $Q = Q_k^*$ the solid Earth tide phase delay ε and the solid Earth tide dissipation rate $\dot{E}_{SE}$. All these quantities are directly related to the real and imaginary part of the complex Love number

$$k^* = k^R + jk^I \tag{27}$$

at tidal frequencies. We will be considering only the main tidal frequency M_2:

Dispersion Δk

$$\Delta k = k - k_0 \tag{28}$$

with

$$k = |k^*|$$

k_0 = unrelaxed Love number (the elastic case!)

Apparent tidal Q

$$Q_k^* = \frac{k^R}{(-k^I)} \tag{29}$$

Solid Earth tide phase delay ε

$$\tan \varepsilon = (Q_k^*)^{-1} = \frac{(-k^I)}{k^R} \tag{30}$$

Solid Earth tide dissipation rate $\dot{E}_{SE}$

$$\dot{E}_{SE} = C_3 k \sin \varepsilon = C_3 k^R \tan \varepsilon = C_3 k^I$$

with

$$C_3 = \tfrac{3}{4} G (C_0 m)^2 \omega (R^5/D^6) \tag{31}$$

Except for (28) all the other equations were already given in Zschau (1978). There we have also explained that the apparent tidal $Q = Q_k^*$ does not have the meaning of an ordinary specific dissipation function in the sense that it determines the ratio between the stored strain energy and the energy dissipated within one harmonic cycle. The index k indicates that Q_k^* is related to the phase angle of the tidal potential perturbation, i.e. to ε. This phase angle should not be mixed up with various other tidal phase angles such as those of tidal gravity and tilt perturbations.

Given a certain elasticity structure at a reference frequency ω_0, a Q structure at a reference frequency ω_1 and an α value for the degree of frequency dependence of Q will now allow us to calculate the above parameters of tidal friction as well as the Chandler wobble period. This is done in the next section for various Q structures at different reference frequencies and for various α values.

V. Results of Model Calculations

In this section we will show that there is practically a linear relationship between the before-mentioned parameters of tidal friction and the Chandler wobble period T, no matter what the Q structure in the Earth's mantle looks like, no matter what the reference frequency ω_1 for the Q structure is chosen to be and no matter what the coefficient α of frequency dependence looks like. Furthermore, it will be demonstrated that as far as the solid Earth dissipation rate is concerned, this linear relationship is practically independent of α for realistic α values, and that all the results for different Q structures and different ω_1 and α values lie approximately on the same line in the $\dot{E}_{SE}$–T diagram. This enables a reliable determination of the solid Earth tidal dissipation rate $\dot{E}_{SE}$ from the wobble period T without making too restrictive assumptions on the Earth's Q structure, the degree of frequency dependence of Q and the reference frequency ω_1.

A. The Earth Models

The Earth models used for calculating complex Love numbers are models for the elasticity and density structure as well as for the Q distribution within the Earth. For the elasticity and density structure we have used model PREM (Dziewonski and Anderson, 1981) which is valid at the reference period $T_0 = 2\pi/\omega_0 = 1$ s.

Different from the elasticity and density structure, the distribution of the quality factor Q within the Earth is not so well known, and there is a variety of proposals for Q especially in the lower mantle. In order to show that the final conclusions are independent of the Q structure, we have chosen a broad spectrum of Q models (see Fig. 4) including even unrealistic constant Q models. $Q = 170$ everywhere in the mantle may be considered as the average Q of near vertical S_CS core reflections for paths through the oceanic mantle and $Q = 450$ for paths through the continental mantle; $Q = 750$ is the highest Q in the lower mantle. It is obtained by model LMS (Smith, 1972). Model SL3 (Anderson and Hart, 1978a), on the other hand,

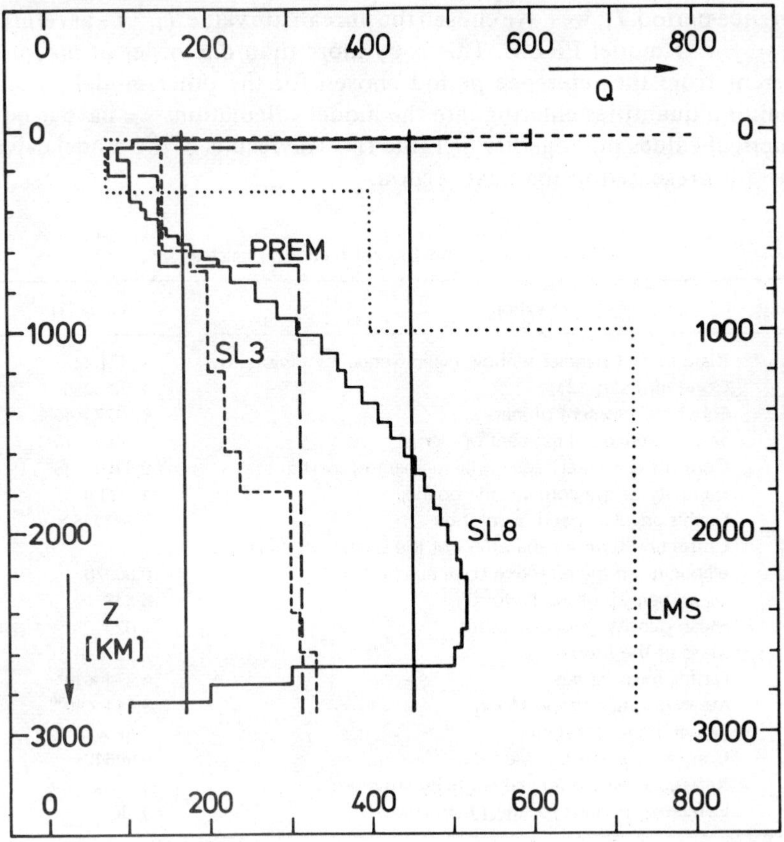

Fig. 4. The Q models used: (1) LMS (Smith, 1972), (2) $Q=450=\text{const}$, (3) SL8 (Anderson and Hart, 1978b), (4) SL3 (Anderson and Hart, 1978a), (5) $Q=170=\text{const}$, and (6) PREM (Dziewonski and Anderson, 1981). PREM is taken to be valid at a reference period of 1 s, whereas for all the other models the reference period is chosen to be 30 s.

has 330 as its highest Q value in the lower mantle. Models SL8 (Anderson and Hart, 1978b) and PREM (Dziewonski and Anderson, 1981) have Q values in-between. Models SL3, SL8 and PREM were constructed from a broad spectrum of data including body wave and surface wave as well as normal mode data. Model LMS was obtained from normal mode data alone.

Except for model PREM we have taken all Q models to be valid at a reference period $T_1 = 2\pi/\omega_1 = 30$ s. Whereas this is a good approximation for S_cS body wave data, the normal mode data are probably centered around some higher period. Smith and Dahlen (1981) used $T_1 = 200$ s. In order to demonstrate that the final conclusions are independent of the reference period T_1, we have chosen the unrealistic value $T_1 = 1$ s as reference period for Q-model PREM. This is by more than one order of magnitude different from the reference period chosen for the other models. For the remaining quantities entering into the model calculations we have used the numerical values put together in Table III. The results of the model calculations are presented in the next section.

TABLE III. Basic parameters for numerical calculations

Symbol	Definition	Value [egs]
$a^{*)}$	Rigid body Chandler wobble eigen frequency divided by Ω	1/304.437
$b^{*)}$	Constant in Eq. (23)	1/287.259
$A^{*)}$	Equatorial moment of inertia	8.0177×10^{44}
$A_M^{*)}$	Mantle equatorial moment of inertia	7.1005×10^{44}
$A_C^{*)}$	Core (inner + outer) equatorial moment of inertia	9.1100×10^{43}
$\varepsilon_b^{*)}$	Ellipticity of the core–mantle boundary	1/393.0
$\Omega^{*)}$	Earth's angular speed of rotation	7.292115×10^{-5}
$\Delta k_1^{*\,*)}$	Correction term for the effect of the Earth's rotation and ellipticity on the effective Love number	0.00070
$\bar{\rho}^{*)}$	Mean density of the Earth	5.517
ρ_w	Mean density of ocean water	1.025
m	Mass of the Moon	7.35×10^{25}
R	Earth's mean radius	6.371×10^{8}
D	Mean distance to the Moon	3.844×10^{10}
G	Gravitational constant	6.67×10^{-8}
C_0	Constant for the M_2-tide	0.90812
T_0	Reference period for the elasticity structure	1
T_1	Reference periods for the Q-structures	1, 30

* From Smith and Dahler (1981).

B. Tidal Friction in the Solid Earth and the Chandler Wobble Period: Numerical Results

The results of tidal friction parameters such as the dispersion of the Love number k, the apparent tidal $Q = Q_k^*$ and the solid Earth dissipation rate are shown in Figs. 5, 6 and 7, respectively. (The results on the solid Earth tide phase delay ε are directly related to the apparent tidal Q [see Eq. (30)] and are discussed in Section III.) In all three figures the tidal friction parameters are plotted against the Chandler wobble period and its lengthening due to inelasticity.

For the wobble period in the elastic case we have 426.7 ± 1.0 days (see Smith and Dahlen, 1981), provided that the pole tide can be considered as

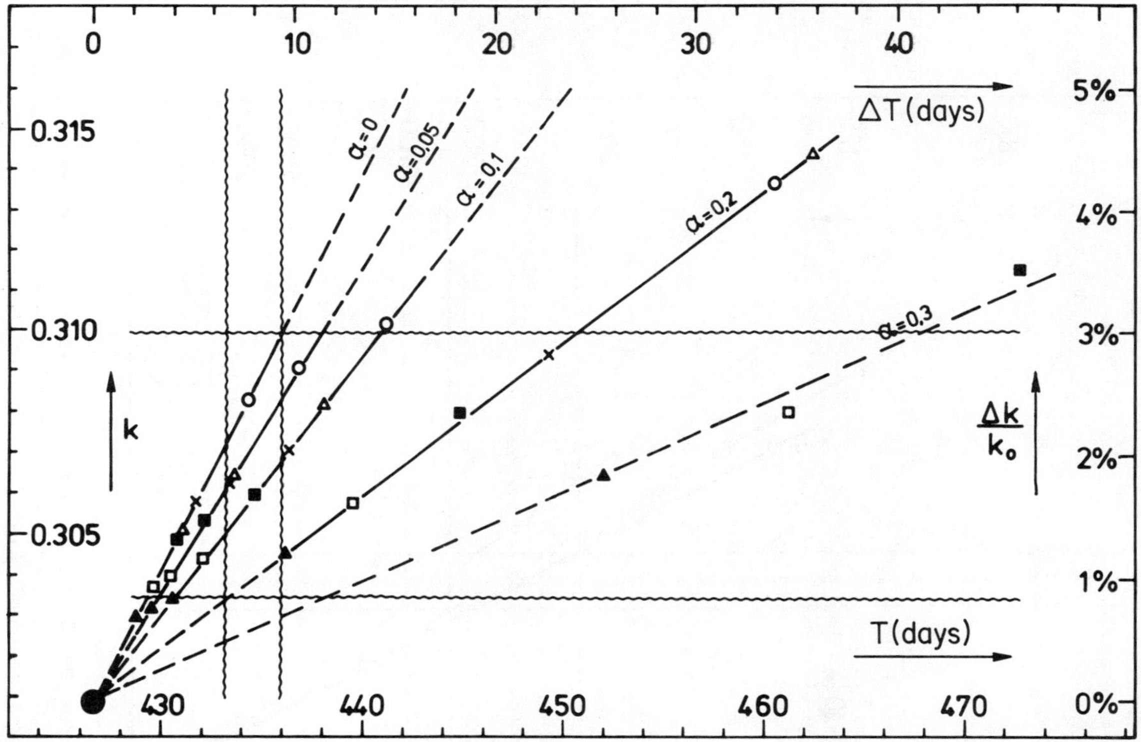

Fig. 5. M_2 tide Love number k as a function of the Chandler wobble period T. $\Delta k/k_0$, the relative increase of the tidal Love number k with respect to the unrelaxed one k_0 on the elastic Earth (effect of dispersion); and ΔT the lengthening of the Chandler wobble period due to inelasticity of the mantle. The results are based on different Q models for the mantle (▲ LMS, □ $Q = 450 =$ const, ■ SL8, × SL3, ○ $Q = 170 =$ const, △ PREM) and on different degrees α of frequency dependence of $Q (Q \sim \omega^{\alpha})$. The wavy lines indicate the spread of observations for the wobble period and the range of Love numbers k, respectively, which is in agreement with these observations.

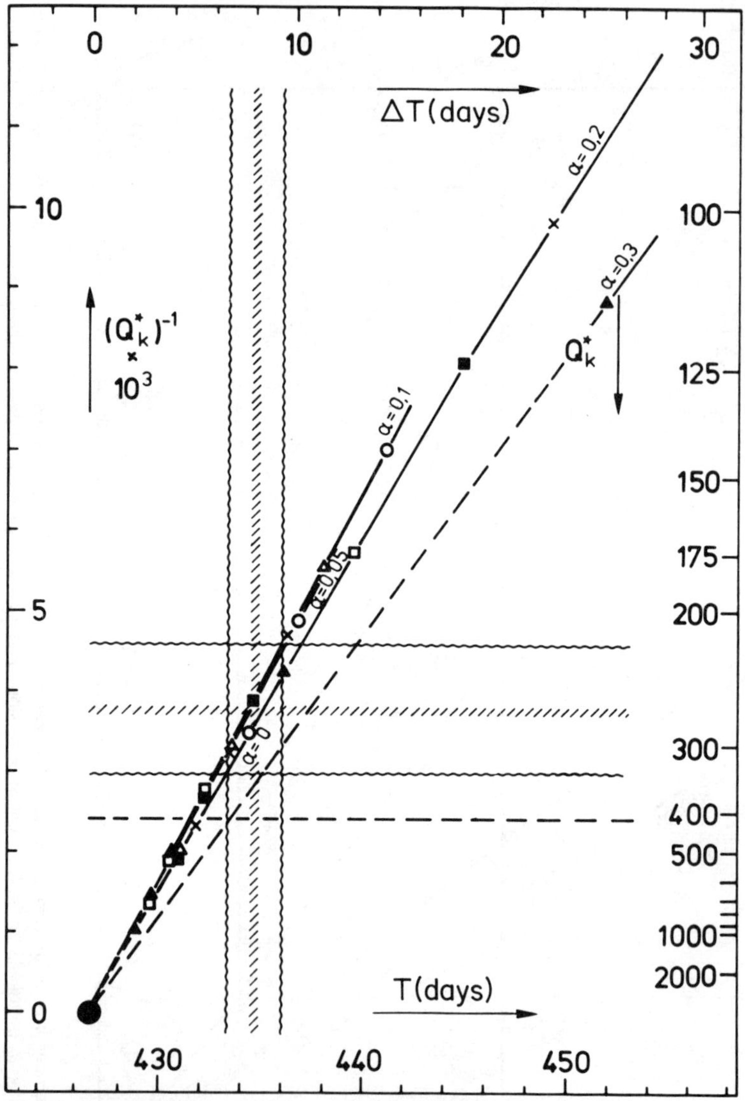

Fig. 6. Apparent M_2 tide $Q = Q_k^*$ as a function of the Chandler wobble period T. ΔT the lengthening of the Chandler wobble period due to inelasticity of the mantle. The results are based on different Q models for the mantle (▲ LMS, □ $Q = 450 = \text{const}$, ■ SL8, × SL3, ○ $Q = 170 = \text{const}$, △ PREM) and on different degrees α of frequency dependence of $Q (Q \sim \omega^\alpha)$. The striped line marks the average observed wobble period and the corresponding result for Q_k^*. The other vertical and horizontal lines indicate the spread of observations for the wobble period and the range of Q_k^* values, respectively, which is in agreement with these observations.

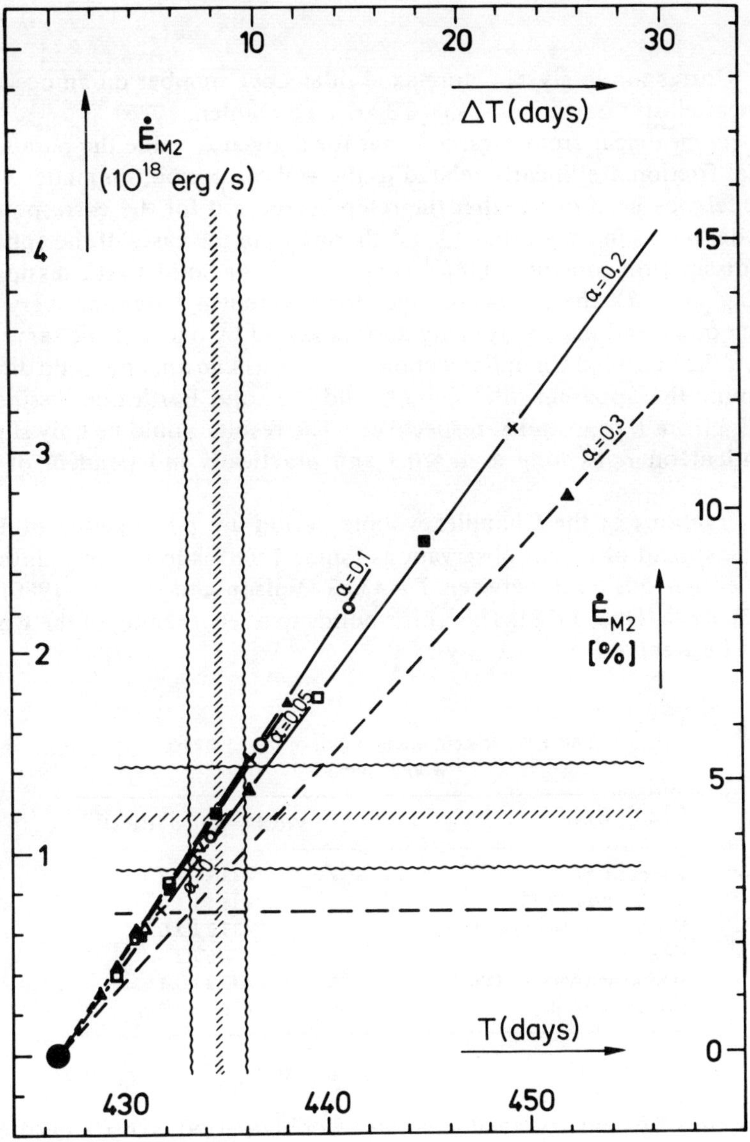

Fig. 7. Solid Earth M_2 tide dissipation rate $\dot{E}_{M2}$ as a function of the Chandler wobble period T. ΔT is lengthening of the Chandler wobble period due to inelasticity of the mantle. The right ordinate gives the percentage with respect to the total dissipation rate (solid Earth and ocean) of 2.7×10^{19} erg/s. The results are based on different Q models for the mantle (▲ LMS, □ $Q = 450 =$ const, ■ SL8, × SL3, ○$Q = 170 =$ const, △ PREM) and on different degrees α of frequency dependence of $Q(Q \sim \omega^{\alpha})$. The striped line marks the average observed wobble period and the corresponding result for $\dot{E}_{M2}$. The other vertical and horizontal lines indicate the spread of observations for the wobble period and the range of $\dot{E}_{M2}$ values, respectively, which is in agreement with these observations.

static. Correspondingly, the unrelaxed tidal Love number on an oceanless spherical elastic Earth is $k = k_0 = 0.3009$ (see Dahlen, 1976).

It becomes clear from Figs. 5–7 that for a given α value the parameters of tidal friction are linearly related to the wobble period, no matter which Q model was used or whether the reference period for the corresponding Q model is 30 s or as short as 1 s. Furthermore, in the cases of the apparent tidal dissipation function $(Q_k^*)^{-1}$ (Fig. 6) and the solid Earth dissipation rate $\dot{E}_{SE}$ (Fig. 7), the regression lines for different α values are very near to each other and may be roughly approximated by one and the same line. Thus, if the observed Chandler wobble period was known, one could directly determine the apparent tidal $Q = Q_k^*$ and the solid Earth tide dissipation rate $\dot{E}_{SE}$ from Figs. 7 and 8, respectively. The results would be only slightly dependent on α as long as $\alpha \leqslant 0.3$ and practically independent of α if $\alpha \leqslant 0.2$.

Observations of the Chandler wobble period are put together in Table IV. The spread of recent observations (since 1968) is less than 3 days. All observed periods lie in-between $T = 433.3$ (Wilson and Vicente, 1980) and $T = 436$ days (Ooe, 1978). This corresponds to a lengthening of the wobble period between 6.6 and 9.3 days.

TABLE IV. Recent observations of the Chandler wobble period

Source	Wobble period [days]
Jeffreys (1968)	434.3 ± 2.2
Currie (1974, 1975)	434.1 ± 1.0
Wilson and Haubrich (1976)	435.2 ± 2.6
Ooe (1978)	436 ± 2
Wilson and Vicente (1980)	433.3 ± 3.1
Höpfner (1983)	435.5 ± 0.7

The spread of the observed Chandler wobble period is also indicated in Figs. 5–7. It is immediately clear that $\alpha \geqslant 0.3$ is in gross disagreement with the observed wobble period. Even the (somewhat unrealistic) high Q model LMS would still give a wobble period of around 452 days if $\alpha = 0.3$. This is about 17 days too high. Realistic changes in the reference period for model LMS would not be sufficient to bring the theoretical period down to the observed one. There is no realistic Q model together with a realistic reference period that would explain the observed wobble period. Thus $\alpha \geqslant 0.3$ can be ruled out. For the same reason even $0.2 < \alpha < 0.3$ seems to be improbable, although not impossible. Both thresholds for α are indicated in the figures.

The main results which can be concluded from Figs. 5–7 are as follows:

(1) Dispersion of tidal Love number k

Maximum dispersion: $\Delta k/k_0 \leqslant 3\%$

Most probable dispersion: $\Delta k/k_0 \approx 2\%$

Minimum dispersion: $\Delta k/k_0 \geqslant \begin{cases} 0.8\% \ (\alpha = 0.2) \\ 0.5\% \ (\alpha = 0.3) \end{cases}$

(2) Apparent tidal $Q = Q_k^*$

Minimum Q_k^*: $Q_k^* \geqslant 220$

Most probable Q_k^*: $Q_k^* \approx 270$

Maximum Q_k^*: $Q_k^* \leqslant \begin{cases} 340 \ (\alpha = 0.2) \\ 420 \ (\alpha = 0.3) \end{cases}$

(3) Solid Earth tide dissipation rate $\dot{E}_{SE}$

Maximum dissipation: $\dot{E}_{SE} \leqslant 1.4 \times 10^{18}$ erg/s $\approx 5.2\%$

Most probable dissipation: $\dot{E}_{SE} \approx 1.2 \times 10^{18}$ erg/s $\approx 4.4\%$

Minimum dissipation: $\dot{E}_{SE} \geqslant \begin{cases} 0.9 \times 10^{18} \text{ erg/s} \approx 3.3\% \ (\alpha = 0.2) \\ 0.7 \times 10^{18} \text{ erg/s} \approx 2.6\% \ (\alpha = 0.3) \end{cases}$

The second quantity is the percentage with respect to the total dissipation rate (solid Earth and ocean) of 2.7×10^{19} erg/s.

The corresponding results for the solid Earth phase delay ε can be directly deduced from the apparent tidal Q by making use of Eq. (30):

(4) Solid Earth phase delay ε

Maximum ε: $\varepsilon \leqslant 0.26°$

Most probable $\varepsilon : \varepsilon \approx 0.21°$

Minimum $\varepsilon : \varepsilon \geqslant \begin{cases} 0.17° \ (\alpha = 0.2) \\ 0.14° \ (\alpha = 0.3) \end{cases}$

All results are valid for the M_2 tide. Other numerical values for the tidal friction parameters than those above would be in gross disagreement with the observed Chandler wobble period.

VI. Discussion and Conclusions

We have examined two different ways of determining parameters of tidal friction in the solid Earth:

(i) by subtracting the effect of ocean tides on close Earth satellites from the observed tidal perturbations of their orbits. The remaining perturbations were attributed to solid Earth tides.

(ii) by making use of the linear relationship between the poorly known parameters of tidal friction and the well-known Chandler wobble period. The linear relationship was found by means of model calculations based on frequency dependent Q models for the Earth's mantle.

The first way turned out to be unsuccessful for several reasons. One of them is the relatively large error in the satellite solutions. For the total dissipation rate, for instance, the error comes up to $\pm 1.3 \times 10^{18}$ erg/s which is about 5% of the total rate. This compares to the solid Earth dissipation rate of only 1.2×10^{18} erg/s as estimated by means of the second method.

The situation does not improve, if instead of the satellite solutions, we take the results of lunar laser ranging, although in this field similar important progress as in the field of satellite observations has been made recently. Lunar laser ranging observations have provided good estimates of the Moon's mean motion $\dot{n}$ which is directly proportional to the total tidal dissipation rate. However, the errors in this method are also not less than 5% at present. Calame and Mulholland (1978) give $\dot{n} = -24.6 \pm 1.6''/100\ \text{yr}^2$; Williams *et al.* (1978), $\dot{n} = -23.8 \pm 4''/100\ \text{yr}^2$; Ferrari *et al.* (1980), $\dot{n} = -23.8 \pm 3.1''/100\ \text{yr}^2$; and Dickey *et al.* (1983), $\dot{n} = -25.2 \pm 1.2''/100\ \text{yr}^2$. These values include all tidal waves, not only M_2. Significant improvements in the laser ranging as well as in the satellite solutions are necessary before they can be used for determining the relatively small amount of solid Earth tide dissipation. However, improvements in the laser ranging and satellite solutions alone are not enough to separate the small amount of solid Earth dissipation rate from the total one, because more crucial than the errors mentioned before are the shortcomings in the ocean tide models. They introduce uncertainties in the value of the solid Earth dissipation rate which may be several times the solid Earth dissipation rate itself.

Finally, the unreliability of the combined satellite-ocean tide solutions is clearly demonstrated by determining the solid Earth phase delay ε and the Love number k separately and plotting the results in an ε-k plot. None of the existing solutions are in agreement with the principle of causality which requires certain relationships between k and ε. Moreover, all the solutions are in gross disagreement with the observed Chandler wobble period.

Because of these obvious shortcomings in the combined satellite–ocean tide solutions, we have concentrated on the before mentioned second method for the determination of solid Earth tidal friction parameters. The model calculations inherent in this method were based on frequency dependent Q models for the Earth's mantle, and they relate the tidal friction parameters to the Chandler wobble period. Using the observed wobble period yields estimates of the tidal Love number dispersion $\Delta k/k_0$ of the apparent tidal $Q = Q_k^*$, of the solid Earth phase delay ε and of the solid Earth tide dissipation rate $\dot{E}_{SE}$, which are much more accurate than those obtained from any other method before. However, the method used here requires

(i) that in between the M_2 tide and the Chandler wobble period the frequency dependence of the quality factor Q is smooth and can be approximated by $Q \sim \omega^\alpha$, and
(ii) that the pole tide is not a dynamic but a static one.

It is not known whether these requirements are fulfilled. However, the first one is probably not too restrictive, because smoothness of Q is only required for a frequency band of less than three decades in width. Considering that a single Debye peak (one single relaxation mechanism!) has already a width at half maximum of more than one decade, it would not need many different relaxation mechanisms to obtain a Q which is approximately smooth in ω within this frequency band. However, deviations from the required smoothness cannot be ruled out.

Dickmann (1979) has shown that the effects of a non-equilibrium pole tide on the Chandler wobble period would be greater than equilibrium theory predicts. In this case part of the observed lengthening of the wobble period would not be due to inelasticity in the Earth's mantle but instead due to the non-equilibrium pole tide. As a consequence our numerical values for the dispersion of the tidal Love number k, for the tidal dissipation function $(Q_k^*)^{-1}$, for the solid Earth phase delay ε, as well as for the solid Earth tidal dissipation rate $\dot{E}_{SE}$ would come out smaller than those obtained with the equilibrium assumption. However, a dynamic pole tide would not affect the main conclusion of this chapter:

> The solid Earth M_2 tide dissipation rate is less than 1.4×10^{18} erg/s, i.e. less than 5.2% of the total M_2 tide dissipation rate (solid Earth and ocean), no matter whether the pole tide is static or dynamic and no matter what the degree of frequency dependence of Q in the Earth's mantle

looks like, as long as the frequency dependence is fairly smooth between the M_2 tide and the Chandler wobble period.

A dynamic pole tide would only reduce but not increase the above value.

Acknowledgments

I would like to thank Kathleen Helbig for critically checking and typing the manuscript. I also wish to thank Dr J. Hinderer for pointing out an error concerning the formulation of Eq. (26).

References

Anderson, D. L. and Given, J. W. (1982). Absorption band Q model for the Earth. *J. Geophys. Res.* **87**, 3893-3904.

Anderson, D. L. and Hart, R. S. (1978a). Attenuation models for the Earth. *Phys. Earth Planet. Int.* **16**, 289-306.

Anderson, D. L. and Hart, R. S. (1978b). The Q of the Earth. *J. Geophys. Res.* **83**, 5869-5882.

Anderson, D. L. and Minster, J. B. (1979). The frequency dependence of Q in the Earth and implications for mantle rheology and Chandler wobble. *Geophys. J. R. Astron. Soc.* **58**, 431-440.

Berckhemer, H., Auer, F. and Drisler, J. (1979). High-temperature anelasticity and elasticity of mantle peridotite. *Phys. Earth Planet. Int.* **20**, 48-59.

Berckhemer, H., Kampfmann, W. and Aulbach, E. (1982). Anelasticity and elasticity of mantle rocks near partial melting. *In* "High Pressure Researches in Geoscience" (W. Schreyer, ed.), pp. 113-132. E. Schweizerbart'sche Verlagsbuchhandlung, Stuttgart.

Calame, O. and Mulholland, D. (1978). Lunar tidal acceleration determined from laser range measures. *Science* **199**, 977-978.

Cazenave, A. and Daillet, S. (1981). Lunar tidal acceleration from Earth satellite orbit analyses. *J. Geophys. Res.* **86**, 1659-1663.

Currie, R. G. (1974). Period and Q_W of the Chandler wobble. *Geophys. J. R. Astron. Soc.* **38**, 179-185.

Currie, R. G. (1975). Period, Q_p and amplitude of the pole tide. *Geophys. J. R. Astron. Soc.* **43**, 73-86.

Dahlen, F. A. (1976). The passive influence of the oceans upon the rotation of the Earth. *Geophys. J. R. Astron. Soc.* **46**, 363-406.

Daillet, S. (1978). Determination par observation des trajectoires de satellites artificiels des andes semi-diurnes de la marée océanique. Application au calcul de l'accélération séculaire de la Lune. *Ann. Géophys.* **34**, 79-87.

Dickey, J. O., Newhall, X. X., Williams, J. G. and Yoder, C. F. (1983). Modulation of the lunar tidal accelration. *Abstr. AGU Spring Meet., EOS* **64**, 204.

Dickman, S. R. (1979). Consequences of an enhanced pole tide. *J. Geophys. Res.* **84**, 5447-5456.

Doornbos, D. J. (1983). Observable effects of the seismic absorption band in the Earth. *Geophys. J. R. Astron. Soc.* **75**, 693-711.

Dziewonski, A. M. and Anderson, D. L. (1981). Preliminary reference Earth model. *Phys. Earth Planet. Int.* **25**, 197-356.

Estes, R. (1977). A computer software system for the generation of global ocean tides including self-gravitation and crustal loading effects. Rep. X-290-77-82-GSFC. Goddard Space Flight Center, Greenbelt, Maryland.

Felsentregger, T. L., Marsh, J. G. and Williamson, R. G. (1978a) Tidal perturbations on the satellite 1967-92A. *J. Geophys. Res.* **83**, 1837–1842.

Felsentregger, T. L., Marsh, J. G. and Williamson, R. G. (1978b). M_2 ocean tide parameters and the deceleration of the Moon's mean longitude from satellite orbit data. *NASA Tech. Mem. 79571.*

Ferrari, A. J., Sinclair, W. S., Sjorgren, W. L., Williams, J. G. and Yoder, C. F. (1980). Geophysical parameters of the Earth–Moon system. *J. Geophys. Res.* **85**, 3939–3951.

Goad, C. and Douglas, B. (1978). Lunar tidal acceleration obtained from satellite-derived ocean tide parameters. *J. Geophys. Res.* **83**, 2306–2310.

Höpfner, J. (1983). Umfassende Analyse der Breitenbestimmungen des geodätischastronomischen Observatoriums Potsdam. Ph. D. thesis, Techn. Univ. Dresden.

Jeffreys, H. (1968). The variation of latitude. *Mon. Not. R. Astron. Soc.* **141**, 255–268.

Kagan, B. A. (1981). Dissipation of gravitational tidal energy in the atmosphere. *Izv. Acad. Sci. USSR Atmos. Oceanic Phys.* (*Engl. Transl.*) **17**, 896–898.

Kjartansson, E. (1979). Constant *Q*-wave propagation and attenuation. *J. Geophys. Res.* **84**, 4737–4748.

Lambeck, K. (1975). Effects of tidal dissipation in the oceans and on the Moon's orbit and the Earth's rotation. *J. Geophys. Res.* **80**, 2917–2925.

Lambeck, K. (1977). Tidal dissipation in the oceans: Astronomical, geophysical and oceanographic consequences. *Philos. Trans. R. Soc. London, A* **287**, 545–594, 1977.

Liu, H. P., Anderson, D. L. and Kanamori, H. (1976). Velocity dispersion due to anelasticity; implications for seismology and mantle composition. *Geophys. J. R. Astron. Soc.* **47**, 41–48.

Lundquist, G. M. and Cormier, V. C. (1980). Constraints on the absorption band model of *Q. J. Geophys. Res.* **85**, 5244–5256.

Minster, J. B. (1980). Anelasticity and attenuation. *Phys. Earth's Interior*, LXXVIII Corso. Soc. Italiana di Fiscia, Bologna, Italy, 153–211.

Müller, G. (1983). Rheological properties and velocity dispersion of a medium with power-law dependence of *Q* on frequency. *J. Geophys.* **54**, 20–29.

Okubo, S. (1982). Theoretical and observed *Q* of the Chandler wobble—Love number approach. *Geophys. J. R. Astron. Soc.* **71**, 647–657.

Ooe, M. (1978). An optimal complex AR,MA model of the Chandler wobble. *Geophys. J. R. Astron. Soc.* **53**, 445–457.

Parke, M. E. (1982). O1, P1, N2 models of the global ocean tide on an elastic Earth plus surface potential and spherical harmonic decompositions for M2, S2 and K1. *Mar. Geod.* **6**, 35–81.

Platzman, G. W. (1984). Planetary energy balance for tidal dissipation. *Rev. Geophys. Space Phys.* **22**, 73–84.

Schwiderski, E. W. (1983a). Atlas of ocean tidal charts and maps. I, The semidiurnal principle lunar tide M_2. *Mar. Geod.* **6**, 219–265.

Schwiderski, E. W. (1983b). The braking of the Earth's rotation. Naval Surface Weapons Center, Dalhgren, Virginia.

Smith, S. W. (1972). The anelasticity of the mantle. *In* "The Upper Mantle" (A. R. Ritsema, ed.). *Tectonophysics* **13**, 601–622.

Smith, M. L. and Dahlen, F. A. (1981). The period and *Q* of the Chandler wobble. *Geophys. J. R. Astron. Soc.* **64**, 223–281.

Williams, J. G., Sinclair, W. S. and Yoder, C. F. (1978). Tidal acceleration of the Moon. *Geophys. Res. Lett.* **5**, 943–946.

Wilson, C. R. and Haubrich, R. A. (1976). Meteorological excitation of the Earth's wobble. *Geophys. J. R. Astron. Soc.* **46**, 707–743.

Wilson, C. R. and Vicente, R. O. (1980). An analysis of the homogeneous ILS polar motion series. *Geophys. J. R. Astron. Soc.* **62**, 605–616.

Zschau, J. (1977). Phase shifts of tidal sea load deformations of the Earth's surface due to low viscosity layers in the interior. *Proc. Int. Symp. Earth Tides, 8th Bonn,* 372–398. Schweizerbart'sche Verlagsbuchhandlung, Stuttgart.

Zschau, J. (1978). Tidal friction in the solid Earth: Loading tides versus body tides. *In* "Tidal Friction and the Earth's Rotation" (P. Brosche, and J. Sündermann, eds.), pp. 62–94. Springer-Verlag, New York.

Zschau, J. (1979). Auflastgezeiten. Habil. thesis, Univ. Kiel.

Zschau, J. (1983). Rheology of the Earth's mantle at tidal and Chandler wobble periods. *Proc. Int. Symp. Earth Tides, 9th, Bonn,* 605–630. Schweizerbart'sche Verlagsbuchhandlung, Stuttgart.

Part 5

Dynamics of the Lithosphere

5.1 Use of Satellite Altimetry in Studies of the Oceanic Lithosphere

A. Cazenave[1] and E. A. Okal[2]

[1] *Groupe de Recherches de Géodésie Spatiale, Centre National d'Etudes Spatiales, Toulouse, France and* [2] *Department of Geological Sciences, Northwestern University, Evanston, Illinois, USA*

I. Introduction

Over the last decade, the collection of marine geoid data over all oceanic areas by the GEOS-3 and SEASAT altimeter satellites has triggered a renewal of interest in the use of gravity information in marine geophysics, and has motivated a number of new studies of the oceanic lithosphere and upper mantle. Several maps of the sea surface height have been produced on both global and regional scales, based either on GEOS-3, SEASAT or both datasets (e.g., Rapp, 1983; Marsh *et al.*, 1985, 1986). In order to obtain maps of the mean sea surface heights over a regular grid from individual

SPACE GEODESY AND GEODYNAMICS
ISBN 0-12-057840-9

satellite profiles, a number of interpolation methods are used after minimizing by least squares procedures the differences at altimeter profile intersections which result from inadequate orbital modeling (e.g. Balmino *et al.*, 1979; Marsh *et al.*, 1980; Rapp, 1983). The raw sea surface topography contains a component of oceanic origin (tides, currents, eddies, etc.), which is known to be less than 1 m in amplitude. Therefore, the sea surface maps closely represent the geoid, and can be used for investigations of density differences within the Earth. The so-called crossing arcs adjustment procedures, applied to remove orbital errors, also remove oceanographic effects so that the precision of the geoid surface approaches that of the measurements themselves, i.e. ~10–20 cm for SEASAT.

As we already know from global geopotential solutions derived from satellite orbit perturbations (e.g. Lerch *et al.*, 1981; Reigber *et al.*, 1985) the geoid is dominated by long anomalies with wavelengths of several thousand kilometers, poorly correlated to surface tectonic features, and usually attributed to density variations in the lower mantle or at the core-mantle interface. The marine geoid presented in Fig. 1 is one of the latest solutions obtained by Marsh *et al.* (1986) with SEASAT data over a $0.5° \times 0.5°$ grid, and confirms these characteristics. In addition, and unlike global geopotential solutions, the altimeter geoid clearly depicts numerous bathymetric features such as deep-sea trenches, volcanic chains or major fracture zones. In order to be used successfully for geophysical interpretation, these tectonic signatures can be enhanced by removing unrelated longer-wavelength geoid undulations; this is currently done by subtracting the cumulative long-wavelength contributions of a global geopotential field; since spherical harmonic expansion on a sphere is analogous to Fourier spectral expansion on a plane, each harmonic of degree n corresponds to a particular wavelength $\simeq 2\pi R/n$ (R being the Earth's radius). Subtracting a global geopotential field up to degree 10 from the altimeter geoid therefore provides a residual geoid surface with undulations of wavelength $\lambda < 4000$ km, on which tectonic signatures are now very apparent, as shown in Figs. 2 and 3.

As evidenced in Fig. 2, the main characteristic of the residual geoid is a clear three-dimensional pattern of geoid highs and lows, with typical wavelengths λ ~2000–3000 km and typical amplitudes 5–10 m. These undulations have been attributed to small-scale convection in the upper mantle (e.g. McKenzie *et al.*, 1980; Marsh *et al.*, 1984; Watts *et al.*, 1985). At smaller scales, marine geoid data offer important constraints on the physical structure of the oceanic lithosphere and have been extensively used in recent studies in addition to bathymetry, gravity, sea floor age, seismic or heat flow data. While it is well known that inversion of gravity for internal density distribution is highly non-unique, it appears that if tectonic structures are isostatically compensated (locally or regionally, statically or dynamically),

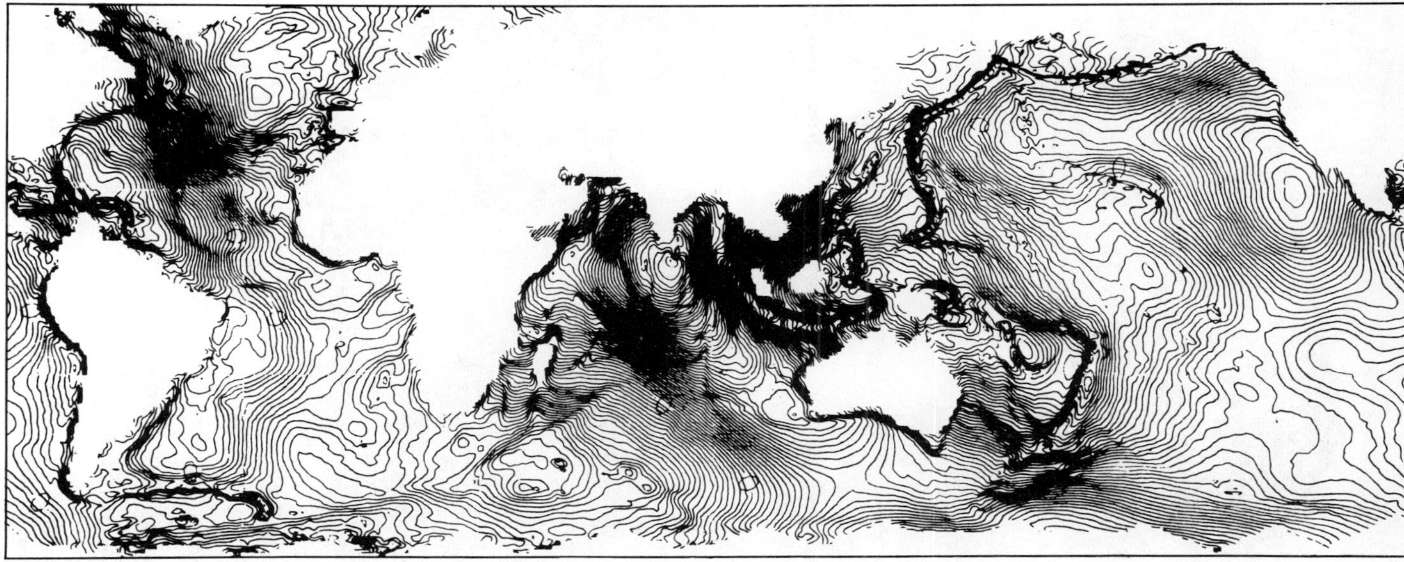

Fig. 1. Marine geoid derived from the SEASAT altimeter data. [After Marsh *et al.* (1985b).]

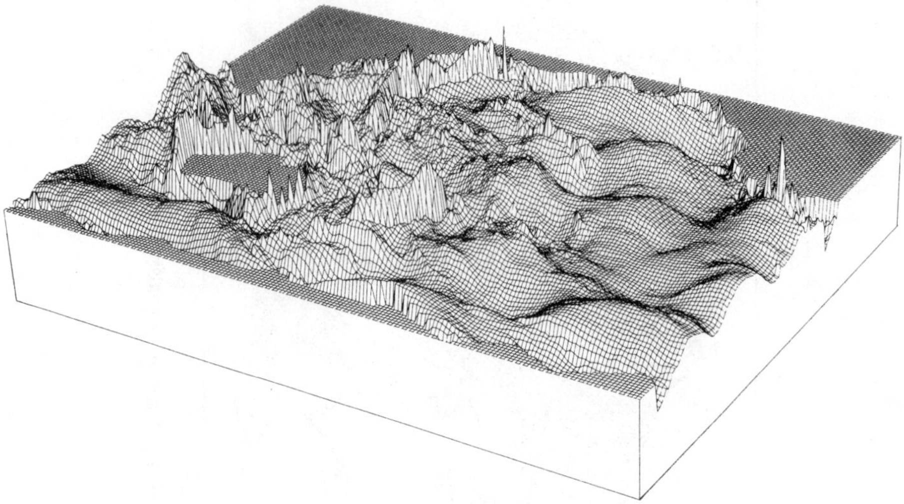

Fig. 2. Three-dimensional view of the geoid over the Pacific Ocean looking northeast, after removal of wavelengths greater than 4000 km.

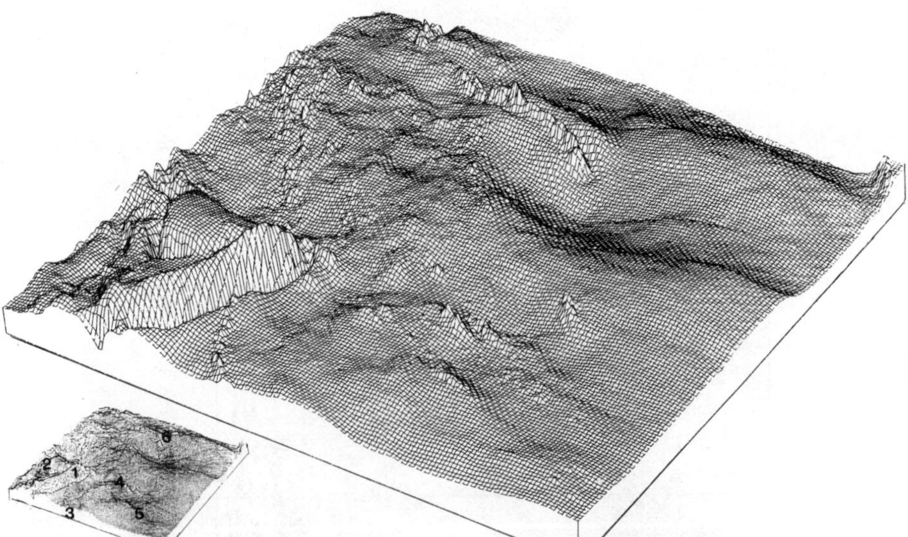

Fig. 3. Three-dimensional view of the geoid over a close-up section of the Pacific Ocean, after removal of the degree 10 geoid. Inset at lower left identifies: (1) Tonga Trench, (2) Vanuatu Trench, (3) Louisville Ridge, (4) Society Islands Chain (Tahiti), (5) Austral Islands Chain, and (6) Hawaiian Islands Chain.

the corresponding geoid anomalies give strong constraints on any proposed theoretical model. This results partly from the fact that each kind of tectonic feature gives a specific signature in the geoid; thus, on the basis of generally accepted models, analyses of the shape, amplitude and wavelength of the associated geoid anomalies can yield information on the internal density structure as well as on its state of isostatic compensation (e.g. Turcotte, Chapter 5.4). The following is a short description of the typical geoid signatures of a few major tectonic features.

1. Volcanic Chains and Isolated Seamounts

These give large positive geoid anomalies, up to 5–6 m amplitude and ~150-km wide, sometimes flanked by small adjacent minima (see Fig. 4). These anomalies result from two opposite effects: the positive geoid anomaly due to the topography and the negative anomaly due to its (often partial) isostatic compensation inside the lithosphere.

2. Symmetrically Spreading Ridges

Over symmetrically spreading ridges, the geoid height anomaly results from the cooling of the lithosphere and decreases from the ridge crest to old basins as sea floor age increases. Typical geoid height changes are 5–10 m over distances of 1000–2000 km. The observed shape of the geoid has been used as a constraint on the cooling models of the lithosphere or on the magnitude of the plates' driving forces (Sandwell and Schubert, 1980; Parsons and Richter, 1980). Figure 5 shows observed geoid height over the East Pacific Rise, decreasing away from the ridge crest with a symmetric behavior. In general, slow-spreading ridges, such as the Mid-Atlantic Ridge, have much stronger geoid signatures than fast-spreading ones, such as the East Pacific Rise.

3. Fracture Zones

Over fracture zones, geoid anomalies show a step-like signal of 1–3 m amplitude, downward from the younger, shallower side to the older, deeper side of the fracture zone. Figure 6 shows examples of this geoid step. It occurs because the portions of the plate on each side of the fracture zone have different thicknesses and density structures, as predicted by thermal models of the evolution of the oceanic lithosphere. Several analyses of geoid

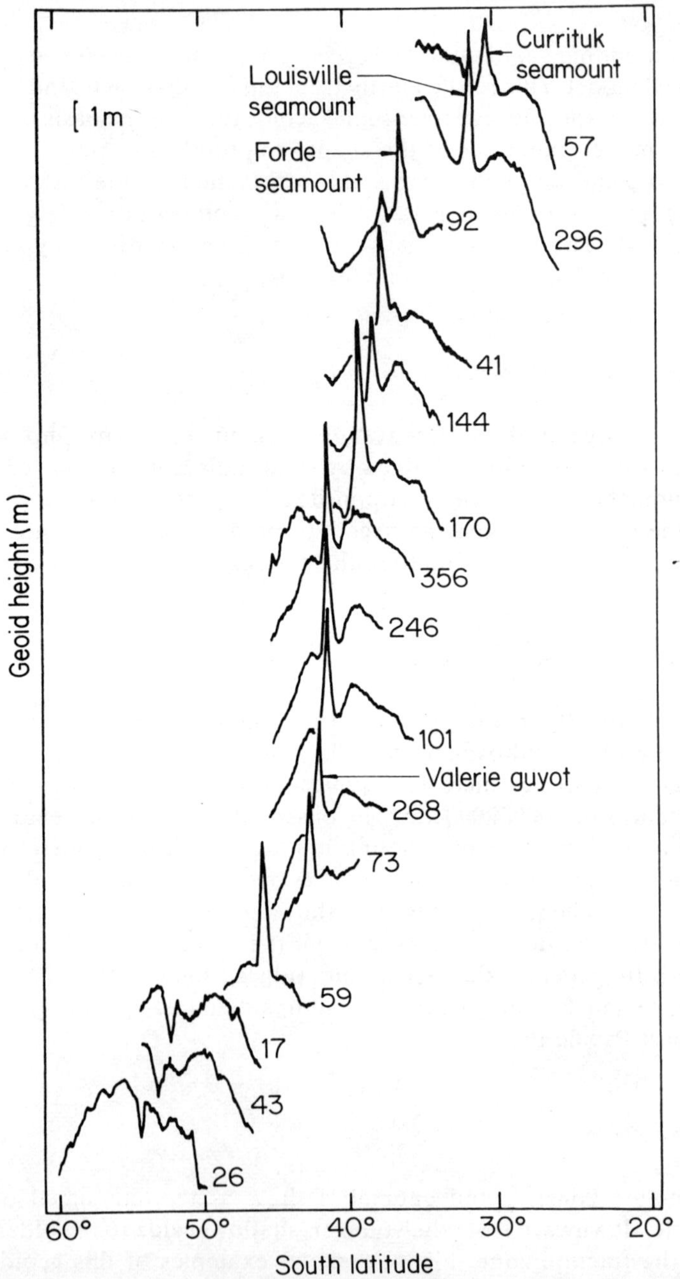

Fig. 4. Descending SEASAT profiles crossing the Louisville seamounts and the Udintsev and Tharp fracture zones. Geoid anomalies are in m: profiles are arranged by increasing southward latitudes.

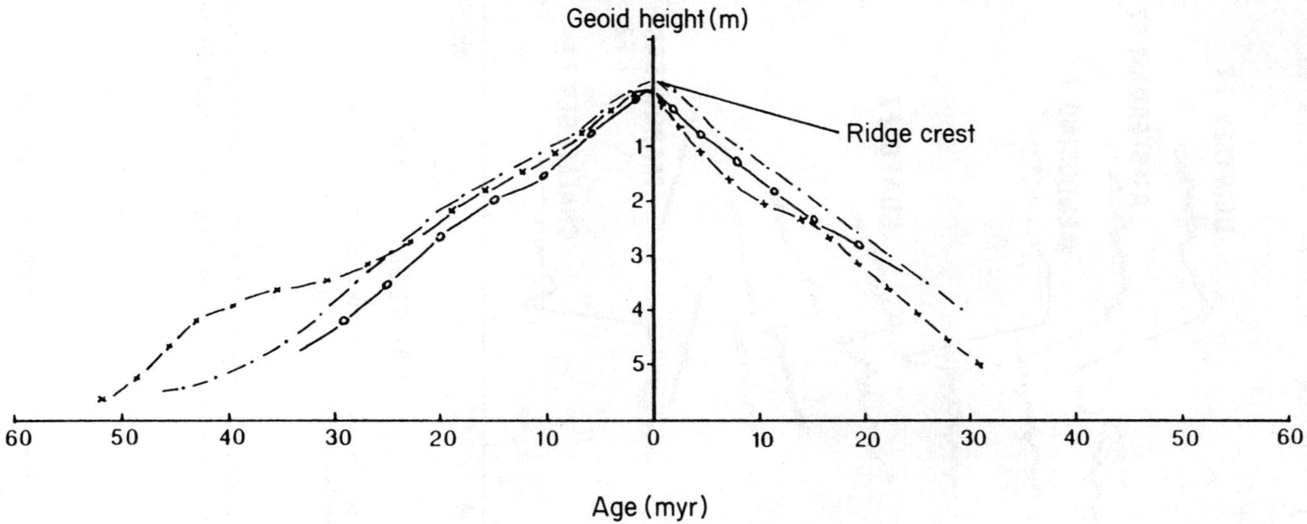

Fig. 5. Geoid profiles across the East Pacific Rise. These are cross-sections perpendicular to the rise, which have been reconstructed from the three-dimensional SEASAT geoid computed over this region.

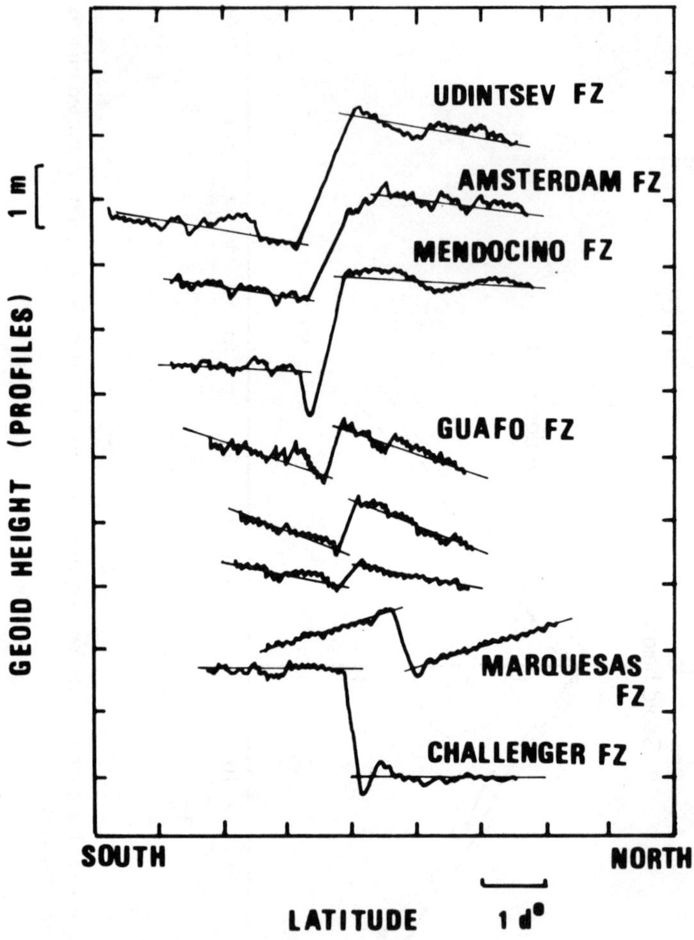

Fig. 6. SEASAT geoid profiles across a selection of fracture zones. Geoid offsets Δh are measured between the parallel solid lines.

anomalies at fracture zones have been conducted to estimate the parameters entering in the models of lithospheric cooling, as we will discuss in more detail in Section III.

4. Deep-sea Trenches

Deep-sea trenches also give rise to a typical geoid signature (Fig. 7). Directly over the trench axis, the geoid presents a narrow trough 5–20 m in depth

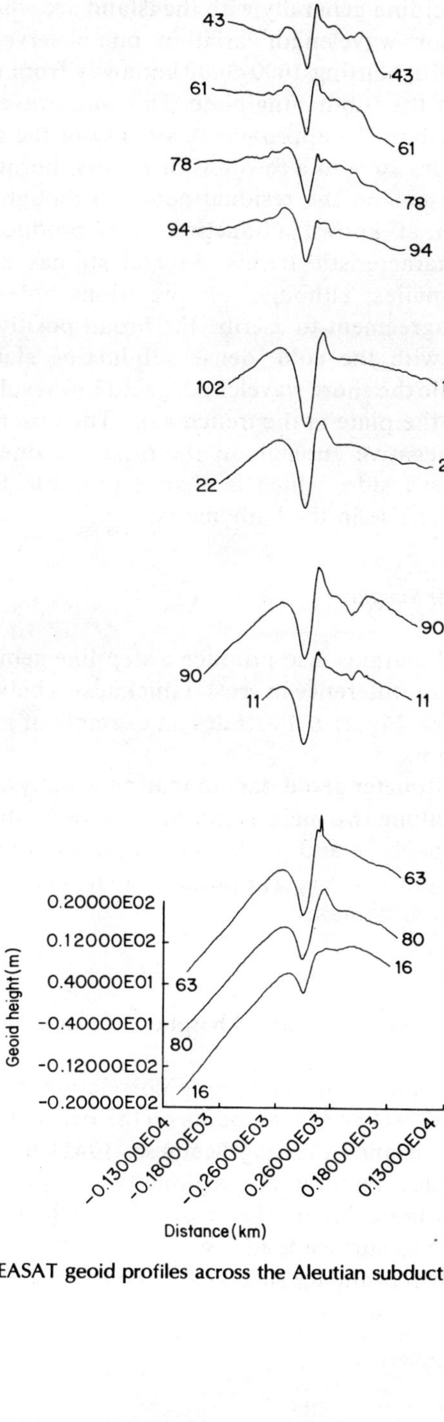

Fig. 7. SEASAT geoid profiles across the Aleutian subduction zone.

and 100–400 km in width. From the trench to the landward side, there is a steep increase coinciding generally with the island arc when present. Superimposed on this short-wavelength variation, one observes a broad positive anomaly in the geoid, starting 1000–3000 km away from the trench axis on the seaward side of the subducting plate. This long-wavelength geoid high reaches its maximum in the approximate vicinity of the trench and typical amplitudes vary from 10 m to a few tens of meters; because of the filtering procedure, it is absent on the residual geoid. Although differing in their detailed expression, all known subduction zones produce geoid anomalies with the above characteristic trends. Several studies have attempted to explain these anomalies; although interpretations differ in some details, there is a general agreement to ascribe the broad positive high to thermal effects associated with the cold, dense subducting slab penetrating the asthenosphere, while the short-wavelength geoid low results from the flexure and subduction of the plate at the trench axis. The observed asymmetry in the shape of the negative anomaly at the trench is due to flexure of the plate on the seaward side, which is also responsible for the occasional presence of an outer rise in the bathymetry.

5. Passive Continental Margins

Passive continental margins also produce a step-like geoid signature which is due in this case to a difference in crustal thicknesses between the continental and oceanic sides. Figure 8 illustrates an example of geoid profile across a continental margin.

Application of altimeter geoid data to marine geophysics have up to now been concentrated along two main avenues: (1) investigation of the rheology of the oceanic lithosphere and its thermal evolution and (2) prediction of bathymetry in uncharted areas. The following sections review some recent work in each of these domains.

II. Geoid Anomalies Over Seamounts: Lithospheric Flexure

We have seen that there exists a strong correlation between local geoid anomalies and the presence of seamounts on the ocean floor. Earlier studies over the Hawaiian islands (Vening Meinesz, 1941) had shown that such localized topographic features are regionally, rather than locally, compensated by lithospheric flexure. Later on, the mechanical response of the oceanic lithosphere to surface loads (such as isolated seamounts, volcanic island chains, sediments and ice sheets) or to bending stresses (in subduction

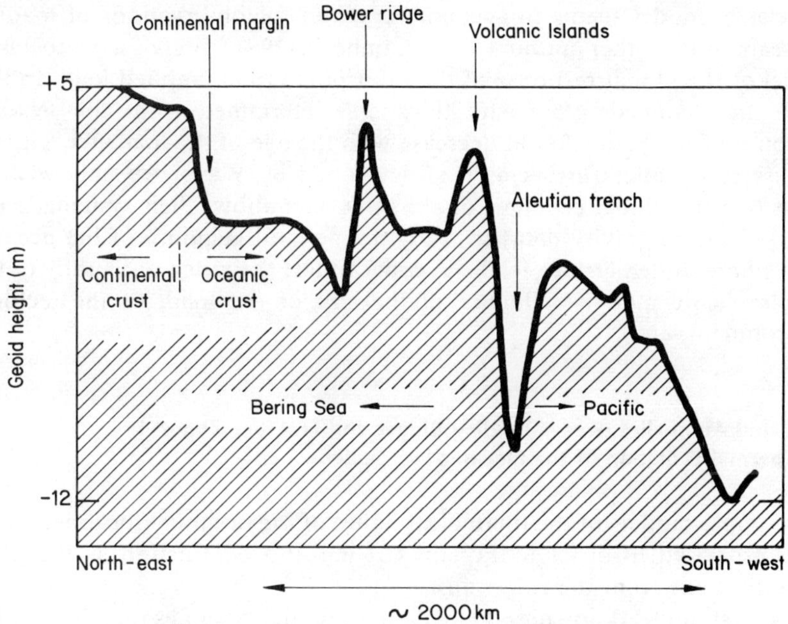

Fig. 8. SEASAT geoid profile across the Bering Sea continental margin.

areas) was studied extensively on the basis of simple models of elastic or viscoelastic flexure, by using data from bathymetry, uplift or subsidence, gravity and more recently, the geoid. The most widely applied model is the elastic one, based on the equilibrium theory of deformation of thin elastic loaded plates overlying a weaker fluid medium. The parameter that characterizes the flexure of the plate is the flexural rigidity D, related to the elastic thickness H by:

$$D = EH^3/[12(1-\nu^2)]$$

where E and ν are Young's modulus and Poisson's ratio, respectively.

The comparison of observed data (gravity, geoid height, uplift, etc.) with theoretical estimates allows the derivation of best-fitting values for D, and thus for H. Watts *et al.* (1982) compiled all available elastic thicknesses, estimated by various authors at different oceanic locations; they showed that the elastic thickness of the oceanic lithosphere increases linearly with the square root of the age of the plate at the time of loading, and concluded that the upper layer of the lithosphere is reasonably well described by an elastic rheology for geological time scales (see also Watts, 1978). Although the elastic model seems to account very well for observations of loading by seamounts, other authors (e.g. Lambeck, 1981) prefer a viscoelastic model of the global response of the lithosphere to an applied load. In this case, the "equivalent" elastic thickness determined from the flexural response of the plate should decrease with the age of the load. We will not discuss this matter further as it is developed by Watts; we only wish to stress that the global coverage of altimeter data allows their systematic use (unlike marine gravity data) to study the flexural response of the oceanic lithosphere under surface loads, in order to test the range of validity of the simple elastic model (influence of the size of the load, of the tectonic environment, etc.).

III. Geoid Anomalies over Spreading Ridges and Fracture Zones: Thermal Evolution of the Oceanic Lithosphere

Two simple models of the thermal structure of the oceanic lithosphere have been proposed in order to explain cooling and contraction as the plate expands away from the ridge crest.

The first model, known as the half-space model, assumes that the cooling plate can be treated as a thermal boundary layer of a large-scale mantle convection cell (Turcotte and Oxburgh, 1967). The base of the plate is then defined as the solid–liquid phase boundary of the material, and thus follows an isotherm (Parker and Oldenburg, 1973; Oldenburg, 1975). A further assumption is the absence of lateral conduction in the plate. In this model, both plate thickness and depth to the seafloor increase linearly with the square root of age, while heat flow decreases as the inverse square root of age.

The second model, known as the plate model, assumes a plate of nearly constant thickness moving away from the ridge crest at a constant velocity. This kind of model was originally developed by McKenzie (1967) to account for heat flow observations, and several variations of the model have later been proposed (e.g. Bottinga, 1974; Bottinga and Allègre, 1976).

The two models predict almost the same behavior for depth to the seafloor, heat flow and geoid in young oceans, but in oceans older than about 70 Ma, the plate model predicts a general tapering of the cooling with depth to the seafloor and heat flow tending asymptotically toward a constant value. This asymptotic behavior is indeed generally observed in the older ocean basins, and thus the plate model is generally preferred to the half-space model. Figure 9 presents the observed ocean depth as a function of age (after Parsons and Sclater, 1977). The solid curve corresponds to the predicted variation in the case of the plate model while the

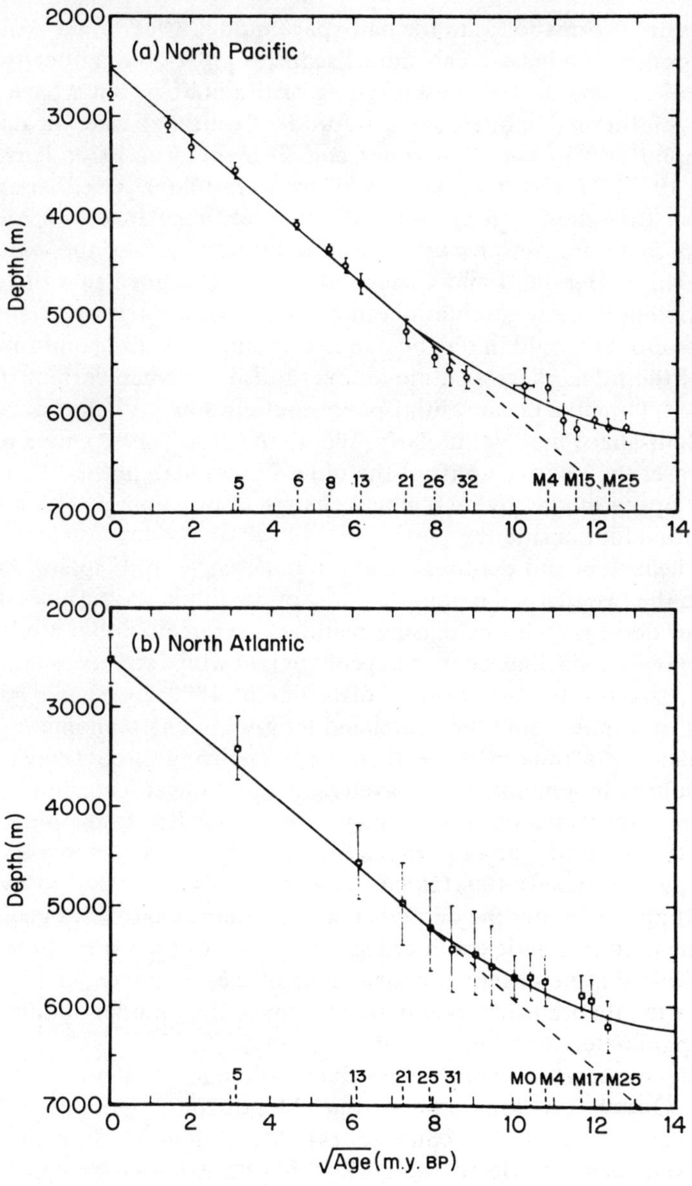

Fig. 9. (a) Plot of mean depth in the North Pacific versus square root of age. Numbers at bottom of the figure refer to selected Cenozoic and Mesozoic magnetic anomalies. (b) Same plot for the North Atlantic. Open circles with error bars give the mean depth and standard deviation, the solid curve gives the theoretical elevation for the plate model, and the dashed curves give the linear $t^{1/2}$ relation. [From Parsons and Sclater (1977).]

dashed curve corresponds to the half-space model. Clearly, the plate model fits the depth data better. This model requires an additional heat source in the mantle to counteract the thickening of the plate in older basins.

Tests of thermal models have traditionally been based on data from topography and heat flow [Sclater and Francheteau, 1970; Parsons and Sclater, 1977]. However, large areas of the ocean floor are either shallower or deeper than predicted by "normal" thermal subsidence, while available heat flow data are not always a sensitive discriminant of the two models. In particular, Heestand and Crough (1981) have argued that the tapering off of the bathymetry in older ocean basins (a major argument in favor of the plate model) could in fact be due to thinning, reheating and subsequent uplift of the lithosphere as it moves over hotspots; when corrected for this bias, they claim that oceanic lithosphere could follow the behavior predicted by the half-space model. Similarly, Wood and Yuen (1983) have proposed to interpret the relative uplift of the older lithosphere in the framework of the half-space model, by involving a change of mineralogy with a negative Clapeyron slope inside the plate.

Like heat flow and depth data, geoid data can in principle discriminate between the two thermal models: cooling of the lithosphere causes the geoid height to decrease with increasing plate age, symmetrically away from the ridge crest. The absolute change in geoid height with age, due to lithospheric cooling, is typically 5–10 m over distances of 1000–2000 km, and thus is difficult to isolate from other unrelated long-wavelength anomalies. On the other hand, at a fracture zone, there exists a strong discontinuity in plate age, resulting in a major short-wavelength geoid offset featuring a step-like signature, downward from the younger, shallower side to the older, deeper side, with amplitude ranging from a few tens of centimeters to a few meters over distances of only 100–150 km. The geoid offset Δh divided by the age offset Δt approximates the derivative with respect to age of the geoid height $h(t)$, due to lithospheric cooling. Thus, if the age of each side of the fracture zone is known, the observed dependence of $\Delta h/\Delta t$ on age can be used to compare the theoretical curves derived from cooling models, allowing some model parameters to be estimated.

We have analyzed geoid height data over many fracture zones of the Pacific Ocean (among others, the Mendocino, Udintsev, Menard, Marquesas, Challenger fracture zones). More than 15 different fracture zones have been considered and ~85 geoid offsets have been collected over a large area extending 60°S–40°N and 80°–160°W, and covering a range of plate ages of ~60 Ma. The quantity $\Delta h/\Delta t$ has been estimated and the observed dependence of $\Delta h/\Delta t$ on age has been derived. A semi-logarithmic plot of this relationship is shown in Fig. 10.

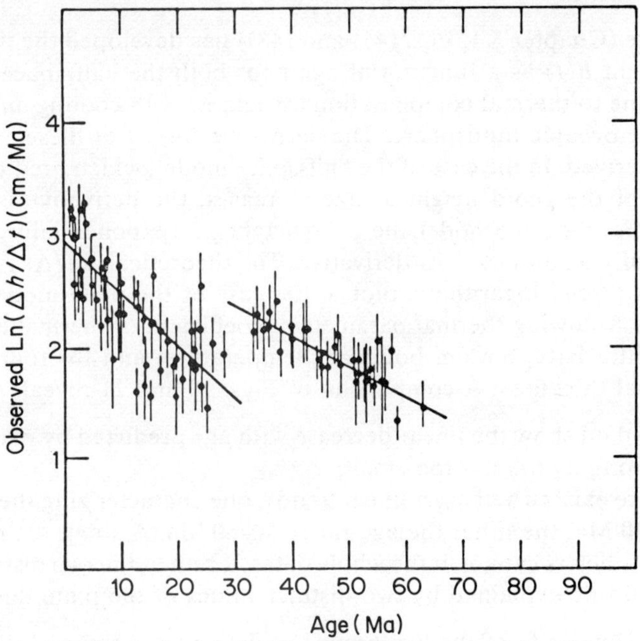

Fig. 10. Observed variation of the ratio of geoid offset to age offset ($\Delta h/\Delta t$) with increasing age.

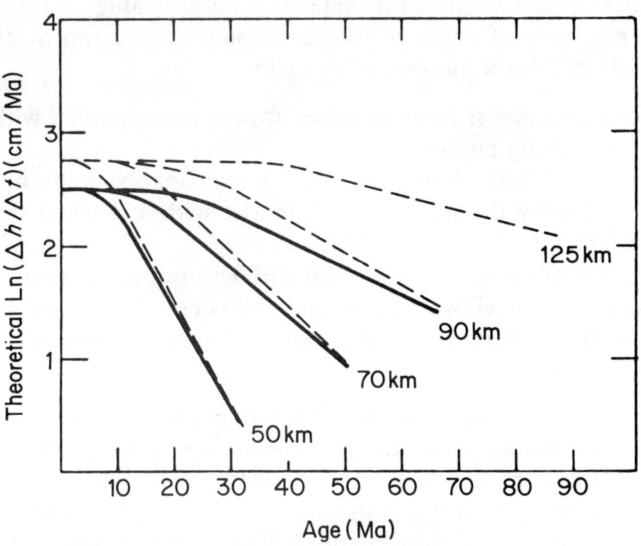

Fig. 11. Theoretical variations of $\text{Ln}(\Delta h/\Delta t)$ with age for the plate model, computed for several values of the thermal thickness H. Solid curves correspond to values of α, κ, T_m best-fitting the geoid data, while dashed curves correspond to values determined by Parsons and Sclater (1977); the top dashed curve, for $H = 125$ km, is their preferred model.

Turcotte (Chapter 5.4; Eqs. (43) and (48)) has developed the variation in geoid height $h(t)$ as a function of age t for both the half-space and plate models, due to thermal compensation associated with cooling and contraction of the oceanic lithosphere. The derivative $\Delta h/\Delta t$ of these expressions is easily derived. In the case of the half-space model, which predicts a linear decrease of the geoid height as age increases, the derivative is simply a constant. For the plate model, the geoid height is an exponentially decreasing function of age, and so is its derivative. The theoretical $\Delta h/\Delta t$ is shown in Fig. 11 in a semi-logarithmic plot in the case of the plate model, for two sets of the following thermal parameters: coefficient of thermal expansion, thermal diffusivity, bottom boundary temperature; and for four values of the thermal thickness. A comparison of Figs. 10 and 11 reveals that:

(i) the data show the linear decrease with age predicted by plate models, as long as t is not too small;

(ii) there exist two distinct linear trends, one characterizing the age range 0–30 Ma, the other the age range 30–60 Ma. A single set of thermal parameters cannot fit the whole dataset; rather, the two distinct trends could be explained by two distinct values of the plate thickness.

A least squares fit of the two groups of data gives a thermal thickness of ~65 km for the age range 0–30 Ma and ~95 km for the age range 30–60 Ma. These values appear significantly smaller than the value (~125 km) determined on the basis of depth to the sea floor (Parsons and Sclater, 1977). These results call for a number of remarks:

(i) geoid data across fracture zones appear incompatible with the half-space cooling model,

(ii) beyond 30 Ma, these data follow reasonably well the behavior predicted by the plate cooling model with a thermal thickness of ~95 km,

(iii) at younger ages, the data could still be interpreted in the context of the plate model with a thermal thickness of ~65 km but do not present the initial constant trend seen in the model curves (compare Figs. 10 and 11).

In view of these results, we are led to conclude that the plate model in its simplest version does not explain the geoid observations for young ages. Many reasons may be invoked. The apparent thinner thermal thickness for young lithosphere indicates that the bottom boundary isotherm of the plate is uplifted, a result of heat perturbations. If the geoid data considered above rise over large positive geoid anomalies associated with hot rising material of small-scale convection cells, we may thus invoke some heat perturbation. Other mechanisms should also be considered, in particular the possibility that the initial geotherm assumed in the plate model may not be an isotherm. Hot spots could represent another source of heat perturbation. This problem needs further investigation; in particular analyses of geoid data in other oceans would eventually help in clarifying the situation.

IV. Use of Altimeter Data in Uncharted Marine Areas: Application to Bathymetric Prediction and Tectonic Reconstitutions

In some parts of the oceans, the spatial coverage of SEASAT exceeds that of surface-ship bathymetry. Altimeter derived geoid data have, therefore, been used to detect and predict bathymetry in uncharted regions. As we have shown previously, each type of tectonic unit gives a specific response in the geoid and a systematic inspection of altimeter data has revealed the presence of numerous unsuspected seamounts, guyots, and in some cases of fracture zones and even subduction zones (e.g., Sailor and Okal, 1983; Ruff and Cazenave, 1985).

A. Seamounts

An example of the discovery of two uncharted seamounts from SEASAT data is shown on Figure 12 along the Louisville Ridge volcanic chain (southeast Pacific). At geographical coordinates (37.5°S, 169.5°W) and (42.5°S, 163°W), the geoid points out to positive anomalies having characteristic seamount signatures in areas with extremely poor shipboard bathymetric coverage. Indeed, the discovery of new seamounts in oceanic areas has become very common since the advent of satellite altimetry (e.g. Lambeck and Coleman, 1982; Lazarewicz and Schwank, 1982; Dixon and Parke, 1983; Sandwell, 1984).

In the past 20 years, it has become apparent that our knowledge of the seamount population on the floor of the oceans is extremely poor. For example it took a strong volcanoseismic swarm to discover Macdonald Seamount (29°S, 140°W) in 1967, when its summit reached only 49 m below sea level (Johnson and Malahoff, 1971). Recently, Jordan *et al.* (1983) have used statistical techniques to extrapolate the known density of seamounts along linear shiptracks, and have revised the estimated density of seamounts in the Pacific by 1 to 2 orders of magnitude, suggesting that they make up as much as 0.4% of the oceanic crustal area, the great majority of them having yet to be discovered. It is clear that SEASAT-type altimetry can be

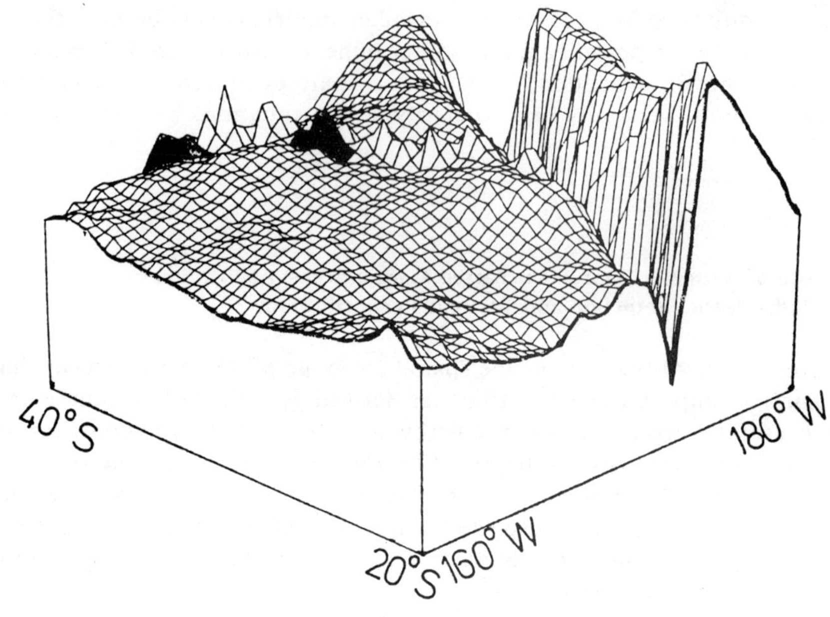

Fig. 12. Three-dimensional close-up of the geoid (with long wavelengths removed) in the area of the Tonga Trench, looking southwest, showing individual seamounts of the Louisville Ridge; the two seamounts shaded were discovered and identified by Cazenave and Dominh (1984).

of major help in starting to make inroads in this endeavor, and procedures have been developed for a systematic and automated scanning of SEASAT tracks for uncharted seamounts (e.g. White èt al., 1983; Dixon et al., 1983).

The interpretation of geoid signals in terms of the tectonic history of the associated bathymetric features is one of the major goals of their study; in the case of *seamounts*, the geoid response results from two opposite contributions: the positive effect of the topography plus the negative effect of its isostatic compensation. It has been found that the flexure of the lithosphere and geoid response to the loading depend critically on the age of the plate at the time of loading: an island or seamount formed on a very young portion of plate is supported only by a thin elastic layer ($H \leq 10$ km), and as a result literally sinks into the plate until substantial compensation is achieved. The resulting geoid signal is very weak if at all detectable. On the other hand, an island forming on sufficiently old (cooler) lithosphere is supported by a thicker elastic layer ($H \simeq 20$–30 km), resulting in only partial isotatic compensation, and a strong geoid anomaly. Thus, the total observed geoid anomaly is smaller for seamounts formed "on-ridge" than for seamounts formed "off-ridge".

Therefore, and in order to predict bathymetry from altimetry in the case of a *previously uncharted seamount*, it is necessary to know the tectonic setting of the topography (i.e. age of the crust and age of the seamounts) as well as the relative density of the load and underlying mantle. Besides this problem of geophysical nature there is a severe problem of coverage of altimeter profiles. Where the satellite track passes adjacent to a seamount or in a saddle between two seamounts, a positive geoid anomaly is recorded which will of course be less than that produced by the actual bathymetry anomaly displaced off-track. Clearly, closely spaced ascending and descending profiles are needed to provide a valuable description of the topography.

Conversely, in the case of *charted seamounts or islands*, the concurrent knowledge of the geoid signature and of the bathymetry makes it possible to obtain the response function of the lithosphere to loading. Watts *et al.* (1980) have used this method to characterize a population of Pacific seamounts as generated either "off-ridge" or "on-ridge". In trying to model quantitatively the elastic response of the lithosphere to a seamount load, several problems arise: first from a methodological point of view, Cazenave and Dominh (1984) have shown that it is necessary to use real-life, two-dimensional modeling, rather than the linear, one-dimensional notion of admittance. Second, in interpreting the derived elastic thickness of the plate in terms of the age of the lithosphere upon formation of the seamount, it is necessary to take into account the effect of reheating (also known as "rejuvenation") of the plate by the mere presence of the magmatic edifice responsible for the island structure. This effect, originally proposed by Crough (1978) has been modeled, most recently by McNutt (1984), who has shown that, although the lithosphere bearing Hawaii is presently 80 Ma, it has the elastic properties of a 20 Ma plate. Similar numbers for Tahiti would be 75 and 15 Ma.

Nevertheless, it is usually possible to derive the basic character ("on-ridge" leading to full compensation, or "off-ridge", leading to elastic support) of an island or seamount for which geoid data and bathymetry are known. An example, taken from Okal and Cazenave (1985) is given in Fig. 13: the set of SEASAT tracks samples in a very similar fashion the flanks of the northern Tuamotu Islands (Tatakoto, Pukaruha) and southern Tuamotu Islands (Acteon, Marutea), as well as of the more recent Gambier chain. While the southern islands have a strong geoid signature, even at distances of 20–30 km, the northern ones are barely if at all detected. Quantification of these results led Okal and Cazenave to propose lithospheric ages of respectively 6–7 Ma and less than 2 Ma upon formation of

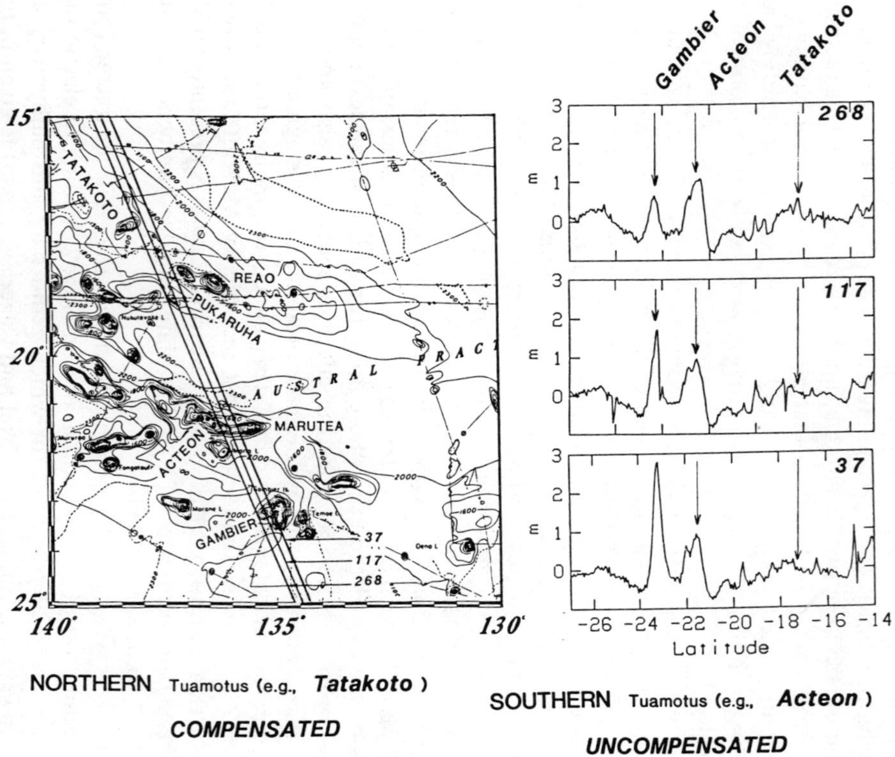

Fig. 13. Examples of compensated and uncompensated atolls in the Tuamotu Islands; see text for discussion. [After Okal and Cazenave (1985).]

the two chains. In the absence of igneous rocks allowing precise radiometric dating, such results are critical to the reconstruction of the tectonic history of oceanic atolls.

B. Fracture Zones

Although most major fracture zones have long been mapped on the basis of correlation of magnetic anomaly offsets, there remain areas of the ocean basins where improved knowledge of the topography of small fracture zones is necessary. Once again, the interpretation of SEASAT data has allowed complementary mapping of the extent of known fracture zones, as well as the discovery of new systems; Sailor and Okal (1983) and later Okal and Cazenave (1985) have used SEASAT data to infer the presence of two small fracture zones in the south central Pacific (see Fig. 14). These features, active about 25 Ma ago, were part of a short-lived pattern of spreading at the eastern boundary of the Pacific plate, which was then involved in an episode of ridge jump and reorientation, believed to be a possible consequence of the collision of North America with the Farallon Ridge. The transient nature of the fracture zone systems means that their present-day expression is confined to a small area, approximately $500 \times 500\ \mathrm{km}^2$, which had hitherto escaped shipboard detection; by using the models for $\Delta h/\Delta t$ described previously, Okal and Cazenave were able to estimate the age gap along these fracture zones. On the other hand, and more surprisingly, these authors have failed to identify the prolongation of a well-known nearby fracture zone (the Austral fracture zone) outside of its segment mapped from shipboard bathymetry (see Fig. 15). They proposed that an episode of rift propagation along the old Farallon ridge resulted in the transfer of the offset along the Austral fracture zone to the transient FZ1, as part of the process of evolution of the Farallon Ridge into the present-day East Pacific Rise.

C. Subduction Zones

As opposed to passive features, such as individual seamounts on the ocean floor or even fossil fracture zones, subduction zones are active agents in the motion of plates at the Earth's surface. It may then sound surprising that satellite altimeter data could be used to *discover and map* subduction zones, whose system should in principle be known.

However, in a recent study, Ruff and Cazenave have used SEASAT data to recognize the Hjort Trench, in the southernmost part of the Macquarie

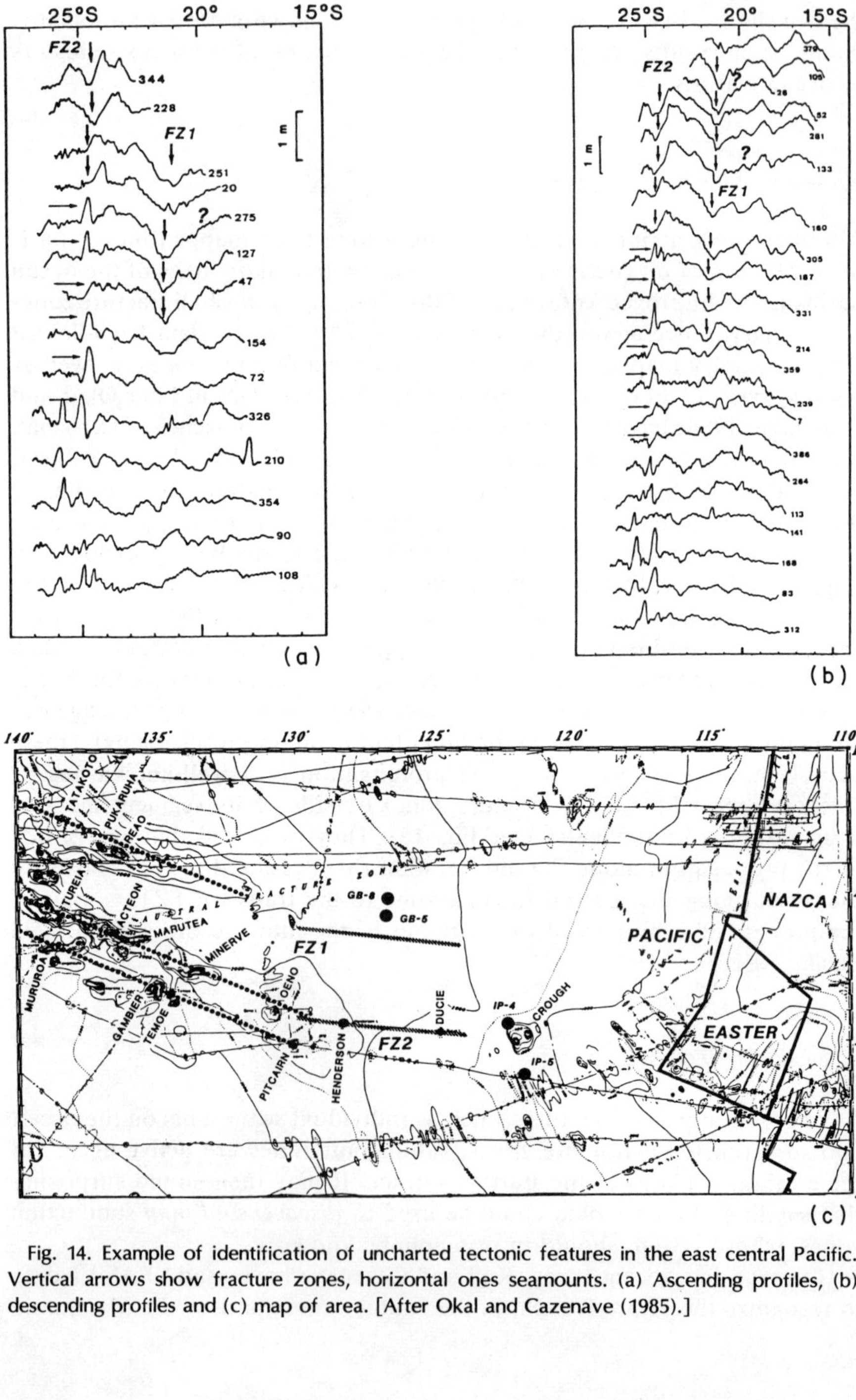

Fig. 14. Example of identification of uncharted tectonic features in the east central Pacific. Vertical arrows show fracture zones, horizontal ones seamounts. (a) Ascending profiles, (b) descending profiles and (c) map of area. [After Okal and Cazenave (1985).]

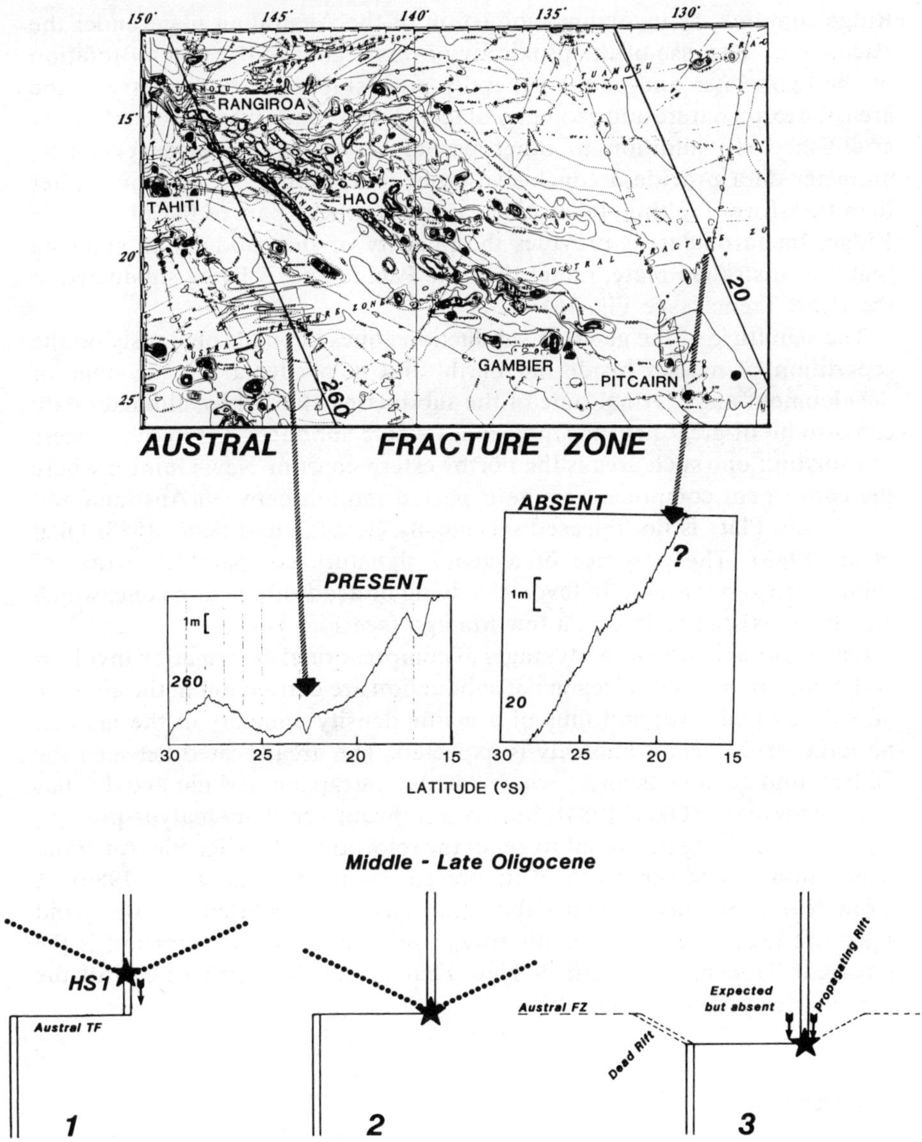

Fig. 15. Disappearance of Austral FZ signature east of 135° W. (top) Shipboard bathymetric map. (center) SEASAT data along tracks 20 and 260; arrows point from topographic location of crossing to corresponding point on SEASAT profile; note strong signal present on the western crossing, but absent from the eastern one (question mark), despite larger scale of figure. (bottom) Interpretation of the absence of the Austral fracture zone, as due to rift propagation of the Mendoza Ridge, upon interaction of the northern Tuamotu Hotspot (HS1) with the Austral transform fault.

Ridge complex, as involving subduction of the Australian plate under the Pacific one. Because of the proximity of this region to the pole of rotation of the two plates and of the generally poor shipboard bathymetry in the area, the exact nature and geometry of this plate boundary was not previously established. In addition to other partial seismic and marine evidence, altimeter data provide a conclusive proof that oblique subduction, rather than transform faulting, takes place at the southern part of the Macquarie Ridge. In particular, it provides the polarity of the subduction, showing that the Australian plate, rather than the Pacific one, is being subducted at the Hjort Trench (see Fig. 16).

The signature of the geoid in subduction zones depends obviously on the repartition of masses inside the Earth, and in particular on the stage of development and thermal state of the subducting slab. Thus, altimeter data can also be of great help to study zones where subduction is in the process of evolving: one such area is the northwestern coast of New Guinea, where the convergent component of the expected motion between Australia and the Pacific Plate is not released seismically (Kaplan and Seno, 1983; Okal *et al.*, 1983). The presence of a geoid signature comparable to that of subduction zones argues in favor of a dying or dead subduction zone, which may have existed as late as a few Ma ago (see Fig. 17).

On the other hand, the early stages of compressional deformation involved at the initiation of a new regime of subduction are carried out in the absence of a developed slab, and thus of a major density anomaly in the mantle; no large geoid height anomaly is expected. The area located between the Gilbert and Samoa Islands, where intense intraplate seismic activity has been recognized (Okal, 1984), but no significant geoid anomaly is present, may be undergoing the initial stages of the relocation of the Pacific–Australia plate boundary to the north of its present position (Okal *et al.*, 1986). A more developed stage, where subduction has clearly started, but the geoid signature lacks the characteristic trough present at mature trenches, is the Puysegur Trench, just south of New Zealand, at the northern end of the Macquarie system (see Fig. 16).

V. Conclusion

Satellite altimeters have given us coverage of the geoid at wavelengths which were previously inaccessible on a worldwide basis. As we have detailed before, a wealth of new geophysical results have been obtained. When combined with bathymetric knowledge, geoid data provide constraints on the Earth's elastic response and, thus, on the cooling processes of the oceanic lithosphere. On the other hand, in the absence of adequate bathymetric

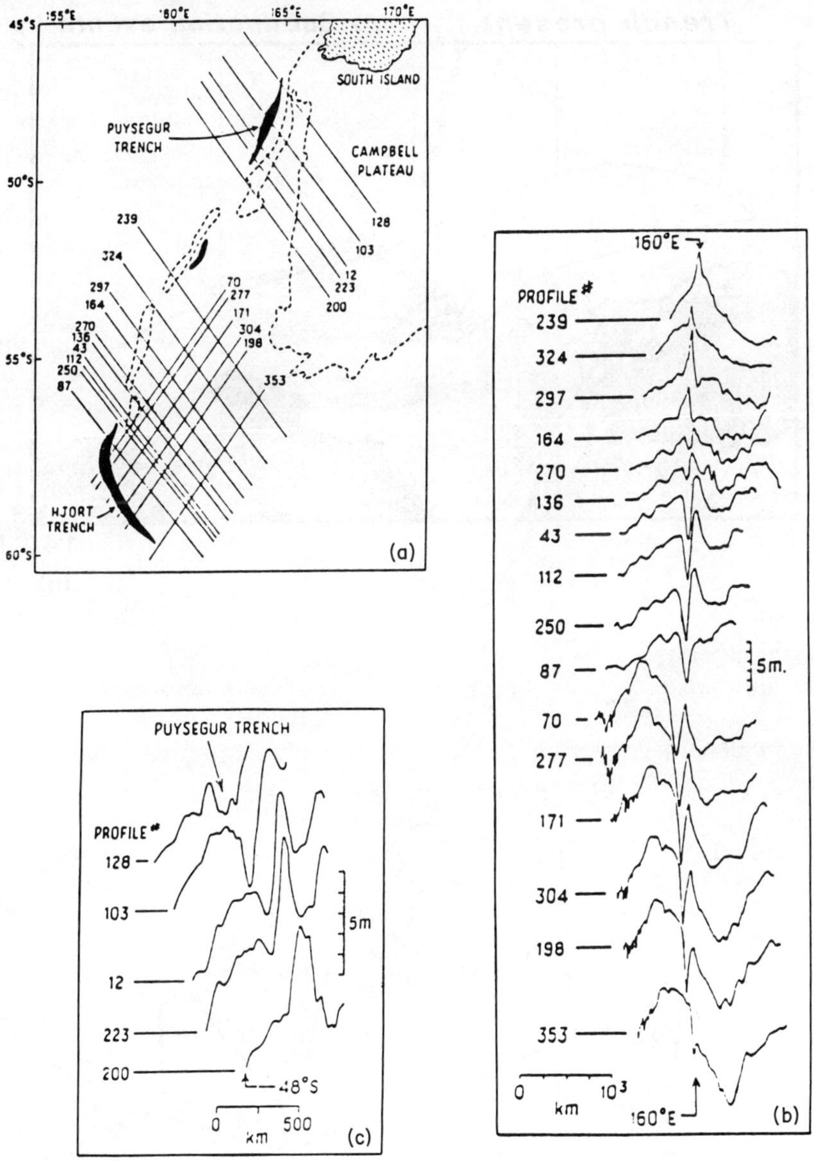

Fig. 16. (a) Index map of SEASAT profiles used in the Macquarie region. Trenches are identified by the 2600-fathom contour and the dashed lines show the 1000-fathom contour. (b) SEASAT profiles across the Hjort Trench region. Note the definite trough present on profiles 43, 112, 250, 87 and the outer rise on tracks 70 and 277. (c) SEASAT profiles across the Puysegur Trench. Note absence of the 5-m deep-trough characteristic of mature subduction and present at the Hjort Trench. [After Ruff and Cazenave (1985).]

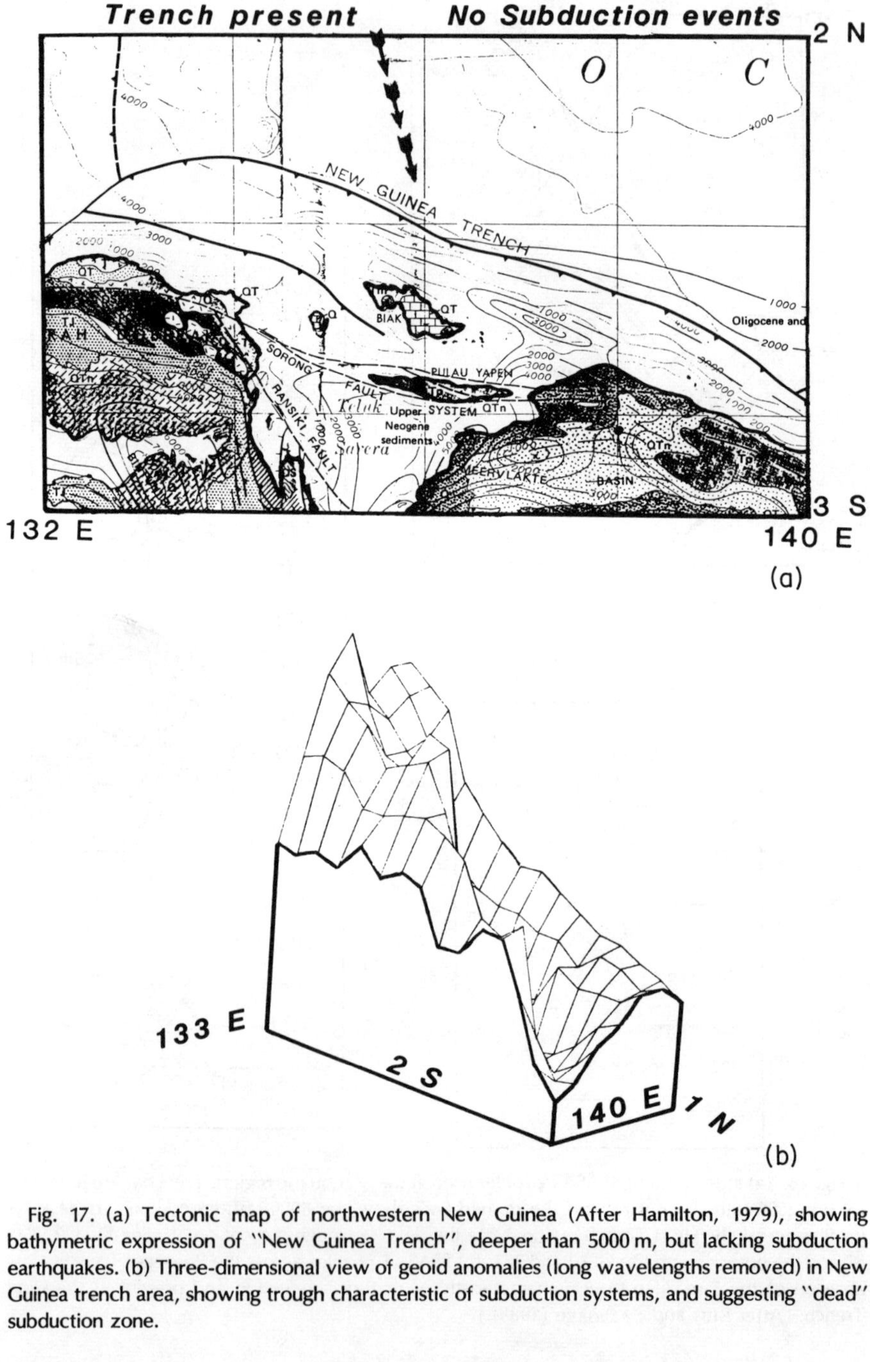

Fig. 17. (a) Tectonic map of northwestern New Guinea (After Hamilton, 1979), showing bathymetric expression of "New Guinea Trench", deeper than 5000 m, but lacking subduction earthquakes. (b) Three-dimensional view of geoid anomalies (long wavelengths removed) in New Guinea trench area, showing trough characteristic of subduction systems, and suggesting "dead" subduction zone.

coverage, satellite altimetry is the most efficient way of mapping uncharted tectonic features. Several problems of importance to present-day plate tectonics have been solved through this technique.

These results are all the more encouraging in the view of the untimely failure of the SEASAT mission after only 105 days in orbit, which resulted in uneven and occasionally sparse gridding of the ocean surface. This in turn has hampered efforts aimed at systematic bathymetric prediction. It would be highly desirable that future planning of altimeter missions be designed to fill the gaps between existing profiles, at least over selected oceanic regions.

References

Balmino, G., Brossier, C., Cazenave, A., Dominh, K., Nouel, F. and Vales, N. (1979). Geoid of the Kerguelen Islands area determined from GEOS-3 altimeter data. *J. Geophys. Res.* **84**, 3827–3831.

Bottinga, Y. (1974). Thermal aspects of sea-floor spreading, and the nature of the suboceanic lithosphere. *Tectonophysics* **21**, 15–38.

Bottinga, Y. and Allègre, C. J. (1976). Geophysical, petrological and geochemical models of the oceanic lithosphere. *Tectonophysics* **32**, 9–59.

Cazenave, A. (1984). Thermal cooling of the oceanic lithosphere: new constraints from geoid height data. *Earth Planet. Sci. Lett.* **70**, 395–406.

Cazenave, A. and Dominh, K. (1984). Geoid heights over the Louisville Ridge (South Pacific). *J. Geophys. Res.* **89**, 11171–11179.

Crough, S. T. (1978). Thermal origin of mid-plate hotspot swells. *Geophys. J. R. Astron. Soc.* **55**, 451–469.

Dixon, T. H. and Parke, M. E. (1983). Bathymetry estimates in the southern oceans from SEASAT altimetry. *Nature* **304**, 406–411.

Hamilton, W. (1979) Tectonics of the Indonesian Region. *U.S. Geol. Surv. Prof. Pap.* **1078**.

Heestand, R. L. and Crough, S. T. (1981). The effect of hotspots on the oceanic age-depth relation. *J. Geophys. Res.* **86**, 6107–6114.

Johnson, R. H. and Malahoff, A. (1971). Relationship of Macdonald Volcano to migration of volcanism along the Austral Chain. *J. Geophys. Res.* **76**, 3282–3290.

Jordan, T. H., Menard, H. W. and Smith, D. K. (1983). Density distribution of seamounts in the eastern Pacific inferred from wide-beam sounding data. *J. Geophys. Res.* **88**, 10508–10518.

Kaplan, D. A. and Seno, T. (1983). Seismotectonics of western New Guinea. *EOS, Trans. Am. Geophys. Un.* **63**, 1023 (abstract).

Lambeck, K. (1981). Lithospheric response to volcanic loading in the southern Cook islands. *Earth Planet. Sci. Lett.* **55**, 482–496.

Lambeck, K. and Coleman, R. (1982). A search for seamounts in the southern Cook and Austral region. *Geophys. Res. Lett.* **9**, 389–392.

Lazarewicz, A. R. and Schwank, D. C. (1982). Detection of uncharted seamounts using satellite altimetry. *Geophys. Res. Lett.* **9**, 385–388.

Lerch, F. J., Putney, B. H., Wagner, C. A. and Klosko, S. M. (1981). Goddard Earth models for oceanographic applications. *Mar. Geod.* **4**, 145–187.

Marsh, B. D., Marsh, J. G. and Williamson, R. G. (1984). On Gravity from SST, geoid from SEASAT, and plate age and fracture zones in the Pacific. *J. Geophys. Res.* **89**, 6070–6078.

Marsh, J. G., Martin, T. V., McCarthy, J. J. and Chovitz, P. S. (1980). Mean sea surface computation using GEOS-3. *Mar. Geod.* **3**, 359–378.

Marsh, J. G., Cheney, R. E., McCarthy, J. J. and Martin, T. V. (1985). Regional mean sea surfaces based upon GEOS-3 and SEASAT. *Mar. Geod.* **8**, 385–402.

Marsh, J. G., Brenner, A. C., Beckley, B. D. and Martin, T. V. (1986). Global mean sea surfaces based upon the SEASAT altimeter data. *J. Geophys. Res.* **91**, 3501–3506.

McKenzie, D. P. (1967). Some remarks on heat flow and gravity anomalies. *J. Geophys. Res.* **72**, 6261–6273.

McKenzie, D. P., Watts, A. B., Parsons, B. and Roufosse, M. (1980). Planform of mantle convection beneath the Pacific Ocean. *Nature* **288**, 442–446.

McNutt, M. K. (1984). Lithospheric flexure and thermal anomalies. *J. Geophys. Res.*, **89**, 11180–11194.

Okal, E. A. (1984). Intraplate Seismicity of the southern part of the Pacific plate. *J. Geophys. Res.* **89**, 10053–10071.

Okal, E. A., and Cazenave, A. (1985). A model for the plate tectonics evolution of the east central Pacific based on SEASAT investigations. *Earth Planet. Sci. Lett.* **72**, 99–117.

Okal, E. A., Woods, D. F. and Lay, T. (1986). Intraplate deformation in the Samoa-Gilbert-Ralik area: a prelude to a change of plate boundaries in the South Pacific? *Tectonophysics* (in press).

Okal, E. A., Heineman, M. and Grall, H. M. (1983). Tectonic implications of the seismicity of northwestern New Guinea, 1910–1979. *EOS, Trans. Am. Geophys. Un.* **64**, 265 (abstract).

Oldenburg, D. W. (1975). A physical model for the creation of the lithosphere. *Geophys. J. R. Astron. Soc.* **43**, 425–451.

Parker, R. L., and Oldenburg, D. W. (1973). Thermal models of ocean ridges. *Nat. Phys. Sci.* **242**, 137–139.

Parsons, B. and Richter, F. M. (1980). A relation between the driving force and geoid anomaly associated with mid-ocean ridges. *Earth Planet. Sci. Lett.* **51**, 445–450.

Parsons, B., and Sclater, J. G. (1977). An analysis of the variation of ocean floor bathymetry and heat flow with age. *J. Geophys. Res.* **82**, 803–827.

Rapp, R. H. (1983). The determination of geoid undulations and gravity anomalies from SEASAT altimeter data. *J. Geophys. Res.* **88**, 1552–1562.

Reigber, C., Muller, H., Bosch, W., Balmino, G. and Moynot, B. (1985). GRIM gravity model improvement using LAGEOS. *J. Geophys. Res.* (submitted).

Ruff, L. J. and Cazenave, A. (1985). SEASAT geoid anomalies and the Macquarie Ridge complex. *Phys. Earth Planet. Inter.* **38**, 59–69.

Sailor, R. V. and Okal, E. A. (1983). Application of SEASAT data in seismotectonic studies of the south central Pacific. *J. Geophys. Res.* **88**, 1572–1580.

Sandwell, D. T. (1984). A detailed view of the South Pacific geoid from satellite altimetry. *J. Geophys. Res.* **89**, 1089–1104.

Sandwell, D. T. and Schubert, G. (1980). Geoid height versus age for symmetric spreading ridges. *J. Geophys. Res.* **85**, 7235–7241.

Sclater, J. G. and Francheteau, J. (1970). The implications of terrestrial heat flow observations on current tectonic and geochemical models of the crust and upper mantle of the Earth. *Geophys. J. R. Astron. Soc.* **20**, 509–542.

Turcotte, D. L. and Oxburgh, E. R. (1967). Finite amplitude convective cells and continental drift. *J. Fluid Mech.* **28**, 29–42.

Vening Meinesz, F. A. (1941). Gravity over the Hawaiian archipelago and over the Madiera area: conclusions about the Earth's crust. *Proc. Kon. Ned. Akad. Wetensch.*

Watts, A. B. (1978). An analysis of isostasy in the world's oceans, 1. Hawaiian–Emperor seamount chain. *J. Geophys. Res.* **83**, 5989–6004.

Watts, A. B., Bodine, J. H. and Ribe, N. M. (1980). Observations of flexure and the geological evolution of the Pacific Ocean basin. *Nature* **283**, 532–537.

Watts, A. B., Karner, G. D. and Steckler, M. S. (1982). Lithospheric flexure and the evolution of sedimentary basins. *Phil. Trans. R. Soc. London, A* **305**, 249–281.

Watts, A. B., McKenzie, D. P., Parsons, B. and Roufosse, M. (1985). The relationship between gravity and bathymetry in the Pacific Ocean. *Geophys. J. R. Astron. Soc.* **83**, 263–298.

White, J. V., Sailor, R. V., Lazarewicz, A. R. and LeSchack, A. R. (1983). Detection of seamount signatures in SEASAT altimeter data using matched filters. *J. Geophys. Res.*, **88**, 1541–1551.

Wood, B. J. and Yuen, D. A. (1983). The role of lithospheric phase transitions on seafloor flattening at old ages. *Earth Planet. Sci Lett.* **66**, 303–314.

5.2 Geodetic and Oceanographic Applications of Satellite Altimetry

R. E. Cheney, B. C. Douglas, D. C. McAdoo and D. T. Sandwell

National Geodetic Survey, Charting and Geodetic Services, National Ocean Service, NOAA, Rockville, Maryland, USA

I. Introduction

The form of the ocean surface, or "sea surface topography", has both constant and time-varying components. The principal constant component is caused by the geopotential, which produces deviations of the sea surface from a best-fitting ellipsoid of revolution of up to 100 m. Tides and geostrophic currents cause additional smaller (~1 m) deviations from the level surface produced by the geopotential. As we shall see, for most marine geophysical applications, the purely oceanographic effects can be neglected and the sea surface assumed level.

Time-varying changes of sea surface topography, although small compared to the permanent undulations created by the geopotential, are large compared to the accuracy (a few centimeters) of a satellite altimeter. In addition, sea height variability can be observed independent of geoid

SPACE GEODESY AND GEODYNAMICS
ISBN 0-12-057840-9

topography either at intersections ("crossovers") of the satellite ground track or on repeating (collinear) tracks. Both measures of variability have been used successfully, but the latter is preferred because a more consistent and uniform time series of observations is realized.

Sea surface profiles are obtained by accurately determining the round trip travel time of the microwave pulse between the satellite and ocean surface. But the pulse is also affected by the sea surface. For example, crests of waves scatter energy, and troughs focus it. Thus the greater the sea state, the more the pulse shape is distorted. The very short-wavelength capillary waves caused by the wind have a different effect, scattering energy and reducing the amplitude of the return pulse as wind speed increases. Suitable algorithms exist that can be used to determine sea state and wind speed from the form and strength of the return pulse. Comprehensive discussions of all applications of altimeter data from GEOS-3 and SEASAT have been collected in several volumes of the *Journal of Geophysical Research* (Vol. 84, no. B8, 1979; Vol. 87, no. C5, 1982; Vol. 88, no. C3, 1983), and a complete bibliography can be found in Brown and Cheney (1983).

In addition to the obvious problems of interpreting altimeter data, there is the geodetic problem of determining the trajectory of the satellite, because altimeter profiles are taken relative to the satellite path which varies in its own unique way. Thus interpreting altimeter data cannot be separated from the problem of satellite tracking and orbit determination. Indeed it is for this reason that there is such a large representation of satellite geodesists in the community that analyzes satellite altimeter data.

II. Geometry of Satellite Altimeter Observations

An altimetric satellite, by the nature of its mission, has severe limitations placed on the parameters of its orbit. It must not be at so low an altitude that atmospheric drag significantly perturbs the orbit (<800 km) nor so high (>1500 km) that an extraordinary amount of power is required by the altimeter. In addition, the inclination of the orbit to the equator must be high to observe as much of the ocean surface as possible, but at the same time low enough that both components of the surface slope are accurately observed. A further complication arises from certain high inclinations where the orbital precession rate leads to aliasing of tidal and other perturbations on the orbit with sought-after frequencies of ocean circulation. These limitations mean that the orbital frequency will be 13–14 rev. per day and the inclination well away from sun-synchronous.

The planned NASA TOPEX (topography experiment) satellite (National Aeronautics and Space Administration, 1981) illustrates well the above considerations. The inclination of the orbit to the equator will be approximately 63.5° to prevent aliasing of solar pressure and tidal perturbations with seasonal variations of the general circulation. An altitude of approximately 1300 km was chosen to place the satellite at a height where drag effects are very small, but low enough for reasonable power requirements for the altimeter.

Another constraint on the orbit of an altimeter satellite occurs because an oceanographic satellite must make repeated measurements of the surface. Although such observations occur at intersections of the groundtrack for any orbit, these "crossovers" do not occur spatially or temporally in a manner suitable for spectral analyses. Intersections occur nearly at integral multiples of an orbital period, and are not uniformly distributed. For example, a high-inclination orbit will produce many intersections in the polar regions, but few at low latitudes. Thus the preferred sampling scheme involves repeating (collinear) groundtracks. However, a tradeoff exists here between temporal and spatial sampling because a rapid time sampling leads to coarse geographic coverage, and vice versa. For example, the 3-day repeat cycle used by SEASAT during its last month of operation produced a groundtrack spacing of 930 km at the equator. The TOPEX groundtrack will repeat every 10 days, yielding 215-km spacing of the tracks.

III. Satellite Orbit Determination

A satellite altimeter mission places extreme requirements on the accuracy to which the path of the satellite must be known. Variations in ocean dynamic height of a few centimeters over wavelength scales from a few tens of kilometers to entire ocean basins need to be observed accurately. Since altimetric profiles of the ocean surface are taken relative to the satellite trajectory, the maximum benefit of a satellite altimeter mission will be obtained only if the trajectory of the satellite is known to an accuracy approaching that of the altimeter itself.

The problem of satellite orbit determination for the SEASAT altimeter satellite was extensively covered in a special issue of the *Journal of the Astronautical Sciences* (Vol. 28. no. 4, 1980). The results of these studies confirmed what has long been known about error in position of a satellite. That is, the error is very systematic, being primarily concentrated at a frequency of one per revolution, with power falling off very rapidly at both lower and higher frequencies. Thus at the wavelength associated with mesoscale oceanographic phenomena or statically supported loads on the lithosphere (a few hundred kilometers), the error in the satellite altitude is insignificant. It is for this reason that the most successful applications of

GEOS-3 and SEASAT data were in computing mesoscale variability from repeated profiles of data, or geoidal undulations at similar small scales. Attempts to compute ocean circulation basin-scale features were less successful, although still interesting, as will be seen later. The SEASAT project computed the radial position (and hence altitude) of the satellite to an accuracy of 60–70 cm (Marsh and Williamson, 1980). Geometric algorithms utilizing the groundtrack crossovers to constrain the trajectory were able to further reduce this figure by at least one-half (Douglas *et al.*, 1984).

The problem of satellite orbit determination is essentially one of reconciling observations made of a *component* of the satellite's position or velocity from a network of stations with the integrated effects of the forces on the satellite. Unlike the situation for aircraft or ships, complete position fixes are never obtained. The orbit determination problem involves computing a continuous trajectory that is simultaneously consistent with the tracking data and satellite equations of motion. The latter equations are of course well known, but some force model parameters used to compute the accelerations of the satellite are in varying degrees uncertain. The principal errors in the force model are uncertainty of the geopotential and atmospheric drag. Significant effects of drag can be eliminated either by placing the altimetric satellite at a sufficiently high altitude or employing surface-force compensation devices on the satellite. The problem of the geopotential is much more difficult. This matter is being addressed by major efforts at several organizations in the United States and abroad. If the flight program envisioned as a part of the NASA Geopotential Research Mission (GRM) is undertaken, the effect of the error of the geopotential on satellite orbit determination will be entirely eliminated. In the case of TOPEX, precision orbits at the decimeter level of accuracy will be obtained by employing a world-wide network of improved TRANET Doppler tracking stations providing very nearly continuous coverage of the orbit.

IV. Marine Gravity and Geophysics

Undulations of the mean sea surface, or the (nearly identical) geoid, are caused primarily by lateral density inhomogeneities within the Earth. These mass anomalies are physically supported by deviatoric stresses arising either from convective motions of the viscous mantle or from static loads on the rigid lithosphere. Mantle-wide convection maintains many of the longer-wavelength ($>10^6$ m), larger-amplitude (>10 m) geoid undulations (Kaula, 1972). Indeed, a number of studies point out the high correlation between geoid undulations and surface manifestations of mantle convection such as spreading ridges, subduction zones and hot-spot swells (Haxby and

Turcotte, 1978; Sandwell and Schubert, 1980; Griggs, 1972; Hager, 1984; Crough and Jurdy, 1980).

Although the geoid is dominated by these longer-wavelength undulations, much information is contained in the shorter-wavelength ($\ll 10^6$ m) signals. These primarily originate from density anomalies within the lithosphere. In all but a few extreme cases, shorter-wavelength geoid undulations are highly correlated with seafloor topography. The high resolution and accuracy of the SEASAT radar altimeter provides a global view of these shorter wavelength geoid undulations.

The most prominent seafloor features reflected in the sea-surface topography are trenches, fracture zones, large seamounts and spreading ridges. Examples of SEASAT profiles across trenches and oceanic fracture zones are presented below along with physical models of these features. Other examples of altimeter profiles over seamounts and spreading ridges are presented in papers by Haxby and Turcotte (1978) and Watts and Ribe (1984).

The shape of the marine geoid at shorter spatial wavelengths is predominantly a smoothed, attenuated version of seafloor topography (see Chapman, 1979). It is, therefore, not surprising that deep-ocean trenches (and the outer rises which flank them) appear in the marine geoid and are readily observed by satellite altimeters. In fact, the deep-ocean trenches represent the largest amplitude undulations in the marine geoid exclusive of very long (>4000 km) wavelengths. The Aleutian trench, for example, contributes 15 m of relief. Trenches manifest themselves so clearly in the geoid because they are paramount bathymetric features and they are not, to a large degree, isostatically compensated (Vening Meinesz, 1964).

Outer rises lying just seaward of most deep-ocean trenches also are isostatically uncompensated features (Watts and Talwani, 1974; McAdoo and Martin, 1984). These have wavelengths (transverse to trench axes) of several hundred kilometers and contribute as much as 6 m of relief to the marine geoid. They are attributable to flexure of the oceanic lithosphere prior to subduction. This flexure is generally modeled by representing the oceanic lithospheric as an end-loaded elastic plate (Fig. 1). This elastic model of lithospheric deflection predicts the topographic shape for the outer rise and outer wall of the trench. It can be extended to predict concomitant geoid undulations. An approximate, analytical expression for these predicted geoid undulations is derived in McAdoo and Martin (1984). This enables predicted geoid undulations to be compared with individual SEASAT or other satellite altimeter passes over outer rises. First, passes are high-pass filtered by removing a reference geoid. By using the method of least squares, elastic model parameters such as flexural wavelength, can be estimated for each altimeter pass. Two such SEASAT passes, an ascending pass over the

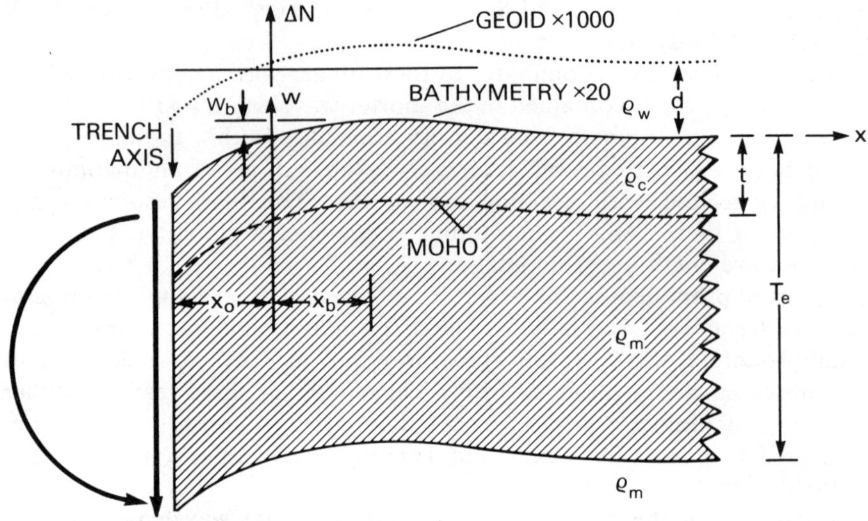

Fig. 1. Model of end-loaded elastic lithospheric plate overlying a fluid-like asthenospheric substrate. Crust and lithosphere deflect as a unit. Model outputs include seafloor deflection profile, $w(x)$ and concomitant profile of geoid heights, $N(x)$ [From McAdoo and Martin (1984).]

southern Mariana Trench and a descending pass across the South Sandwich Trench, are shown in Fig. 2. Note that available bathymetric data across these trenches are not suitable for studies such as this. Estimates of flexural wavelength can be converted to estimates of lithospheric thickness. In McAdoo and Martin (1984) and McAdoo *et al.*, (1985), SEASAT altimeter data over various outer rises are used to demonstrate that the mechanical lithosphere thickens with age in a manner consistent wth established thermal models (e.g. Parsons and Sclater, 1977), and can achieve a thickness as great as 60 or 70 km. So, SEASAT data significantly augment available information concerning the shapes of outer rises and regional variations in the thickness of the lithosphere.

In general, satellite altimeter data are particularly suited to geophysical studies of long, linear features like trenches, outer rises or fracture zones because these data enjoy the advantage of geoidal representation in which unmodeled, local, short-wavelength effects such as those of seamounts are attenuated.

Fracture zones are linear scars in the seafloor produced by transform faulting (Wilson, 1965). Topography along their inactive segments consists of long ridges, troughs and scarps separating regions of different depth (Menard and Atwater, 1969). Figure 3 shows how fracture zones (FZs) are formed as part of the seafloor spreading process. Spreading ridges generally

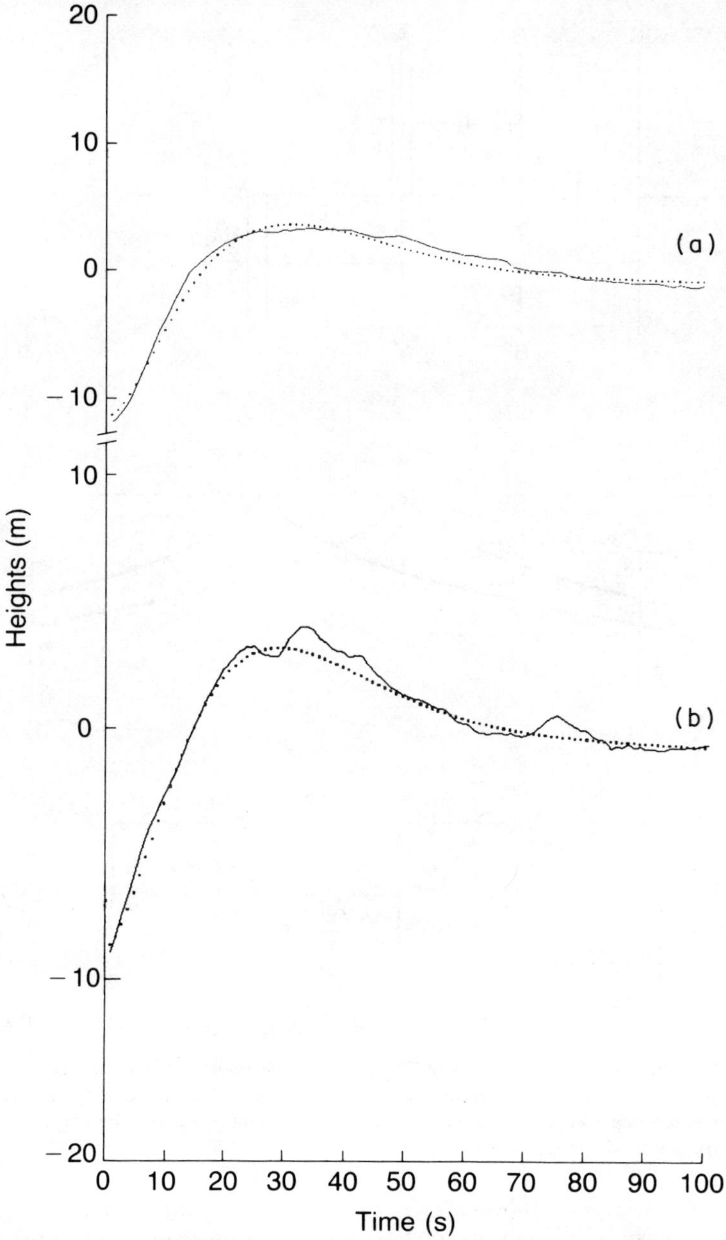

Fig. 2. Individual SEASAT passes (low-cut filtered) across two trenches. Shown as dotted curves are best-fitting geoid profiles generated by an elastic lithosphere model. (a) South Sandwich Trench, $\phi = -55.60°$, $\lambda = 333.85°$, Rev = 468, $\alpha = 105.5 \pm 4.3$, $W_0 = 1975 \pm 94$. (b) Southern Mariana Trench, $\phi = 10.85°$, $\lambda = 140.92°$, Rev = 829, $\alpha = 103.1 \pm 5.3$, $W_0 = 1683 \pm 10$.

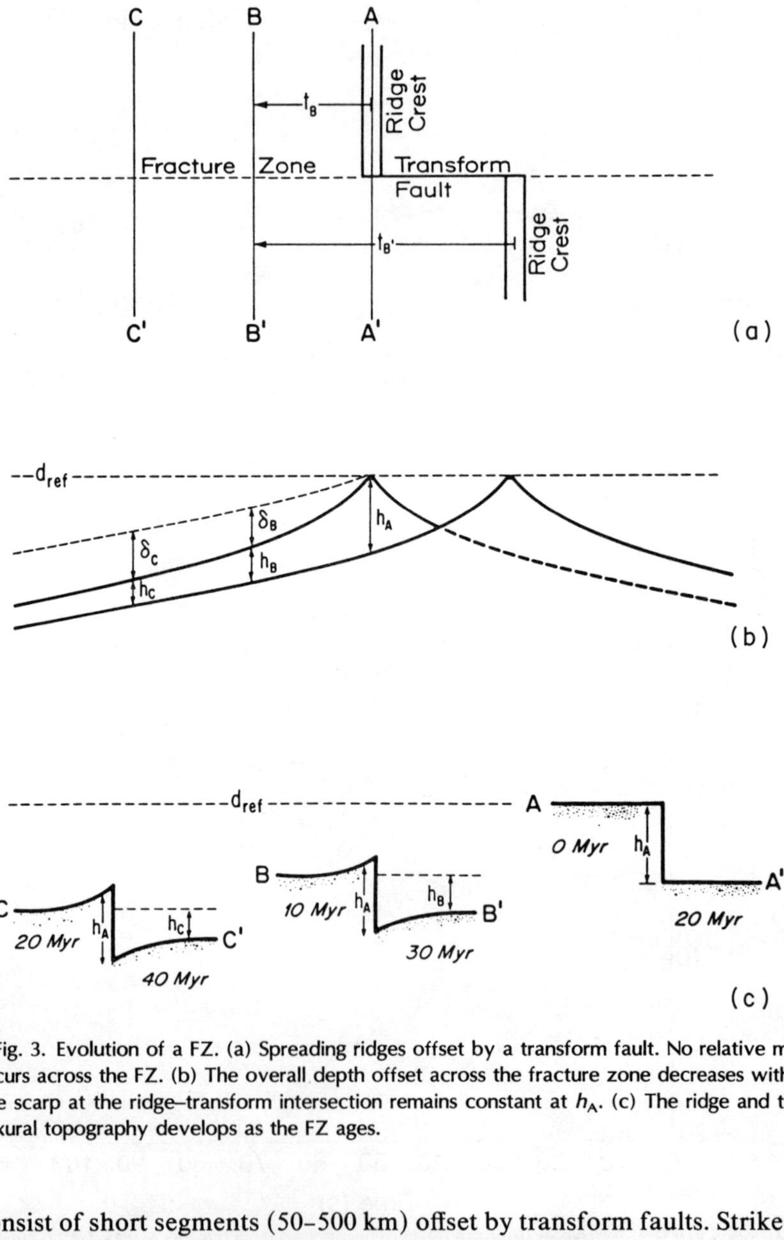

Fig. 3. Evolution of a FZ. (a) Spreading ridges offset by a transform fault. No relative motion occurs across the FZ. (b) The overall depth offset across the fracture zone decreases with age. The scarp at the ridge–transform intersection remains constant at h_A. (c) The ridge and trough flexural topography develops as the FZ ages.

consist of short segments (50–500 km) offset by transform faults. Strike-slip motion occurs along the transform fault as shown in the top of Fig. 3, producing fractures in the seafloor. Once the lithosphere migrates beyond the intersection of the spreading ridge with the transform fault, the two sections of lithosphere move at the same rate and eventually become welded together. The fractured seafloor remains along this older section of the FZ although no new fractures develop.

The general increase in seafloor depth with increasing seafloor age produces additional topography in the vicinity of a FZ. As shown in Fig. 3 (Sandwell and Schubert, 1982), a FZ separates newly created lithosphere (0 Myr) from older lithosphere (20 Myr) (see A′–A′). As these two segments of lithosphere move away from the ridge crest from A–A′ to B–B′, the depth far from the FZ on the younger lithospheric segment increases at a higher rate than the depth far from the FZ on the older segment (Menard and Atwater, 1969; Delong *et al.*, 1977; Sibuet and Mascle, 1978). This effect is shown by the two subsidence curves in the middle of Fig. 3. Since the lithosphere is welded at the ridge–transform intersection, the FZ does not slip during its evolution. Therefore the height of the scarp will remain at its initial value h_A. The lithosphere must flex to satisfy both of these requirements as shown in the lower portion of Fig. 3.

Results from a physical model for the thermomechanical evolution of a FZ are shown in Fig. 4 (Sandwell, 1984). Figure 4(a) shows the development of the ridge and trough topography across a large age-offset FZ (i.e. 20 Ma age offset). Since the younger lithosphere subsides at a higher rate than the older lithosphere, the ridge and trough signature develops in just a few million years. This flexural topography extends to about 100 km on either side of the FZ.

The same model is used to predict the geoid step (Fig. 4(c)) and the geoid slope (Fig. 4(b)) across a FZ. The geoid step has an amplitude of 3 m while maximum geoid slopes (i.e. deflections of the vertical) are 80 μrad. As the FZ evolves the geoid step becomes complex, reflecting the development of the ridge and trough topography. The deflection of the vertical, rather than the geoid profile, is compared with the data because differentiation of the SEASAT profile removes the long-wavelength, large-amplitude components of the geoid that are associated with density anomalies in the mantle.

Large values of the deflection of the vertical (~50 μrad = 10 arcsec) occur along the Udintsev FZ because of its large age offset (~18 Ma, Weissel *et al.*, 1977). Examples of along-track deflection of the vertical profiles, computed by differentiating a subset of descending SEASAT altimeter passes, are shown in Fig. 5. On this projection, the Heezen, Tharp and Udintsev transform faults are heavy horizontal lines whereas segments of the Pacific–Antarctic spreading ridge are vertical heavy lines. Deflections of the vertical along the Udintsev FZ (i.e. filled profiles in Fig. 5) show a simple systematic pattern. At the midpoint between the two spreading ridges, a positive peak lies to the north of the transform fault while a negative peak lies to the

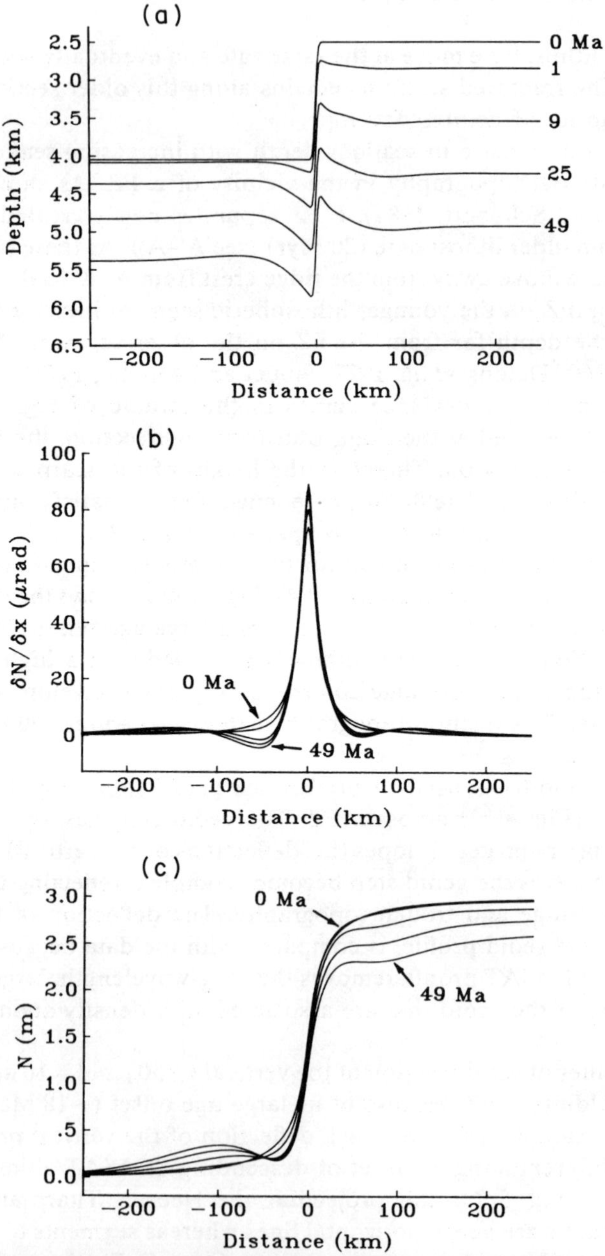

Fig. 4. Evolution of a FZ (20 Ma age offset) from a thermomechanical model. (a) Topography within a few million years flexural ridges and troughs develop. (b) Deflection of the vertical peak retains its high initial value reflecting the persistence of the bathymetric scarp. (c) Geoid step changes in shape but not in overall amplitude as the FZ ages. [Sandwell (1984).]

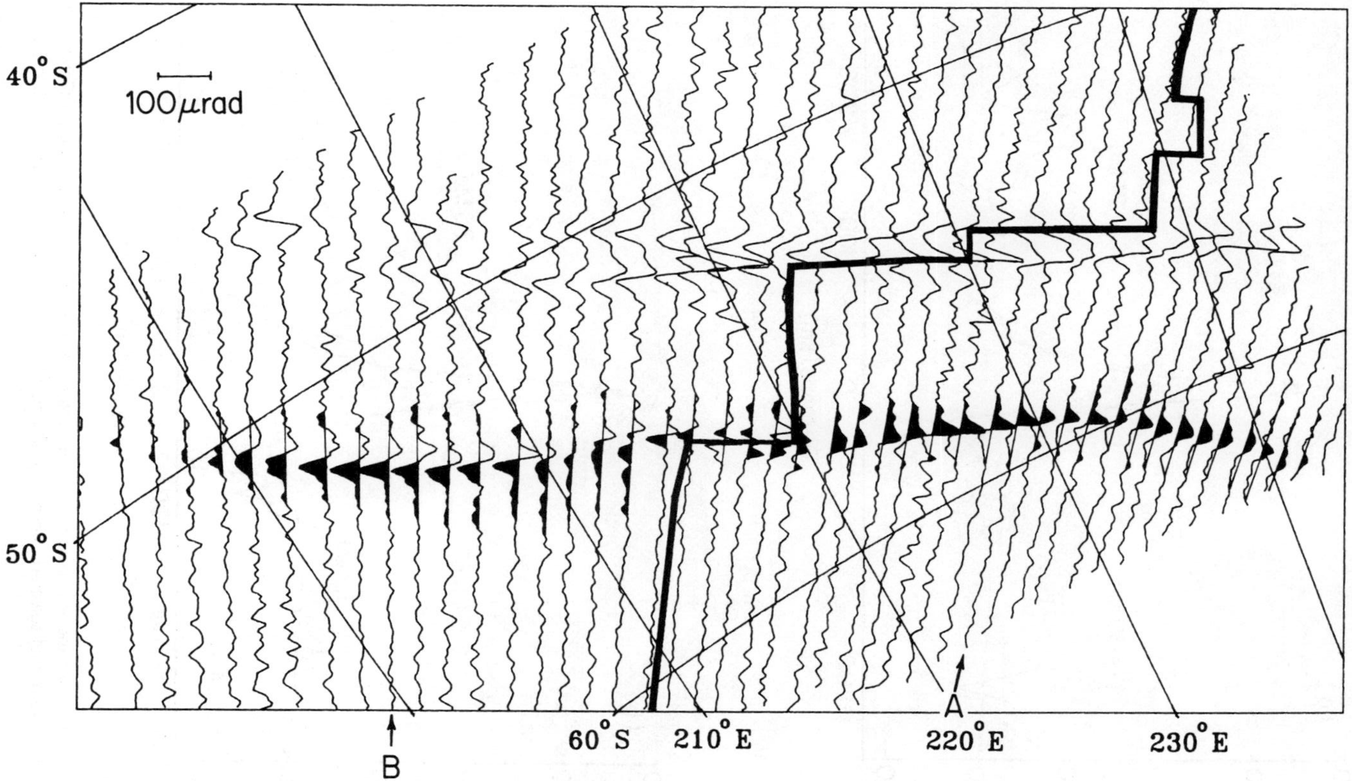

Fig. 5. Along-track deflection of the vertical profiles from differentiated SEASAT altimeter passes. Map projection is oblique Mercator where pole is the relative rotation axis between the Pacific and Antarctic plates. The plate boundary (heavy line) consists of spreading ridges (vertical lines) and transform faults (horizontal lines). Highlighted peaks reveal the characteristic signature of the major age-offset Udintsev FZ.

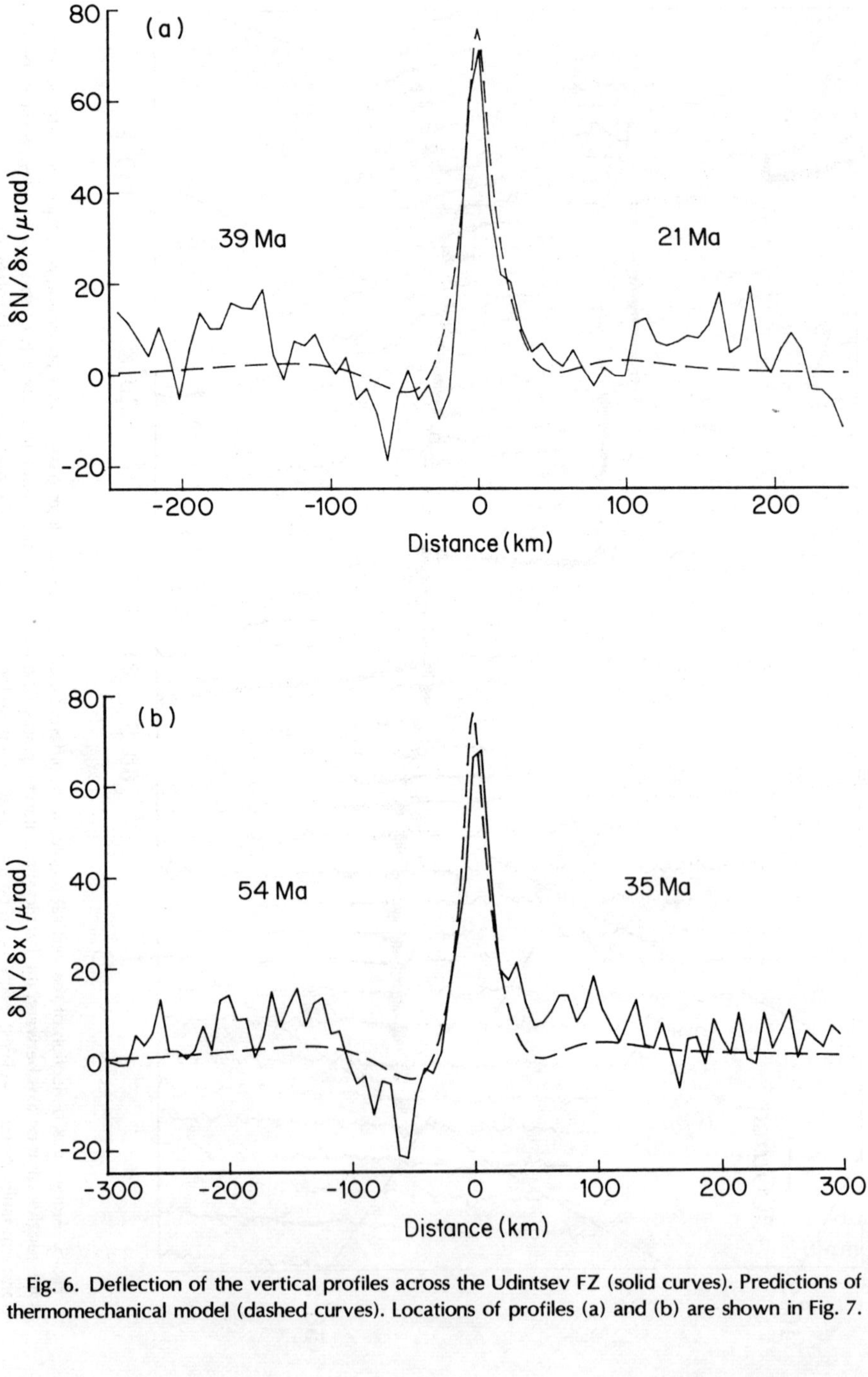

Fig. 6. Deflection of the vertical profiles across the Udintsev FZ (solid curves). Predictions of thermomechanical model (dashed curves). Locations of profiles (a) and (b) are shown in Fig. 7.

south. At the left ridge–transform intersection there is only a positive peak. Further to the left, along the inactive portion of the FZ, there is a positive peak centered above the FZ with a negative sidelobe on the older side. An analogous pattern occurs along the right inactive FZ segment. Thus, along all inactive portions of the Udintsev FZ there is a prominent sidelobe on the older side with a small or absent sidelobe on the younger side.

Two features of the deflection profiles across the Udintsev FZ support the flexure model of a FZ. The first is the persistence of the deflection of the vertical signature along older portions of the Udintsev FZ. This signal persists to at least 60 Ma where it diminishes because of a decrease in age offset (Weissel, personal communication, 1983). The second confirmation of the flexure model is the apparent asymmetry in deflection of the vertical.

Quantitative agreements between model profiles and observed profiles are shown in Fig. 6. The locations of the two profiles are marked in Fig. 5. Profile A was inverted and replotted (solid curve) along with the model prediction (dashed curve). For both profiles, the model matches the data fairly well although there are short-wavelength variations in the data that cannot be fit by the model. The model agrees equally well with about half of the other profiles shown in Fig. 5. Disagreements of the model with the other profiles could be due to crustal thickness variations or deviations of the thermal structure from the boundary layer cooling model.

V. Mesoscale Variability

Geostrophic surface currents are maintained by horizontal pressure gradients, and manifested as sea surface slopes relative to the geoid. While it is difficult to separate this dynamic component of sea height from the static geoid signal, an altimeter can readily detect time-dependent variations of height. The most energetic fluctuations have typical scales of 100–300 km and are attributable to the ocean eddy field (meandering of narrow currents and migration of detached vortices). Study of the oceanic eddy field is important because eddying motions are believed to be the dominant mechanism for transferring energy and momentum. Obtaining a basic description of global mesoscale eddy variability is thus fundamental to an improved understanding of ocean dynamics.

Computing mesoscale variability from altimeter data of course requires solving the problem of satellite orbit error. The method demonstrated here employs relatively short (1000–2000 km) profiles obtained along repeated ground tracks. Because orbit error has dominant scales many times larger than this, the relative error between two or more repeated passes can be effectively removed as a linear trend. Analysis of profiles having identical tracks also eliminates any contribution from the geoid, enabling determination of time-variable sea height changes alone. An example is shown in Fig. 7. Two passes of GEOS-3 data were taken along the same track in the North Atlantic but at different times. A linear trend has been removed from each profile to eliminate the long-wavelength orbit error. Each pass contains the same geoid profile, as demonstrated by the close agreement over the seamount. However, at the location of the Gulf Stream there is a difference of 120 cm due to a meander or shift in the current's position. Mesoscale variability can therefore be detected with pairs of altimeter passes having the same track.

This procedure is most applicable to the GEOS-3 data which were collected in great volume in certain areas such as the North Atlantic. Even though GEOS-3 did not have an orbit which repeated exactly at specific intervals, it did produce many pairs of profiles along virtually the same ground tracks. Figure 8 is a map of sea height variability in the Gulf of Mexico and western North Atlantic derived from about 1000 collinear pairs of GEOS-3 data collected over 3.5 years. These results show peaks of variability created by the Loop Current in the Gulf of Mexico and by the

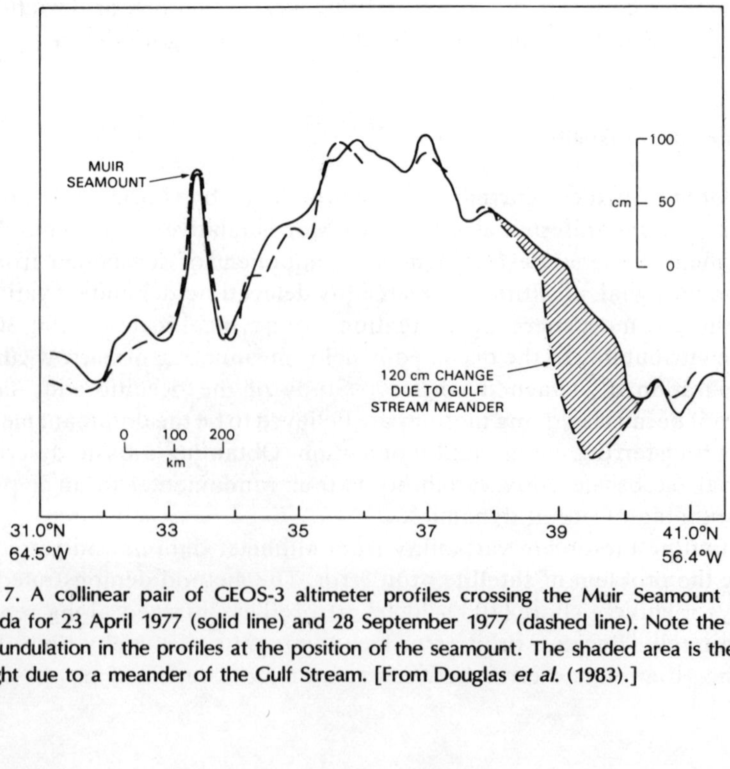

Fig. 7. A collinear pair of GEOS-3 altimeter profiles crossing the Muir Seamount north of Bermuda for 23 April 1977 (solid line) and 28 September 1977 (dashed line). Note the identical geoid undulation in the profiles at the position of the seamount. The shaded area is the change in height due to a meander of the Gulf Stream. [From Douglas *et al.* (1983).]

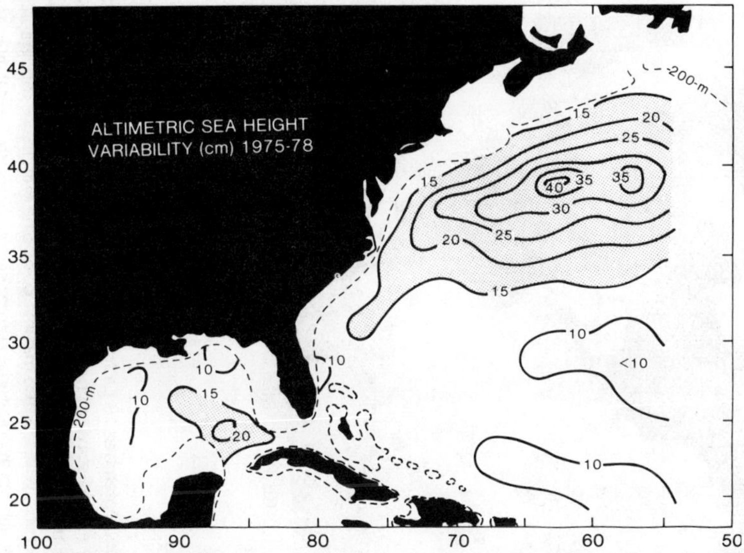

Fig. 8. Mesoscale sea surface height variability in centimeters computed from repeated pairs of GEOS-3 altimeter profiles during 1975–1978. Maximum values are associated with the Loop Current in the Gulf of Mexico and the Gulf Stream meander region downstream of Cape Hatteras. [From Douglas *et al.* (1983).]

Gulf Stream east of Cape Hatteras. The least energetic area is south of 35°N. A more detailed discussion of these results, including comparisons with similar maps derived from standard oceanographic data, has been published separately (Douglas *et al.*, 1983).

The technique of collinear differences has also been used to derive the global mesoscale variability from SEASAT altimeter data (Cheney *et al.*, 1983). In this case, statistics were derived not from pairs, but from nests of collinear profiles accumulated during a 25-day period when the ground track repeated at 3-day intervals. Variability was derived by computing the standard deviation of sea height every 7 km along the track from the series of eight or nine observations. Because of the relatively coarse cross-track grid described by the ground tracks (approximately 600 km at mid-latitudes), interpolation was required to compute values on a uniform global grid for contouring. The resulting map (Fig. 9) displays patterns consistent with our present understanding of the surface ocean circulation. In the North Pacific and the North Atlantic, the Kuroshio and Gulf Stream systems are represented as variability maxima extending 3000–4000 km into mid-ocean. The smoothing and interpolation process has damped their amplitudes down to about 10 cm from their actual along-track values of 20–30 cm, but they

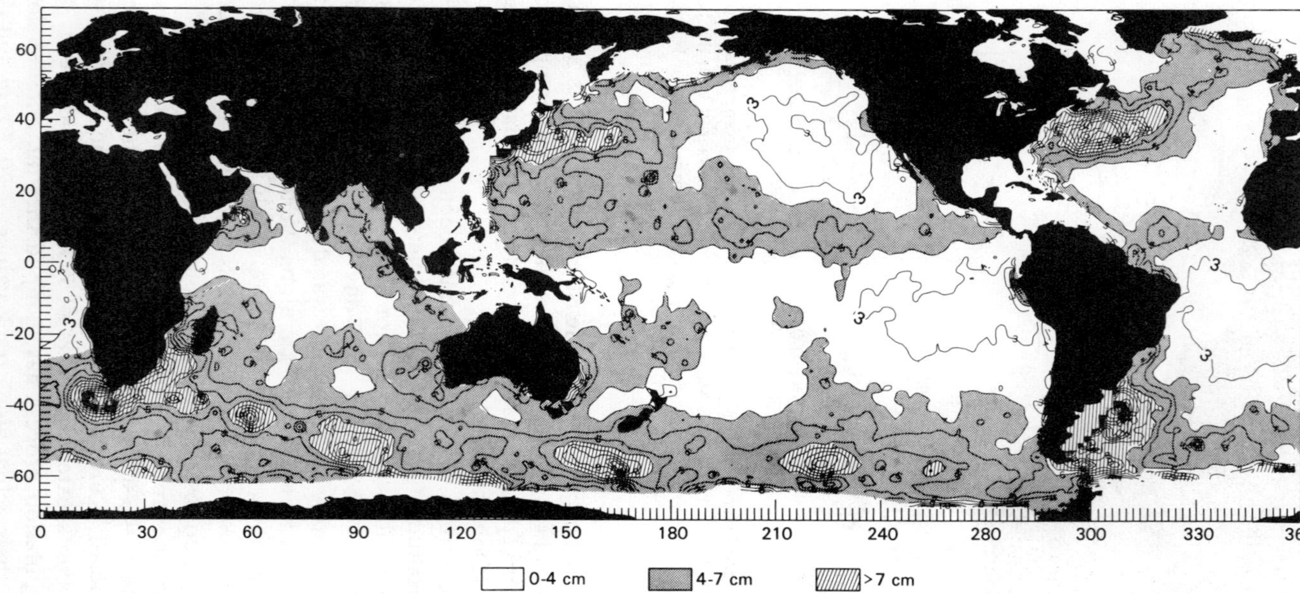

Fig. 9. Global mesoscale sea height variability measured by the SEASAT altimeter, 15 September–10 October, 1978. The North Atlantic and North Pacific are dominated by the highly energetic Gulf Stream and Kuroshio systems that extend seaward nearly 4000 km. In the Southern Hemisphere the Agulhas Current below Africa and the Falkland Brazil/Current confluence off South America are clearly apparent. High variability due to the Antarctic Circumpolar Current extends in a nearly continuous band through the polar oceans, with isolated maxima coinciding with major topographic ridges and plateaus. Owing to the predominance of values less than 4 cm in mid-ocean, the north equatorial current systems in both the Atlantic and Pacific can be seen as zonal bands of higher variability. [From Cheney *et al.* (1983).]

are still the dominant features of the Northern Hemisphere oceans. The North Equatorial Current systems are also associated with zonal bands of higher variability. The most prominent features in the Southern Hemisphere are the Agulhas Current off the tip of South Africa, the confluence of the Falkland and Brazil Currents east of Argentina and the Antarctic Circumpolar Current, represented as a series of variability maxima surrounding Antarctica. Because of westward intensification of the wind-driven ocean currents, higher variability is most prevalent in the western parts of the oceans. This is best demonstrated in the Atlantic where the Northern and Southern Hemispheres are nearly mirror images of each other. There is also a marked contrast between the highly variable western North Pacific and the quieter eastern basin, with the dividing line occurring near the Emperor seamount chain. This line of seamounts apparently acts as an efficient barrier to the high variability generated by the Kuroshio in the west. East–west asymmetry is characteristic of all oceans with values as small as 2–3 cm in the eastern parts of the major basins. Of all global variability values computed, 70% were less than 5 cm, 26% were in the 5–10 cm range, and only 4% were greater than 10 cm. It is this broad background of low variability that permits some of the equatorial currents to be detected even though their dynamic height signatures are relatively small (25 cm).

VI. Basin-scale Variability

The relationship of the ocean to climate is a key issue in contemporary oceanography. The ocean has the capacity to store large amounts of heat more readily than the atmosphere. We therefore need to understand the dynamics which influence the ocean–atmosphere balance, especially in the tropics where heat storage is large and time scales are short.

The El Niño–Southern Oscillation is one of the most dramatic air–sea interaction phenomena. During El Niño there is a collapse of the trade winds, resulting in sea-level changes of several decimeters over vast regions of the equatorial Pacific. In recent years, these sea-level changes have been well documented in the western Pacific by using island tide-gauge data (Wyrtki, 1979). This network has provided clear evidence of the very large scales of sea-level response during El Niño, particularly the pronounced 1982–1983 event when 40-cm monthly mean anomalies were observed. In contrast, El Niño signal is poorly observed in the eastern half of the Pacific where few islands are found.

No altimeter data are available from past El Niño years. However, SEASAT data can be used in simulations to test the ability of collinear altimeter data tracks to monitor El Niño events. The approach is similar to that used for mesoscale variability except that longer profiles are required to detect the larger scales associated with El Niño.

El Niño sea-level signal extends along the equatorial ocean with meridional scales of a few thousand kilometers and zonal scales of up to 10,000 km. If we restrict our attention to individual altimeter tracks, we need only be concerned with detection of the relatively shorter-meridional signal since high-inclination satellite orbits cross the equator in a predominantly north–south direction.

As mentioned earlier in this chapter, the principle problem of satellite altimetry is error in determining the satellite's radial position. For SEASAT the magnitude of this error was 1–2 m with a dominant wavelength equal to the Earth's circumference. A secondary problem is removal of the tidal signal from altimetric profiles. At the scale of mesoscale variability, both effects can be eliminated simply by removing linear trends from profiles 1000–2000 km long. To observe sea-level changes with the much longer scales of El Niño, longer arcs are required, making the problem more difficult. The tide must be subtracted explicitly by using modeled heights, and orbit error must be approximated by a quadratic or higher-order trend.

In order to obtain a realistic test in the presence of radial orbit error, our simulation was performed using two adjacent 3-day arcs of SEASAT altimeter data taken in September, 1978. During this time the satellite groundtrack repeated every 3 days. Collinear altimetric profiles of sea surface topography 3 days apart should not show any large differences because the time scale is too small compared to any large-scale oceanographic phenomena other than the tides.

The first test was designed to evaluate how well orbit error and tidal signals could be removed. Data in the Pacific were used between 40°N and 40°S, yielding passes approximately 9500-km long (Fig. 10). To determine the sea-level change in 3 days, members of collinear pairs were subtracted to form difference profiles. Figure 11 shows an example for a collinear pair in the central Pacific. The upper profile is the difference between the two altimeter passes after removing only the modeled tide (Schwiderski, 1979). The large scale and amplitude of this curve are clearly indicative of radial orbit error. In the lower illustration, results of removing a quadratic trend from the difference profile are shown. The apparent change in sea level has been reduced by an order of magnitude to approximately 10 cm. (In computing the quadratic trend of the difference profile, values within 15° of the equator were not used. This was done to reduce damping of El Niño signal in the subsequent simulation.)

This procedure was followed for each of the Pacific collinear pairs, and mean sea height differences were accumulated in 10° blocks. Removal of the tide and a quadratic trend from each pair successfully reduced all block

Fig. 10. SEASAT ground tracks during 23–28 September 1978. Two altimeter profiles, separated in the cross-track direction by only 0.4 km, were obtained along each track.

values to 10 cm or less, demonstrating that orbit error and tidal effects could be minimized over long arcs spanning the Pacific.

The El Niño simulation was performed based on modeled height anomalies. One of the 3-day sets of altimeter data was modified by the values shown in Fig. 12 to see if the signal could be recovered in the presence of 1-m orbit error. The signal was given a maximum amplitude of 40 cm at the equator, falling off to zero at 20°N and 20°S latitude. This is analogous to conditions that existed in December 1982 during the peak of the most pronounced El Niño event on record (Wyrtki, 1985). Of course, El Niño has never been accurately mapped in the eastern Pacific. In our model, we simply assumed a signal centered on the equator and symmetric about 160°W.

Figure 13 shows results of the simulation described previously. The model signal, present in one of the 3-day arcs, but not the other, was recovered virtually intact when the arcs were differenced. Peak amplitude was reduced by a few centimeters, and only 5 cm of noise was introduced in mid-latitude regions. The simulation demonstrates that orbit error can be removed from long arcs of altimeter data without aliasing El Niño signal. This technique can be made even more effective by incorporating tide-gauge observations in the solution as constraints wherever they are available. These form an important source of "surface truth" and strengthen the altimetric results.

Monitoring of basin-scale variability in the Pacific by using the method of collinear pairs should become a reality in 1986. The U.S. Navy altimeter

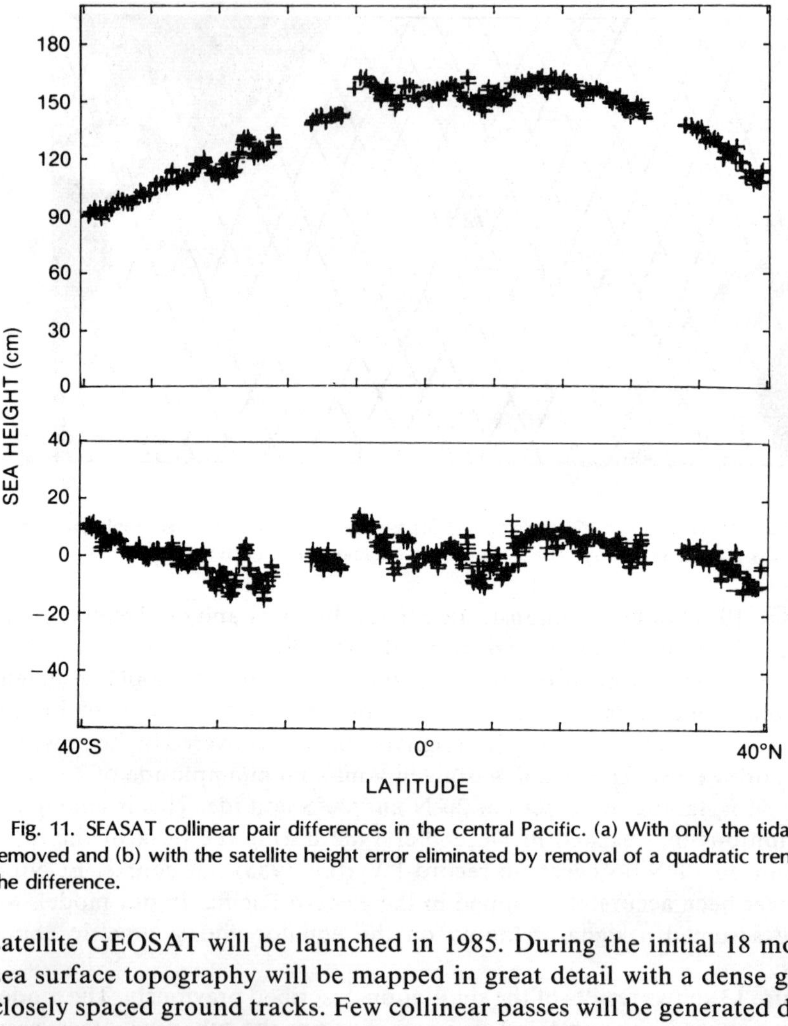

Fig. 11. SEASAT collinear pair differences in the central Pacific. (a) With only the tidal signal removed and (b) with the satellite height error eliminated by removal of a quadratic trend from the difference.

satellite GEOSAT will be launched in 1985. During the initial 18 months, sea surface topography will be mapped in great detail with a dense grid of closely spaced ground tracks. Few collinear passes will be generated during this portion of the mission, but after completion of the geodetic survey, the satellite will be placed in a 17-day repeat orbit. This will yield a spatial resolution of approximately 150 km globally and should provide a unique look at temporal sea-level variations on a variety of scales.

VII. Determining the Mean Circulation

Ultimately it is believed that satellite altimetry will contribute to determination of the general circulation of the oceans (the long-term mean movement of water). In order to accomplish this, advances will be required in gravity

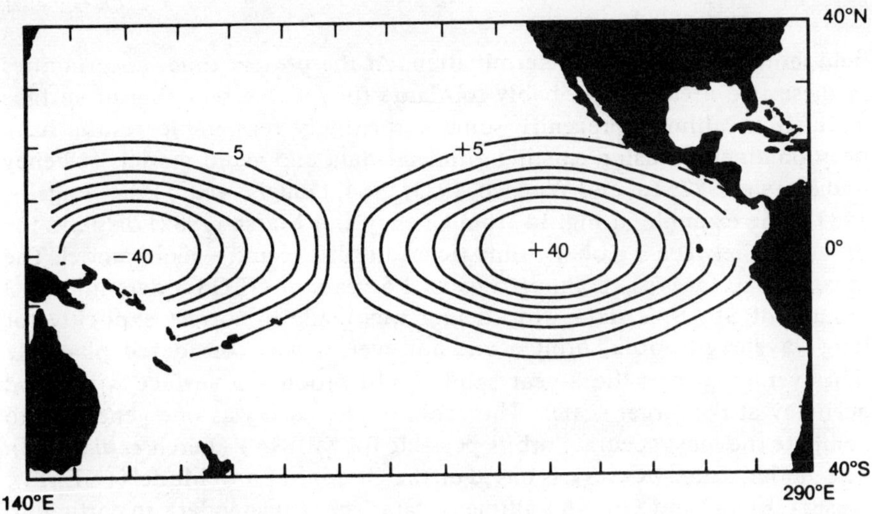

Fig. 12. Model of El Niño sea-level signal based on the 1982–1983 event. Values were assigned in 10° blocks and represent anomalies (in centimetres) from the long-term mean.

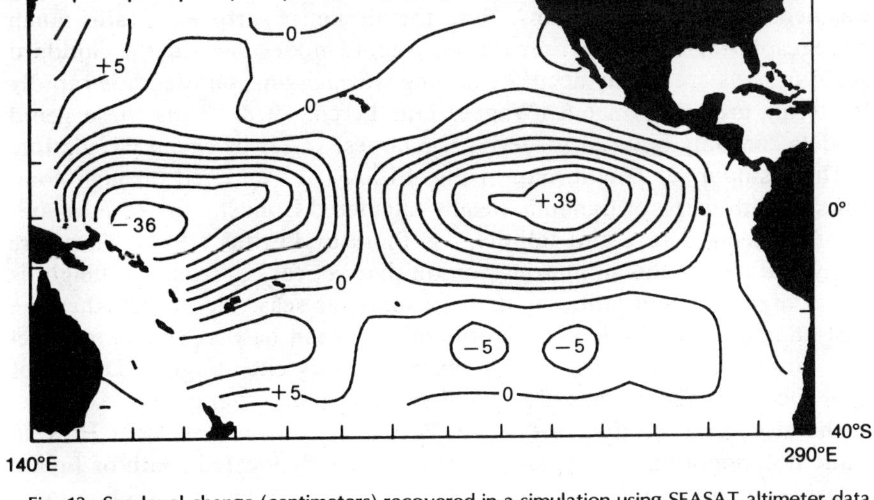

Fig. 13. Sea-level change (centimeters) recovered in a simulation using SEASAT altimeter data modified by model heights. Tides and orbit error were successfully removed over 9000-km arcs without significantly altering El Niño signal.

field modeling and orbit determination. At the present time, uncertainties in these two areas are probably too large for reliable mapping of surface circulation, although recently some surprisingly reasonable results have been obtained by using existing altimeter data and geoid models (Cheney and Marsh, 1982; Tai and Wunsch, 1983, 1984; Douglas *et al.*, 1984; Engelis, 1983). The example in Fig. 14 from Cheney and Marsh (1982) displays the difference between a global altimetric mean surface and a geoid model. The mean sea surface was computed using 1.5 years of GEOS-3 data and all 3 months of SEASAT data. No attempt was made to correct explicitly for long-wavelength radial orbit error; however, it was considered plausible that averaging over the 2-year span would produce a surface with good accuracy at the larger scales. The geoid model used was one generated to compute the most accurate orbits possible for SEASAT (Lerch *et al.*, 1982). The model, called PGS-S4, is based on the Goddard Earth Models (GEMS); it uses GEOS-3 and SEASAT altimeter data. The altimeter data, in particular, were weighted so that the higher-degree and -order terms of the geoid were improved (to 36,36). Sea heights for the global mean surface and the geoid model were both computed on a 1° grid.

The residual map in Fig. 14 was derived by subtracting the geoid from the altimetric surface. The residuals were then smoothed so that only features with scales greater than a few thousand kilometers were retained. (The filtering scheme has been described by Marsh and Martin, 1982. The 1° resolution surfaces were recomputed on a 5° grid by fitting a biquadratic surface to each point based on data within a 20° radius.) This smoothing was necessary for two reasons. First, the altimetric surface contains much more resolution and detail than the 36,36 geoid model. Second, the Goddard geoid models are highly accurate at long wavelengths, but degrade rapidly for terms greater than 6,6 (Wagner and Lerch, 1978). Thus these geoid models can only be used to derive the largest scales of ocean circulation.

The residual sea height map in Fig. 14 may be evaluated through comparison with maps of dynamic height calculated from shipboard measurements of temperature and salinity, e.g. those by Levitus (1982). There are many features of the altimetric solution that appear realistic. Sea height is 1–2 m higher at low latitudes than in subpolar seas. This establishes the well-known anticyclonic gyres in the major ocean basins. The simple fact that these gyres have the correct sense is highly encouraging. Details of their locations and magnitudes are of secondary importance. For example, in the altimetric solution, the North Pacific gyre is centered near Hawaii, while oceanographic charts show it more properly located south of Japan. Considering orbit and geoid uncertainties, such discrepancies are not unexpected. The altimetric solution does contain a strong, continuous current in the southern oceans much like the Antarctic Circumpolar Current. The

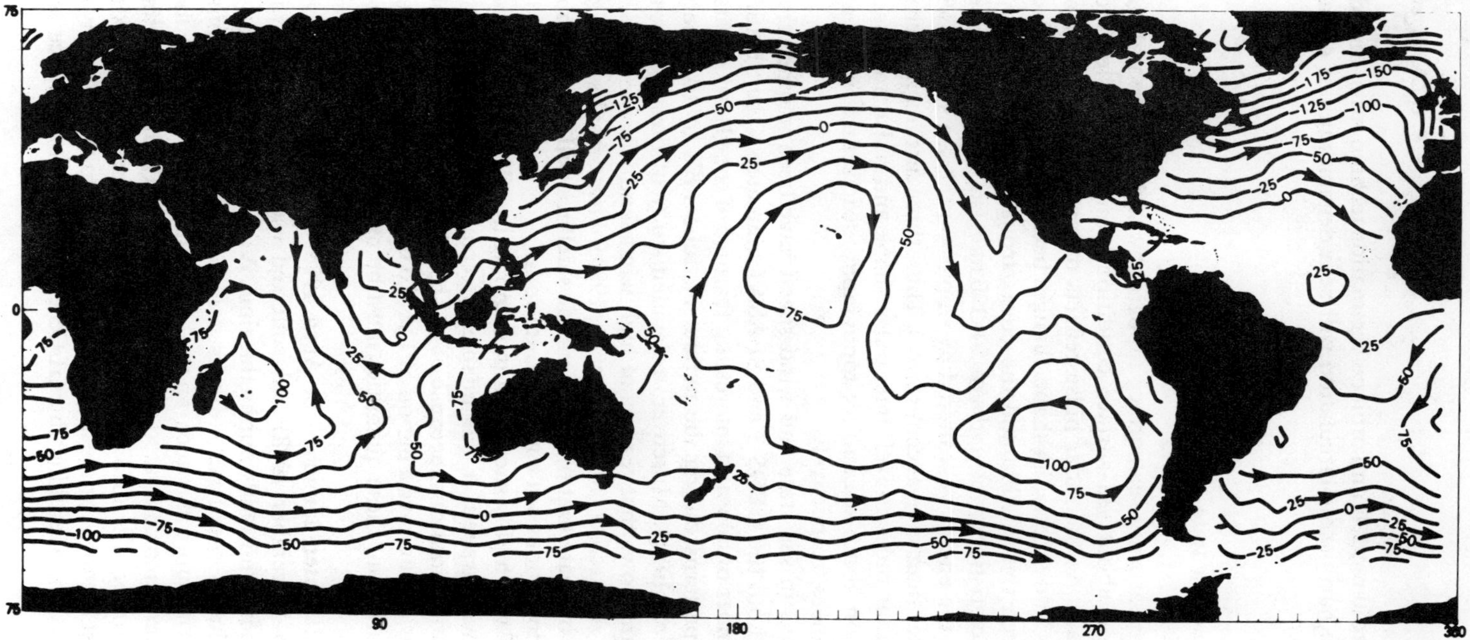

Fig. 14. Dynamic topography determined from altimetry. Contours in centimeters represent the difference between a long-term mean sea surface and a global geoid model. The surface consists of 1.5 years of GEOS-3 data plus all existing SEASAT altimetry. The geoid model, called PGS-S4 is based on a combination of satellite observational data, surface gravity, and also altimeter data. Both surfaces were smoothed before performing the difference to reveal only the large-scale (>10,000 km) circulation features. Although certain aspects of the map are probably unrealistic, based on comparison with dynamic height maps from hydrography, gyres with the correct sense are found in the major ocean basins. A strong Antarctic Circumpolar Current is also apparent. [From Cheney and Marsh (1982).]

ability of future geoid models and orbit determination techniques to improve this picture of the basin-scale circulation will be a demanding test of their absoute accuracy.

VIII. Wind Speed and Sea State

In addition to measuring sea surface topography and its temporal variations, information about sea surface conditions can be extracted from radar altimeter data. As the radar pulse reflects off an area of sea surface several kilometers in diameter, its shape changes because of the differences in distance between the satellite and the crests/troughs of surface waves which have characteristic wavelengths of 50–300 m. Also, during the reflection, a portion of the energy is scattered by ripples on the sea surface with wavelengths greater than several centimeters. Thus, three distinct properties of the reflected radar pulse, travel time, pulse shape and pulse amplitude, provide information about sea surface topography, sea surface waves and sea surface roughness, respectively.

Spatial variations in sea state and wind speed have been mapped from data acquired by both the GEOS-3 and SEASAT altimeters. The SEASAT maps reveal a number of new and interesting features of the global sea state and wind speed patterns during the period, 28 June 1978 to 10 October 1978 (Chelton *et al.*, 1981). Moreover, a detailed study of these data over the southern ocean shows that regions of high winds and waves migrate in longitude around Antarctica (Mognard *et al.*, 1983). These results from the SEASAT mission indicate that future altimetric satellites will provide more detailed information about sea surface conditions.

While SEASAT was highly successful in mapping the global distribution of sea state and wind speed, it was unable to recover much information about their temporal variations, because of its short lifespan. Undoubtedly, the largest time variations have a period of 1 yr. To obtain a reliable estimate of the seasonal variation, these quantities must be measured for a number of years.

The GEOS-3 altimeter, launched in April 1975, operated for nearly 4 years (April 1975–December 1978) and collected wave height and wind speed information intermittently during this time. A limitation of the GEOS-3 altimeter was that it could not store its data between times of ground communication. Therefore the geographical distribution of GEOS-3 data is very non-uniform when compared with the later SEASAT altimeter which carried a tape recorder. The GEOS-3 data density is highest near tracking stations and lowest in more remote areas of the world. Despite this relatively poor data distribution, averaging the GEOS-3 data over both space and time reveals significant seasonal variation in wind speed and sea state. Moreover, the maps derived from the GEOS-3 data (Sandwell and Agreen, 1984) show substantial hemispherical asymmetry in the seasonal variations in these quantities. Most of the seasonal variation occurs in the Northern Hemisphere while winds and waves in the Southern Hemisphere are fairly steady.

During the austral winter (June–September) the largest wind speeds of 8–12 m/s occur in the Southern Hemisphere between latitudes of 40° and 60°. This band of high wind speed, which also appears on the SEASAT-derived maps (Chelton *et al.*, 1981; Mognard *et al.*, 1983), surrounds the Antarctic continent and is associated with mid-latitude winter storms. Highest average wind speeds (>10 m/s) generally occur away from continents and are centered over the Antarctic Circumpolar Current. Other more localized areas of high wind speed appear in the North Atlantic, the Gulf of Alaska, south of the Baja California Peninsula and in the Bay of Bengal. The latter feature is associated with monsoons. As noted by Chelton *et al.* (1981), both the North Atlantic and North Pacific are characterized by light winds during boreal summer.

The patterns of wind speed change drastically between summer and winter. As expected, wind speeds are very high in the North Atlantic and North Pacific during their winter. Indeed, the highest average wind speeds (>14 m/s) in the world's oceans occur in the northern North Atlantic during this period. It is interesting that there is not a complementary decrease in Southern Hemisphere wind speeds between the austral winter and summer.

GEOS-3 wind speed maps show zonal wind patterns that sailors have known about for centuries. These are the westerlies, the horse latitudes, the trades and the doldrums. To enhance these zonal wind speed patterns and their seasonal variations, the data were averaged into latitudinal zones. The results are shown in Fig. 15, where average wind speed for each seasonal period is plotted as a function of latitude. The zonal wind patterns are evident as local maxima and minima in this figure. The 5°–10° shift in the zonal patterns between the two seasonal periods is also apparent. The most striking feature of the zonal wind patterns is the large asymmetry in the summer-to-winter wind speed variation between the two hemispheres. In particular, the westerlies in the Northern Hemisphere increase by approximately 50% between summer and winter whereas the corresponding increase in the Southern Hemisphere is only 10%.

The global distribution of significant wave height and its seasonal variation is perhaps less well known, especially in the southern ocean. Since waves are generated by winds, one would expect a high correlation between wind speed and $H_{1/3}$. During boreal summer, average wave heights are low (2–3 m) in both the North Atlantic and North Pacific. In the southern oceans

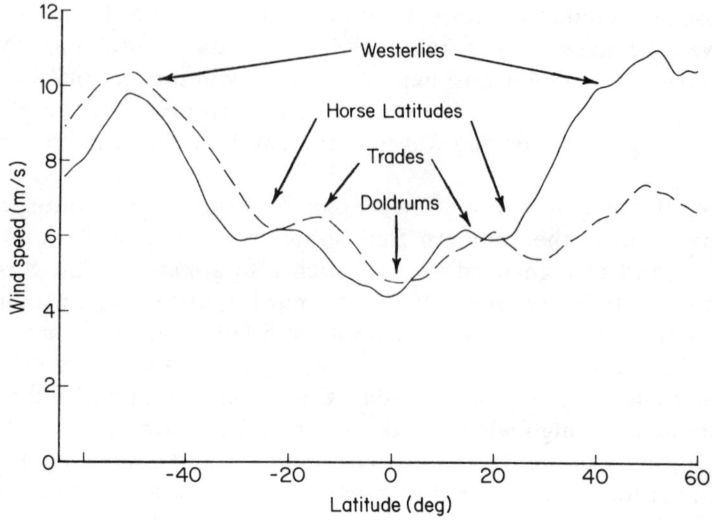

Fig. 15. Wind speed averaged over longitude for two seasonal periods, June–Sepember (dashed curve) and December–March (solid curve). The major zonal wind patterns shift between the two seasons.

significant wave heights are highest (4–5.5 m) during this period. Wave heights are generally low in equatorial regions because of low wind speeds in these areas.

The significant wave height map (see Sandwell and Agreen, 1984) for boreal winter shows a dramatic increase in $H_{1/3}$ in the northern ocean areas and only a small decrease in the southern oceans. This reflects the seasonal variation in wind speed. Like the average wind speed, the highest waves (>5.5 m) occur in the North Atlantic during its winter. Localized wave height maxima occur over the Kuroshio and Gulf Stream as well.

The zonal wave patterns are illustrated in Fig. 16 where longitudinally averaged significant wave height is plotted against latitude. The dominant zonal wave patterns are associated with the westerlies and therefore occur at mid-latitudes. Again note the dramatic seasonal change in the Northern Hemisphere while the seasonal variation in the Southern Hemisphere is relatively small. There is no pronounced local minimum corresponding to the horse latitudes although a hint of minimum appears at +20° on the December–March profile. The trade winds do not produce a local maximum in the zonal wave heights. There is, however, a wave height minimum corresponding to the doldrums. Moreover, this minimum shifts by about 5–10° between the two seasonal periods reflecting the shift in the zonal wind patterns.

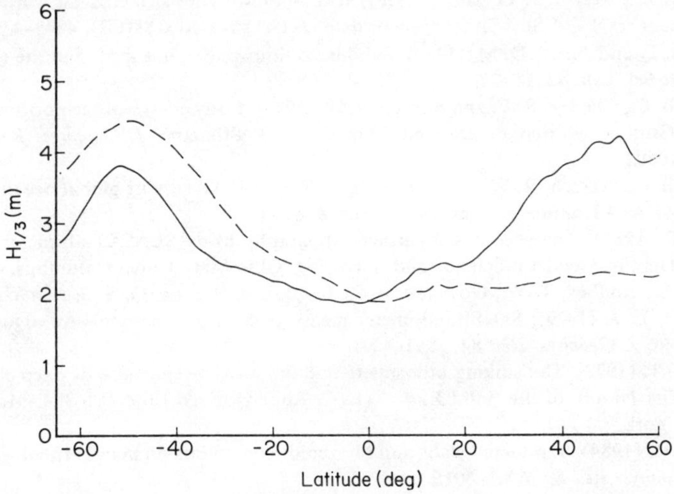

Fig. 16. Significant wave height averaged over longitude for the two seasonal periods, June–September (dashed curve) and December–March (solid curve). Wave propagation smooths the zonal wave patterns with respect to the zonal wind patterns.

The smoothness of the zonal wave patterns as compared with the zonal wind patterns is due to wave propagation. Waves generated by storms in the mid and high latitudes propagate to lower latitudes where they are detected by the altimeter (Mognard, 1982). This propagation effect is most apparent on the December–March map where the central ocean areas, with their large aperture of exposure, have higher waves than coastal areas where waves can approach from only one side.

Acknowledgments

The authors gratefully acknowledge the U.S. Government to reproduce the figures in this chapter.

References

Brown, O. B. and Cheney, R. E. (1983). Advances in satellite oceanography, *Rev. Phys. Space Phys.* **21** (5), 1216–1230.

Chapman, M. E. (1979). Techniques for interpretation of geoid anomalies, *J. Geophys. Res.* **84**, 3793–3901.

Chelton, D. B., Hussey, K. J. and Parke, M. E. (1981). Global satellite measurements of water vapor, wind speed and wave height, *Nature* **294**, 529–532.

Cheney, R. E. and Marsh, J. G. (1982). Global ocean circulation from satellite altimetry. *Trans. Am. Geophys. Un.* **63**, 997.

Cheney, R. E., Marsh, J. G. and Beckley, B. D. (1983). Global mesoscale variability from collinear tracks of SEASAT altimeter data. *J. Geophys. Res.* **88**(C7), 4343–4354.

Crough, S. T. and Jurdy, D. M. (1980). Subducted lithosphere, hot spots and the geoid. *Earth Planet Sci. Lett.* **48**, 15–22.

Douglas, B. C., Cheney, R. E. and Agreen, R. W. (1983). Eddy energy of the northwest Atlantic and Gulf of Mexico determined from GEOS-3 altimetry. *J. Geophys. Res.* **88**(C14), 9595–9603.

Douglas, B. C., Agreen, R. W. and Sandwell, D. T. (1984). Observing global ocean circulation with SEASAT altimeter data. *Mar. Geod.* **8**, 67–83.

Engelis, T. (1983). Analysis of sea surface topography using SEASAT altimeter data. Rep. 343, Dep. of Geodetic Science and Surveying, Ohio State Univ., Columbus, Ohio.

Fedor, L. S., Godbey, T. W., Gower, J. F. R., Guptill, R., Hayne, G. S., Rufenach, C. L. and Walsh, E. J. (1979). Satellite altimeter measurements of sea state—An algorithm comparison. *J. Geophys. Res.* **84**, 3991–4001.

Griggs, D. T. (1972). The sinking lithosphere and the focal mechanisms of deep earthquakes. In "The Nature of the Solid Earth" (E. C. Robertson, ed.) pp. 361–384, McGraw-Hill, New York.

Hager, B. H. (1984). Subducted slabs and the geoid: constraints on mantle rheology and flow. *J. Geophys. Res.* **89**, 6003–6015.

Haxby, W. F., and Turcotte, D. L. (1978). On isostatic geoid anomalies. *J. Geophys. Res.* **83**, 5473–5478.

Lerch, F. J., Marsh, J. G., Klosko, S. M., and Williamson, R. G. (1982). Gravity model improvement for SEASAT. *J. Geophys. Res.* **87**, 3281–3296.

Levitus, S. (1982). Climatological atlas of the world ocean. NOAA Professional Paper 13 Rockville, Maryland.

Marsh, J. G. and Martin, T. V. (1982). The SEASAT altimeter mean sea surface model. *J. Geophys. Res.* **87**, 3269–3280.

Marsh, J. G. and Williamson, R. G. (1980). Precision orbit analyses in support of the SEASAT saltimeter experiment. *J. Astron. Sci.* **28**, (4), 345–369.

McAdoo, D. C. and Martin, C. F. (1984). SEASAT observations of lithospheric flexure seaward of trenches. *J. Geophys. Res.* **81**, 3201–3210.

McAdoo, D. C., Martin, C. F. and Poulose, S. (1985). SEASAT observations of flexure: Evidence for a strong lithosphere. *Tectonophysics.* **116**, 209–222.

Menard, Y. (1983). Observations of eddy fields in the northwest Atlantic and northwest Pacific by SEASAT altimeter data. *J. Geophys. Res.* **86**, 8022–8030.

Mognard, N. M. (1982). Applications of satellite radar altimetry to the determination of sea state. Thesis, University Paul Sabatier de Toulouse, Toulouse, France.

Mognard, N. M., Campbell, W. J., Cheney, R. E. and Marsh, J. G. (1983). Southern ocean mean monthly waves and surface winds for winter 1978 by SEASAT radar altimeter. *J. Geophys. Res.* **88**, 1736–1744.

National Aeronautics and Space Administration (1981). Satellite altimetric measurements of the ocean: Report of the TOPEX science working group. Jet Propulsion Laboratory, Pasadena, California.

Parsons, B. and Sclater, J. G. (1977). An analysis of the variation of ocean floor bathymetry with age. *J. Geophys. Res.* **82**, 803–827.

Sandwell, D. T. (1984). Thermomechanical evolution of oceanic fracture zones. *J. Geophys. Res.* **89**, 11401–11413.

Sandwell, D. T. and Agreen, R. W. (1984). Seasonal variation in wind speed and sea state from global satellite measurements, *J. Geophys. Res.* **89**.

Sandwell, D. T. and Schubert, G. (1982). Lithospheric flexure at fracture zones. *J. Geophys. Res.* **87**, 4657–4667.

Schwiderski, E. W. (1979). Ocean Tides, Part I: Global ocean tidal equations. *Mar. Geod.* **3**, 161–217.

Tai, C. K. and Wunsch, C. (1983). Absolute measurement by satellite altimetry of dynamic topography of the Pacific Ocean. *Nature* **301**, 408–410.

Tai, C. K. and Wunsch, C. (1984). An estimate of global absolute dynamic topography, *J. Phys. Oceanogr.* **14** (2), 457–463.

Vening-Meinesz, F. A. (1964) "The Earth's Crust and Mantle", p. 124. Elsevier, New York.

Wagner, C. A. and Lerch, F. J. (1978). The accuracy of geopotential models. *Planet. Space Sci.* **26**, 1081–1140.

Watts, A. B. and Ribe, N. W. (1984). On geoid heights and flexure of the lithosphere at seamounts. *J. Geophys. Res.* **89**, 11152–11170.

Watts, A. B. and Talwani, M. (1974). Gravity anomalies seaward of deep-sea trenches and their tectonic implications. *Geophys. J. R. Astron. Soc.* **36**, 57–90.

Wilson, J. T. (1965). A new class of faults and their bearing on continental drift. *Nature* **207**, 343–347.

Wyrtki, K. (1979). Sea level variations; monitoring the breath of the Pacific. *EOS*, **60**(3), 25–27.

Wyrtki, K. (1985). Monthly maps of sea level in the Pacific during the El Niño of 1982 and 1983. *IOC Tech. Ser. 30*, UNESCO, Paris.

5.3 Seismological Studies of the Deformation of Oceanic Lithosphere

Seth Stein and Emile A. Okal

Department of Geological Sciences, Northwestern University, Evanston, Illinois, USA

I. Introduction

Global earthquake seismology and plate tectonics have developed in a parallel and closely coupled manner during the last 20 years. Seismological instrumentation and techniques evolve sufficiently rapidly that seismology continues to provide new insight on tectonic questions of current interest.

During the first decade of plate tectonics, 1965–1975, many investigations focused on present-day plate *kinematics*, defining the location and nature of motion on plate boundaries. Introduction of the World Wide Standardized Seismograph Network in 1963 provided reliable earthquake locations

SPACE GEODESY AND GEODYNAMICS
ISBN 0-12-057840-9

which helped delineate mid-ocean ridge systems and downgoing plates at subduction zones. The large dataset that was obtained from reliably calibrated seismic stations with adequate global distribution yielded earthquake focal mechanisms demonstrating extension at spreading centers, subduction at convergent boundaries and confirming the predicted strike-slip motion on transform faults (an adequate listing of the relevant papers would involve several hundred; see Cox (1973) for a review). The geometry of earthquake mechanisms, along with spreading rates from magnetic anomalies and transform azimuths, formed the basis for models of global relative plate motions (Chase, 1978; Minster and Jordan, 1978).

With the general kinematics of present plate motions established, increasing attention has been devoted to three general issues:

(i) dynamics of plate tectonic processes,
(ii) deformation within plates, and
(iii) evolution of plate boundaries.

These topics are coupled and cannot be discussed separately: for example, intraplate deformation often reflects either forces applied to plates or plate boundary evolution. As these questions came to the fore, by about 1975, new methods in seismology provided important insights. The purpose of this chapter is to briefly review some of these results for a nonseismological audience. We first illustrate some seismological techniques, and then discuss plate scale processes, plate boundary processes and localized intraplate processes. Due to length limitations, we restrict our attention to the oceanic lithosphere.

II. Techniques

The before mentioned results were obtained by analyzing most exclusively the polarity (or direction of motion) of the first arrival ("*P*" wave) on seismic records; following advances in theoretical seismology, newer methods now allow general use of longer portions of the waveform. While their goal remains the traditional one of determining earthquake depth, fault orientation and direction and history of slip (Fig. 1), these techniques provide considerably better resolution than was previously possible.

The first method, *body wave modeling*, matches the first minute or two of the *P*-wave seismogram. This approach uses an algorithm which computes the seismogram as a sum of the direct P wave and of near-surface reflections and conversions (Fukao, 1970; Langston and Helmberger, 1975; Kroeger and Geller, 1986). The radiation pattern of the earthquake source, reflection coefficients at interfaces and the delay of the different reflections are included. The resulting series of impulses with the correct amplitudes and

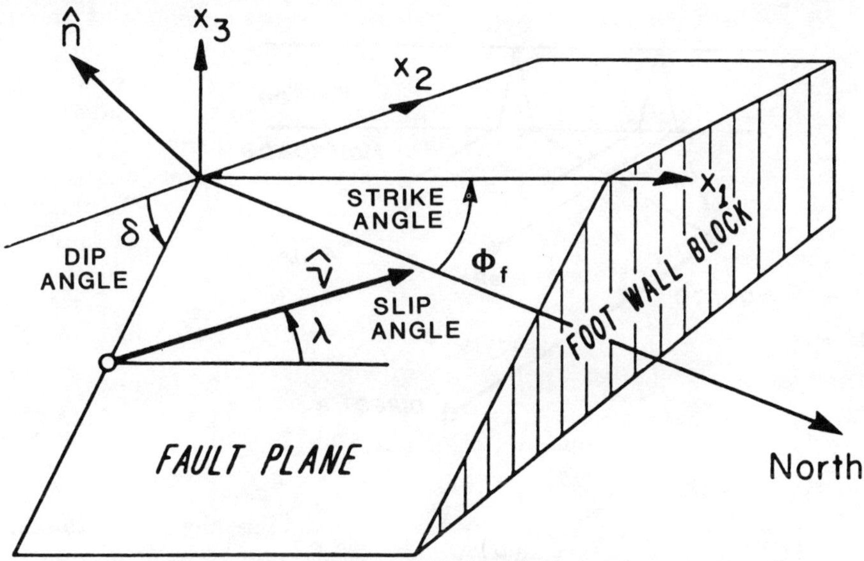

Fig. 1. Fault geometry used in seismological studies. The fault is treated as planar and characterized by strike ϕ, dip δ, and slip λ angles.

delays is convolved with operators representing the pulse generated by the earthquake (source time function), attenuation along the ray path and the seismometer transfer function (Fig. 2). Synthetic seismograms are generated for a range of possible parameters and the best fit to the data found by direct matching or a formal error criterion.

This method defines a best-fitting fault geometry consistent with first motion data, and constrains both focal depth and source time function. It is especially useful for constraining focal depth from the delay of the free surface reflections pP and sP, and of the water bottom reflection pwP, relative to the direct P pulse (Fig. 2). P waveforms also depend on source parameters other than depth; fortunately, the time separation between arrivals, which provides most of the depth information, is relatively insensitive to focal mechanism and source time function. The accuracy of depth estimates relying on travel-time data alone can be as poor as 30 km, especially in remote oceanic locations with no regional station coverage; while they correctly define the brittle lithosphere as a shallow layer (e.g. in a subducting plate), such estimates cannot resolve earthquake depths within the lithosphere. Thus body wave modeling, with depth resolution of several kilometers, proves very valuable for tectonic studies. The depth defined by this process is known as the centroid depth, the average depth of a fault surface which may extend over some depth range.

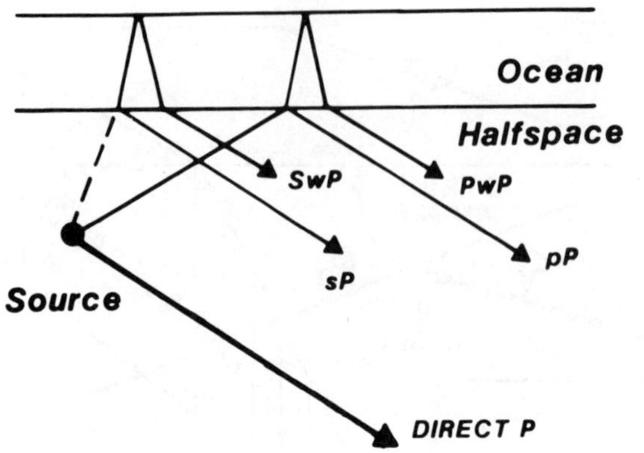

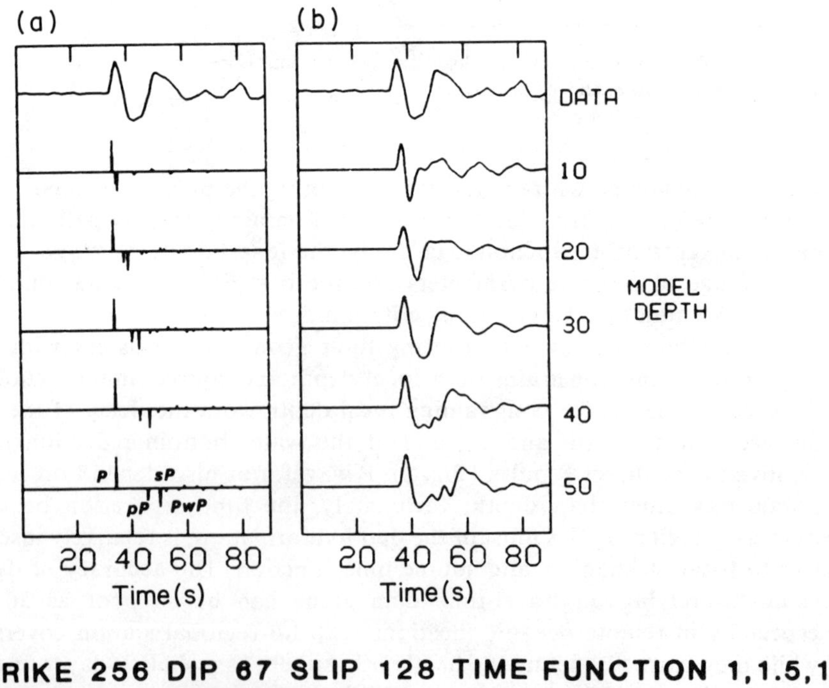

Fig. 2. Body-wave modeling procedure for depth determination. (a) Synthetic seismograms for an assumed fault geometry and (b) including the effects of the seismometer and attenuation calculated for various depths. The data are best fit by a depth near 30 km.

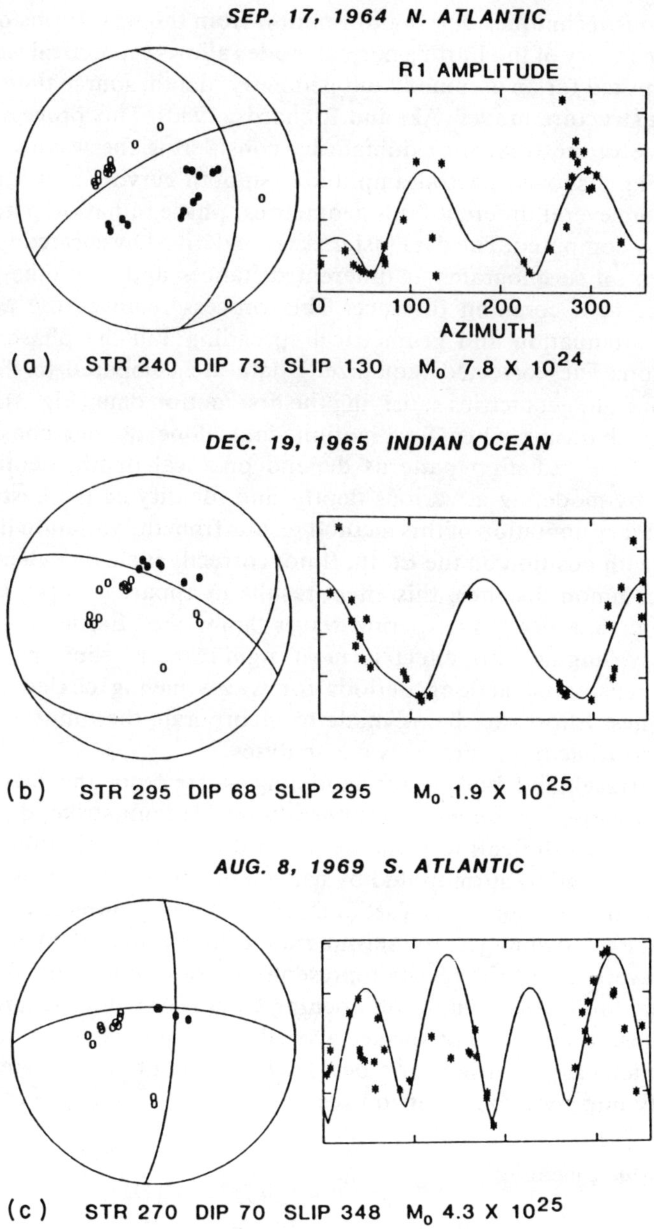

Fig. 3. Surface wave focal mechanism study for (a) thrust, (b) normal, and (c) strike–slip fault earthquakes. The theoretical Rayleigh wave amplitude radiation patterns shown yield the best fit to the data consistent with the first motions (Wiens and Stein, 1984).

Analogous techniques derive information from the waveforms of *surface waves.* The theory of the Earth's normal modes allows theoretical waveforms to be computed for an assumed fault geometry, depth, source time function and Earth structure model (Aki and Richards, 1980). This process is easier to visualize in the frequency domain by considering the wave's radiation pattern; Fig. 3 shows spectral amplitude (smooth curves) as a function of azimuth for several different fault geometries; phase radiation patterns can be similarly computed. The data (asterisks) are derived by correcting spectral amplitudes on seismograms at different distances and azimuths from the earthquake to a common distance. This process removes the amplitude effects of attenuation and geometrical spreading and the phase effect of propagation. The corrected (equalized) data are compared to theoretical spectra for fault geometries satisfying the first motion data (Fig. 3), yielding reliable mechanisms when first motion data alone do not constrain the solution. Since radiation patterns depend on focal depth, depth can be estimated by modeling at various depths and identifying the best fit.

The primary limitation of this method results from the variation in velocity structure with position on the Earth. If not correctly included when equalizing to a common distance, this effect results in apparent noise, especially for phase data. Since longer-period waves "integrate" Earth structure over longer wavelengths, this effect is most significant at short periods, but becomes substantial at long periods for waves having circled the Earth several times. Efforts are being made to incorporate the influence of such *lateral heterogeneity* in surface wave analyses.

Both surface- and body-wave modeling suffer from the fact that the observed seismogram depends nonlinearly on the fault strike, dip and slip angles. It is thus difficult to formally invert data for a best fitting solution. This difficulty can be surmounted by using a moment tensor representation which describes the source as a set of forces and force couples. The moment tensor has two advantages: the seismogram is linear in its elements and can thus be inverted and the tensor represents sources more general than slip on a planar fault (for example, an opening crack) (Aki and Richards, 1980). Moment tensor inversion of surface wave data is conducted for an assumed depth, which can be varied for a best fit. Inclusion of lateral heterogeneity in velocity improves the inversion results.

III. Plate-wide Processes

A. Earthquake Depths and Rheology of the Lithosphere

The thickness of the lithosphere, the upper layer of strong material which deepens as the plate cools, has been inferred from several techniques with different results. The differences result primarily, as discussed later, from the different characteristic times of the forces applied to the plates. *Seismic thickness*, the depth to the low-velocity zone, is measured from surface wave dispersion studies (Forsyth, 1975; Yu and Mitchell, 1979); *elastic thickness*, estimated from flexure below mid-plate loads (Bodine *et al.*, 1981), is found to be much less than seismic thickness. The depths of oceanic intraplate earthquakes provide another measure as the maximum depth of seismicity increases with lithospheric age. Comparison with predicted thermal structures (Fig. 4) indicates that the maximum depth of seismicity corresponds approximately to the 750°C isotherm (Chen and Molnar, 1983; Wiens and Stein, 1983, 1984). The thickness of the seismically active portion of the lithosphere is equal to or slightly greater than the elastic thickness, but about one-half the seismic (surface wave) thickness.

Lithospheric deformation is thought to be controlled by the variation with depth of strength (the maximum difference between principal stresses that rock can support). The earthquake depth data suggest that above about 750°C the lithosphere cannot support stresses needed for seismic failure. Experimental data (Brace and Kohlstedt, 1980; Kirby, 1980) can be used with a geotherm to compute strength as a function of depth (Fig. 5): at shallow depths brittle fracture occurs and strength increases linearly with depth. At greater depths strength is controlled by ductile flow laws, which vary with mineral type; since strength decreases exponentially with temperature, rocks weaken rapidly with depth. Thus, as the plate ages and cools, the strong portion of the lithosphere deepens (Fig. 5).

In the ductile domain, strength depends on strain rate: thus flexural thickness from long-term loads (at low strain rates) is less than the thickness measured with surface waves, corresponding to shorter-term loads or high strain rates. This explains why the responses of the plate to loading by a new volcanic island (which may take typically 1 Ma to grow) and to the passage of a surface wave (of typical period 100 s) differ drastically. Thc strain rates for intraplate deformation shown on Fig. 5 were estimated from the rate of seismic moment release of oceanic intraplate earthquakes: the lower value, $10^{-18}\,s^{-1}$, is for the oceanic lithosphere as a whole; the upper value, $10^{-15}\,s^{-1}$, is for a very active zone of intraplate deformation, the Ninetyeast Ridge area (Stein and Okal, 1978).

The relationship between results derived in the laboratory (and extrapolated to geological strain rates) and the actual behavior of materials in the Earth is somewhat unclear. Nonetheless, it is useful to be able to relate the depth of seismicity to strength envelopes that can be calculated for different thermal and mechanical situations and compared between tectonic environments.

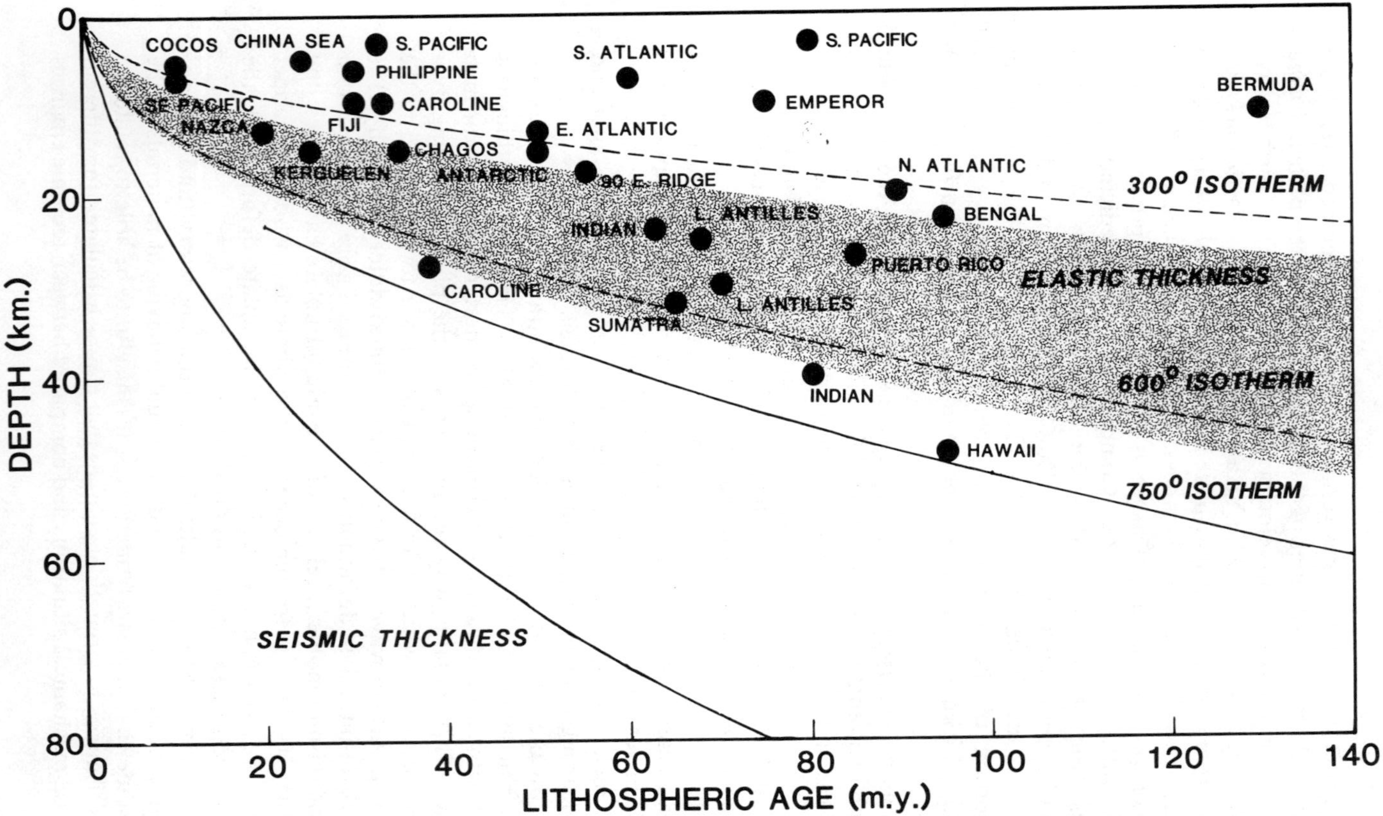

Fig. 4. The maximum depth of intraplate earthquakes deepens as the oceanic lithosphere cools (Wiens and Stein, 1983). The isotherms shown are calculated from a lithospheric cooling model (Parsons and Sclater, 1977). Stippled region denotes range of estimates of the flexural elastic thickness. The seismic thickness is taken from Rayleigh wave dispersion data. The failure limit is the lower limit at which 20 MPa (200 bar) deviatoric stress can be sustained for a dry olivine rheology and a strain rate of 10^{-18} s^{-1}.

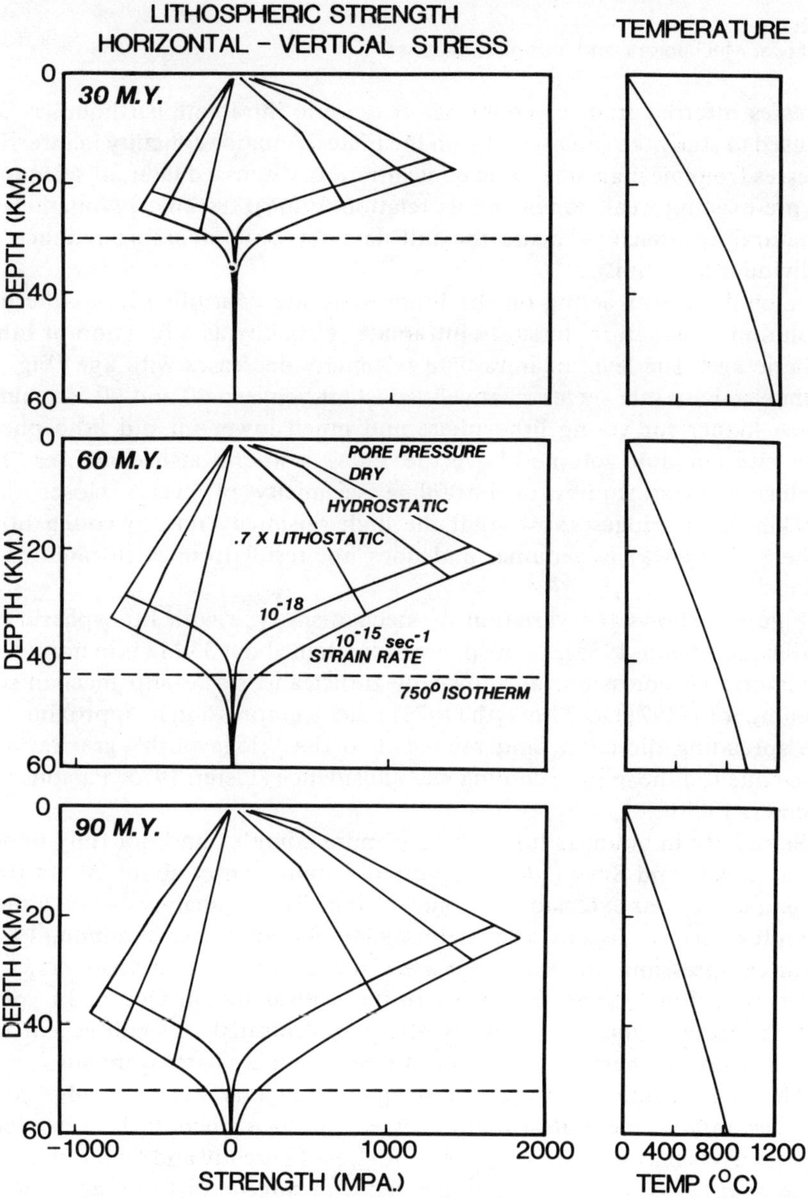

Fig. 5. Strength envelopes showing maximum stress difference as a function of depth calculated using temperatures from the Parsons and Sclater (1977) plate cooling model (Wiens and Stein, 1983). At shallow depth strength is controlled by brittle fracture; at greater depth ductile flow laws predict rapid weakening. The 750°C isotherm (dashed line) is the approximate lower bound for intraplate seismicity. The strong region deepens as the plate cools.

B. Focal Mechanisms and Intraplate Stresses

Stresses inferred from mechanisms of oceanic intraplate earthquakes can be used to study the forces acting on the plates. A major difficulty in inferring stresses from mechanisms is the concentration, discussed later, of seismicity on pre-existing weak zones and its relation to local tectonics. Nonetheless, a natural approach is to use the full data set while avoiding reliance on individual mechanisms.

Since the forces acting on the lithosphere are controlled by its thermal evolution, it is natural to study intraplate seismicity as a function of lithospheric age. The level of intraplate seismicity decreases with age (Fig. 6). The seismicity rate seems relatively constant between 40 and 90 Ma, but is much higher for young lithosphere and much lower for old lithosphere. The rate per unit volume above the 750°C isotherm also decreases. The exclusion from studies of intraplate seismicity of events closer than 100 km to the ridges shows that the high seismicity rate in young lithosphere (5–15 Ma) is genuine, and does not result from mislocated ridge events.

Figure 7 shows the variation of mechanism type with lithospheric age (Wiens and Stein, 1985). Lithosphere older than about 35 Ma is in horizontal deviatoric compression, as shown by thrust and strike-slip mechanisms. Mendiguren (1971) and Forsyth (1973) noted compression in approximately the spreading direction, and related it to the "ridge push" gravitational force due to lithospheric cooling and subsidence (Lister, 1975; Parsons and Richter, 1980).

Seismicity in younger lithosphere is more complex and not fully understood. Sykes and Sbar (1974) suggested a transition at about 20 Ma from spreading-related extension to compression. Recent analyses show a more complex picture (Wiens and Stein, 1984; Bergman and Solomon, 1984). Both compression and extension are observed in young lithosphere, with most extensional events occurring in the central Indian Ocean. In young lithosphere elsewhere, faulting is often concentrated in small seismically active regions, where mechanisms can be extremely heterogeneous.

The high seismicity in young lithosphere suggests that near-ridge stress sources, rather than plate-wide processes, produce much of this seismicity. These effects may include thermoelastic stress (Turcotte and Oxburgh, 1973; Bratt *et al.*, 1985) and stresses associated with ridge–transform intersections (Fujita and Sleep, 1978). However, some information about plate-wide stresses can be extracted from earthquakes in young lithosphere. In lithosphere younger than 35 Ma, tensional axes of normal fault events are oriented at large angles to the spreading direction (Fig. 8); these earthquakes thus *cannot* indicate ridge-related extension. In addition, no age boundary

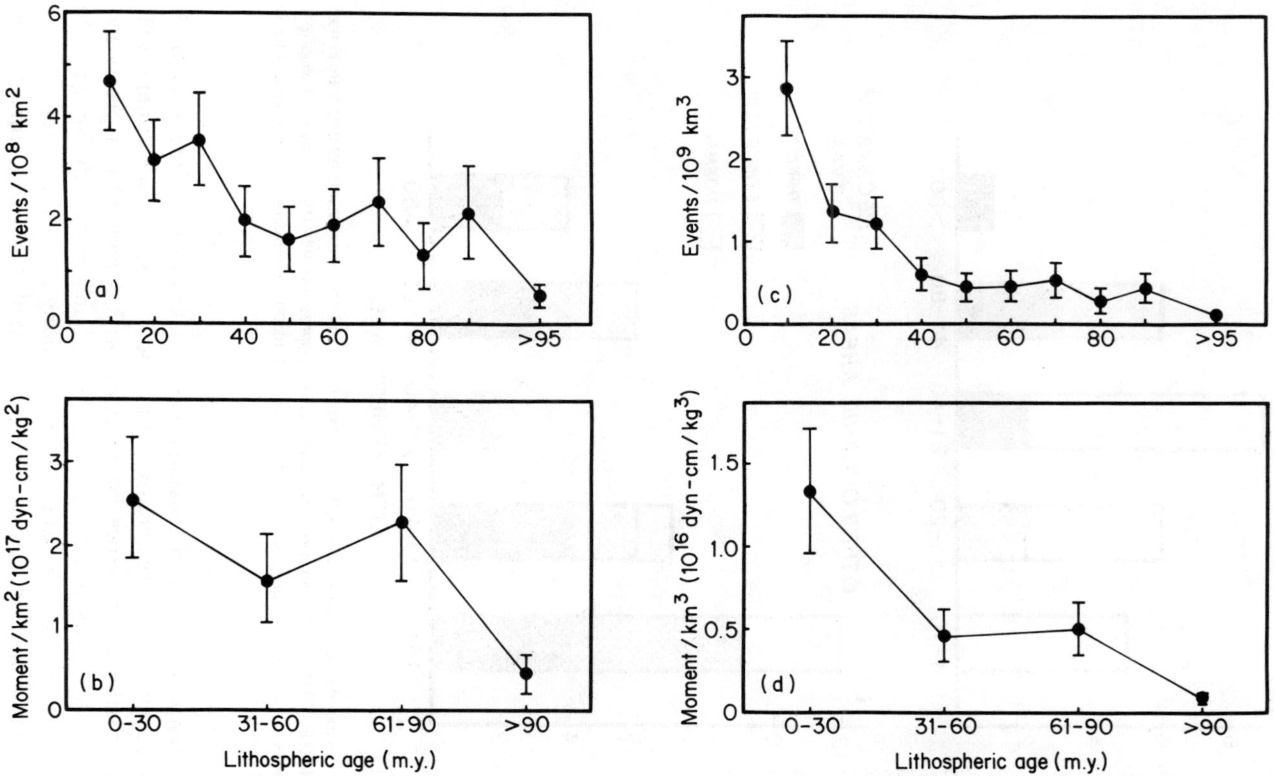

Fig. 6. Decrease in oceanic intraplate seismicity with lithospheric age for the period 1964–1979 (Wiens and Stein, 1983). (a) Number of oceanic intraplate earthquakes with m_b above 5.0 per surface area versus age. (b) Number of intraplate earthquakes per volume of strong lithosphere (that above the 750° isotherm) versus age. Cumulative seismic moment of oceanic intraplate seismicity per area (c) and volume (d) as a function of age.

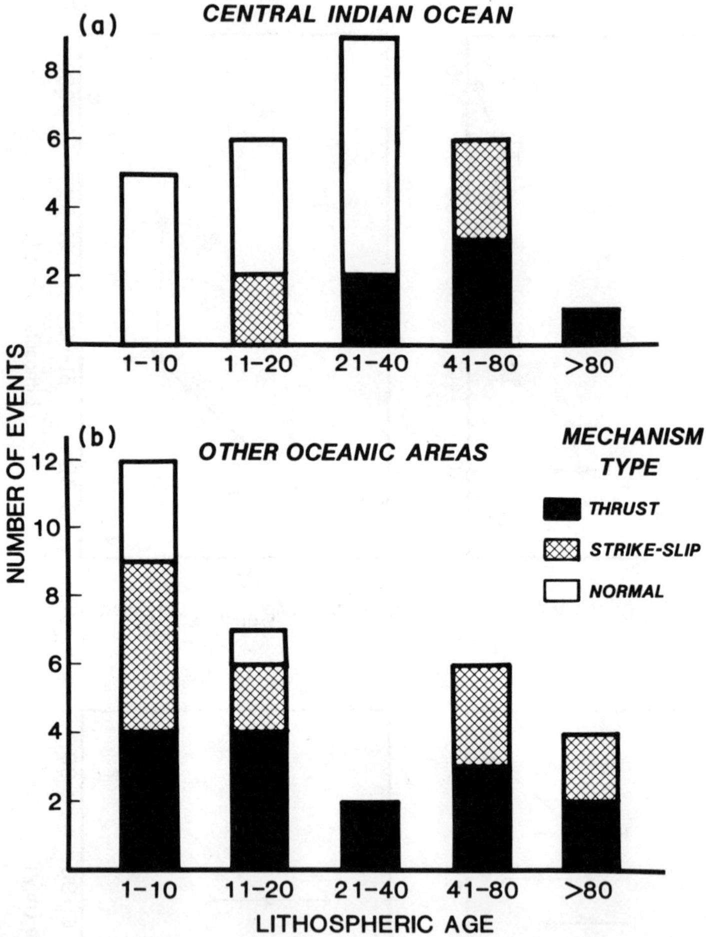

Fig. 7. Focal mechanism type as a function of lithospheric age for oceanic intraplate earthquakes (Wiens and Stein, 1984). Older oceanic lithosphere is in compression, whereas younger lithosphere has both extensional and compressional mechanisms. Extensional events are located primarily in the Central Indian Ocean.

between compressional and tensional earthquakes can be drawn, as both types can coexist in any area younger than about 35 Ma. A two-dimensional "average" plate can therefore be modeled with a general compressional stress in the spreading direction for all ages. Such models do not include the tensional stress parallel to the ridge in young (≤ 35 Ma) areas.

Assuming that the plate is in equilibrium, a simple two-dimensional model shows that stress is determined by the ridge-push force and by boundary

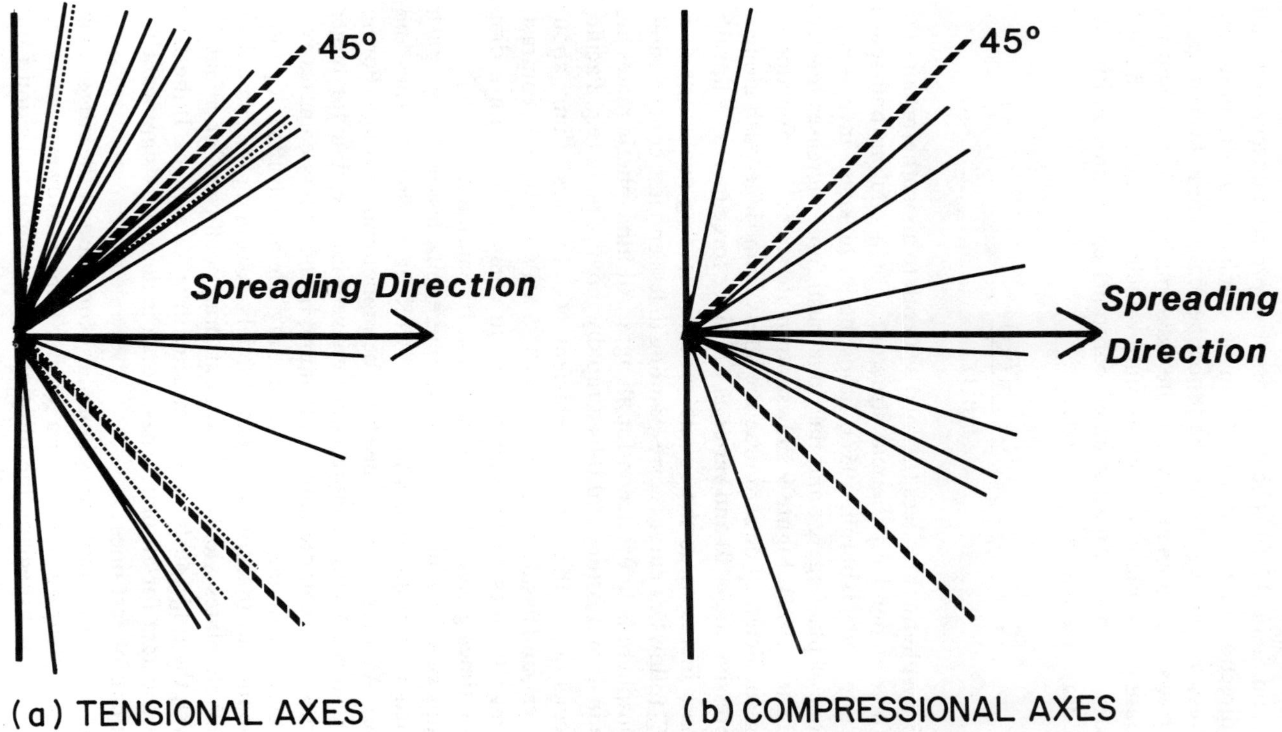

Fig. 8. Angle between the horizontal component of principal stress and the spreading direction for normal faulting (a) and thrust faulting (b) intraplate earthquakes in young oceanic lithosphere (Wiens and Stein, 1984). Tensional axes of normal faulting events, most of which are located in the Central Indian Ocean, generally show extension oblique to the spreading direction, while compressional axes of thrust faulting events show a weak preferred orientation in the spreading direction [Central Indian Ocean (——); Other regions (·····)].

conditions at the ridges and bottom of the plate (Mendiguren and Richter, 1978; Wiens and Stein, 1985). The deviatoric compression observed for all ages in the direction of spreading (x) can be compared to the predicted deviatoric stress $\overline{\sigma_{xx}}$ averaged over the lithospheric thickness $h(t)$ at age t. This stress shows the balance between the ridge-push force $F(t)$ and the drag at the base of the plate. For an absolute velocity ν (taken as the half spreading rate), the basal drag (assumed constant) equals the product of velocity and drag coefficient C, so that:

$$\overline{\sigma_{xx}}(t) = \frac{C\nu^2 t - F(t)}{h(t)}$$

Both the lithospheric thickness (defined as depth to an isotherm) and the ridge-push force depend on thermal structure. For a cooling half-space model (Turcotte and Oxburgh, 1967), lithospheric thickness increases as the square root of plate age $\sqrt{t}$ and the ridge-push force increases linearly with age t (Lister, 1975). Figure 9 shows intraplate stress as a function of age and drag coefficient. For zero drag the stress from ridge-push alone is purely compressive ($\overline{\sigma_{xx}} < 0$) and varies as $\sqrt{t}$, since force increases linearly while thickness increases as the square root. If the drag coefficient C is increased, $\overline{\sigma_{xx}}$ follows $\sqrt{t}$ curves corresponding to less and less compression, until the lithosphere is in extension for all ages. All lithospheric plates are known to be in compression, and thus a rapidly moving plate (e.g. Pacific, 10 cm/yr) constrains the drag coefficient C to less than about 3.5 MPa/(m/yr), and basal stress to less than 0.35 MPa (3.5 bar). In contrast, a plate moving 10 times slower can remain in compression with a drag coefficient 100 times greater and a basal stress 10 times greater.

For a cooling plate model, both the lithosphere thickness and ridge-push force approach asymptotic values with age (McKenzie, 1967; Parsons and Richter, 1980). As a result, stress need not be monotonic with age. For no drag ($C = 0$), stress becomes more compressive with age, but for larger drag, stress reaches a compressional maximum, becomes less compressive, and eventually extensional. The maximum drag coefficient allowing compression is about half that for half-space models. Assuming the same drag coefficient for all plates yields a maximum drag coefficient of about 2–4 MPa/(m/yr); the ridge-push force dominates drag for all ages. If the drag coefficient is constant for all plates, basal drag is much less significant for slow plates than for fast ones.

This model assumes a zero stress boundary condition at the ridge, or in other words, that the ridge axis has no tensile strength. However, stress in a young lithosphere, especially the location of a possible transition from compression to extension in the direction of spreading, would be sensitive

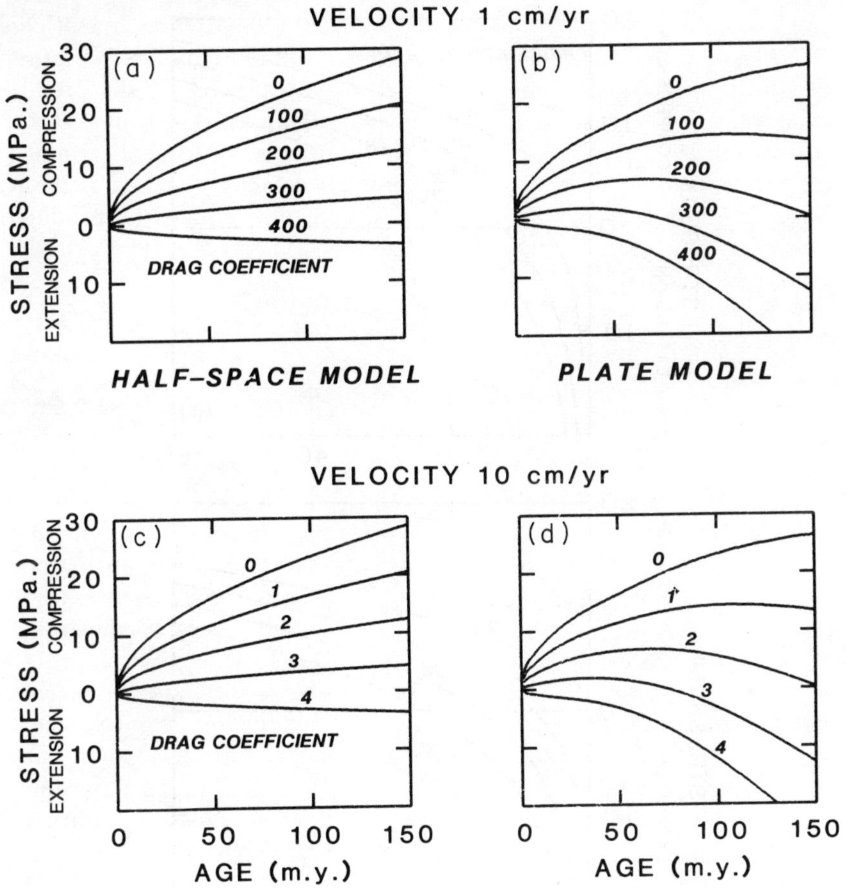

Fig. 9. Intraplate stress as a function of lithospheric age and assumed basal drag coefficient for slow moving (a, b) and fast moving (c, d) plates (Wiens and Stein, 1985). Figures (a) and (c) assume a half-space thermal model; Figures (b) and (d) assume a plate model. The observation of compressional stresses in oceanic lithosphere places an upper bound on the drag coefficient of 2–4 MPa/(m/yr).

to the strength of the ridge (Fig. 10). Alternative models with a tensile stress boundary condition, i.e. substantial strength in the axial region (Fleitout and Froidevaux, 1983), would predict a wide band of intraplate extension in the spreading direction. Since such a zone of normal faulting is not observed, the axial region must have little tensile strength, a useful constraint on ridge models (Sleep and Rosendahl, 1979). Figure 11 summarizes the different tectonic constraints.

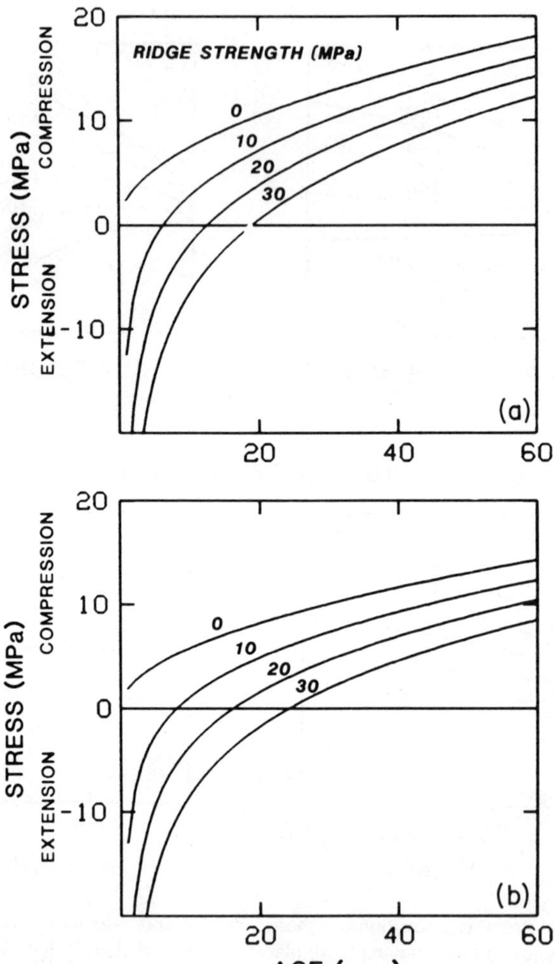

Fig. 10. Vertically averaged stress in the spreading direction as a function of lithospheric age computed for a plate velocity of 5 cm/yr and several values of ridge strength and basal drag (Wiens and Stein, 1985). The age of the transition from ridge-normal extension to compression increases with the strength of the ridge [(a) No basal drag and (b) drag coefficients 3 MPa/(m/yr)].

C. Constraints on Mantle Viscosity

Earthquake mechanisms, though not a primary tool for estimation of mantle viscosity, provide interesting inferences under certain assumptions. The analysis assumes a common model of plate motions driven largely by thermal buoyancy forces acting on the lithosphere with mantle return flow balancing

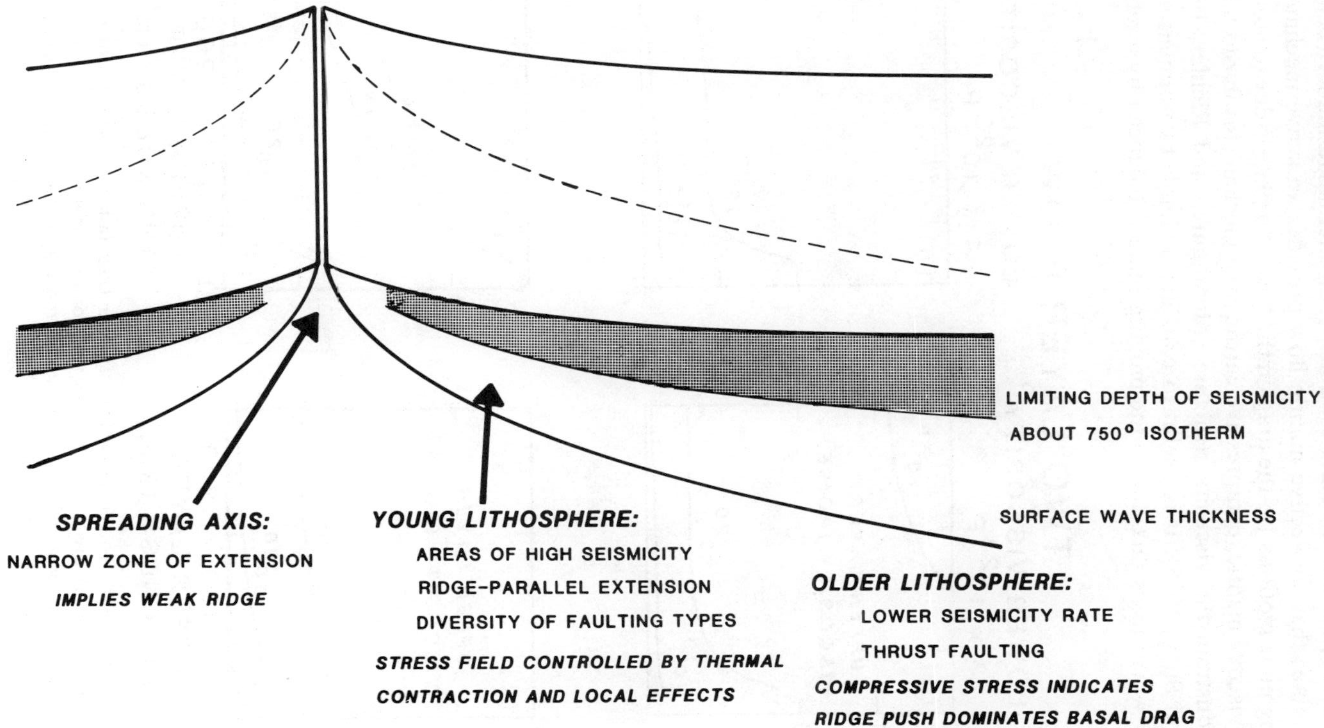

Fig. 11. Schematic diagram relating seismicity observations to the tectonics of the oceanic lithosphere (Wiens and Stein, 1985).

the mass transport due to plate motions. [An alternative model (Peltier, 1980) assumes that convective mantle flow provides an active mechanism for driving plate motions.] If the drag on the base of a plate is due to motion over the viscous mantle, compressive earthquake mechanisms in old lithosphere constrain the viscosity structure (Mendiguren and Richter, 1978; Wiens and Stein, 1985). It is useful to consider a simple two-dimensional geometry where mass flux due to the moving plate is balanced by a return

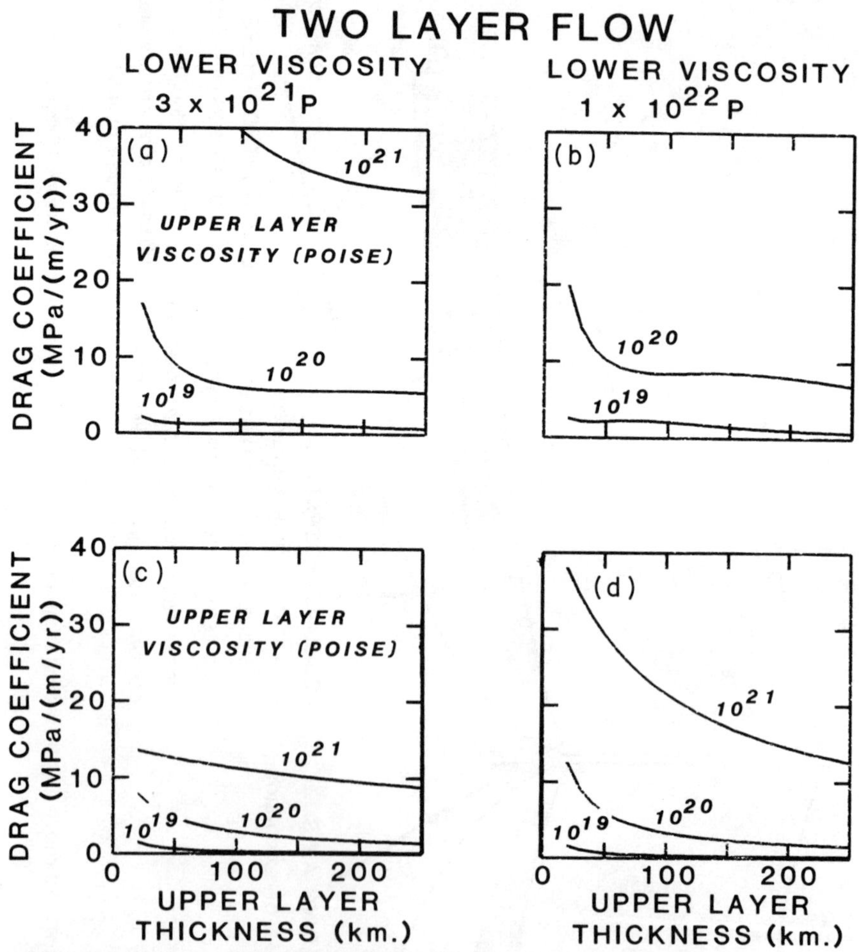

Fig. 12. Basal drag coefficients as a function of upper layer (asthenosphere) thickness and viscosity for two values of lower layer viscosity and total flow depths of 700 km (a, b) and 2900 km (c, d). The upper bounds on drag coefficients required by the focal mechanism data constrain the mantle viscosity. (Wiens and Stein, 1985).

flow at depth. The drag coefficient is proportional to viscosity and inversely proportional to flow depth. For the simplest case, in which the underlying fluid has uniform Newtonian viscosity, acceptable drag values require an average mantle viscosity significantly lower than estimated from glacial rebound, Earth rotation and satellite orbits.

This discrepancy can be reconciled by assuming that the plate is underlain by a thin low-viscosity asthenosphere. Such a model satisfies constraints from gravity (Richter and McKenzie, 1978), driving forces (Hager and O'Connell, 1979) and glacial isostasy (Cathles, 1975; Peltier, 1983). In this model the low-viscosity layer, in which only a fraction of the return flow occurs, decouples the plates from the underlying mantle. For a mantle (lower layer) with viscosity 3×10^{21} P, acceptable drag coefficient values occur for asthenospheric (upper layer) viscosities less than about 6×10^{19} P for upper mantle return flow and 2×10^{20} P for the whole mantle case (Fig. 12). For a more viscous mantle, the maximum acceptable asthenospheric viscosities decrease. The low-viscosity layer significantly reduces the effect of the total flow depth so that either upper mantle or whole mantle flow is acceptable. These viscosity bounds are for oceanic regions and may not be comparable to those for continental regions.

These simple calculations provide only general estimates of viscosity structure. In particular, the assumed two-dimensional flow is reasonable for large plates overlying a low-viscosity asthenosphere but less valid for smaller plates or a higher-viscosity atmosphere (Hager and O'Connell, 1979). Even if the drag coefficient is spatially uniform, the apparent drag would vary when the return flow is not antiparallel to the plate's absolute motion. Additionally, the temperature dependence of viscosity should lead to structures more complicated than simple layers. Nonetheless, intraplate stresses derived from earthquake focal mechanisms can be explained by conventional ideas of mantle viscosity and support a low-viscosity layer below the oceanic plates. Such decoupling is consistent with the lack of correlation between oceanic plate area and absolute velocity (Forsyth and Uyeda, 1975).

IV. Plate Boundary Processes

A. Ridge–Transform Systems

Earthquake seismology has also considerably improved our knowledge of oceanic spreading center systems. Early studies (Sykes, 1967) identified normal faulting on some mid-ocean ridges; such earthquakes are now known to be confined to slow-spreading ridges (less than about 6 cm/yr full rate), also characterized by axial valleys. Figure 13 shows a decrease in seismic

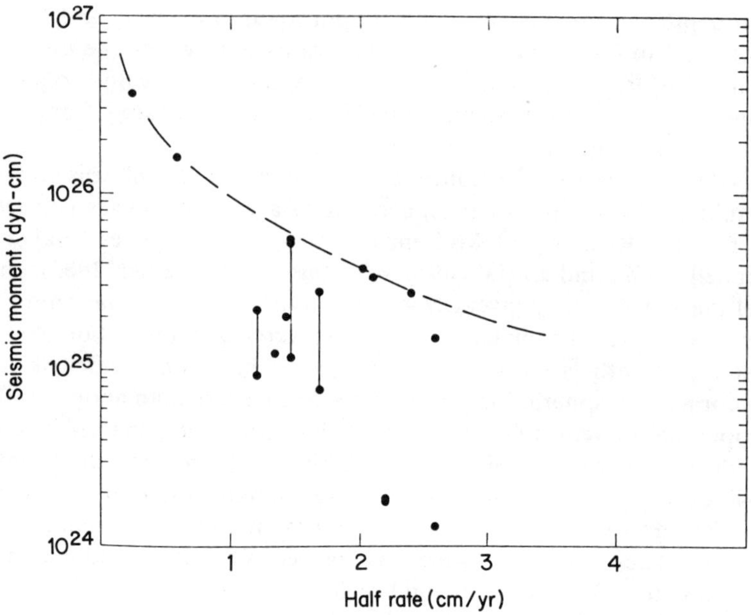

Fig. 13. Decrease in maximum seismic moment of ridge-crest normal fault earthquakes with spreading rate (Solomon and Burr, 1979).

moment of ridge-crest normal faulting earthquakes with spreading rate. This relation is consistent with faulting which is limited by temperature; the depth to a given isotherm and hence the area available for faulting decreases with spreading rate (Solomon and Burr, 1979). Fast-spreading ridges show an axial high, rather than a valley, and thus no normal faulting events.

Despite the presence of topography suggestive of vertical tectonism at transforms [notably anomalously high features (Bonatti, 1978)], detailed focal mechanism studies (Fig. 14) show almost no deviations from the transform-parallel strike–slip motion on steeply dipping fault planes predicted by Wilson (1965). The only exceptions occur at ridge–transform intersections (Fig. 15), which show a variety of tectonic anomalies due to the cooling of the ridge as it approaches an intersection (Fox and Gallo, 1984; Forsyth and Wilson, 1984).

The thermal structure of a transform fault should be essentially the average of the expected temperature on the two sides; coolest at the transform midpoint and hottest at either end (Fig. 16). As expected from the area available for faulting, the maximum seismic moment for transform earthquakes decreases with spreading rate (Fig. 17) (Burr and Solomon,

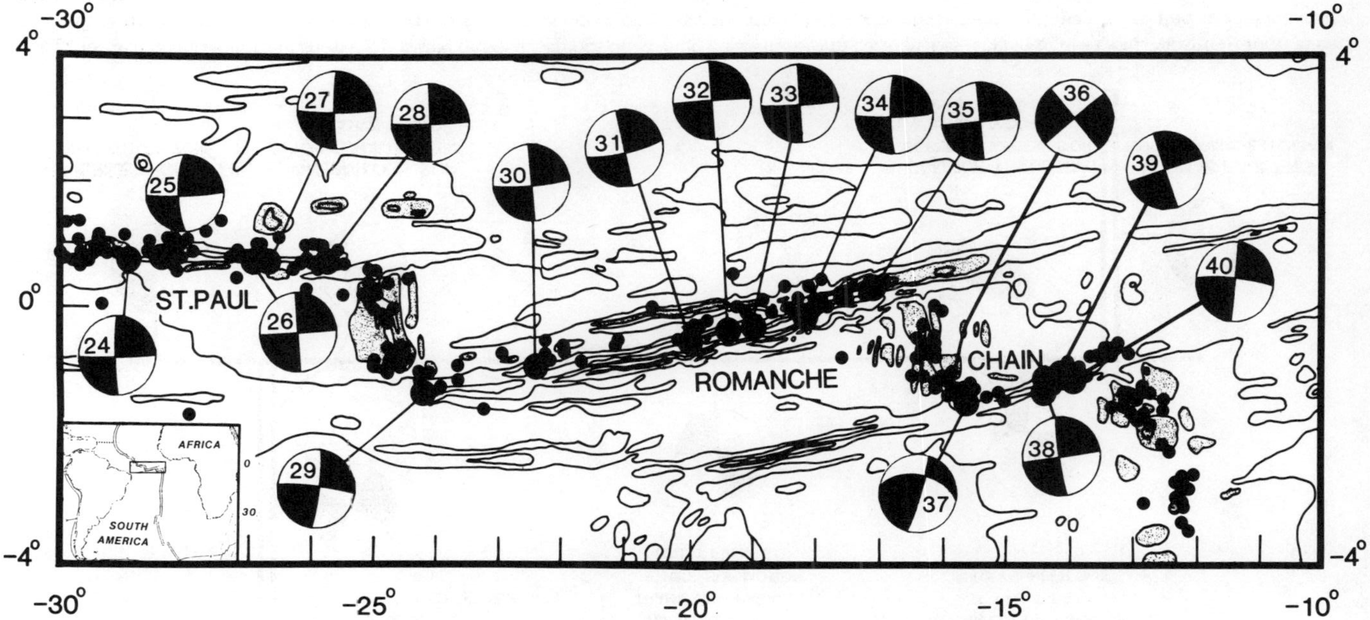

Fig. 14. Earthquake focal mechanisms on equatorial Atlantic transform faults (Engeln *et al.*, 1986). The vast majority of events show the expected transform–parallel strike–slip faulting on a steeply dipping plane. Note anomalous mechanism events near the west end of the Chain transform

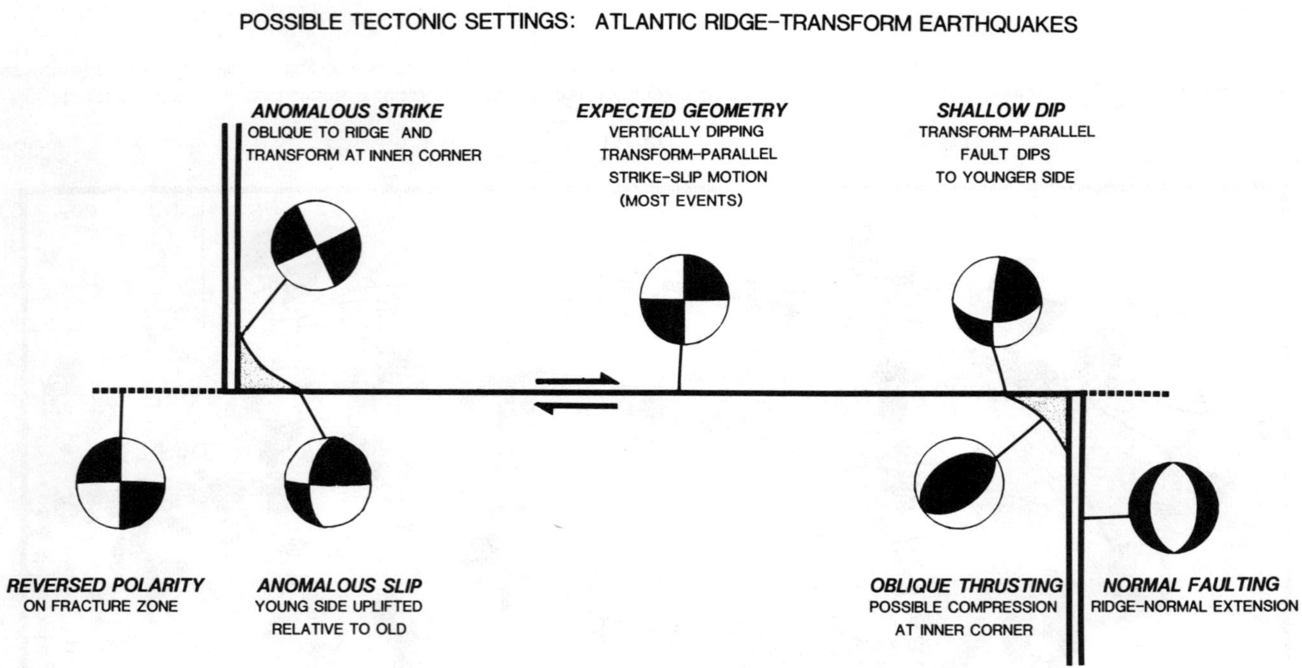

Fig. 15. Schematic transform showing possible tectonic settings of Atlantic ridge–transform earthquakes (Engeln *et al.*, 1986). Most transform events show the expected geometry. Anomalous geometry events occur only near the ridge–transform intersections and may be related to faulting at the inner corner.

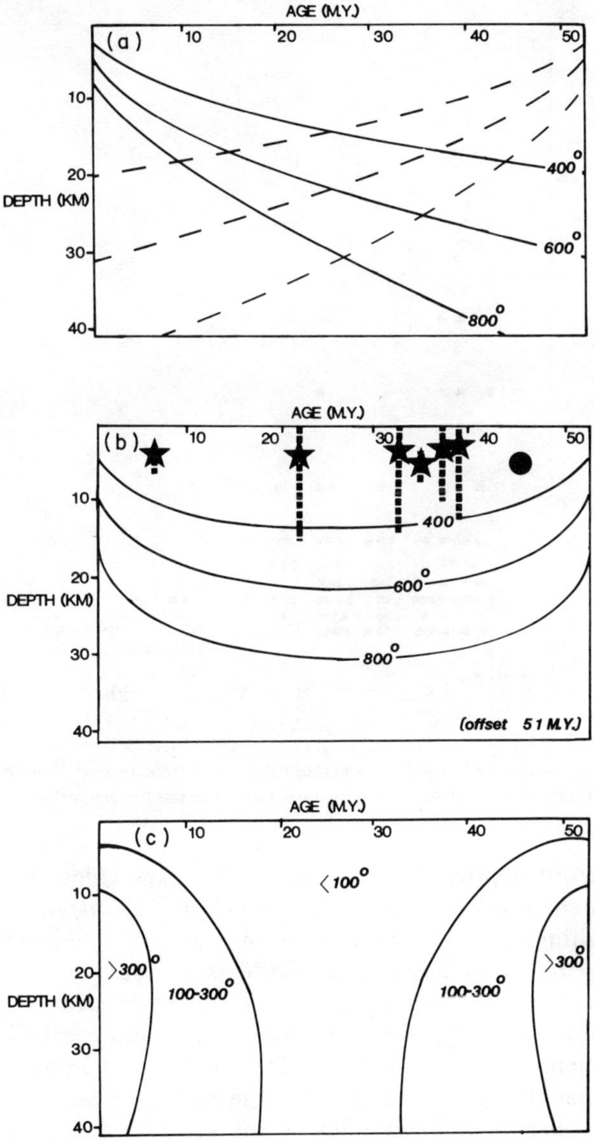

Fig. 16. Thermal structure of the Romanche Transform. (a) Temperatures on either flank predicted by the cooling half-space model. (b) Average temperature distribution along the transform. Stars represent the locations and depths of the well-constrained earthquakes; dots represent events with less reliable depths. Dashes represent the depth range of faulting estimated from the seismic moments. All centroid depths are above the 400° isotherm predicted by the thermal model and most faulting is above the 600°C isotherm. (c) Difference between transform and cooling half-space temperatures (Engeln *et al.*, 1986).

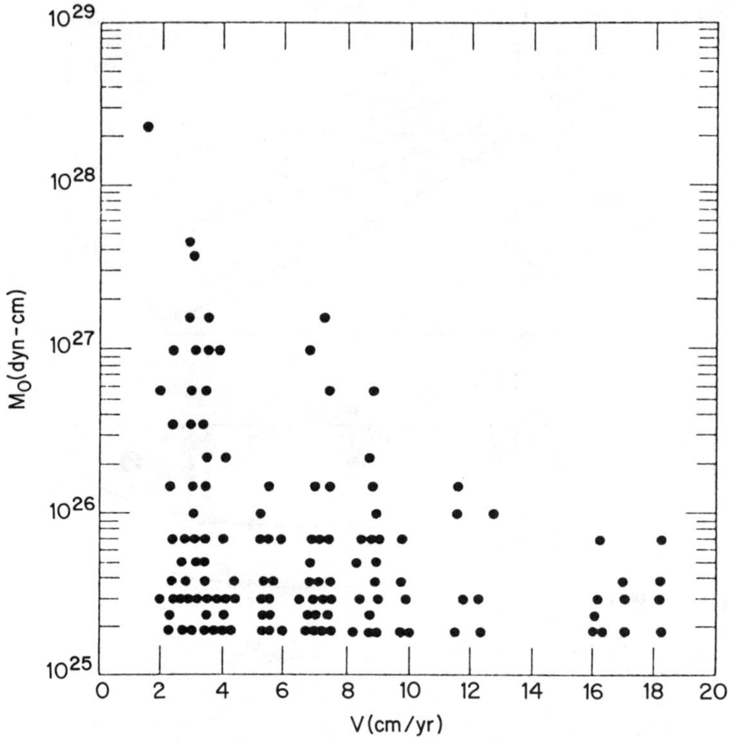

Fig. 17. Seismic moment versus spreading rate for oceanic transforms (Solomon and Burr, 1979). The maximum moment decreases with spreading rate as expected from thermal considerations.

1978). Centroid depths of transform earthquakes (Figs. 16 and 18) are limited to a shallow zone above the predicted 400°C isotherm. The depth range of faulting estimated from the seismic moments suggests that faulting extends approximately down to the 600°C isotherm.

The depth of seismicity has interesting implications for the slip process on transforms. Given the depth of faulting, slip rates can be inferred from seismic moment release and compared to those predicted by plate motions. Assuming that the areas above either the 400°C or 600°C isotherms fail seismically, seismic moment releases for six major Atlantic transforms covering the period 1920–1979 account for varying fractions of the slip predicted by the plate motion (Fig. 19). Some aseismic slip must be occurring if the time period sampled is long enough to be representative. An earlier analysis, predating precise constraints on depths of transform earthquakes, which assumed that all slip was seismic, implied faulting to an average depth approximately that of the 150°C isotherm (Burr and Solomon, 1978). In

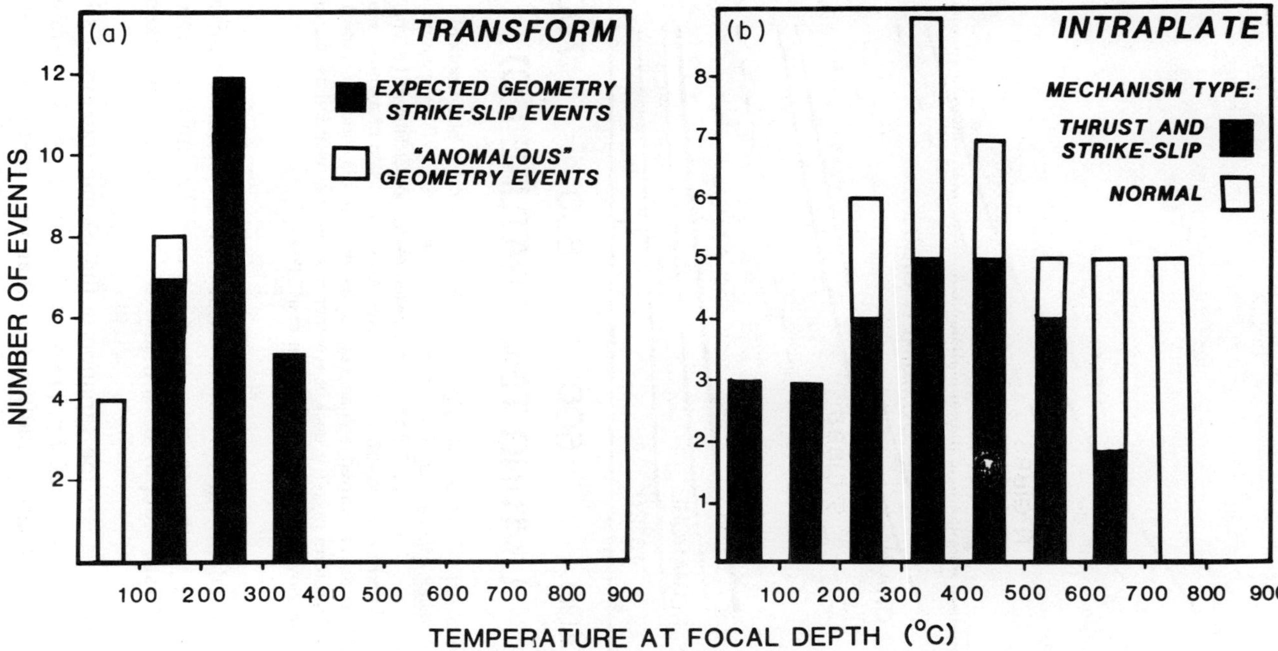

Fig. 18. Comparison of transform (a) and intraplate (b) earthquake depths (Engeln *et al.*, 1986). Centroid depths along transforms are confined to a zone above the 400°C isotherm (a) and concentrated between 100° and 300°C. In contrast, oceanic intraplate seismicity (b) occurs primarily between the 300°C and 600°C isotherms and is limited by the 750°C isotherm (Wiens and Stein, 1983, 1984).

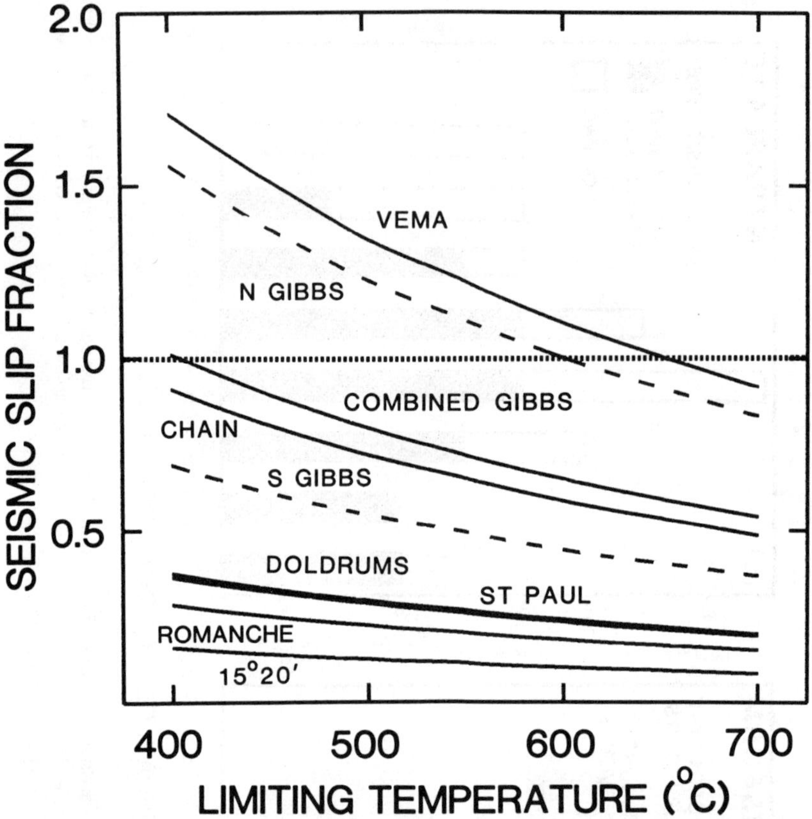

Fig. 19. Fraction of seismic slip (seismic slip rate divided by predicted plate motion) versus limiting isotherm for major Atlantic transforms. Five of the seven have significant aseismic slip if the 400°C isotherm is used. If the 600°C isotherm suggested by the lower limit of faulting (Figs. 16 and 18) is used, all transforms except the Vema have some proportion of aseismic slip. The Vema may have a larger than estimated fault area due to tectonic complications. Because a short offset in the Gibbs Transform makes it unclear whether it should be regarded as one or two transforms, results are shown for both cases (Engeln *et al.*, 1986).

the absence of focal depth interpretation, these are two essentially equivalent ways of viewing the same data, since there is a complete trade-off between the assumed "average" depth of slip and the fraction of slip taken up seismically.

It is unclear what causes the variation in the seismic fraction of slip or equivalently in depth extent of slip. The Vema Transform is a clear example; inferred seismic slip is substantially greater than predicted unless faulting extends to the 650°C isotherm. Transforms may differ substantially; the

Vema may have an unusually long recurrence time or be mechanically anomalous due to its complex tectonic history.

Full understanding of the transform slip process requires knowledge of thermal and mechanical structure. If centroid depths reflect strength as a function of depth, these can offer some insight (Fig. 18). In contrast to transform events, centroid depths for oceanic intraplate earthquakes extend down to approximately the 750°C isotherm. Transforms have much higher strain rates ($10^{-13}\,s^{-1}$) than intraplate areas ($10^{-18}\,s^{-1}$), so if both were characterized by the same dry olivine rheology, the limiting isotherm for transforms should be higher, approximately 1000°C. Alternatively, if a weaker hydrated olivine rheology was more appropriate for transforms, the higher strain rate should cancel the weakening effect and yield a limiting isotherm similar to that for intraplate regions.

Thus, transforms may either have much higher temperatures than expected and/or be much weaker than expected. In the former case, the required anomalously high temperatures, perhaps due to intrusion at depth below transforms, predict measurable heat flow anomalies. The second possibility, unusual rheological weakness, is reminiscent of the unexplained weakness of the San Andreas transform, where heat flow data exclude anomalously high temperatures (Lachenbruch and Sass, 1980). Several lines of inquiry such as teleseismic earthquake studies, ocean bottom seismometer surveys and adequate heat flow should help advance our understanding of the thermal structure and mechanical process on oceanic transforms.

B. Evolution of Ridge–Transform Systems

Seismological studies of rapidly changing ridge–transform systems offer some insights into the process by which plate boundaries evolve. For example, along the East Pacific Rise (EPR) near the location of the fastest spreading currently observed, bathymetric, seismic and magnetic data suggest the existence of two small plates between the Nazca and Pacific Plates (Herron, 1972; Forsyth, 1972). Anderson *et al.* (1974) suggested that the southern boundaries of the northern microplate, known as the Easter Plate, were stable spreading centers joined to the EPR on the north by long transforms. In contrast Handschumacher *et al.* (1981) interpreted the eastern boundary as a propagating rift (Hey, 1977), growing northward at the expense of the west ridge.

The concentration of seismicity along the boundaries suggests a rigid platelet, rather than a diffuse deformation zone. Focal mechanism data can be combined with marine geophysical results to suggest the nature of its boundaries. The east ridge spreading rapidly shows the low seismicity expected for a fast ridge. The southwest boundary is the most active seismically; its normal faulting events suggest slow spreading, by analogy to the Mid-Atlantic Ridge, whereas the strike-slip events suggest transform motion. The northeast boundary, on which both thrust and strike-slip motions occur, is unclear and may be a more diffuse deformation zone.

Treating the Easter Plate as rigid, relative motions between the Easter, Nazca and Pacific Plates can be determined by inversion, using Minster and Jordan's (1978) algorithm (Fig. 20). The proximity of the Easter-Nazca and Easter-Pacific poles allow significant variation in relative motion along the boundaries. The model predictions are generally good; spreading rates decrease rapidly to the north along the East ridge. The Easter-Pacific pole fits the observed rate on the northwest boundary and predicts slow spreading on the southwest boundary. The transform direction implied by the strike-slip events on the southwest boundary is fit well and spreading is predicted to be slightly oblique. The northwest boundary (or diffuse boundary zone) remains unclear, with the predicted motions changing from oblique convergence to strike-slip to oblique divergence. Recent detailed magnetic and bathymetric surveys (Naar and Hey, 1986) of the east ridge should provide the basis for a more accurate model.

The Easter Plate situation may be a model for the evolution of an area where one spreading center is being born and another dies. Figure 21 illustrates the schematic evolution of such a plate for a simple geometry similar to the Easter Plate. As a new spreading center develops, the old one adjusts to keep the total spreading rate and direction constant between the two large plates. Since the propagating rift spreads orthogonally, the dying ridge slows and changes spreading direction. Relative motion across the original transform also changes during propagation. Since motion across the original transform is no longer parallel to this boundary, the transform evolves into a "leaky" transform (Menard and Atwater, 1968) and then to a ridge spreading nearly orthogonally.

Comparison of this model with the Easter and Galapagos situations suggests that the dominant factor determining the ultimate fate of the platelet is its size. The more lithosphere is present, the more difficult it is for this area to be sheared. Thus, while small-scale ridge overlaps appear to have extensively sheared regions between ridges (Macdonald and Fox, 1983), large-scale propagation will tend to produce rigid plates. Microplates have been identified in magnetic anomalies when major boundary reorganizations shifted material from one large plate to another (Menard, 1978; Mammerickx *et al.*, 1980; Cande *et al.*, 1982). If the Easter analogy is valid, such microplates may be the mechanism by which large-scale ridge boundary reorganizations occur.

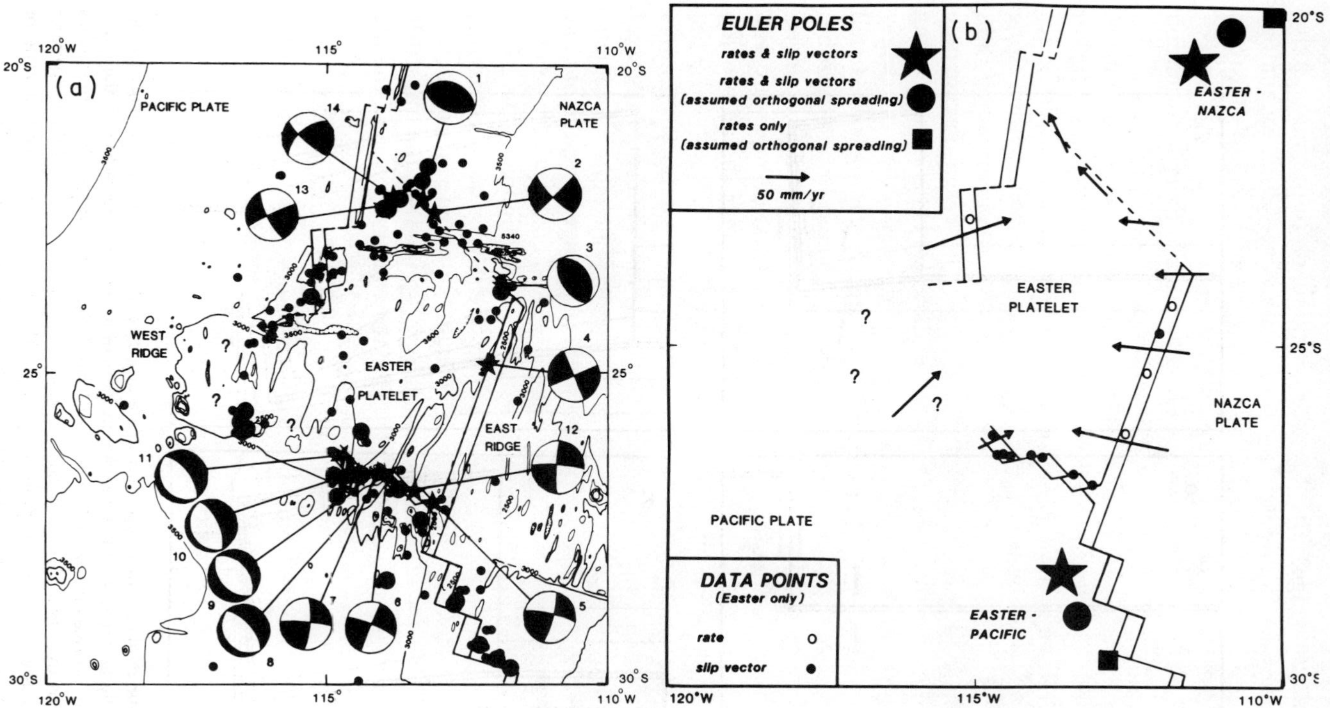

Fig. 20. (a) Bathymetry, seismicity and tectonics of the Easter Plate. Dots represent earthquake epicenters for the period 1963–1978; larger dots indicate events with $m_b > 5.1$. The stars represent earthquakes whose mechanisms have been determined. The East Ridge, the northern West Ridge and the southwest boundary are constrained by geophysical or seismological data; the other boundaries are less well known. (b) Rigid plate analysis of the Easter Plate system showing the Euler poles for different subsets of the data. The motions of the Easter Plate relative to the Nazca and Pacific Plates predicted using all of the data are indicated by the arrows. The open and closed circles represent the locations of rate or slip vector measurements (Engeln and Stein, 1984).

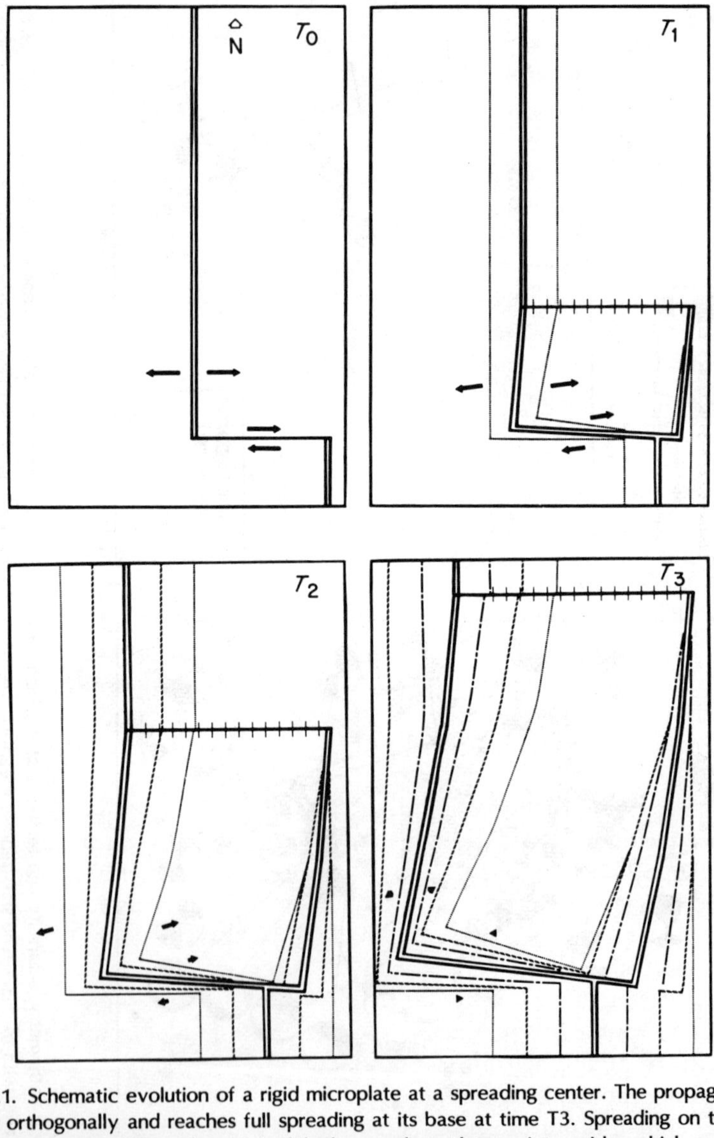

Fig. 21. Schematic evolution of a rigid microplate at a spreading center. The propagating rift spreads orthogonally and reaches full spreading at its base at time T3. Spreading on the dying ridge becomes more oblique (arrows) while the transform changes into a ridge which approaches orthogonal spreading over time. The difference in spreading rate at the base of the propagator offsets the propagating rift from the preexisting ridge. The northern boundary is a zone of compression which migrates along with the propagating tip (Engeln and Stein, 1984).

C. Subduction Zones

One major contribution of seismology to the early developments of plate tectonics was the recognition of the Benioff–Wadati zones of intermediate and deep seismicity and their interpretation as cold downgoing slabs, now interpreted as part of the subduction and ultimate recycling of the oceanic plates. The overthrusting geometry in major interplate earthquakes at subduction zones was confirmed early on (Kanamori, 1971a). A number of secondary problems have brought substantial insight into the mechanics of the subduction process, as well as the elastic structure of the lithosphere itself.

Comparison of the rates of seismic moment release over sufficiently long periods of time with those predicted from published plate motions shows strong differences in behavior between various subduction zones (Fig. 22) (Kanamori, 1977). For major subduction events, the task is made easier by the truly gigantic character of the earthquakes (magnitude M_w reaching 9.5) and the usual proximity of at least a few island stations, allowing the three-dimensional mapping of the fault area through the study of directivity and aftershocks.

While certain subduction zones (most notably Chile, Peru and Alaska) show a generally good quantitative agreement between seismicity and predicted plate motion, others (most notably the Marianas, Vanuatu and Tonga) feature a strong seismic deficiency, which can reach several orders of magnitude in the seismic moment release rate (for example, no earthquakes of magnitude 7.5 are known in the Marianas), indicating that most of the subduction in these areas is taken up by *aseismic creep.* Trenches such as Northern Japan and the Kuriles would be of an intermediate character with 50–60% deficiency. As reviewed by Uyeda and Kanamori (1979), the aseismic character of a subduction zone correlates strongly with extensional back-arc tectonics and development of a marginal sea, with a steeper dip of the Benioff slab, with the absence of a flexural bulge and a gravity high seaward of the trench, and with the generally more andesitic character of the arc volcanics.

All these effects can be ascribed to a general decoupling of the two colliding plates, a model confirmed by studies of the rupture process of subduction zone earthquakes. Body-wave modeling is used to retrieve the time history of the seismic source; some events show a simple rupture propagating rapidly and smoothly along the fault, others a complex series of independent ruptures separated in time, possibly triggering one another. These individual seismic sources along the fault plane have been described by Kanamori (1981) as resulting from the failure of zones of increased strength known as "asperities". In general, earthquakes in the more coupled

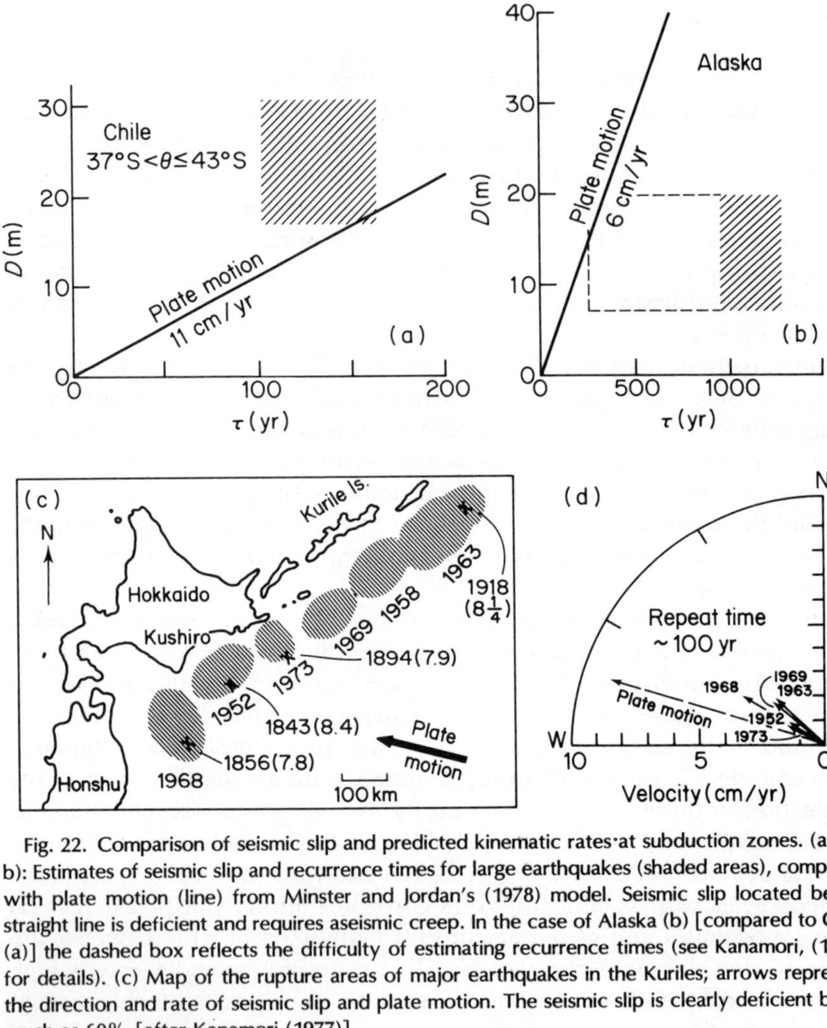

Fig. 22. Comparison of seismic slip and predicted kinematic rates at subduction zones. (a and b): Estimates of seismic slip and recurrence times for large earthquakes (shaded areas), compared with plate motion (line) from Minster and Jordan's (1978) model. Seismic slip located below straight line is deficient and requires aseismic creep. In the case of Alaska (b) [compared to Chile (a)] the dashed box reflects the difficulty of estimating recurrence times (see Kanamori, (1977) for details). (c) Map of the rupture areas of major earthquakes in the Kuriles; arrows represent the direction and rate of seismic slip and plate motion. The seismic slip is clearly deficient by as much as 60% [after Kanamori (1977)].

subduction zones, such as Alaska or Chile, represent the rupture of a single large asperity, whereas earthquakes in partially decoupled areas, such as the Kuriles, involve several distinct asperities (Ruff and Kanamori, 1983). In this interpretation, the areas between asperities yield at lower strains, in the form of foreshocks or aseismically. When the density of asperities becomes very low, they can no longer trigger each other, and seismic events are limited to the rupture of an isolated small asperity, as in the case of the Mariana Trench.

Studies of the maximum known size of earthquakes at subduction zones suggest that the degree of coupling at a trench is a trade-off between the velocity of convergence of the two plates and the age of the subducting lithosphere (Ruff and Kanamori, 1980). On the one hand, a higher collision velocity may favor stronger coupling; on the other hand, the older the subducting plate, the denser it is, and the stronger the *slab-pull* on the plate. This negative buoyancy force tends to sink the subducting plate, and thus may favor decoupling (Fig. 23). Other parameters, such as the development of back-arc spreading and the state of stress in the subducting plate are also controlled by this trade-off. In situations where either a young plate subducts slowly (e.g. Alaska) or an old one subducts fast (e.g. Japan or Tonga), local compressional stresses in the plate can give rise to double Benioff zones (Fujita and Kanamori, 1981; Kawakatsu, 1985).

Seaward of the trench, constraints on the mechanical state of the lithosphere can be obtained from the study of intraplate earthquakes identified by Stauder (1968) as due to the bending of the plate. Later, Chen and Forsyth (1978) identified a pattern of tensional faulting in the upper part of the plate (to a depth of 25 km) and thrusting in its lower part (between 40 and 50 km). Chapple and Forsyth (1979) and Bodine *et al.* (1981) use rheological models with an upper elastic layer and a lower plastic layer to fit the observed depth distribution of flexural events (Fig. 24).

Occasionally, the trench areas are the sites of great earthquakes featuring normal faulting (e.g. Sanriku, 1933, $M_w = 8.4$; Peru, 1970, $M_w = 7.9$; Indonesia, 1977, $M_w = 8.3$). There has been some controversy as to whether to interpret these earthquakes as bending events in the framework of Chen and Forsyth's (1978) "upper" earthquakes or as rupture and decoupling of the entire lithosphere under slab pull. The shallow centroid depth and very fast moment release of the 1977 Indonesian earthquake (Silver and Jordan, 1983) both favor the bending model; on the other hand, for the 1933 Sanriku event, Kanamori (1971b) has argued that a fault width of 100 km is necessary to account for near-field and tsunami amplitudes. In the singular case of the deeper 1977 Tonga event, both aftershock distribution, and the slower rate of seismic moment release suggest vertical rupture extending possibly over 100 km in depth, favoring the decoupling model (Silver and Jordan, 1983).

V. Localized Intraplate Processes

Frequently, seismicity can be used to identify and study regions where intraplate deformation appears to be controlled by localized causes. Many such cases exist; we discuss a few particularly significant or interesting ones.

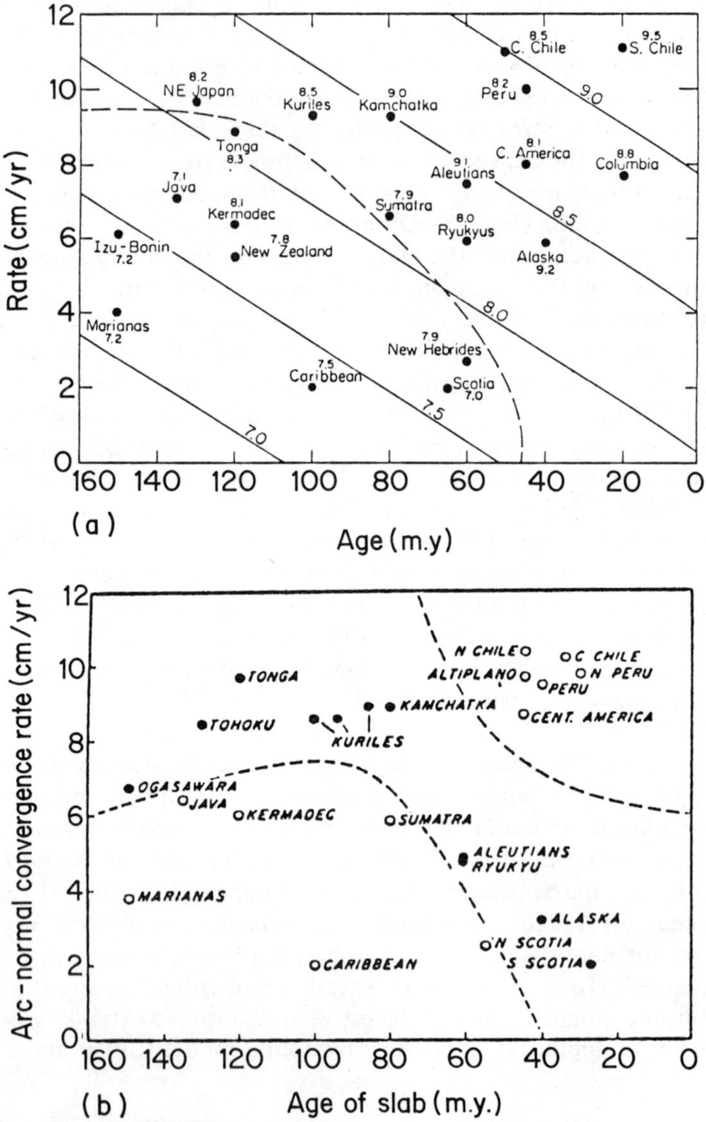

Fig. 23. Variation of coupling at subduction zones as a function of convergence rate and slab age. (a) Maximum known earthquake magnitude M_w. The dashed line outlines the range of back-arc spreading and basin development, and the straight lines are an empirical model for the data [After Ruff and Kanamori, (1980)]. (b) State of stress in the downgoing slab [After Fujita and Kanamori (1981)]. The dashed lines outline the region of compressional stress in which double Benioff zones can exist; ● indicates compressional and ○, tensional stress.

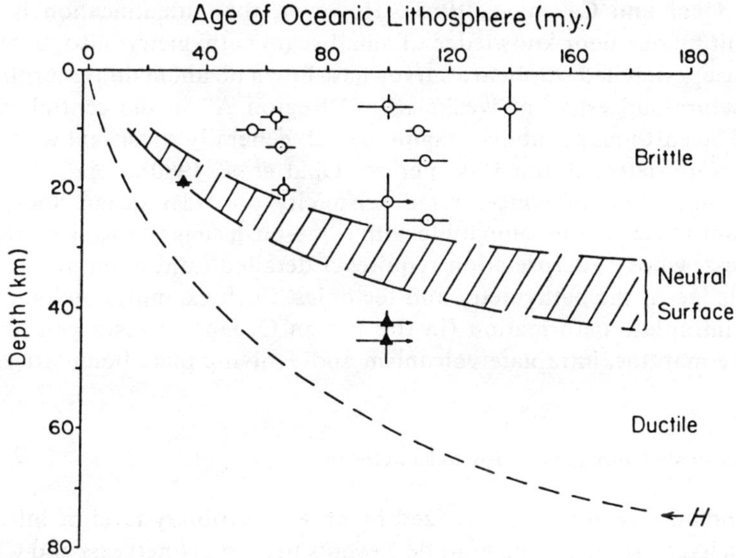

Fig. 24. Focal depths of outer rise flexural earthquakes seaward of trenches (Chapple and Forsyth, 1979) and predictions of a flexural model with a yield strength rheology (Bodine *et al.*, 1981). The plate mechanical thickness *H* increases with age as expected from thermal models (Figs. 4 and 5). Tensional (○) events occur above the neutral surface while compressional (▲) events occur below.

A. Concentration of Seismicity: Local Stresses or Plate Weakness?

Oceanic intraplate earthquakes occur primarily at sites of preferential seismic release, rather than at random in the plates. Although, in principle, this could be an artifact of the limited time span of instrumental seismology, the occasional occurrence of repeat earthquakes (Okal, 1981) argues against this possibility. The preferential character of intraplate seismic sites has been confirmed in both teleseismic and regional investigations (Okal *et al.*, 1980). The question arising is whether these locations are seismic due to preferential release of platewide stresses at zones of local weakness or due to stress concentrations of local origin. The general correlation of stresses inferred from focal mechanisms over plate-wide distances and the success of simple models of their origin, argue in favor of local weakness. Also, a recent experiment has shown agreement between seismic stress release and the orientation of *in situ* stress measured in a borehole (Newmark *et al.*, 1984). Zones of weakness and preferential seismic release include healed fracture zones (Sykes, 1978), lines of age discontinuity (Okal, 1981) and areas with seamounts and other fossil tectonic features (Stein and Okal,

1978; Okal and Cazenave, 1985). However, their identification is made difficult by our poor knowledge of small-scale bathymetry; also, in at least one case, a detailed shipboard survey has shown no anomalous morphological feature suggestive of weakness at "Region A" in the central Pacific, where 86 earthquakes above magnitude 3.2, generally consistent with ridge-push, were detected in a 15-yr period (Okal *et al.*, 1980).

In some areas of concentrated seismicity on older ocean floor, focal mechanisms clearly incompatible with ridge-push suggest localized regimes of stress, whose interpretation requires a detailed (and often unavailable) knowledge of the bathymetry and tectonics. Such examples include large-scale intraplate deformation (in the Indian Ocean), stresses generated at passive margins, intraplate volcanism and evolving plate boundaries.

B. Large-scale Deformation: The Indian Ocean

The Indian Ocean is characterized by an extraordinary level of intraplate seismicity, as shown by magnitude 7 events near the Ninetyeast and Chagos "aseismic" ridges (Fig. 25). Indian Ocean intraplate seismicity has been studied extensively (Stein and Okal, 1978; Stein, 1978; Bergman *et al.*, 1984; Bergman and Solomon, 1985; Wiens, 1985) but is very poorly understood. Earthquake mechanisms along the Ninetyeast Ridge (which is as active seismically as portions of the San Andreas Fault) are generally consistent with left lateral strike–slip motion, and can be interpreted as resulting from a boundary zone between portions of the plate. Stein and Okal suggested that this relative motion results from the Indian (west) side encountering resistance due to the Himalayan collision, while the Australian (east) side subducts smoothly at the Sumatra Trench. The deformation may be localized as the Ninetyeast Ridge marks the location of a former transform boundary. In stark contrast, the region along the Chagos–Laccadive Ridge shows normal faulting.

The region between the two ridges is seismically active, generally showing thrust and strike–slip faulting and characterized by anomalies in heat flow (Geller *et al.*, 1983), gravity and bathymetry (Weissel *et al.*, 1980). Furthermore, as discussed earlier, the region of young lithosphere in the central Indian Ocean is an unusual zone of normal faulting.

This intraplate deformation is sufficient to provide statistical significance to the modeling of the Indian plate as two plates in inversions of global relative plate motions (Minster and Jordan, 1978; Stein and Gordon, 1984). It is still unclear whether this separation is real, what it implies and how it relates to the variety of deformation styles over the region. Additional relative plate motion data for the area should provide some help. Further

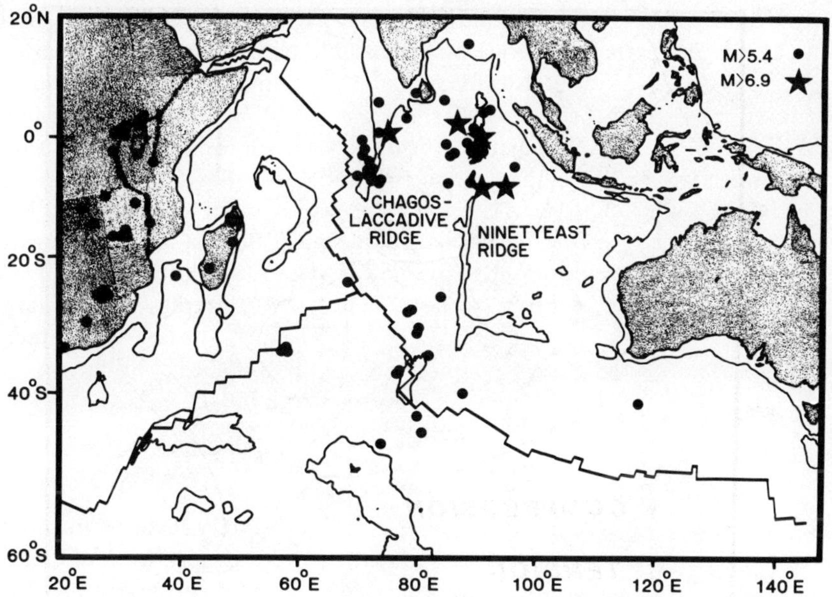

Fig. 25. Intraplate seismicity of the Indian Plate, the most seismically active area of the ocean basins. Magnitude-seven earthquakes occur near the Ninetyeast and Chagos–Laccadive Ridges.

insight may be derived from tectonic models such as the finite element calculation of the theoretical stress field for the Indian Plate (Cloetingh and Wortel, 1985) (Fig. 26). The predicted stresses in the Ninetyeast Ridge area are very large and consistent with the focal mechanisms. Finally, the Indian Ocean example points out the difficulty in distinguishing intense intraplate deformation from a slow (1 cm/yr) plate boundary: the distinction can be rather arbitrary.

C. Seismicity at Passive Margins

An interesting type of intraplate seismicity occurs at passive margins, where continental and oceanic lithosphere join. Although these areas are in general tectonically inactive, major earthquakes can occur. The type example is the active seismic zone on the eastern coast of North America (Stein *et al.*, 1979), which includes the 1929 Grand Banks ($M_s = 7.2$) and 1933 Baffin Bay ($M_s = 7.3$) earthquakes. These events are thought to involve stresses related to the removal of Pleistocene glacial loads extending on to the continental shelf, which reactivated faults remaining from the original

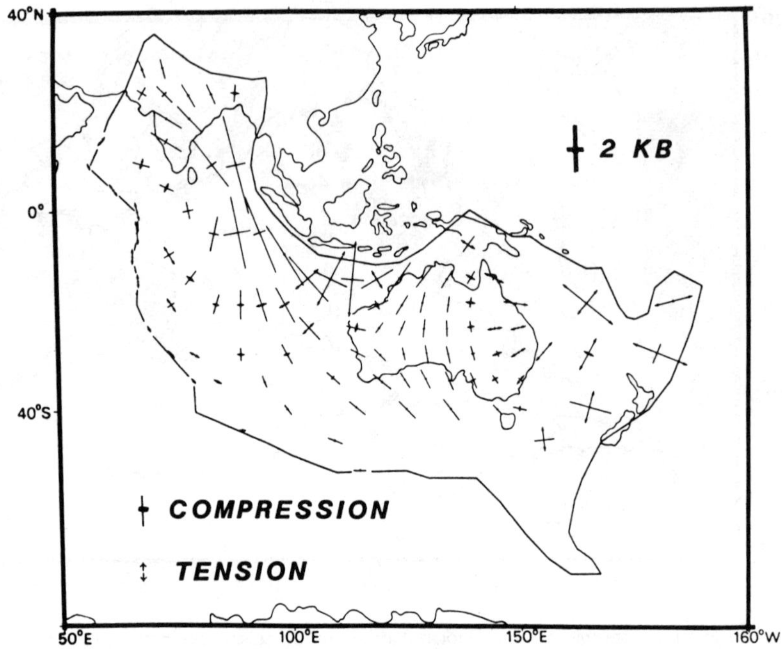

Fig. 26. Theoretical stress model for the Indian Plate (Cloetingh and Wortel, 1985). The predicted high stresses near the Ninetyeast Ridge are consistent with the earthquake focal mechanisms.

rifting. Simple flexure calculations predict stresses of 100–150 bar, adequate to trigger earthquakes (Stein *et al.*, 1979; Quinlan, 1984). While all passive margins are subject to a stress field due to the differing densities of continental and oceanic lithosphere (Bott and Dean, 1972) the effect of deglaciation appears to be to trigger earthquakes, as such events are observed primarily on glaciated margins, such as eastern Canada, Fennoscandia and Greenland.

D. Volcanism

Intraplate volcanism, expressed most spectacularly in the Hawaiian-type island chains, is thought to be a general property of the seafloor (Batiza, 1982). Statistical arguments suggest that it contributes as much as 6% to the total surface of the ocean floor (Jordan *et al.*, 1983). Most of these seamounts are still to be discovered due to poor knowledge of small-scale bathymetry.

In areas of documented volcanism (i.e. reaching above sea level), a clear geographic and often temporal correlation exists between volcanic and seismic activities, and earthquake mechanisms have been successfully interpreted in the framework of the magmatic evolution of the volcano. Examples include deep fracture under increased magma pressure (Eaton and Murata, 1960), horizontal faulting upon magma intrusion (Furumoto and Kovach, 1979), and post-eruptive caldera collapses (Kaufman and Burdick, 1980). Because of the particular mechanical conditions of the magmatic bodies and surrounding rock, volcanic earthquakes feature several singular characteristics.

Volcanic earthquakes, responding to strain rates as high as 10^{-12}/s, can extend deeper into the lithosphere and have been located under Hawaii down to 55 km (Butler, 1982). In addition, the generally weaker character of the rock leads to rupture at low levels of strain, and preferential concentration of seismicity at lower magnitudes, as confirmed from rock mechanics experiments (Mogi, 1963). Finally, the mechanism of earthquake rupture can depart significantly from the simple model of slip on a fault plane. Significant second double-couples and isotropic sources have been documented in volcanic areas (Dziewonski *et al.*, 1983), and on a much smaller scale, resonant magma oscillation inside cracks leads to high-frequency tremor observed prior to and during volcanic eruptions (Aki and Koyanagi, 1981).

In underwater areas with no record of volcanism, these characteristics have been used to identify and discover active submarine volcanoes [e.g. Macdonald (Norris and Johnson, 1969) and Teahitia (Talandier and Okal, 1984)]. However, recognition of a seismic swarm as volcanic is made difficult by such factors as poor depth control at regional distances and rapid spatial attenuation of tremors. Most importantly, reduced detection capabilities can leave most of a volcanic swarm undetected, with only a few events recorded. Thus, the possibility that a major single earthquake recorded in a remote oceanic area may be part of an unsuspected episode of active volcanism cannot always be discarded, and the interpretation of such an event in terms of large-scale tectonic stresses may be erroneous. For example, Okal (1984) has speculated that a major normal faulting event in 1955, on the flank of Crough Seamount in the east central Pacific, may be related to volcanic activity.

E. Evolving Plate Boundaries

A characteristic of plate boundary systems is often their stability over extended periods of time; e.g. the Mid-Atlantic Ridge has kept an approximately constant orientation and spreading rate for the past 160 Ma. There exist however cases when plate boundaries *must* change, primarily because of

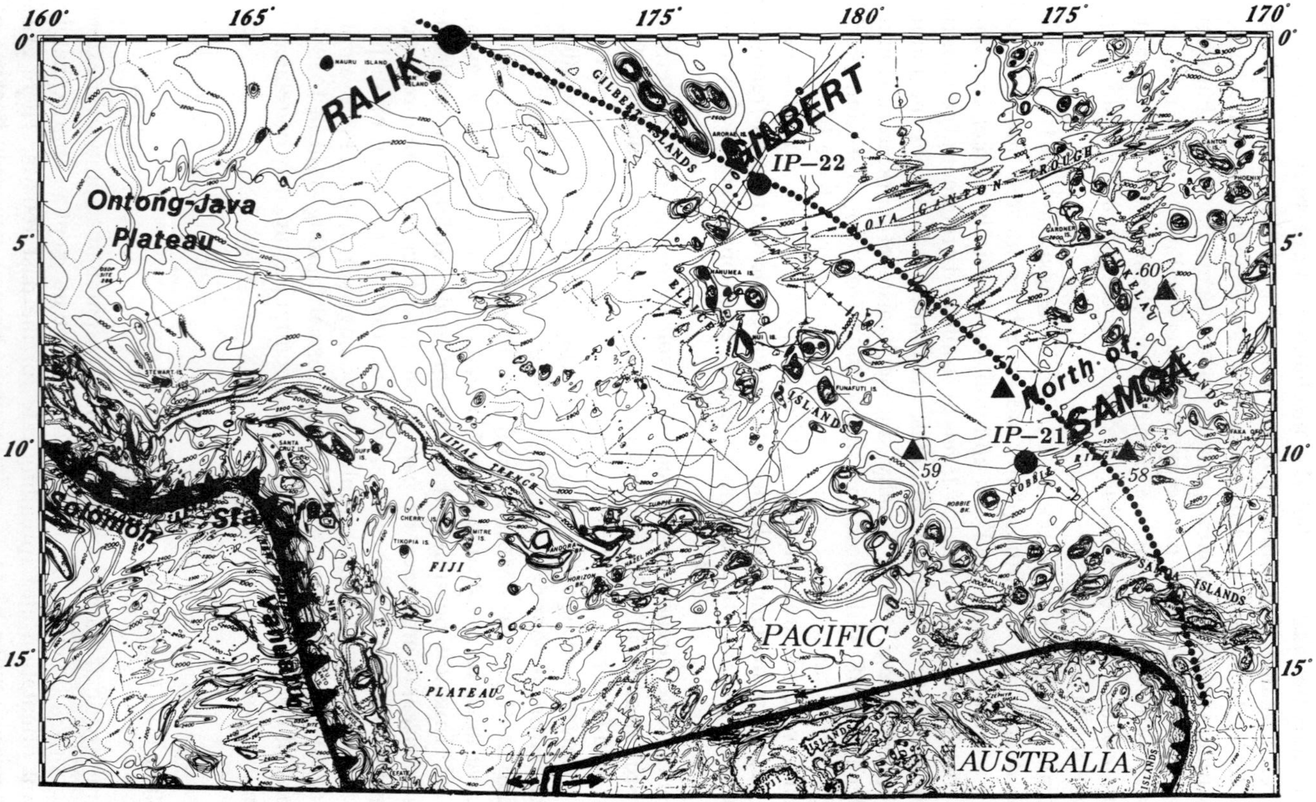

Fig. 27. Map of the Ralik–Gilbert–Samoa Island area, with intraplate seismicity [individual dots and triangles; see Okal (1984)]. The dotted line indicates the potential new plate boundary.

the unsubductible character of continents and of very young oceanic lithosphere. A well-known example is the cessation of subduction at the western edge of North America in the Eocene (Atwater, 1970). The process involved in the reorganization of the plate system is a complex one, which can feature the transient existence of a number of platelets (Menard, 1978). One such platelet is believed to be the Caroline Plate, northwest of New Guinea, where subduction under the Australian Plate has ceased (Kaplan and Seno, 1982). Seismicity in the Caroline Plate is characterized by comparatively deep earthquakes (down to 28 km at only 40 Ma), and generally north–south compressional stresses. Both features can be due to underthrusting of the Pacific Plate under the Caroline Plate, possibly accompanied by shortening (Weissel and Anderson, 1978; Wiens and Stein, 1983).

Another region of interest is the area extending north of Samoa to the Gilbert Islands and beyond, where at least three seismic foci have been identified (see Fig. 27) (Lay and Okal, 1983; Okal, 1984; Walker, 1984). Focal mechanisms at the Gilbert site require a horizontal compressional stress oriented N20°E, more or less at right angles to ridge-push, and clearly of a regional nature. One possible explanation would interpret the Samoa–Gilbert–Ralik seismic line as an incipient subduction zone, which in the future will take up the convergence between the Pacific and Australian Plates: the present subduction system at the Vanuatu (ex-New Hebrides), Santa Cruz and South Solomon Trenches is about to collide with buoyant structures such as New Caledonia, the Norfolk and d'Entrecasteaux Ridges and the Loyalty and Rennell Islands, All this material is light and cannot subduct and the same applies to the overthrusting material on the Pacific side (Fiji and Ontong–Java plateaus), barring a simple reversal of the polarity of the subduction. Thus, the subduction zone will have to jump in search of subductible material, possibly along the Samoa–Ralik seismic line. Although speculative, this interpretation explains the compressional nature of the Gilbert events, and predicts an eventual and general realignment of the southern boundary of the Pacific Plate to the north of the unsubductible material making up the Caroline Plate and the Ontong–Java and Fiji Plateaus.

Acknowledgments

The research effort of the authors described in this review was supported by NSF grants EAR 84-07510, EAR 82-06381 (SS), and EAR 81-06106 (EAO), under ONR Contracts N-00014-79-C-0292 and N-00014-84-C-0616 (EAO), NASA Crustal Dynamics Contract NAS5-27238 (SS) and a Cotrell Research Grant (SS). Acknowledgment is made to the donors of the Petroleum Research Fund, administered by the American Chemical Society, for partial support of this research (SS).

References

Aki, K. and Koyanagi, R. Y. (1981). *J. Geophys. Res.* **86**, 7095-7109.
Aki, K. and Richards, P. G. (1980). "Quantitative Seismology", 2 Vols., Freeman, San Francisco, California.
Artyushkov, E. V. (1973). *J. Geophys. Res.* **78**, 7675-7708.
Anderson, R. N., Forsyth, D. W., Molnar, P. and Mammerickx, J. (1974). *Earth Planet. Sci. Lett.* **24**, 188-202.
Atwater, T. (1970). *Geol. Soc. Am. Bull.* **81**, 3513-3536.
Batiza, R. (1982). *Earth Planet. Sci. Lett.* **60**, 195-206.
Bergman, E. A. and Solomon, S. C. (1980). *J. Geophys. Res.* **85**, 5389-5410.
Bergman, E. A. and Solomon, S. C. (1984). *J. Geophys. Res.* **89**, 11415-11441.
Bergman, E. A. and Solomon, S. C. (1985). *Phys. Earth Planet. Int.* **40**, 1-23.
Bergman, E. A., Nábělek, J. L. and Solomon, S. C. (1984). *J. Geophys. Res.* **89**, 2425-2443.
Bodine, J. H., Steckler, M. S. and Watts, A. B. (1981). *J. Geophys. Res.* **86**, 3695-3707.
Bonatti, E. (1978). *Earth Planet. Sci. Lett.* **37**, 369-379.
Bott, M. H. P. and Dean, D. S. (1972). *Nature* **235**, 23-25.
Brace, W. F. and Kohlstedt, D. L. (1980). *J. Geophys. Res.* **85**, 6248-6252.
Bratt, S. R., Bergman, E. A. and Solomon, S. C. (1985). *J. Geophys. Res.* **90**, 10249-10260.
Burr, N. C. and Solomon, S. C. (1978). *J. Geophys. Res.* **83**, 1193-1205.
Butler, R. G. (1982). *Geophys. J. R. Astron. Soc.* **69**, 173-186.
Cande, S. C., Herron, E. M. and Hall, B. R. (1982). *Earth Planet. Sci. Lett.* **57**, 63-74.
Cathles, L. M. (1975). "The Viscosity of the Earth's Mantle," Princeton Univ. Press, Princeton, New Jersey.
Chapple, W. M. and Forsyth, D. W. (1979). *J. Geophys. Res.* **84**, 6729-6749.
Chase, C. G. (1978). *Earth Planet. Sci. Lett.* **37**, 355-368.
Chen, T. and Forsyth, D. W. (1978). *J. Geophys. Res.* **83**, 4995-5004.
Chen, W.-P. and Molnar, P. (1983). *J. Geophys. Res.* **88**, 4183-4214.
Cloetingh, S. and Wortel, W. (1985). *Geophys. Res. Lett.* **12**, 77-80.
Cox, A. V. ed. (1973). "Plate Tectonics and Geomagnetic Reversals," Freeman, San Francisco, California.
Dziewonski, A. M., Friedman, A., Giardini, D. and Woodhouse, J. H. (1983). *Phys. Earth Planet Int.* **33**, 76-90.
Eaton, J. P. and Murata, K. J. (1960). *Science* **132**, 925-938.
Engeln, J. F. and Stein, S. (1984). *Earth Planet Sci. Lett.* **68**, 259-270.
Engeln, J. F., Wiens, D. A. and Stein, S. (1986). *J. Geophys. Res.* **91**, 548-577.
Fleitout, L. and Froidevaux, C. (1983). *Tectonics* **2**, 315-324.
Forsyth, D. W. (1972), *Earth Planet. Sci. Lett.* **17**, 189-193.
Forsyth, D. W. (1973). *Nature* **243**, 78-79.
Forsyth, D. W. (1975). *Geophys. J. R. Astron. Soc.* **43**, 103-162.
Forsyth, D. W. and Uyeda, S. (1975). *Geophys. J. R. Astron. Soc.* **43**, 163-200.
Forsyth, D. W. and Wilson, B. (1984). *Earth Planet. Sci. Lett.* **70**, 355-362.
Fox, P. J. and Gallo. D. G. (1984). *Tectonophysics* **104**, 205-242.
Fujita, K. and Kanamori, H. (1981). *Geophys. J. R. Astron. Soc.* **66**, 131-156.
Fujita, K. and Sleep, N. H. (1978). *Tectonophysics* **50**, 207-221.
Fukao, K. (1970). *Bull. Earthquake Res. Inst. Univ. Tokyo* **48**, 707-727.
Furumoto, A. S. and Kovach, R. L. (1979). *Phys. Earth Planet. Int.* **18**, 197-208.
Geller, C. A., Weissel, J. K. and Anderson, R. N. (1983). *J. Geophys. Res.* **88**, 1018-1032.
Hager, B. H. and O'Connell, R. J. (1979). *J. Geophys. Res.* **84**, 1031-1048.

Handschumacher, D. W., Pilger, R. H., Jr., Forman, J. A. and Campbell, J. F. (1981). *Geol. Soc. Am. Mem.* **154**, 63–76.
Herron, E. M. (1972). *Nature* **240**, 35–37.
Hey, R. N. (1977). *Earth Planet. Sci. Lett.* **37**, 321–325.
Jordan, T. H., Menard, H. W. and Smith, D. K. (1983). *J. Geophys. Res.* **88**, 10508–10518.
Kanamori, H. (1971a). *Tectonophysics* **12**, 187–198.
Kanamori, H. (1971b). *Phys. Earth Planet. Int.* **4**, 289–300.
Kanamori, H. (1977). *In* "Island Arcs, Deep Sea Trenches and Back Arc Basins" (M. Talwani and W. C. Pitman, III, eds.) pp. 163–174. Maurice Ewing Series, **1**. Am. Geophys. Un., Washington, D.C.
Kaplan, D. A. and Seno, T. (1982). *EOS, Trans. Am. Geophys. Un.* **63**, 1023 (abstract).
Kaufman, K. and Burdick, L. J. (1980). *Bull. Seismol. Soc. Am.* **70**, 1759–1770.
Kawakatsu, H. (1985). *Nature* **316**, 53–55.
Kirby, S. H. (1980). *J. Geophys. Res.* **85**, 6353–6363.
Kroeger, G. C. and Geller, R. J. (1986). *J. Geophys. Res.* (in press).
Lachenbruch, A. H. and Sass, J. H. (1980). *J. Geophys. Res.* **85**, 6185–6223.
Langston, C. A. and Helmberger, D. V. (1975). *Geophys. J. R. Astron. Soc.* **42**, 117–130.
Lay, T. and Okal, E. A. (1983). *Phys. Earth Planet. Int.* **33**, 284–303.
Lister, C. R. B. (1975). *Nature* **257**, 663–665.
Macdonald, K. C. and Fox, P. J. (1983). *Nature* **302**, 55–58.
Mammerickx, J., Herron, E. M. and Dorman, L. M. (1980). *Geol. Soc. Am. Bull.* **91**, 263–271.
McKenzie, D. P. (1967). *J. Geophys. Res.* **72**, 6261–6273.
Menard, H. W. (1978). *J. Geology* **86**, 99–110.
Menard, H. W. and Atwater, T. (1968). *Nature* **219**, 463–467.
Mendiguren, J. A. and Richter, F. M. (1978). *Phys. Earth Planet. Int.* **16**, 318–326.
Minster, J. B. and Jordan, T. H. (1978). *J. Geophys. Res.* **83**, 5331–5354.
Mogi, K. (1963). *Bull. Earthq. Res. Inst. Univ. Tokyo* **41**, 615–658.
Naar, D. F. and Hey, R. N. (1986). *J. Geophys. Res.* **91**, 3425–3438.
Newmark, R. L., Zoback, M. D. and Anderson, R. N. (1984). *Nature* **311**, 424–428.
Norris, R. A. and Johnson, R. H. (1969). *J. Geophys. Res.* **74**, 650–664.
Okal, E. A. (1981). *Earth Planet. Sci. Lett.* **52**, 397–409.
Okal, E. A. (1984). *J. Geophys. Res.* **89**, 10053–10071.
Okal, E. A. and Cazenave, A. (1985). *Earth Planet Sci. Lett.* **72**, 99–116.
Okal, E. A., Talandier, J., Sverdrup, K. A. and Jordan, T. H. (1980). *J. Geophys. Res.* **85**, 6479–6495.
Parsons, B. and Richter, F. M. (1980). *Earth Planet. Sci. Lett.* **51**, 445–450.
Parsons, B. and Sclater, J. G. (1977). *J. Geophys. Res.* **82**, 803–827.
Peltier, W. R. (1980). Physics of the Earth's interior. *Proc. Int. Sch. Phys. "Enrico Fermi"*. (A. M. Dziewonski and E. Boschi, eds.), pp. 362–431. North-Holland, Amsterdam.
Peltier, W. R. (1983). *Nature* **304**, 434–436.
Quinlan, G. (1984). *Can. J. Earth Sci.* **21**, 1018–1023.
Richter, F. M. (1977). *Tectonophysics* **38**, 61–88.
Richter, F. M. and McKenzie, D. P. (1978). *J. Geophys.* **44**, 441–471.
Ruff, L. J. and Kanamori, H. (1980). *Phys. Earth Planet. Int.* **23**, 240–252.
Ruff, L. J. and Kanamori, H. (1983). *Phys. Earth Planet. Int.* **31**, 202–230.
Sclater, J. G. and Fisher, R. L. (1974). *Geol. Soc. Am. Bull.* **85**, 683–702.
Silver, P. G. and Jordan, T. H. (1983). *J. Geophys. Res.* **88**, 3273–3293.
Sleep, N. H. and Rosendahl, B. R. (1979). *J. Geophys. Res.* **84**, 6831–6839.
Solomon, S. C. and Burr, N. C. (1979), *Tectonophysics* **55**, 107–126.
Stauder, W. J. (1968). *J. Geophys. Res.* **73**, 7693–7701.

Stein, S. (1978). *Geophys. J. R. Astron. Soc.* **55**, 577–588.
Stein, S. and Gordon, R. G. (1984). *Earth Planet. Sci. Lett.* **69**, 401–412.
Stein, S. and Okal, E. A. (1978). *J. Geophys. Res.* **83**, 2233–2245.
Stein, S., Sleep, N. H., Geller, R. J., Wang, S.-C. and Kroeger, G. C. (1979). *Geophys. Res. Lett.* **6**, 537–540.
Sykes, L. R. (1967). *J. Geophys. Res.* **72**, 2131–2153.
Sykes, L. R. (1978). *Rev. Geophys. Space Phys.* **16**, 621–688.
Sykes, L. R. and Sbar, M. L. (1974). In "Geodynamics of Iceland and the North Atlantic area" (L. Kristjansson, ed.) pp. 207–224. Reidel Publ., Boston, Massachusetts.
Talandier, J. and Okal, E. A. (1984). *J. Geophys. Res.* **89**, 11216–11234.
Turcotte, D. L. and Oxburgh, E. R. (1967). *J. Fluid Mech.* **28**, 29–42.
Turcotte, D. L. and Oxburgh, E. R. (1973). *Nature* **244**, 337–339.
Uyeda, S. and Kanamori, H. (1979). *J. Geophys. Res.* **84**, 1049–1061.
Walker, D. A. (1984). *OPA, Newsletter of the Ocean P Alliance*, Univ. Hawaii, Honolulu, 15 November 1984.
Weissel, J. K. and Anderson, R. N. (1978). *Earth Planet. Sci. Lett.* **41**, 143–158.
Weissel, J. K., Anderson, R. N. and Geller, C. A. (1980). *Nature* **287**, 284–291.
Wiens, D. A. (1985). *Earth Planet. Sci. Lett.* **76**, 350–360.
Wiens, D. A. and Stein, S. (1983). *J. Geophys. Res.* **88**, 6455–6468.
Wiens, D. A. and Stein, S. (1984). *J. Geophys. Res.* **89**, 11442–11464.
Wiens, D. A. and Stein, S. (1985). *Tectonophysics* **116**, 143–162.
Wilson, J. T. (1965). *Nature* **207**, 343–347.
Yu, G.-K. and Mitchell, B. J. (1979). *Geophys. J. R. Astron. Soc.* **57**, 311–341.

5.4 Properties of the Lithosphere and Asthenosphere Deduced from Geoid Observations

Donald L. Turcotte

Department of Geological Sciences, Cornell University, Ithaca, New York, USA

I. Introduction

The gravitational potential and acceleration at any point on the Earth can be determined if the distribution of densities within the Earth are prescribed. However, the surface field cannot be uniquely inverted to give the distribution of densities. The determination of the gravitational potential and acceleration at any point is quite difficult because it often requires an integral over a considerable volume before convergence is achieved.

Under some conditions the surface gravitational potential or acceleration can be related to the one-dimensional distribution of density beneath the point of measurement. One example of such a relationship is the Bouguer

SPACE GEODESY AND GEODYNAMICS
ISBN 0-12-057840-9

formula for the surface gravity anomaly Δg

$$\Delta g(x, y) = 2\pi g\sigma(x, y) \tag{1}$$

where G is the gravitational constant and the surface source density distribution is

$$\sigma(x, y) = \int_0^h \Delta\rho(x, y, z)\, dz \tag{2}$$

where $\Delta\rho$ is the density anomaly, z is the vertical coordinate and h is the depth of compensation. The Bouguer formula is valid if the horizontal scale of the density variation is large compared with the vertical scale h and $h \ll a$ where a is the radius of the Earth.

By using the technique of matched asymptotic expansions, Ockendon and Turcotte (1977) have derived a power series expansion for the gravitational potential caused by slowly varying density changes. The leading term of the expansion is the Bouguer formula. By using this expansion Haxby and Turcotte (1978) found that if the near-surface density distribution is in isostatic equilibrium, then the surface gravitational potential anomaly ΔU is given by

$$\Delta U = -2\pi G\delta(x, y) \tag{3}$$

where the surface dipole density distribution is

$$\delta(x, y) = \int_0^h z\, \Delta\rho(x, y, z)\, dz \tag{4}$$

The conditions for the validity of (3) are the same as for the Bouguer formula (1) with the additional isostatic constraint. For these conditions isostasy requires that the surface source distribution be zero, $\sigma = 0$; it follows from (1) that the gravity anomaly given by the Bouguer formula is zero.

The surface gravitational potential anomaly ΔU can be related directly to the geoid anomaly ΔN. The potential anomaly is defined by

$$\Delta U = U_{m0} - U_0 \tag{5}$$

where U_{m0} is the measured potential at the location of the reference geoid and U_0 is the reference value of the potential that defines the reference geoid. Assuming $\Delta N \ll a$ it follows that

$$U_0 = U_{m0} + \left(\frac{\partial U}{\partial r}\right)_{r=r_0} \Delta N \tag{6}$$

where r_0 is the radial position of the reference geoid. However

$$g_0 = \left(\frac{\partial U}{\partial r}\right)_{r=r_0} \tag{7}$$

where g_0 is the gravitational acceleration on the reference geoid. Combining (5), (6) and (7) gives

$$\Delta U = -g_0 \, \Delta N \tag{8}$$

and from (3)

$$\Delta N = (2\pi G/g_0)\delta(x, y). \tag{9}$$

Thus the geoid anomaly can be directly related to the dipole distribution of density beneath the point of measurement if the condition of isostasy is applicable and if the geoid anomaly is caused by near-surface density variations.

Measurements of sea surface heights from the GEOS-3 and SEASAT missions have directly provided accurate determinations of geoid anomalies over the oceans. Many of these geoid anomalies can be explained directly using (9). The positive geoid anomalies associated with the mid-ocean ridge system are consistent with thermal isostasy (Haxby and Turcotte, 1978; Cazenave *et al.*, 1983; Sandwell and Schubert, 1980). The geoid offsets across fracture zones can also be explained in terms of thermal isostasy and the age difference across the fracture zones (Crough, 1979; Sandwell and Schubert, 1982; Cazenave *et al.*, 1982). Systematic correlations of geoid and bathymetry have also been found for aseismic ocean ridges and plateaus (Angevine and Turcotte, 1980; 1983) and for oceanic swells (Haxby and Turcotte, 1978; Crough, 1978, 1982). Geoid anomalies can also be used to study passive continental margins (Haxby and Turcotte, 1978) and to determine the thickness of the continental lithosphere (Turcotte and McAdoo, 1979).

II. Analysis

We will next show how the relationship between the gravitational potential anomaly (geoid anomaly) and the dipole density distribution can be derived. We will also test the range of validity of (3) and (9).

A. Disk Approximations

We first consider a circular disk of thickness h and radius R as shown in Fig. 1. The density of the disk is a function of the vertical coordinate z, $\rho(z)$, but not of r. The contribution to the gravitational acceleration of each element of mass in the disk is (Turcotte and Schubert, 1982)

$$d\mathbf{g} = -\frac{G\mathbf{s}\, dM}{|s|^3} \tag{10}$$

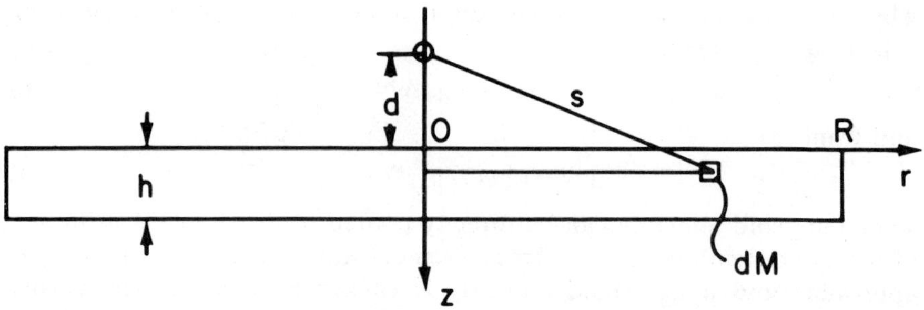

Fig. 1. Illustration of the circular disk formulation.

Integrating over the volume of the disk to obtain the vertical components of the gravitational acceleration on the axis at a distance d above the disk yields

$$g_z = 2\pi G \int_0^h \int_0^R \frac{(d+z)r\rho(z)\, dr\, dz}{[r^2+(d+z)^2]^{1/2}} \tag{11}$$

First by integrating with respect to r and then taking the limit $R \to \infty$ gives

$$g_z = 2\pi G \int_0^h \rho(z)\, dz \tag{12}$$

which is the Bouguer formula previously given in (1).

The gravitational potential due to each element of mass is

$$dU = \frac{G\, dM}{|s|} \tag{13}$$

Integrating over the volume of the disk to obtain the potential at a height d on the axis of the disk yields

$$U = 2\pi G \int_0^h \int_0^R \frac{r\rho(z)\, dz\, dr}{[r^2+(d+z)^2]^{1/2}} \tag{14}$$

Integration with respect to r and expanding for large R gives

$$U = 2\pi G \left[R \int_0^h \rho(z)\, dz - \int_0^h (d+z)\rho(z)\, dz + \frac{1}{2R} \int_0^h (d+z)^2 \rho(z)\, dz + O(R^{-3}) \right] \tag{15}$$

First applying the condition of isostasy, i.e.

$$\int_0^h \rho(z)\,dz = 0$$

and then taking the limit $R \to \infty$ yields

$$U = -2\pi G \int_0^h z\rho(z)\,dz \tag{16}$$

which is the formula previously given in (3).

B. Mass-layer Approximations

For mass anomalies confined to thin layers it is useful to integrate (10) over the cylindrical volume illustrated in Fig. 2. Gauss's theorem may then be used to convert one of the volume integrals to a surface integral with the result

$$\int\int \mathbf{g} \cdot d\mathbf{s} = -4\pi G \int\int\int \rho\,dv \tag{17}$$

In the limit $h \to 0$ these integrals can be evaluated to yield

$$g_z^+ - g_z^- = 4\pi G\sigma \tag{18}$$

However by symmetry

$$g_z^+ = -g_z^- \tag{19}$$

so that

$$\bar{g}_z = 2\pi G \int_0^h \rho\,dz \tag{20}$$

which is the Bouguer formula (1).

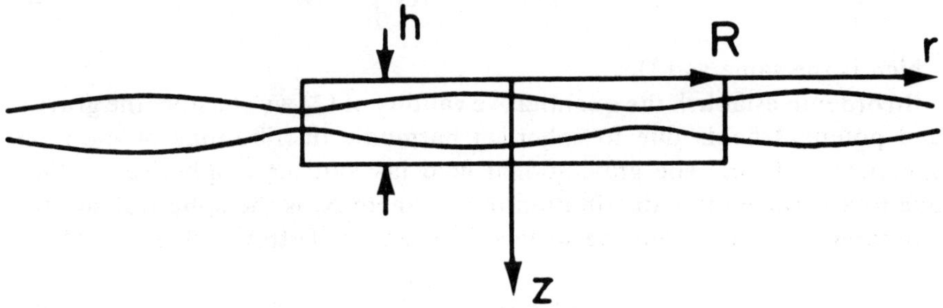

Fig. 2. Illustration of the application of Gauss's theorem to a thin layer of mass anomalies.

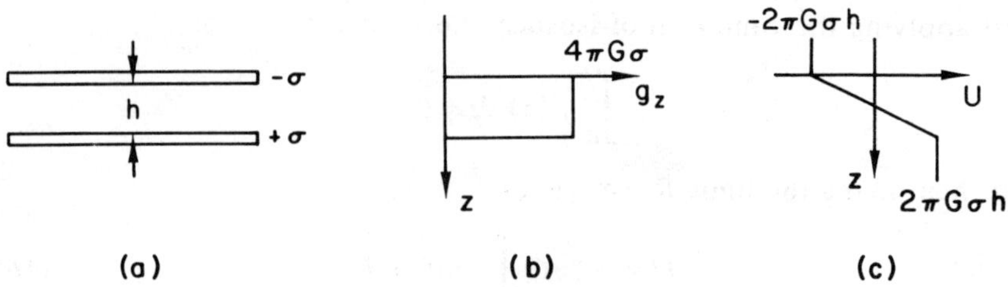

Fig. 3. Gravitational acceleration (b) and potential (c) associated with a dipole mass distribution (a).

A surface dipole mass distribution is obtained by taking two equal surface mass distributions of opposite sign, $\sigma_1 < 0$ on $z = 0$ and $\sigma_2 = -\sigma_1$ on $z = h$ (Fig. 3(a)); the limits $\sigma_2 \to +\infty$, $\sigma_1 \to -\infty$ and $h \to 0$ are taken such that

$$\int_0^h z\rho\, dz = \sigma_2 h = \delta$$

is finite. It follows from (18) and (19) that

$$\begin{aligned} g_z &= 0, \qquad z > h, \qquad z < 0 \\ g_z &= 4\pi G\sigma_2, \qquad 0 < z < h \end{aligned} \tag{21}$$

By using the relationship between the gravitational field and potential $g_z = \partial U/\partial z$, the difference in the potential across the dipole layer is

$$U^+ - U^- = 4\pi G\sigma_2 h = 4\pi G\delta \tag{22}$$

We choose our origin for U such that

$$U^+ = -U^- \tag{23}$$

so that the distribution of U illustrated in Fig. 3(c) is obtained and

$$U^- = -2\pi G\delta = -2\pi G \int_0^h z\rho\, dz \tag{24}$$

which is the same as (3).

In order to establish the quantitative validity of (3) we consider the gravity and potential fields due to spherical harmonic distributions of mass on spherical surfaces. The gravitational field just outside a spherical surface due to a surface mass distribution $\sigma_n S_n$ (where S_n is the spherical surface harmonic of order n) on the surface is given by (Jeffreys, 1976, p. 234)

$$g = 4\pi G\left(\frac{n+1}{2n+1}\right)\sigma_n S_n \tag{25}$$

The wavelengths of the surface mass distribution can be related to the order of the harmonic n by

$$\lambda = \frac{2\pi a}{n} \tag{26}$$

where a is the radius of the sphere (of the Earth). For short wavelength distributions, we take the limit $n \to \infty$ in (25) with the result

$$g = 2\pi G \sigma_n S_n \tag{27}$$

which is the Bouguer formula.

In order to represent a dipole distribution of mass, we consider a spherical harmonic distribution of surface mass on a sphere of radius $r = a - h$ with amplitude $K_n \sigma_n S_n$ in addition to the distribution of surface mass $\sigma_n S_n$ on the sphere $r = a$. The resulting gravitational potential just outside the outer sphere is given by Jeffreys (1976, p. 237)

$$U = \frac{-4\pi G}{(2n+1)} \sigma_n S_n a \left[1 + K_n \left(1 - \frac{h}{a} \right)^{n+2} \right] \tag{28}$$

The condition of isostasy requires and equal mass defect on the inner sphere to that on the outer sphere. By allowing for the difference in area, we require (Jeffreys, 1976, p. 237)

$$K_n = -\frac{a^2}{(a-h)^2} \tag{29}$$

Substitution of (29) into (28) yields

$$U = \frac{-4\pi G}{(2n+1)} \sigma_n S_n a \left[1 - \left(1 - \frac{h}{a} \right)^{n} \right] \tag{30}$$

Taking the limit $h/a \to 0$ in (30) gives

$$U = \frac{-4\pi n G \sigma_n S_n h_n}{2n+1} \tag{31}$$

Next taking the limit $n \to \infty$, we find

$$U = -2\pi G \sigma_n S_n h \tag{32}$$

and noting that $\sigma_n S_n h$ is the surface dipole distribution of mass, this is the same as (3).

By using (30) we can compare the results for finite depths of compensation h with the limiting solution given in (32). This is done in Fig. 4. The ratio of the surface potential U from (30) to the value $2\pi G \sigma_n S_n h$ is given as a function of $1/n$ for $h = 25$, 50, 100, 200 and 400 km. The corresponding

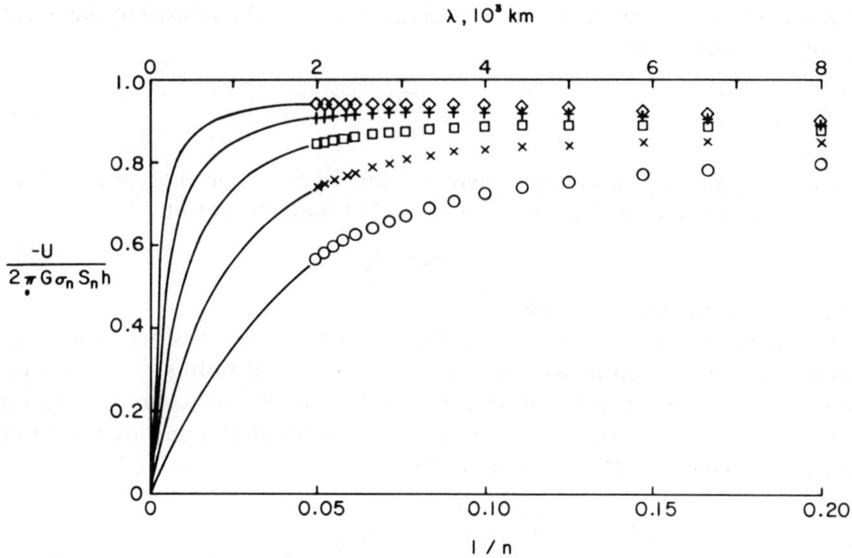

Fig. 4. The ratio of the surface potential U on the Earth from (30) to the value for a surface dipole layer (32) as a function of $1/n$ for various values of the depth of compensation (◇, 25 km; +, 50 km; □, 100 km; ×, 200 km; ○, 400 km). Also included are the values of the wavelength λ corresponding to the value of $1/n$ from (26).

wavelengths from (26) are also included. We can see that the approximation breaks down as expected when the wavelength is of the same order as the depth of compensation, i.e. as $\lambda/h \rightarrow 1$. For depths of compensation of 50 km or less, the error in using (32) is 10% or less over a wide range of wavelengths. It should be emphasized that the case of two mass layers is an extreme case of compensation at depth. The realistic case of distributed mass with depth will lead to lower errors than those given in Fig. 4 if the mass differences are limited to a depth h.

C. Mechanisms of Compensation

1. Airy Compensation

In many mountain ranges elevated topography is primarily compensated by a thick crustal roof. This is Airy isostasy and is illustrated in Fig. 5(a). The density of the crust ρ_c and the mantle ρ_m are assumed to be constant. The thickness of continental crust with zero elevation is H; crust with an

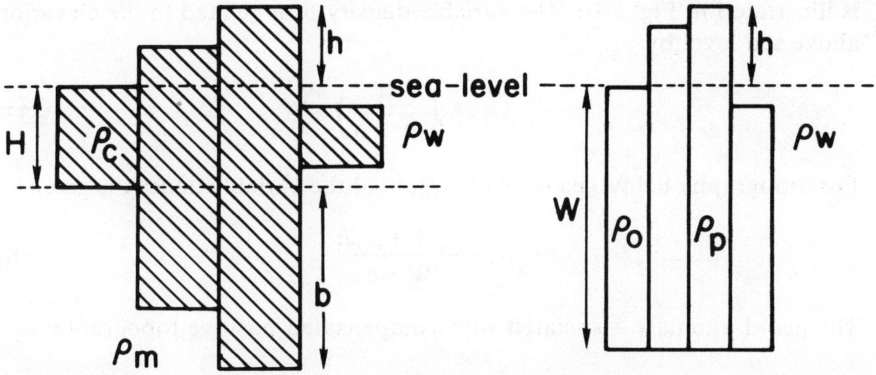

Fig. 5. Isostasy diagram for Airy and Pratt compensation. (a) Airy isostasy in which a crustal block of elevation h above sea level is compensated by a root of depth b below a zero elevation reference crust of thickness H. (b) Pratt isostasy in which surface topography is compensated by mass columns of variable density ρ_p; ρ_0 is the reference density of a zero elevation block and W is the depth of compensation to which the variable densities extend.

elevation h has a crustal root of thickness b. Isostasy requires

$$b = \frac{\rho_c h}{\rho_m - \rho_c} \tag{33}$$

If the height of the topography is negative and is covered with water

$$b = \left(\frac{\rho_c - \rho_w}{\rho_m - \rho_c}\right) h \tag{34}$$

Taking continental crust with zero elevation as reference, the geoid anomaly associated with positive topography from (3) and (4) is

$$\Delta N = \frac{\pi G}{g} \rho_c \left[2Hh + \frac{\rho_m}{(\rho_m - \rho_c)} h^2 \right] \tag{35}$$

For topography below sea level (h negative) the geoid anomaly is given by

$$\Delta N = \frac{\pi G}{g} (\rho_c - \rho_w) \left[2Hh + \frac{(\rho_m - \rho_w)}{(\rho_m - \rho_c)} h^2 \right] \tag{36}$$

2. Pratt Compensation

Another classical compensation mechanism uses horizontal variations in density over a prescribed depth. This is known as Pratt compensation and is illustrated in Fig. 5(b). The variable density ρ_p is related to the elevation above sea level by

$$\rho_p = \rho_0 \left(\frac{W}{W+h} \right) \tag{37}$$

For topography below sea level (h negative) the variable density is given by

$$\rho_p = \frac{\rho_0 W + \rho_w h}{W+h} \tag{38}$$

The geoid anomaly associated with compensated positive topography is

$$\Delta N = \frac{\pi G}{g} \rho_0 W h \tag{39}$$

and with negative topography is

$$\Delta N = \frac{\pi G}{g} (\rho_0 - \rho_w) W h \tag{40}$$

3. *Thermal Compensation*

Thermal compensation is associated with the thickening thermal boundary layer defining the oceanic lithosphere. We will consider two models for the thermal structure of the oceanic lithosphere. The first, the half-space model, assumes that the cooling plate is the thermal boundary layer of a large-scale mantle convection cell (Turcotte and Oxburgh, 1967). The second, the plate model, considers the cooling of a prescribed plate of thickness H (Parsons and Sclater, 1977). For the half-space model, the lithosphere continues to thicken at all ages, but for the plate model the plate thickness is the maximum thickness of the lithosphere.

For the half-space model the temperature distribution in the lithosphere is

$$T - T_0 = (T_m - T_0)\,\mathrm{erf}[(z/2)(\kappa t)^{-1/2}] \tag{41}$$

where T_0 is the surface temperature, T_m the mantle temperature, κ the thermal diffusivity, t the age of the sea floor and erf is the error function. By assuming isostasy, the subsidence relative to the ocean ridge crest is

$$w = 2\rho_m \alpha (T_m - T_0)(\rho_m - \rho_w)^{-1} \left(\frac{\kappa t}{\pi} \right)^{1/2} \tag{42}$$

where α is the coefficient of thermal expansion. The corresponding geoid anomaly relative to the ocean ridge crest as obtained from (3) and (4) is

$$\Delta N = -\frac{2\pi G\rho_{\mathrm{m}}\alpha(T_{\mathrm{m}}-T_0)\kappa}{g}\left[1+\frac{2\rho_{\mathrm{m}}\alpha(T_{\mathrm{m}}-T_0)}{\pi(\rho_{\mathrm{m}}-\rho_{\mathrm{w}})}\right]t \tag{43}$$

The predicted geoid anomaly is a linear function of the age of the seafloor.

For the plate model the equations are more complicated. The temperature in the slab is given by

$$T(z,t) = T_{\mathrm{m}}\left(\frac{z}{H}+\frac{2}{\pi}\sum_{n=1}^{\infty}\frac{1}{n}\left(\sin\frac{n\pi z}{H}\exp-\frac{\beta_n ut}{H}\right)\right) \tag{44}$$

where

$$\beta_n = \left[\left(\frac{uH}{2\kappa}\right)^2+n^2\pi^2\right]^{1/2}-\frac{uH}{2\kappa} \tag{45}$$

and the subsidence is given by

$$w(t) = \alpha\rho_{\mathrm{m}}\frac{(T_{\mathrm{m}}-T_0)}{2}(\rho_{\mathrm{m}}-\rho_w)^{-1}\left[1-\frac{8}{\pi^2}\sum_{n=1}^{\infty}\frac{\exp(-\beta_{2n-1}\,ut/H)}{(2n-1)^2}\right] \tag{46}$$

where u is the half-spreading velocity. Integration yields

$$\Delta N(t) = -2\pi\frac{G}{g}H^2\left[(\rho_{\mathrm{m}}-\rho_{\mathrm{w}})\frac{w(t)^2}{2}+\alpha\rho_{\mathrm{m}}(T_{\mathrm{m}}-T_0)\left(\frac{1}{6}+\frac{2}{\pi^2}\sum_{n=1}^{\infty}Q_n\right)\right] \tag{47}$$

where

$$Q_n = \frac{(-1)^n}{n^2}\exp\left(-\beta_n\frac{ut}{H}\right) \tag{48}$$

Changing the velocity has relatively little effect on the geoid anomaly. For this model $\Delta N(t)$ asymptotically approaches a constant value at large ages and $\Delta N/\Delta t$ is approximately constant for very young ages.

4. Compensation with Crustal and Lithospheric Thinning

A relatively simple model for continental rifting and mountain formation can be developed by introducing a crustal thinning factor $\beta_{\mathrm{c}} \equiv y_{\mathrm{c}}/y_{\mathrm{c0}}$. Where y_{c} is the crustal thickness after rifting or mountain building and y_{c0} is the reference crustal thickness, as well as a lithospheric thinning factor $\beta_{\mathrm{L}} \equiv y_{\mathrm{L}}/y_{\mathrm{L0}}$ where y_{L} is the lithosphere thickness after rifting or mountain building and y_{L0} is the reference lithosphere thickness. This model is illustrated in Fig. 6. This approach follows that given by McKenzie (1978) who introduced a stretching factor (the reciprocal of the thinning factor) and assumed that

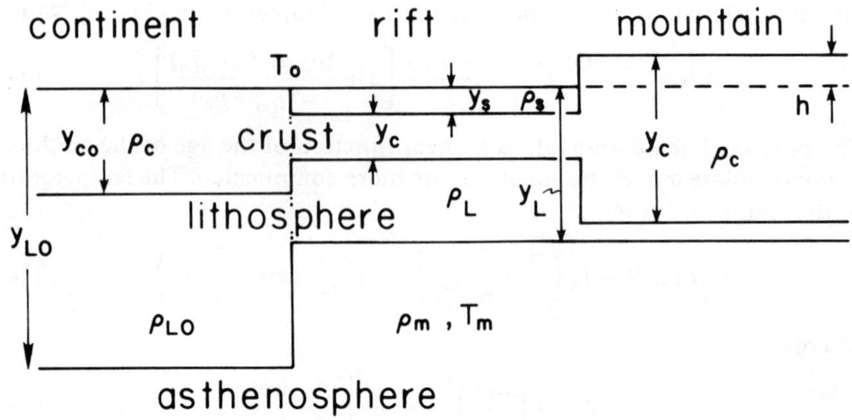

Fig. 6. Idealized structure of the crust and lithosphere in a normal continental region, a continental rift, and a mountain belt.

both crust and lithosphere are stretched together. Although mechanical stretching may be applicable to the crust, it is doubtful that it is applicable to the lithosphere. The model adopted here follows that of Royden and Keen (1980) and Turcotte (1983).

Assuming isostasy, the depth of sediments in a rift is related to the crustal and lithospheric thinning factors by

$$y_s = \frac{(\rho_m - \rho_c)}{(\rho_m - \rho_s)} y_{c0}(1-\beta_c) - \frac{\rho_m \alpha (T_m - T_0)}{2(\rho_m - \rho_s)} y_{L0}\,(1-\beta_L) \tag{49}$$

And the height of the topography in a mountain belt is related to the crustal and lithospheric thinning factors by

$$h = \frac{(\rho_m - \rho_c)}{\rho_m} y_{c0}(\beta_c - 1) + \tfrac{1}{2}\alpha (T_m - T_0) y_{L0}(1-\beta_L) \tag{50}$$

Taking $\rho_m = 3.3\ \mathrm{g/cm^3}$, $\rho_c = 2.8\ \mathrm{g/cm^3}$, $\rho_s = 2.5\ \mathrm{g/cm^3}$, $y_{c0} = 35$ km, $y_{L0} = 175$ km, $T_m - T_0 = 1200$ K and $\alpha = 3 \times 10^{-5}\ \mathrm{K^{-1}}$ the dependences of y_s and h on β_c for several values of β_L are given in Fig. 7.

The geoid anomaly for an isostatically compensated rift is given by

$$\begin{aligned}\Delta N = \frac{2\pi G}{g}\Bigg[&-\frac{(\rho_m - \rho_c)(\rho_c - \rho_s)}{2(\rho_m - \rho_s)} y_{c0}^2 (1-\beta_c^2) \\ &+ \tfrac{1}{2}\rho_m \alpha (T_m - T_0) y_{L0}^2 \left[\tfrac{1}{3}(1-\beta_L^2) - \frac{(\rho_m - \rho_c) y_{c0}}{(\rho_m - \rho_s) y_{L0}} (1-\beta_L) \right] \\ &+ \frac{\rho_m \alpha^2 (T_m - T_0)^2 y_{L0}^2}{8(\rho_m - \rho_s)} (1-\beta_L^2) \Bigg]\end{aligned} \tag{51}$$

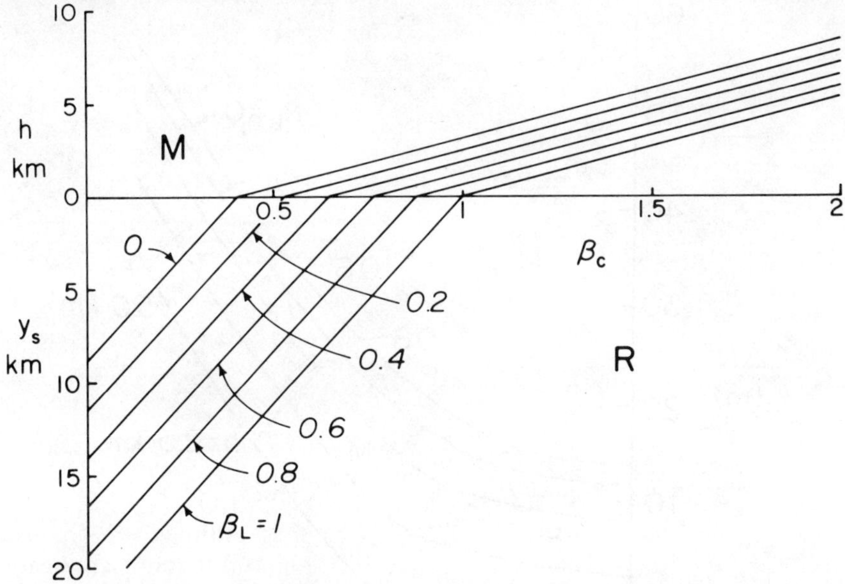

Fig. 7. Dependence of the topographic elevation of an isostatically compensated mountain belt (M) and the depth of an isostatically compensated rifted basin (R) on the crustal thinning factor for several values of the lithospheric thinning factor.

And the geoid anomaly for an isostatically compensated mountain range is

$$\Delta N = \frac{2\pi G}{g}\left[\frac{(\rho_m - \rho_c)\rho_c}{2\rho_m} y_{L0}^2(\beta_c^2 - 1) + \tfrac{1}{6}\rho_m \alpha (T_m - T_0) y_{L0}^2 (1 - \beta_L^2) \right.$$
$$+ \tfrac{1}{2}(\rho_m - \rho_c)\alpha(T_m - T_0) y_{L0} y_{c0}(\beta_c \beta_L - 1) \tag{52}$$
$$\left. + \tfrac{1}{8}\alpha^2 (T_m - T_0)^2 y_{L0}^2 (1 - \beta_L^2)\right]$$

Taking the parameter values given above the geoid anomalies associated with rifts and mountain belts are given in Fig. 8 as a function of β_c for several values of β_L.

III. Applications

A. Ocean Ridges

Ocean ridges are a primary example of thermal subsidence. A GEOS-3 altimeter track across the mid-Atlantic ridge at 44.5°N is shown in Fig. 9(a). The observed geoid anomaly is compared with the calculated anomaly from

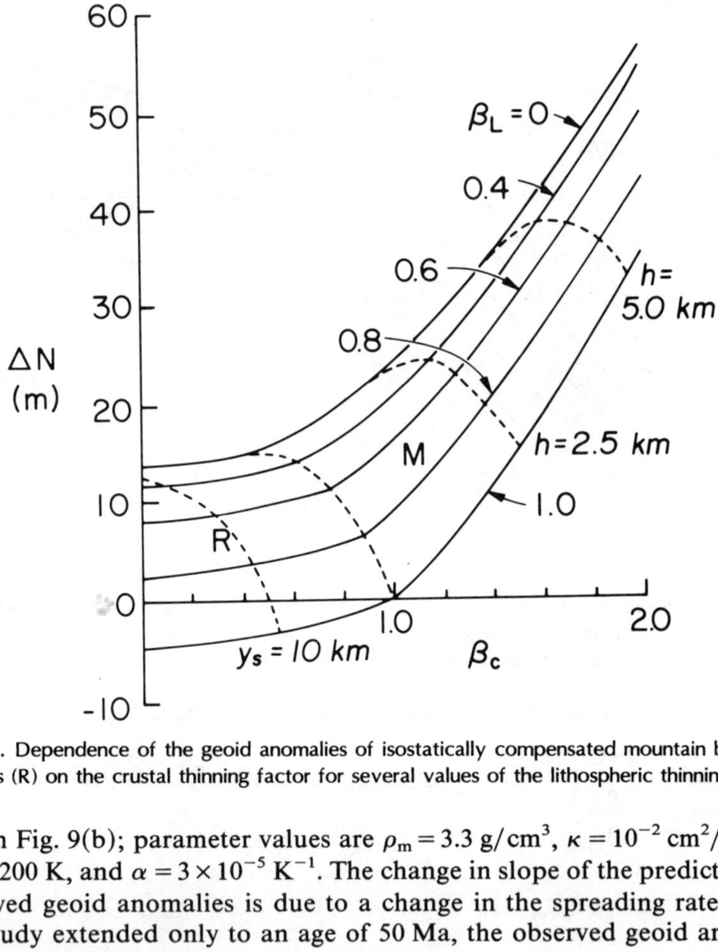

Fig. 8. Dependence of the geoid anomalies of isostatically compensated mountain belts (M) and rifts (R) on the crustal thinning factor for several values of the lithospheric thinning factor.

(42) in Fig. 9(b); parameter values are $\rho_m = 3.3\ \text{g/cm}^3$, $\kappa = 10^{-2}\ \text{cm}^2/\text{s}$ $T_m - T_0 = 1200$ K, and $\alpha = 3 \times 10^{-5}\ \text{K}^{-1}$. The change in slope of the predicted and observed geoid anomalies is due to a change in the spreading rate. Since this study extended only to an age of 50 Ma, the observed geoid anomaly increases linearly with time.

The primary difficulty with this type of direct study is the noise introduced by the regional anomalies associated with geoid anomalies of deeper origin. This is a particularly serious problem for fast-spreading ridges. In order to overcome this difficulty Sandwell and Schubert (1980) averaged the geoid anomalies on the two sides of symmetrical spreading ridges relative to the ridge crest. Their results in terms of the dependence of the geoid slope $(\Delta N/\Delta t)$ on time are given in Fig. 10 for the North Atlantic, South Atlantic and Indian Oceans. Obviously there is still considerable scatter in the data. Also included in Fig. 10 are the theoretical results for plate thickness of 50, 75, 100 and 125 km from (47). Because of the large scatter it is difficult to distinguish between the alternative models.

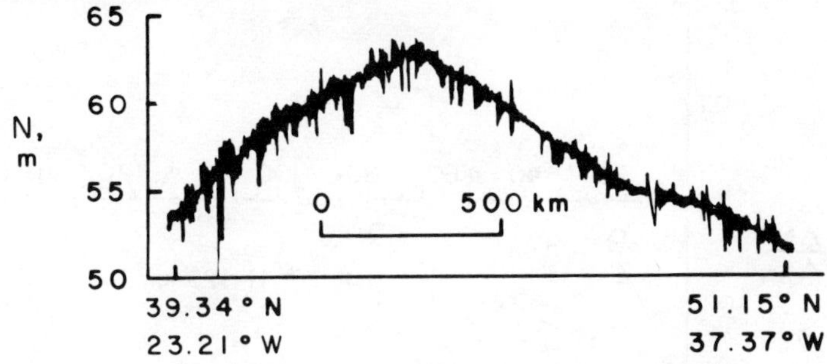

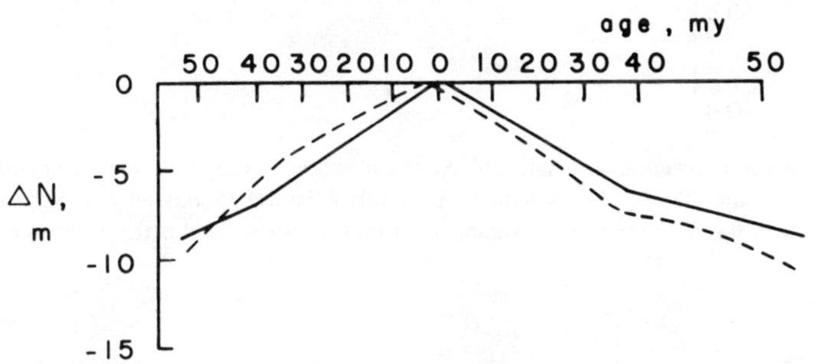

Fig. 9. (a) Geoid anomaly across the Mid-Atlantic Ridge at 44.5 obtained by the altimeter on the GEOS-3 satellite. (b) Measured geoid anomaly from (a) as a function of seafloor age compared with the predicted anomaly from (43) (observed indicated by dashed line and theory by solid line).

Another approach to the study of geoid anomalies associated with ocean ridges is to utilize the geoid changes across major fracture zones. Major fracture zones are generated by transform faults on the ocean ridge system. Thus the ocean floors on the two sides of a fracture zone have age differences that are related to the offsets in the ridge at the transform fault that generates the fracture zone. The age difference can be determined from the seafloor magnetic anomalies. The geoid change across the Mendocino fracture zone as obtained by Cazenave *et al.* (1982) is given as a function of the age of the younger side in Fig. 11. Also included in this figure is the predicted geoid change for the plate model as given by (47) assuming an age difference of 26 Myr and plate thicknesses of 50, 75, 100 and 125 km. The data appear to favor a plate thickness of about 75 km. The discrepancy at ages greater than 25 Ma can be attributed to a change in the age difference across the fracture zone.

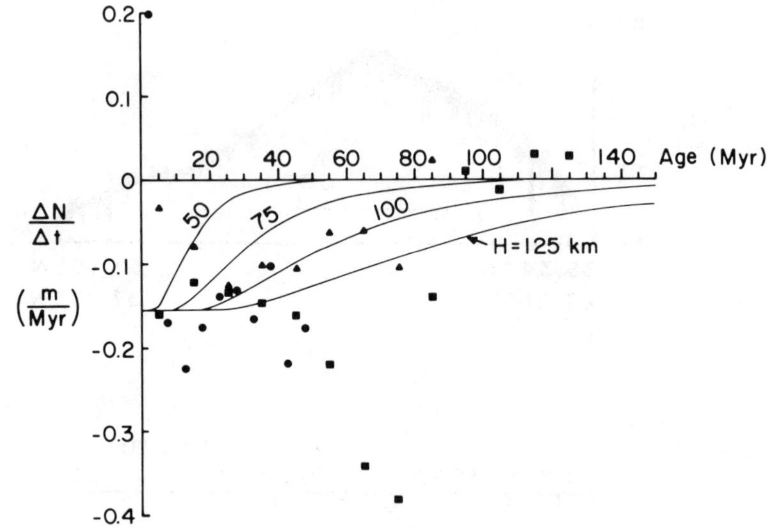

Fig. 10. Slope of the geoid anomaly $\Delta N/\Delta t$ as a function of age for several ocean ridge systems (■, North Atlantic; ●, Southeast Indian; ▲, South Atlantic) [Sandwell and Schubert (1980)] compared with the predicted geoid anomalies from (47) for several plate thicknesses H.

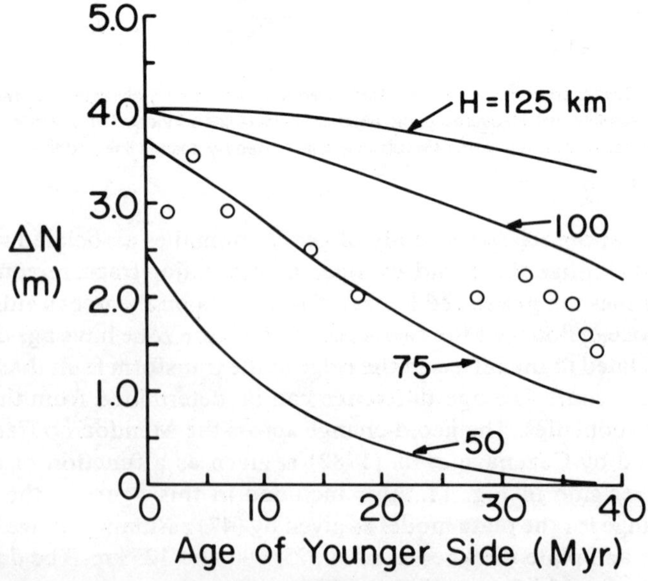

Fig. 11. The difference in geoid height across the Mendocino fracture zone as a function of the age of the younger side (Cazenave *et al.*, 1982) compared with the predicted geoid difference from (47) for several plate thicknesses H assuming an age difference of 26 Ma.

B. Ocean Swells

Ocean swells are near-circular regions of anomalously shallow bathymetry with a radius of 500–1000 km. Examples include the Hawaiian Swell that surrounds the active volcanic center at the end of the Hawaiian Islands and seamount chain and the Bermuda Swell. A GEOS-3 altimeter track across the Bermuda swell is given in Fig. 12. Also included in Fig. 12 is the bathymetry across the swell along the same track. It is seen that there is a good correlation between bathymetry and geoid anomaly. The dependence of geoid anomaly on the depth anomaly at several points along this track is given in Fig. 13. Also included in Fig. 13 are data for several points on the Hawaiian Swell (Crough, 1978) and the Cape Verde Rise (Crough, 1982). It is seen that the three features correlate well with each other with relatively little scatter. Included for comparison are the predicted Pratt correlations from (40) with $W = 75$, 100 and 125 km. Good agreement is obtained for Pratt compensation with $W = 100$ km.

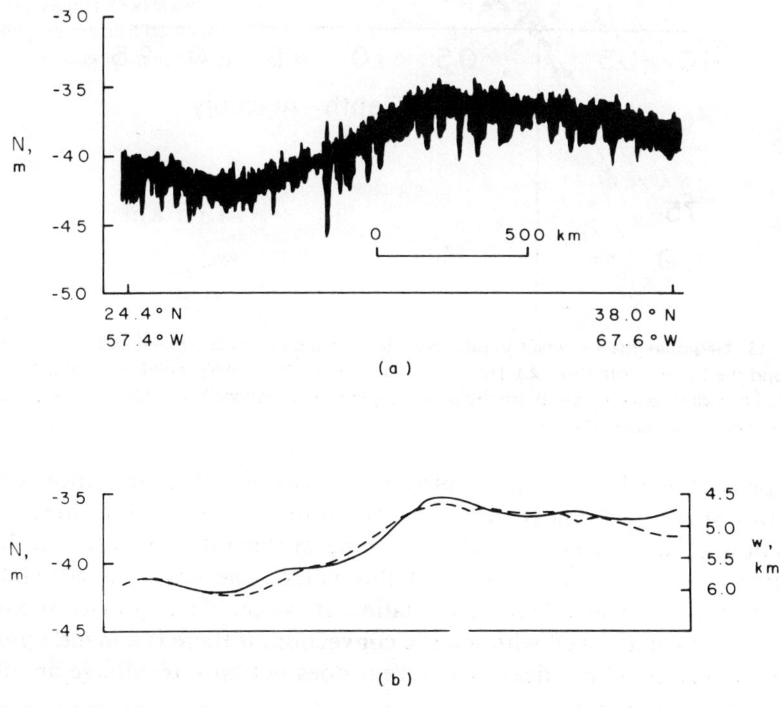

Fig. 12. (a) Geoid anomaly across the Bermuda Swell obtained by the altimeter on the GEOS-3 satellite. (b) Bathymetry along the same track (dashed line indicates geoid and solid line, bathymetry).

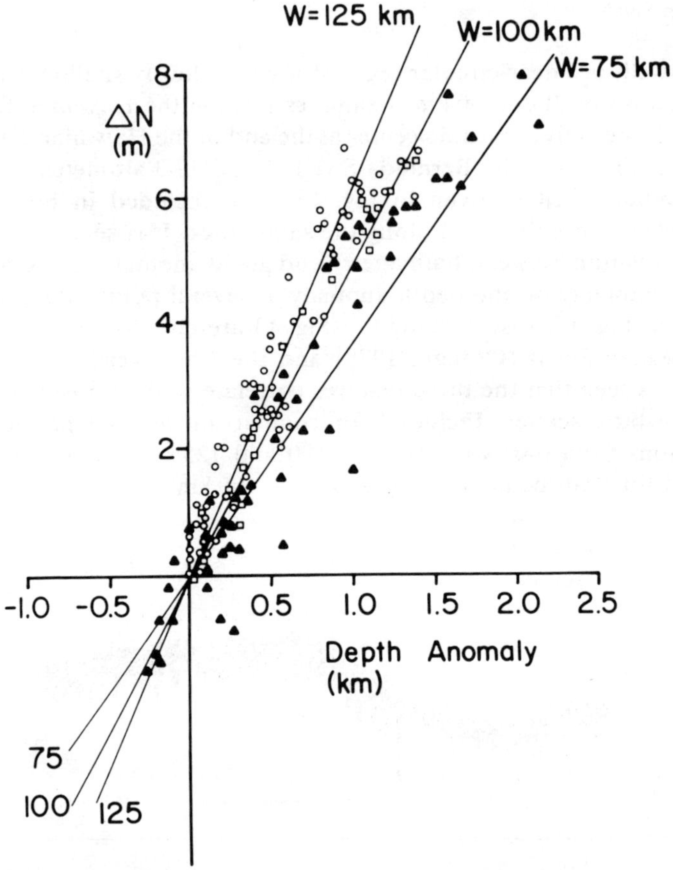

Fig. 13. Geoid anomalies versus depth anomalies for the Bermuda Swell (□), Hawaiian Swell (○), and the Cape Verde Rise (▲). [Data is from Crough (1978, 1982) and Haxby and Turcotte (1978).] Included for comparison are the predicted Pratt correlations from (40) for several values of the depth of compensation *W*.

A physical mechanism that would appear to explain this correlation would be lithospheric thinning (Crough, 1978). Normal oceanic lithosphere has a thickness of about 100 km so that thinning of this lithosphere would have a depth of compensation of about this value. The observed correlation would appear to preclude any association of the geoid and gravity anomaly over the Hawaiian swell with mantle convection. If there is a mantle plume beneath Hawaii with a deep structure it does not appear to have an effect on the geoid anomaly.

Although it is not surprising that the Cape Verde Rise correlates so well with the Hawaiian Swell since it is also a recognized intraplate hot spot, the correlation with the Bermuda swell is somewhat unexpected. We would conclude that the Bermuda swell is also the result of lithospheric thinning. Also, in order to have persisted to the present this thinning must have occurred within the last 50 Myr or so.

C. Aseismic Ocean Ridges and Plateaus

Aseismic ocean ridges and plateaus occur throughout the ocean basins. These regions of shallow topography do not have the near-circular geometry of the ocean swells. Also, they are not associated with active intraplate hot spots. Anomalously shallow topography is now being created by the Mid-Atlantic Ridge in the vicinity of the Iceland hot spot. It is natural to associate the aseismic ocean ridges and plateaus with this process.

Correlations of geoid anomalies with bathymetric anomalies for the Walvis Ridge in the South Atlantic Ocean (Angevine and Turcotte, 1980) and the Agulhas Plateau in the southwestern Indian Ocean (Angevine and Turcotte, 1983) are given in Fig. 14.

Again these features are seen to correlate with each other with relatively little scatter. Included for comparison are the predicted Pratt correlations from (40) with $W = 20$, 30 and 40 km. Good agreement is obtained for Pratt compensation with $W = 30$ km.

There is strong evidence that the Walvis Ridge is anomalously thick ocean crust formed at the Mid-Atlantic Ridge. However, the geoid anomaly obtained for a simple crustal thickening model is much smaller than the observed values. An anomalously light upper mantle is required. It is quite possible that this lighter upper mantle resulted from the emplacement of the thicker oceanic crust at the ridge crest. Depleted mantle rock is lighter than undepleted mantle rock (Angevine and Turcotte, 1983). Thus we do not attribute the anomalous bathymetry of the Walvis Ridge to lithospheric thinning. The excellent correlation between the Walvis Ridge and Agulhas Plateau, illustrated in Fig. 13, is strong evidence that the same process is responsible for the formation of these features.

D. Mid-Atlantic Ridge Crest

There is also a positive correlation between geoid anomalies and topography along the crest of the Mid-Atlantic Ridge. The geoid anomaly on the ridge crest is given as a function of latitude in Fig. 15(a). The anomalous bathymetry on the ridge crest is given as a function of latitude in Fig. 15(b). The dependence of geoid anomaly on the depth anomaly at several points along

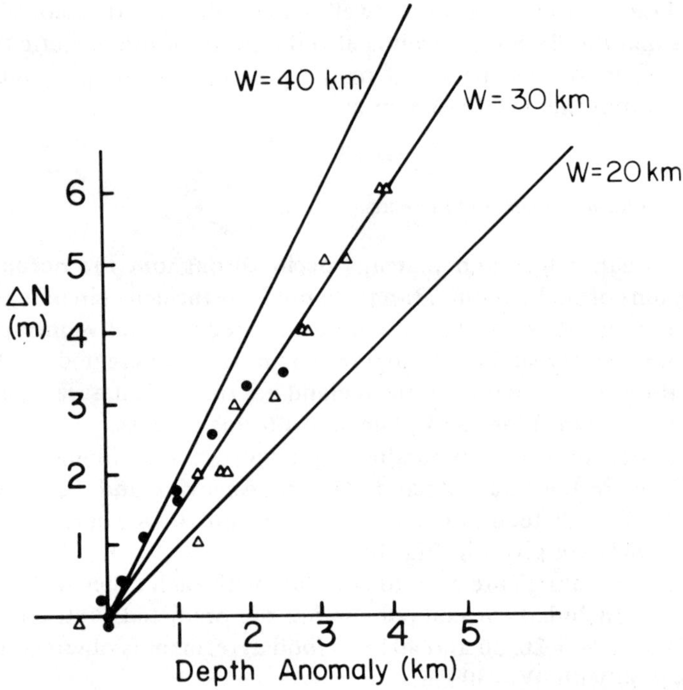

Fig. 14. Geoid anomalies versus depth anomalies for the Agulhas Plateau (●) and Walvis Ridge (△). [Data is from Angevine and Turcotte (1980, 1983).] Included for comparison are the predicted Pratt correlations from (40) for several values of the depth of compensation *W*.

the ridge crest is given in Fig. 16. Although there is considerable scatter there is a strong correlation of positive geoid anomaly with shallow bathymetry. Also included in Fig. 16 are the predicted Pratt correlations from (40) with $W = 500$, 750, 1000 and 1500 km. A best fit would be with $W \approx 1000$ km. This is evidence of a deep origin for the geoid anomaly associated with the Icelandic hot spot. It is interesting to note that this was not found for the Hawaiian hot spot.

E. Thickness of the Continental Lithosphere

A GEOS-3 altimeter profile across the Atlantic margin of North America at 40.5°N is shown in Fig. 17(a). The corresponding bathymetry along the profile is given in Fig. 17(b). Clearly there is a strong correlation between the geoid anomaly and the seafloor topography. In order to calculate a geoid anomaly we assume that the bathymetry is isostatically compensated,

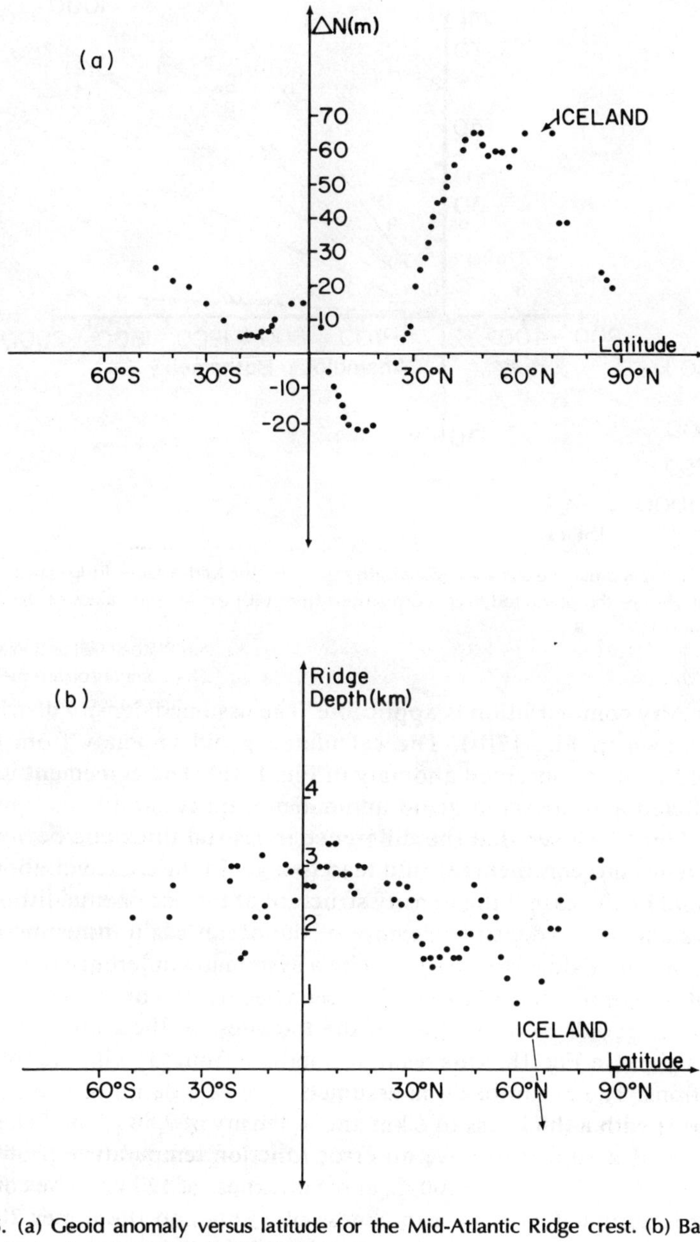

Fig. 15. (a) Geoid anomaly versus latitude for the Mid-Atlantic Ridge crest. (b) Bathymetry versus latitude along the Mid-Atlantic Ridge crest.

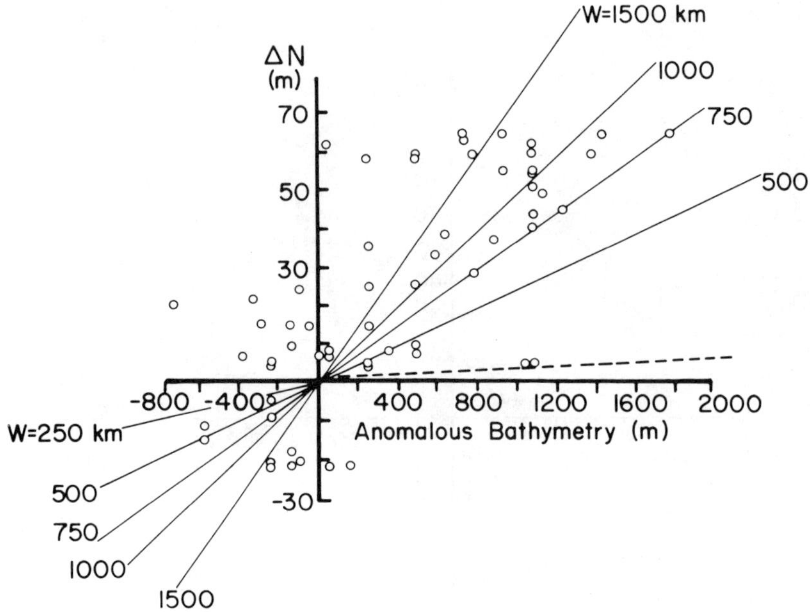

Fig. 16. Geoid anomaly versus anomalous bathymetry for the Mid-Atlantic Ridge crest. Included for comparison are the predicted Pratt correlations from (40) for several values of the depth of compensation *W*.

and that Airy compensation is applicable. The assumed density distribution is also shown in Fig. 17(b). The calculated geoid anomaly from (40) is compared with the observed anomaly in Fig. 17(c). The agreement between the predicted and observed geoid anomalies is quite good.

From Fig. 17 we see that the difference in crustal thickness between the ocean basins and continents would lead to a geoid difference of about 6 m. This would be the case if the density structure of the continental lithosphere was the same as the density structure of the ocean basin lithosphere.

However, there does not appear to be a systematic difference in the geoid height of the continents relative to the ocean basins. The predicted difference between the geoids as a function of the thickness of the continental lithosphere is given in Fig. 18. This result is obtained from (3) with the following assumptions. The ocean basin is assumed to have a depth of 6 km with an ocean crust with a thickness of 6 km and a density of 2.88 g/cm^3. The ocean lithosphere is assumed to have an error function temperature profile with $\alpha = 3 \times 10^{-5}\ \mathrm{K}^{-1}$, $T_m - T_0 = 1300°\mathrm{C}$, and a thickness of 120 km. The continental crust is assumed to have a thickness of 35 km and the upper 20 km to have a density of 2.70 g/cm^3. The density of the lower crust is varied to

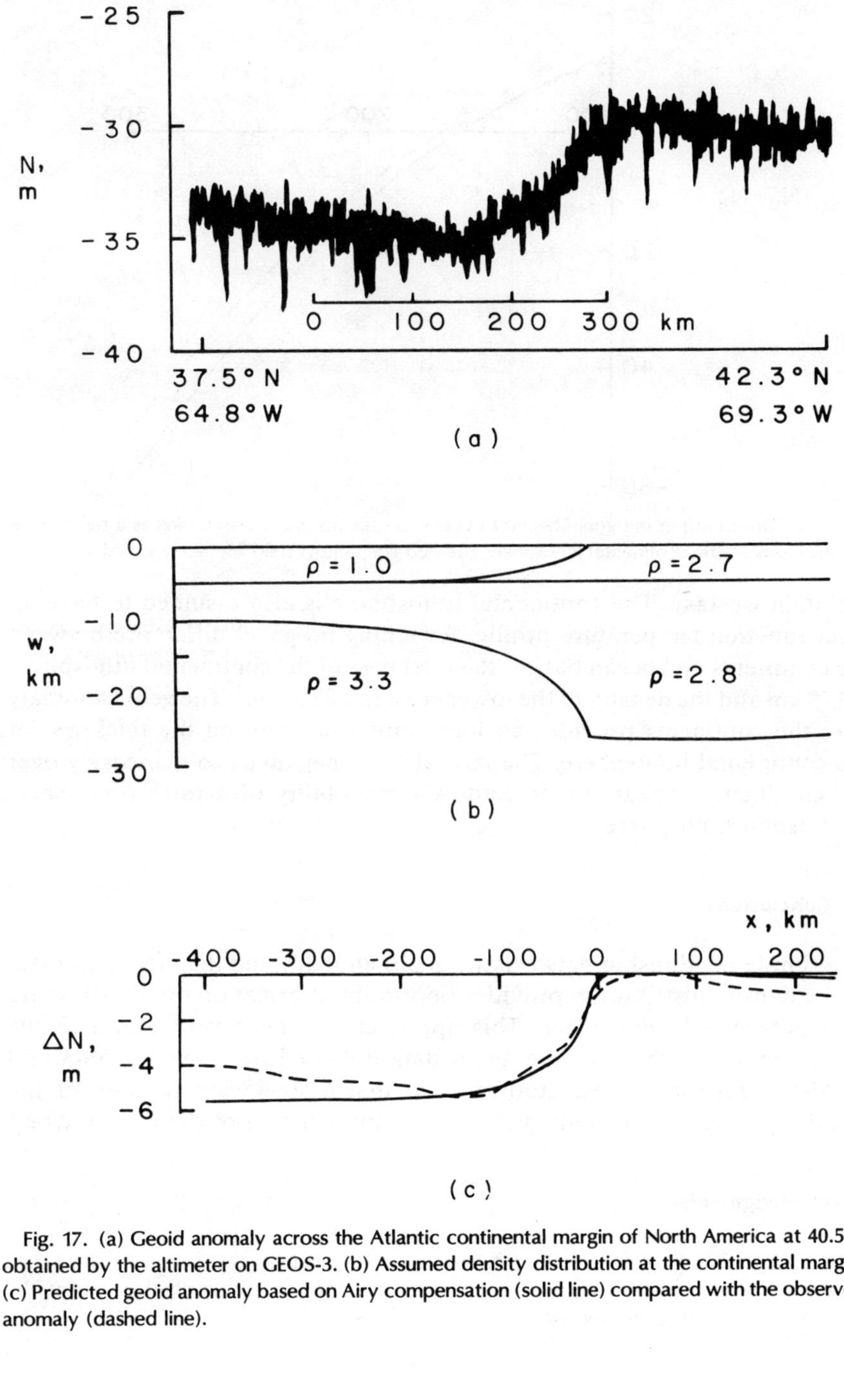

Fig. 17. (a) Geoid anomaly across the Atlantic continental margin of North America at 40.5°N obtained by the altimeter on GEOS-3. (b) Assumed density distribution at the continental margin. (c) Predicted geoid anomaly based on Airy compensation (solid line) compared with the observed anomaly (dashed line).

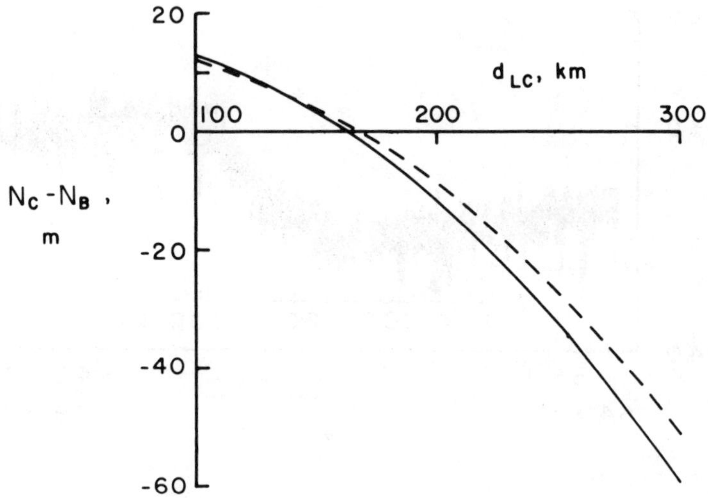

Fig. 18. Difference in the geoid height between continents and ocean basins as a function of the thickness of the continental lithosphere (dashed line, linear; solid line, error function).

maintain isostasy. The continental lithosphere is also assumed to have an error function temperature profile. Assuming no geoid difference between the continents and ocean basins, the thickness of the continental lithosphere is 175 km and the density of the lower crust is 2.83 g/cm^3. The geoid anomaly over the continents provides an important constraint on the thickness of the continental lithosphere. The lack of large negative geoid anomaly over the continents appears to preclude the possibility of a thick (>250 km) continental lithosphere.

IV. Conclusions

The simple relationship between the geoid anomaly and the dipole moment of the density distribution provides important information on the structure of compensated lithosphere. This approach has been very useful in the interpretation of the accurate geoid data obtained from the GEOS-3 and SEASAT missions. These studies show that a significant fraction of the geoid signal can be attributed to the density structure of the lithosphere.

Acknowledgments

The preparation of this paper and much of the research included was supported by the Office of Naval Research under Contract No. N00014-79-C-0569. It is contribution 806 of the Department of Geological Sciences, Cornell University.

References

Angevine, C. L. and Turcotte, D. L. (1980). On the compensation mechanism of the Walvis Ridge. *Geophys. Res. Lett.* **7**, 447–479.

Angevine, C. L. and Turcotte, D. L. (1983). Correlation of geoid and depth anomalies over the Agulhas Plateau. *Tectonophysics* **100**, 43–53.

Cazenave, A., Lago B. and Dominh K. (1982). Geoid anomalies over the northeast Pacific fracture zones from satellite altimeter data. *Geophys. J. R. Astron. Soc.* **69**, 15–31.

Cazenave, A., Lago, B. and Dominh K. (1983). Thermal parameters of the oceanic lithosphere estimated from geoid height data. *J. Geophys. Res.* **88**, 1105–1118.

Crough, S. T. (1978). Thermal origin of mid-plate, hot-spot swells. *Geophys. J. R. Astron. Soc.* **55**, 451–469.

Crough, S. T. (1979). Geoid anomalies across fracture zones and the thickness of the lithosphere. *Earth Planet. Sci. Lett.* **44**, 224–230.

Crough, S. T. (1982). Geoid height anomalies over the Cape Verde Rise. *Mar. Geophys. Res.* **5**, 263–271.

Haxby, W. F. and Turcotte, D. L. (1978). On isostatic geoid anomalies. *J. Geophys. Res.* **83**, 5473–5478.

Jeffreys, H. (1976). "The Earth". Cambridge Univ. Press, London and New York.

McKenzie, D. (1978). Some remarks on the development of sedimentary basins. *Earth Planet. Sci. Lett.* **40**, 25–32.

Ockendon, J. R. and Turcotte, D. L. (1977). On the gravitational potential and field anomalies due to thin mass layers. *Geophys. J. R. Astron. Soc.* **48**, 479–492.

Parsons, B. and Sclater, J. G. (1977). An analysis of the variation of the ocean floor bathymetry and heat flow with age. *J. Geophys. Res.* **82**, 803–827.

Royden, L. and Keen, C. E. (1980). Rifting process and thermal evolution of the continental margin of eastern Canada determined from subsidence curves. *Earth Planet. Sci. Lett.* **51**, 343–361.

Sandwell, D. and Schubert, G. (1980). Geoid height versus age for symmetric spreading ridges. *J. Geophys. Res.* **85**, 7235–7241.

Sandwell, D. T. and Schubert, G. (1982). Geoid height-age relation from SEASAT altimeter profiles across the Mendocino fracture zone. *J. Geophys. Res.* **87**, 3949–3958.

Turcotte, D. L. (1983). Mechanisms of crustal deformation. *J. Geol. Soc., London* **140**, 701–724.

Turcotte, D. L. and McAdoo, D. C. (1979). Geoid anomalies and the thickness of the lithosphere. *J. Geophys. Res.* **84**, 2381–2387.

Turcotte, D. L. and Oxburgh, E. R. (1967). Finite amplitude convection cells and continental drift. *J. Fluid Mech.* **28**, 29–42.

Turcotte, D. L. and Schubert, G. (1982). "Geodynamics". Wiley, New York.

Index